Joachim Knoch

Nanoelectronics

Also of Interest

Solid State Physics
Siegfried Hunklinger, Christian Enss, 2022
ISBN 978-3-11-066645-8, e-ISBN 978-3-11-066650-2

Nanochemistry
From Theory to Application for In-Depth Understanding of Nanomaterials
Edited by Xuan Wang, Sajid Bashir, Jingbo Liu, 2023
ISBN 978-3-11-073985-5, e-ISBN 978-3-11-073987-9

Carbon for Micro and Nano Devices
Swati Sharma, 2024
ISBN 978-3-11-062062-7, e-ISBN 978-3-11-062063-4

Handbook of Nanoethics
Edited by Gunjan Jeswani, Marcel Van de Voorde, 2021
ISBN 978-3-11-066923-7, e-ISBN 978-3-11-066928-2

Expansion of Physics through Nanoscience
What is Time at the Basic Level?
Wolfram Schommers, 2020
ISBN 978-3-11-052460-4, e-ISBN 978-3-11-052559-5

Joachim Knoch

Nanoelectronics

From Device Physics and Fabrication Technology to Advanced Transistor Concepts

2nd edition

DE GRUYTER

Author
Prof. Dr. Joachim Knoch
Institute of Semiconductor Electronics
RWTH Aachen University
Sommerfeldstr. 24
52074 Aachen
Germany
knoch@iht.rwth-aachen.de

ISBN 978-3-11-105424-7
e-ISBN (PDF) 978-3-11-105442-1
e-ISBN (EPUB) 978-3-11-105501-5

Library of Congress Control Number: 2023951185

Bibliographic information published by the Deutsche Nationalbibliothek
The Deutsche Nationalbibliothek lists this publication in the Deutsche Nationalbibliografie; detailed bibliographic data are available on the Internet at http://dnb.dnb.de.

Cover image: blackdovfx / E+ / Getty images
Typesetting: VTeX UAB, Lithuania
Printing and binding: CPI books GmbH, Leck

www.degruyter.com

To my family

Preface

Nanoelectronics devices have become ubiquitous in recent years, penetrating almost every sector of modern life. This has been made possible by the research work of thousands of scientists and engineers. I still remember when I first came across nanoelectronics devices: as an intern in a large research center, my task was to characterize two-dimensional electron gases in III–V heterostructures by fabricating simple van der Pauw structures and measure them at cryogenic temperatures. What intrigued me and still does is the fact that research on nanoelectronics is truly interdisciplinary, including physics, electrical engineering and chemistry. Indeed, working with and understanding the phenomena observed in nanoelectronics devices requires a solid theoretical basis as well as knowledge of the fabrication techniques, since often the technology used is decisive for the functionality of a device. Part of the fun is also the fact that research on nanoelectronics involves groundbreaking, fundamental work as well as dealing with applied, real-life problems. The bottom line is that research on nanoelectronics provides a broad playground for curiosity-driven, explorative, innovative, disruptive, highly relevant work, which is great!

While the breadth and depth of nanoelectronics are very appealing to me, it turned out to be a curse when writing the present introductory textbook. I realized that one can easily get lost in so many details that it appeared sometimes difficult to decide between more and less important topics. During the writing, I often thought that things were getting too broad and at the same time that other parts were getting too detailed and that I should remove or add content. But then again, I have been working long enough in the field of research to witness the transition from 3D to 2D systems, the excitement about 1D semiconductors followed by the hype regarding 2D materials and the advent of 3D integration. Interestingly, while there have been many newspaper articles about yet another 2D wonder-material, recently you also find papers claiming that "silicon comes back" and reports about how a big manufacturer bets it can turn silicon into a wonder material for quantum computing. So, in the end instead of focusing too much on a certain topic or material, I tried to find a balance because I am convinced that one needs to be able to handle the various aspects related to the materials, the fabrication, the theoretical understanding and simulation of nanoelectronics devices when one wants to contribute to the field. Ultimately, restrictions regarding time and book pages as well as a personal selection of topics determined the content of the book. Furthermore, the book emerged out of the courses that I teach at RWTH Aachen University in Germany, and hence the selection of topics is also strongly impacted by the interaction with many students. For instance, a larger section of the second chapter is devoted to tight-binding calculations, which I consider to be essential for understanding a substantial part of the literature on device simulations. And such calculations, while more common to physics students, are sometimes covered only scarcely in electrical engineering curricula. Another example is Chapter 3 that is devoted to the fabrication technology. Since there is a vast literature available on the subject and because a thorough treatment of the subject

https://doi.org/10.1515/9783111054421-201

justifies a book on its own, I touched on many of the techniques only slightly and rather provided more details on aspects that are relevant to a student's lab work. In addition, I found it important to include a chapter on fabrication to make the book self-contained and provide details on the technology used to realize the experimental devices presented in the book. Chapter 3 also covers a personal selection of process technologies that I find particularly appealing, useful and worth mentioning. As such, the content of the book is broad, yet I hope that it enables students to start their own, more in-depth work in a certain direction of this multidisciplinary field of research.

Research and teaching have always been a great pleasure for me and I consider myself privileged to be able to do so. I certainly owe a great deal of this privilege to inspiring and supporting teachers, advisors, colleagues and friends. The most influential persons in this respect are Professor Joerg Appenzeller, with whom I have been collaborating for more than 20 years. Furthermore, my PhD advisor, Professor Bruno Lengeler impressed me with his dedication to teaching and research. During my career, I was fortunate to work under the guidance of PIs like Professor Siegfried Mantl, who supported and fostered me with a perfect mix of direction and great freedom while I was post-doc at Forschungszentrum Jülich; or Dr. Walter Riess who enabled me to work in a great team of researchers during my time at IBM Research in Zurich. Furthermore, I am also indebted to Professor Jesus del Alamo for the opportunity to work at the Microsystems Techonology Laboratories at MIT and to Professor Hans Lüth who always supported me.

A substantial fraction of the material that I am presenting here is the result of the dedicated and diligent work of a number of PhD students as well as post-docs and it emerged from collaborations with quite a few colleagues. Particularly, I would like to thank Dr. Marcel Müller, Dr. Thomas Grap, Dr. Felix Riederer, Dr. Birger Berghoff, Noel Wilck, Bin Sun, Lena Hellmich, Benjamin Richstein, Jan Klos, Liu Minshang, Alexander Gumprich, Thorben Frahm, Dr. Stefan Scholz and Dr. Karl Wolter. Moreover, many thanks go to the following colleagues and collaborators: Dr. Mikael Björk, Heinz Schmid, Dr. Heike Riel, Dr. Dirk König, Professor Qing-Tai Zhao, Dr. Dan Buca, Professor Yordan Georgiev, Dr. Lars Schreiber, Professor Jörg Schulze, Professor Zhihong Chen, Professor Walter Weber and the late Dr. Klaus Kallis. Most of all, I am indebted to my family for their patience and understanding when I was mentally absent during the last weeks of intensive writing to finish the book.

Aachen, May 2020.

Preface to the Second Edition

When I was contacted by DeGruyter and asked for a second edition of the book, I was very grateful to get the opportunity to remove quite a few mistakes in the text, in equations and figures that crept into the first edition no matter how careful I read the manuscript. My favorite is dropping a factor of e, i. e., the elementary charge. For instance, when coding a simulation I regularly equate V and eV and then insert negative signs and factors at the end as needed to fix units, etc. While this pragmatic way works very well for me, it not necessarily leads to the optimum way of presenting the material. Therefore, I tried to remove as many ambiguities as possible.

Apart from removing mistakes, the second edition of the book comes with multiple updates in the various chapters and additional material. Furthermore, I have added a novel chapter on cryogenic electronics to the current edition.

Finally, the new material that I presented in this second edition is again the result of the diligent work of quite a few collaborators, PhD and Master students. In addition to the ones mentioned in the preface of the first edition, I would like to thank Michail Michailow, Michael Frentzen, Patrick Liebisch, Yujie Li, Nazmus Sakib, Yichao Yang, Eike Ecking and Professor Christoph Stampfer. Moreover, special thanks go to the technical staff of the Central Laboratory for Micro and Nanotechnology at RWTH Aachen University, in particular to Jochen Heiss.

Aachen, January 2024

https://doi.org/10.1515/9783111054421-202

How to Use the Book

In our increasingly digital world, writing a book instead of producing a YouTube video may appear somewhat strange. However, I am convinced that digital formats such as videos alone, do not yield a long-lasting, in-depth understanding of complicated material. Personally, I learned the most when sitting at home trying to solve a problem which sometimes required several hours of thinking, i. e. it required effort. Everyone who plays a musical instrument knows exactly that the ability to play the instrument does not improve by watching videos or listening to the music played by somebody else; this can at best provide instructions, give guidelines or be stimulating. Making improvements ultimately requires practice. Reading a book requires effort, because it is an active process and stimulates thinking about the content. In contrast, watching a movie is a rather passive process. In fact, I frequently see my kids writing messages on their cell phones while watching a movie on TV but I never see this when they read a book because they are completely absorbed in the latter case. Being actively involved in the learning process strongly helps understanding and memorizing the content. From this point of view, a textbook is ideal since the content requires effort and one must be actively involved. On the other hand, every student knows that textbooks can also be highly frustrating in that sometimes the solutions to the problem sets are either missing or is provided in such a reduced way that it is difficult or near impossible to follow. Here, watching a video with step by step instructions is extremely helpful. Furthermore, an appropriate animation is often more elucidating than many pages of written explanations.

The contents of the present book is therefore digitally enhanced in that video explanations and supporting material as well as animations are provided through QR codes. Moreover, there are tasks integrated into the text that are provided with video solutions. As mentioned above, learning is most effective if you try to solve these problems yourself before watching the solution. To support doing so, the videos start with providing hints regarding the solution before going into the details.

Furthermore, I have also added exercises at the end of each chapters. To save space, both exercises and solutions are accessible via QR codes. The difference between the tasks and the exercises is that ideally you should try solving the tasks while reading the particular section they are embedded into. The exercises, on the other hand, can be made anytime after you have gone through the material as they also require knowledge of the material of the preceding chapters.

The last bit of digital enhancement are simulation tools that allow to study the material in more depth. Be aware of the fact that the tools only qualitatively reproduce experimental results since a number of approximations have been made in order to allow a quick simulation on a regular PC. The simulation tools are accessible via QR codes. Finally, a list of the QR codes is added to ease finding specific content.

https://doi.org/10.1515/9783111054421-203

List of QR Codes

https://doi.org/10.1515/9783111054421-204

Contents

1 Introduction

Within the last few decades, micro/nanoelectronics has undergone an enormous evolution and has become an ubiquitous technology that has an amazing impact on our lives.

From my point of view, the evolution can be subdivided into three major phases. The first phase mainly dealt with the miniaturization of the same device structure (the planar metal-oxide-semiconductor field-effect transistor based on bulk silicon) realized with (almost) the same set of materials. In retrospective, this phase is sometimes called the “era of happy scaling”: from an industrial engineering point of view, the “happy” certainly refers to the fact that it was basically clear what to do and often also how to do it. Today this era may better be characterized by the notion of “more Moore.” Currently, we are in the middle of the second phase that is sometimes called “more than Moore” alluding to an extension and diversification of the mere scaling approach of integrated circuits. The diversification of micro/nanoelectronics is characterized by the integration of novel materials, new device structures and novel functionalities. In particular, the advent of one- and two-dimensional materials has spurred investigations for new ways of realizing materials with tailored properties and the realization of smart materials that consist of ultracompact devices buried into three-dimensional material stacks. Furthermore, the connection between micro/nanoelectronics and biological systems will become one of the next big things. Finally, convergence is one of the buzzwords in particular of electronics and photonics systems. The third phase is currently about to lift-off. This phase will be marked by novel computing paradigms such as neuromorphic and quantum computing and their combination with traditional von Neumann architectures.

From a scientists point of view, being somewhere in between all these phases makes nanoelectronics a tremendously fascinating field of research. Mastering the nanotechnology required in order to manipulate materials on the few-nanometer scale, investigating and exploiting the quantum mechanical principles behind materials and device innovations is truly exciting. Particularly interesting is the fact that the development of micro and nanoelectronics led to increasingly powerful computers that have been used to obtain a deeper understanding and enable a further exploration of materials and device properties and concepts through simulations. With neuromorphic and quantum computers being available in the foreseeable future, this self-reinforcing development will be strongly accelerated.

However, nanoelectronics is also an enormously challenging field of research. In recent years, it has evolved into a truly multidisciplinary field of research. Apart from mastering the engineering challenges in terms of the fabrication of appropriate nanoscale structures scientist are faced with quantum physics, material science and chemistry. This can only be handled with a firm basis regarding the materials and device physics aspects as well as a thorough understanding of the technological aspects of manufacturing processes. In addition, a fundamental understanding of device simulations is neces-

https://doi.org/10.1515/9783111054421-001

sary to successfully contribute to the field of research. The present textbook provides an introduction to nanoelectronics covering solid-state physics aspects, semiconductor fabrication, device physics and device simulations as well as novel materials and device concepts. Obviously, each of these topics easily fills entire bookshelves and the present book can therefore only cover selected topics in limited breadth and depth. However, the present book is intended to give graduate students and newcomers alike an integral access to the topic covering the various aspects of the topic on an equal footing.

The book starts with the relevant solid-state and quantum physics fundamentals. A particular focus will be on tight-binding calculations in order to enable students to carry out their own, independent materials exploration. The following chapter covers semiconductor fabrication aspects followed by a chapter on the central ingredient of micro/nanoelectronic devices. Chapters 5 and 6 provide an introduction into the device physics aspects of nanoscale field-effect transistors as well as their simulation using the nonequilibrium Green's function formalism. Equipped with these fundamentals, nanoelectronics devices based on alternative concepts, employing novel materials and operated at cryogenic temperatures will be discussed in the remainder of the book.

2 Solid-State Physics Foundation

The extremely small dimensions of today's nanoelectronics devices, reaching the scale of a few nanometers only, make a quantum mechanical treatment of the phenomena observed in such devices mandatory. Tunneling of charge carriers, the manipulation of the electronic properties of semiconductors by exerting strain, by combining different materials into heterostructures, etc. requires a firm basis in solid-state physics. Therefore, this book starts with a summary of the most important concepts needed when working with nanoelectronics devices. Emphasis is put on quantum confinement and on how this leads to low-dimensional systems. Moreover, a major focus is on tight-binding calculations, since this enables the computation of material parameters of semiconductors providing a good starting point for elaborate device simulations. Finally, electronic transport through nanostructures is discussed.

2.1 Schrödinger Equation

Within a single-particle, quantum mechanical treatment of charge carriers, electrons are described by the wave function $\Psi(\vec{r}, t)$, which is obtained as a solution of the time-dependent Schrödinger equation given by

$$i\hbar\frac{\partial}{\partial t}\Psi(\vec{r}, t) = \underbrace{(\mathcal{H}_{\text{kin}} + \mathcal{H}_{\text{pot}})}_{\mathcal{H}}\Psi(\vec{r}, t), \tag{2.1}$$

where the kinetic term $\mathcal{H}_{\text{kin}}$ and the potential term $\mathcal{H}_{\text{pot}}$ of the Hamiltonian are explicitly given in Equation (2.2). In a static potential, the time dependence of the wave function $\Psi(\vec{r}, t)$ reduces to a time-dependent phase factor, and consequently, the wave function can be written in the form $\Psi(\vec{r}, t) = \psi(\vec{r})e^{-iEt/\hbar}$. Inserting this into Equation (2.1) yields the time-independent Schrödinger equation,

$$\underbrace{-\frac{\hbar^2}{2m}\left(\frac{\partial^2}{\partial x^2} + \frac{\partial^2}{\partial y^2} + \frac{\partial^2}{\partial z^2}\right)}_{\mathcal{H}_{\text{kin}}}\psi(\vec{r}) + \underbrace{V(\vec{r})}_{\mathcal{H}_{\text{pot}}}\psi(\vec{r}) = E\psi(\vec{r}). \tag{2.2}$$

In the present book, we will mostly consider (quasi)static potentials, and hence, we only have to deal with the time-independent Schrödinger equation. Next, we are going to find solutions to Equation (2.2) for various scenarios.

2.1.1 Schrödinger Equation in 1D

Let us start by solving the time-independent Schrödinger equation for a free particle in one dimension (1D). For simplicity, the potential is considered to be at the constant value

https://doi.org/10.1515/9783111054421-002

V_0. For the solution, we can then make the following plane wave ansatz:

$$\psi(x) = Ae^{ikx} + Be^{-ikx}. \tag{2.3}$$

Inserting this into the Schrödinger equation (2.2), it is easy to show that the second derivative of the kinetic term simply yields $-k^2\psi(x)$. As a result, one obtains

$$\frac{\hbar^2 k^2}{2m}\psi(x) + V_0\psi(x) = E(k)\psi(x), \tag{2.4}$$

from which the dispersion relation, i. e., the relation between energy E and the wave number k is obtained as $E(k) = \frac{\hbar^2 k^2}{2m} + V_0$, shown in Figure 2.1 for positive k-values (green parabola). Taking the derivative of the dispersion relation with respect to k and dividing by $\hbar$, one obtains

$$\frac{1}{\hbar}\frac{\partial E(k)}{\partial k} = \frac{1}{\hbar}\frac{\hbar^2 k}{m} = \frac{\hbar k}{m} = \frac{p}{m} = v(k) \tag{2.5}$$

where the de Broglie relation $\hbar k = p$ has been used. The resulting velocity $v(k)$ is the group velocity of a charge carrier.

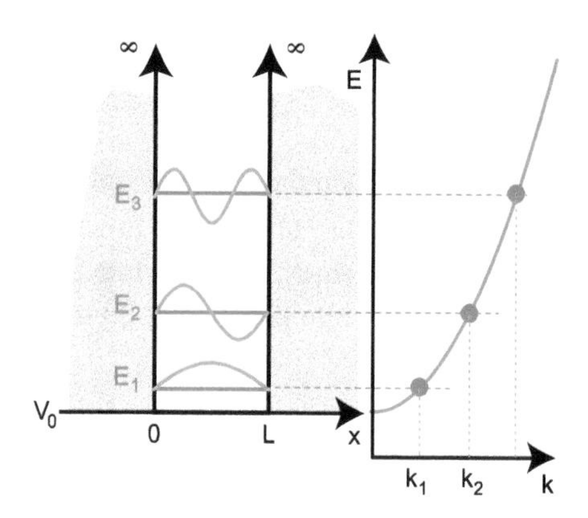

Figure 2.1: Quantization of a particle-in-the-box yields discrete, equidistant k_n, and hence discrete energy eigenvalues E_n.

2.1.2 The Particle-in-the-Box

When the free, one-dimensional motion of a particle is restricted to a region in-between $x = 0$ and $x = L$ with infinitely high potential barriers at the boundaries, the famous particle-in-the-box (PIB) system is obtained, which is illustrated in Figure 2.1 (left). The infinite potential barriers impose fixed boundary conditions, meaning that the wave function is identically zero for $x \leq 0$ and $x \geq L$. Using the ansatz wave function Equation (2.3), the boundary condition at $x = 0$ yields $A+B = 0$ resulting in $\psi(x) = 2iA\sin(kx)$; the boundary condition $\psi(x = L) = 0$ requires that $k \cdot L = n\pi$ with n being a positive integer. As a result, fixed boundary conditions lead to quantized values for k with $k_n = n\frac{\pi}{L}$, and hence discrete energy eigenvalues $E_n = \frac{\hbar^2\pi^2}{2mL^2}n^2 + V_0$.

Since the probability of finding the particle between x and $x+dx$ is given by $|\psi(x)|^2 dx$ and because the probability of finding the particle anywhere in-between 0–L must be unity, we get the following normalization condition that allows one to determine the prefactor A:

$$\int_0^L dx|\psi_n(x)|^2 = 4A^2 \int_0^L dx \sin^2\left(\frac{n\pi}{L}x\right) \stackrel{!}{=} 1 \rightarrow \psi_n(x) = \sqrt{\frac{2}{L}}\sin\left(\frac{n\pi}{L}x\right). \tag{2.6}$$

Solving the integral in Equation (2.6), one finally arrives at $A = \frac{1}{\sqrt{2L}}$. The wave functions of the particle-in-the-box problem are shown in Figure 2.1 (orange lines), which are multiples of a half-wave similar to the string of a guitar or violin. Since the wave is completely reflected at the infinite potentials (with a shift of the phase by π) in the stationary case it bounces back and forth infinitely many times resulting in a complete destructive interference at all energies except at the exact eigenenergies E_n of the PIB where constructive interference occurs (see QR code #1).

2.2 Free Electrons in Various Dimensions

In the preceding section, the 1D Schrödinger equation and the particle-in-the-box were discussed as a “warm-up” for quantum mechanics. Semiconductor devices, on the other hand, are made of three-dimensional solids, and thus we have to extend our considerations of quantum mechanics to the three-dimensional case. On the other hand, nanoelectronics devices have become so small that confinement of carriers within these devices leads to a reduction of the dimension due to quantization. In the following, it will therefore be discussed how quantization leads to low-dimensional carrier transport in semiconductor nanostructures.

Consider a particle in three-dimensional (3D) space and again set the potential energy to a constant, i. e., $V(x,y,z) = V_0$. Solving the 3D Schrödinger equation in this case is straightforward using the separation ansatz $\psi(x,y,z) = \phi(x) \cdot \varphi(y) \cdot \eta(z)$ for the wave function, which results in three 1D Schrödinger equations (see Task 1 for details) giving rise to the following dispersion relation:

$$E(k_x, k_y, k_z) = \frac{\hbar^2 k_x^2}{2m_x} + \frac{\hbar^2 k_y^2}{2m_y} + \frac{\hbar^2 k_z^2}{2m_z} + V_0 \tag{2.7}$$

where the fact has been included that the (effective) masses in x-, y- and z-directions may all be different. In the case $m_x = m_y = m_z$, the constant energy surfaces are spheres as becomes obvious when setting the energy in Equation (2.7) constant. If two effective masses are equal, a rotationally symmetric ellipsoid is obtained (see Task 1).

Task 1.
3D dispersion relation: Show that the dispersion relation of a free particle in 3D in a constant potential $V(x, y, z) = V_0$ is given by Equation (2.7). Use a separation ansatz for the wave function $\psi(x, y, z) = \phi(x) \cdot \varphi(y) \cdot \eta(z)$.

2.2.1 From 3D to 2D Systems

When considering the PIB, it became clear that confinement of carriers in one direction leads to quantized eigenenergy values. In real solid-state systems, carrier confinement can in principle be obtained by, e. g., etching a thin slab out of a 3D, bulk semiconductor. From the solutions of the PIB, it can be inferred that this involves extremely thin slabs with a thickness of a few nanometers only. Such thin slabs can be fabricated for instance with ultrathin silicon-on-insulator (see Figure 3.5 as an example). A very elegant way of realizing quantization is the use of a heterostructure consisting of III–V compound semiconductors: Epitaxial growth allows the insertion of a semiconductor with a small band gap such as InGaAs, into two barrier layers with larger band gap (for instance InAlAs). If an appropriate alignment of the conduction and valence bands can be realized (more on this topic will be discussed in Section 4.7) a quantum well can be formed as schematically shown in the left panel of Figure 2.2. In this case, confinement leads to the quantization of k_z. Neglecting the detailed potential distribution within the quantum well (the curvature of the band) and assuming that the confinement potential is high enough, the quantum well can be interpreted as a particle-in-the-box with infinite potential barriers and a constant potential of V_0 within the box yielding for the k_z-component the energy eigenvalues E_n stated above. As a result,

$$E(k_x, k_y, k_z) \longrightarrow E(k_x, k_y, n) = \frac{\hbar^2 k_x^2}{2m_x} + \frac{\hbar^2 k_y^2}{2m_y} + \frac{\hbar^2 \pi^2}{2m_z L_z^2} n^2 + V_0. \tag{2.8}$$

Confinement leads to a reduction in dimension giving rise to the formation of two-dimensional (2D) subbands, as shown in Figure 2.2, right panel. In this case, the dispersion relations $E(k_x, k_y, n)$ can be plotted with the quantized $k_z^n = n\frac{\pi}{L_z}$ as a parameter yielding rotationally symmetric paraboloids if $m_x = m_y$. Here, each paraboloid indexed with n “starts” at the energy $V_0 + \frac{\hbar^2\pi^2}{2mL_z^2} n^2$. If only $k_{x,y}$-states of the first 2D subband are occupied with electrons, the system is two-dimensional and a so-called two-dimensional electron gas (2DEG) is formed.

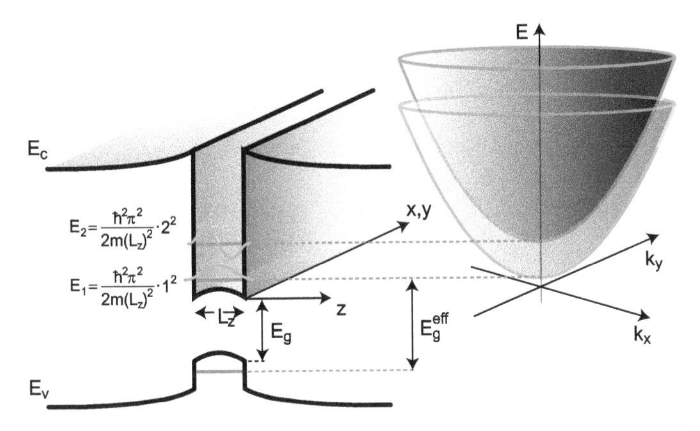

Figure 2.2: Quantization in one spatial direction (here, z-direction) leads to the formation of two-dimensional subbands with free motion within the x–y plane. The dispersion of each 2D subband is a paraboloid, as shown on the right-hand side. If all carriers reside only in the first subband, a two-dimensional electron gas (2DEG) is obtained. A 2DEG can be realized, e. g., with a heterostructure based on III–V compound semiconductors as illustrated in the left panel; here, E_c and E_v refer to the conduction and valence bands, respectively.

With a similar derivation, the impact of quantization on the valence band can be carried out leading to a corresponding set of paraboloids (upside down in this case and with possibly different masses $m^v_{x,y,z}$). One important implication of carrier confinement is that the band gap E_g of the quantized semiconducting material is increased to $E_g^{\text{eff}} = E_g + \frac{\hbar^2\pi^2}{2m_z L_z^2} + \frac{\hbar^2\pi^2}{2m_z^v L_z^2}$ as illustrated in Figure 2.2. Hence, a quadratic increase of E_g^{eff} is expected when L_z is decreased. However, in Section 4.4, ultrathin silicon quantum wells are studied as an example showing that simple carrier confinement may be inadequate to reproduce E_g^{eff} (see Figure 4.10(b)).

2.2.2 From 2D to 1D Systems

If confinement is imposed onto a bulk semiconductor in two spatial directions, e. g., in x- and z-directions, a quasi-1D system such as a nanowire is obtained. Nanowires can either be etched out of a volume material using appropriate lithography and etch steps (top-down fabrication) or can be grown bottom-up with chemical vapor deposition (bottom-up fabrication); Chapter 3 provides details on processing technology discussing different possibilities how to fabricate appropriate nanowires.

Consider a nanowire object where carriers are confined to $x = 0, \ldots, L_x$ and $z = 0, \ldots, L_z$ as depicted in Figure 2.3. Using the quantized eigenenergies of the PIB in x- and z-directions, it is obvious that for sufficiently small L_x and L_z and/or small carrier mass, quantization yields one-dimensional, energetically well-separated subbands. In Figure 2.3, two paraboloids are exemplarily shown that are the result of quantiza-

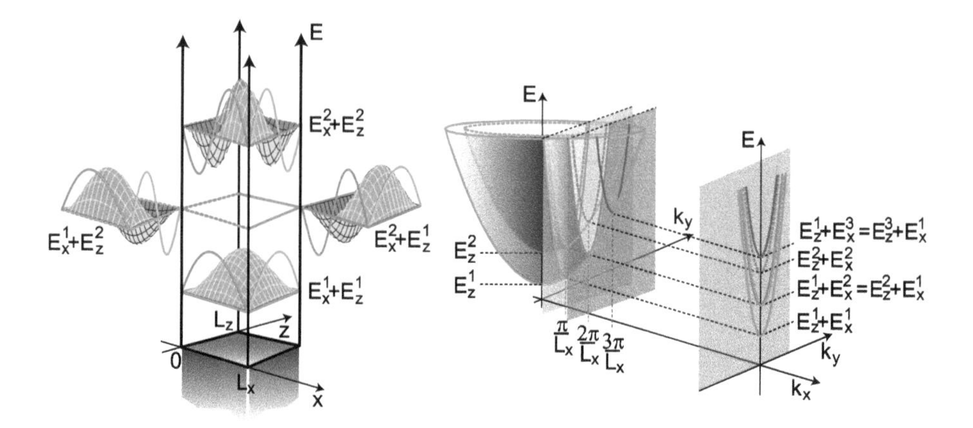

Figure 2.3: Quantization in two spatial directions (x- and z-directions in the given example) leads to a set of one-dimensional subbands that are a result of cuts between the paraboloids (due to k_z quantization) and the plane $k_x^r = r\frac{\pi}{L_x}$ as illustrated on the right-hand side. In the case, $L_x = L_z$ some of the 1D subbands can be degenerate (purple straight and dashed red lines).

tion along the z-direction (the same as shown in Figure 2.2). Now, if we assume that the nanowire can be considered as a 2D particle-in-the-box then $k_x^r = r\frac{\pi}{L_x}$ as well as $k_z^n = n\frac{\pi}{L_z}$ and the dispersion relation becomes

$$E(r, k_y, n) = \frac{\hbar^2 k_y^2}{2m_y} + \underbrace{\frac{\hbar^2 \pi^2}{2m_x L_x^2} r^2}_{E_x^r} + \underbrace{\frac{\hbar^2 \pi^2}{2m_z L_z^2} n^2}_{E_z^n} + V_0, \tag{2.9}$$

where r and n are integer numbers. This means that additional quantization along the x-direction yields k-states on cuts of the planes given by $k_x^r = r\frac{\pi}{L_x}$ = const. with the paraboloids. In turn, these cuts result in parabolas that are projected onto the E–k_y plane; both paraboloids (and also higher ones) provide 1D subbands plotted as straight and dashed lines in Figure 2.3, right panel. The first three energy levels at $E_x^1 + E_z^1 = E^{11}$, $E_x^1 + E_z^2 = E^{12}$ and $E_x^2 + E_z^2 = E^{22}$ and their respective wave functions are displayed in Figure 2.3, left panel. Note that $L_x = L_z$ and $m_x = m_z$ was assumed here, and as a result the energy values for $n = 1, r = 2$ and $n = 2, r = 1$ are degenerate since $E^{12} = E^{21}$.

Task 2.
Carrier confinement and degeneracy: Consider a nanowire with rectangular cross-section with $L_x = 2L_z$. The carrier mass is considered to be equal to m in all three spatial directions. Suppose that for $V_0 = 0$ the first subband lies at $E^{11} = 0.1$ eV. Compute the energy of the next three one-dimensional subbands. Are they degenerate?

2.2.3 From 1D to 0D Systems

If a system is made sufficiently small in all three spatial directions, quantization leads to the formation of fully discrete energy levels. Figure 2.4 shows the one-dimensional parabolas obtained from the quantization of x- and z-directions as detailed in the preceding section (note that the second parabola is twofold degenerate). If in addition, k_y is quantized we obtain a quantum dot. If $L_{x,y,z} = L$ and $m_{x,y,z} = m$, the eigenenergies of the quantum dot are obtained as

$$E^{rln} = \frac{\hbar^2\pi^2}{2mL^2}(r^2 + l^2 + n^2) + V_0 \tag{2.10}$$

where we assumed again a particle-in-the-box quantization with infinite potential. In Equation (2.10), n, r and l are integer numbers representing the indices in the three directions of quantization. The eigenenergies for the lowest indices are displayed in the left panel of Figure 2.4 with some of them being threefold degenerate. Looking at the constant contour plots of the respective eigenfunctions displayed in the right panel of Figure 2.4 it becomes clear that the quantum dot can be interpreted as artificial atom: the lowest eigenenergy E^{111} represents an s-like wave function similar to a hydrogen atom. The next higher energy value with threefold degeneracy ($E^{211} = E^{121} = E^{112}$) is equivalent to the $p_{x,y,z}$ orbitals of a hydrogen atom, etc. The degeneracy follows directly from Equation (2.10) and can also be extracted from the left panel of Figure 2.4.

Task 3.
Quantization in a quantum dot: Consider a quantum dot as displayed in Figure 2.4 that is etched out of a bulk material. The dot is quantized in all three spatial directions with $L_x = L_y = L_z$ but different masses $m_x = 2m_y = 4m_z$. What are the eigenenergies (assume $V_0 = 0$ eV) of the first five levels of the quantum dot? Are they degenerate?

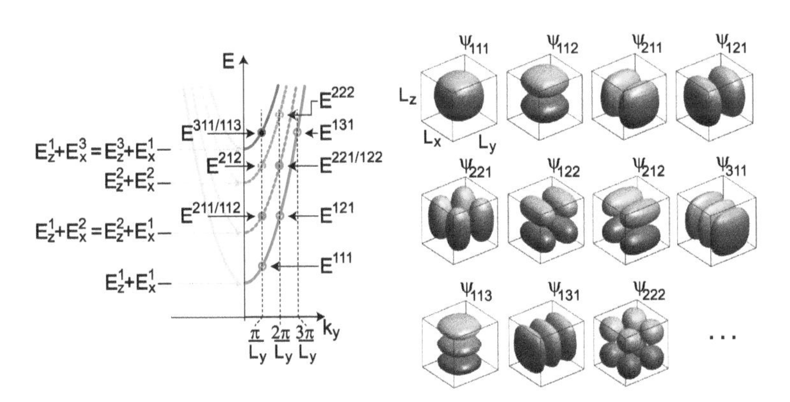

Figure 2.4: Discrete energy eigenvalues of a quantum dot due to quantization in x-, y- and z-directions. Note that because of the particle-in-the-box quantization, only positive k-values are allowed. Right panel: constant contour plots of the appropriate wave functions belonging to the displayed eigenenergies.

2.3 Electrons in Solids—Bravais Lattice, Wigner–Seitz Cell, Brillouin Zone and Related Concepts

Up to now, we considered the electrons as being free particles. However, nanoelectronics devices are made of semiconductors, i. e., crystalline solids and we therefore need to incorporate the solid into our considerations. This implies that the simple quadratic dispersion relation $E(k_x, k_y, k_z)$ of a free particle will change into a more complicated band structure. But since the solid is a crystal the electrons experience a periodic potential that determines how the band structure eventually will look like. In order to understand this, we need a number of concepts from crystallography related to the periodic nature of the crystal and its symmetries. These concepts will only be mentioned briefly but they are necessary for the band structure calculations.

Crystalline semiconductors (or in general solids) are made of a small portion of the crystal called basis that is repeated periodically on a lattice. The lattice represents the underlying periodic structure spanned by linearly independent lattice vectors $\vec{a}_{1,2,3}$ such that each point on the lattice can be reached by the translational vector $\vec{R} = n\vec{a}_1 + m\vec{a}_2 + k\vec{a}_3$ with n, m, k being integer numbers. The lattice structures that can be constructed this way are classified according to their symmetry properties and are called Bravais lattices.

In 1D, only one Bravais lattice exists (the linear chain), but in two dimensions, five Bravais lattices exist that are displayed in Figure 2.5 and in the 3D case, the 14 Bravais lattices displayed in Figure 2.6 are obtained. The area in the 2D case(volume in 3D) spanned by the basis vectors $\vec{a}_{1,2,3}$ is called the unit cell of the lattice. However, as can be inferred from Figure 2.5 the choice of the unit cell is not unique: all (dark and light) blue-shaded areas in the figure represent unit cells of the respective Bravais lattice. The unit cell with the smallest area(volume) is called primitive unit cell. Even this does not have to

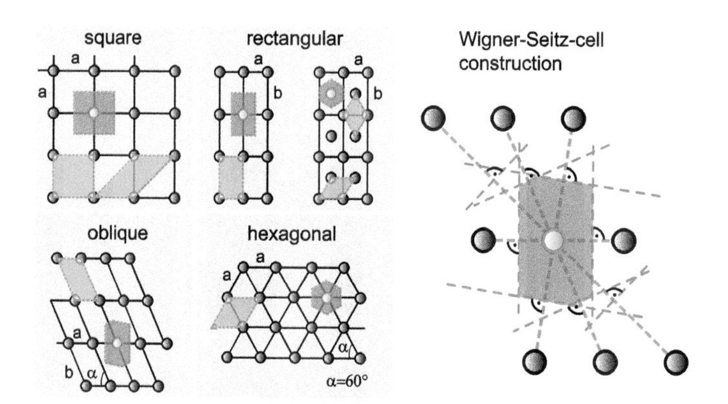

Figure 2.5: Left panels: Possible Bravais lattices in two dimensions. The choice of a unit cell is not unique: the dark and light blue-shaded unit cells show possible examples each containing a single atom. The dark blue unit cell is the Wigner–Seitz cell of the respective lattice. The right panel shows the construction of a Wigner–Seitz cell (dark blue-shaded area) of an oblique 2D lattice.

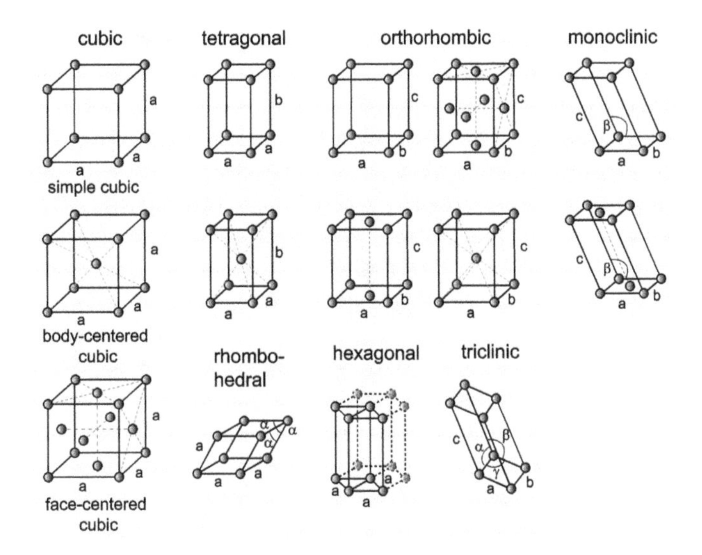

Figure 2.6: The 14 Bravais lattices in three-dimensional space.

be unique as the examples of the square and rectangular lattices in Figure 2.5 show. In order to find a unique unit cell, we can use the approach put forward by Wigner and Seitz [270], illustrated in Figure 2.5, right panel: From a central atom, tie lines (dashed green lines in Figure 2.5, right panel) are drawn to the nearest and next nearest neighboring atoms. At the midpoints of these tie lines, perpendicular lines (in a 2D lattice) or perpendicular mini-planes (in a 3D lattice) are drawn. The minimal area(volume) enclosed by the perpendicular lines(planes) and their intercepts is the so-called Wigner–Seitz cell (dark blue-shaded area in Figure 2.5, right panel). Wigner–Seitz cells for the five two-dimensional Bravais lattices are displayed in Figure 2.5. In the case of a Wigner–Seitz cell, a single lattice point always sits in the center of the cell.

In three dimensions, the cubic lattices and especially the so-called face-centered cubic (fcc) lattices are of particular interest since the diamond lattice structure of silicon, germanium and many III–V compound semiconductors consists of two interwoven fcc lattices (see Figure 2.24). It might not be immediately obvious that, e. g., for the body-centered or the face-centered cubic lattice a primitive unit cell containing only a single lattice point can be found. Based on the unit cells displayed in Figure 2.6 (left column) the bcc lattice contains two and the fcc contains four lattice points. However, Figure 2.7 shows in the case of a fcc lattice that an appropriate unit cell indeed only contains a single lattice point: either the blue-shaded parallelepiped with basis vectors given by $\vec{a}_1 = (a/2, a/2, 0)$, $\vec{a}_2 = (a/2, 0, a/2)$ and $\vec{a}_3 = (0, a/2, a/2)$ can be chosen which is illustrated in Figure 2.7 in the bottom right panel. Alternatively, the Wigner–Seitz cell can be constructed around the blue marked atom leading to the unit cell shown in the right panel of Figure 2.7; it will be used in the next section to compute the band structure of silicon with the tight-binding formalism.

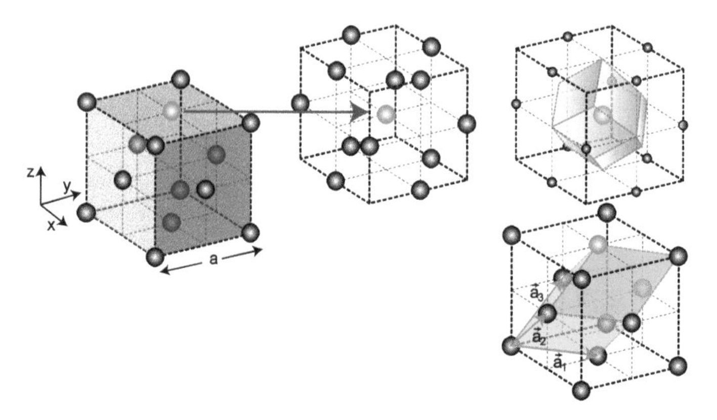

Figure 2.7: Face-centered cubic (fcc) lattice (left). Shifting the cube along the z-direction by $a/2$ allows constructing the Wigner–Seitz cell (right panel) with the blue atom in the center. The bottom right panel shows an alternative unit cell containing a single atom. The primitive lattice vectors $\vec{a}_1$, $\vec{a}_2$ and $\vec{a}_3$ can be constructed exploiting the symmetry of the cubic lattice.

Eventually, a crystal is formed by placing the basis of the crystal at each point of the Bravais lattice. This implies that the primitive unit cell contains as many atoms as the basis. In Section 2.6.2, this will be made clear when discussing the silicon crystal structure. The basis can be rather complicated and may contain a large number of atoms, as will be discussed when computing the band structure of carbon nanotubes in Section 2.9. The importance of the considerations concerning the crystal structure of solids becomes apparent when the tight-binding calculations of the band structure of solids are discussed in Section 2.4.

Task 4.
Wigner–Seitz cell: Consider the two-dimensional crystals displayed in the figures provided through the QR code #5 at the page margin. Find the primitive unit cells of the underlying lattices. What does the basis look like? Construct the Wigner–Seitz cell of the lattices.

2.3.1 Reciprocal Space

The concept of a crystal consisting of basis and lattice is important for the solution of the Schrödinger equation since the resulting wave function can be split into a factor due to the basis atoms[1] within the unit cell, multiplied with a phase factor $e^{i\vec{k}\vec{r}}$ that yields a plane wave with the periodicity of the underlying Bravais lattice (so-called Bloch waves). For each real lattice, there exists a so-called reciprocal lattice spanned by reciprocal basis lattice vectors $\vec{b}_{1,2,3}$. The points of the reciprocal lattice $\vec{K}$ are defined by the condition

1 This will be a linear combination of the atomic orbitals of the atoms of the basis.

that $e^{i\vec{K}(\vec{r}+\vec{R})} \stackrel{!}{=} e^{i\vec{K}\vec{r}}$, which requires that $e^{i\vec{K}\vec{R}} = 1$. In turn, this means that $\vec{K} \cdot \vec{R}$ must be an integer multiple of 2π. It can be shown that the basis vectors of the reciprocal lattice are given by

$$\vec{b}_1 = 2\pi \frac{\vec{a}_2 \times \vec{a}_3}{\vec{a}_1 \cdot (\vec{a}_2 \times \vec{a}_3)}, \quad \vec{b}_2 = 2\pi \frac{\vec{a}_3 \times \vec{a}_1}{\vec{a}_1 \cdot (\vec{a}_2 \times \vec{a}_3)}, \quad \vec{b}_3 = 2\pi \frac{\vec{a}_1 \times \vec{a}_2}{\vec{a}_1 \cdot (\vec{a}_2 \times \vec{a}_3)}. \tag{2.11}$$

If we expand $\vec{k}$ into the basis vectors of the reciprocal lattice, i. e., $\vec{K} = \tilde{n}\vec{b}_1 + \tilde{m}\vec{b}_2 + \tilde{r}\vec{b}_3$ and $\vec{R} = n\vec{a}_1 + m\vec{a}_2 + r\vec{a}_3$ into the basis vectors of the real lattice $\vec{a}_{1,2,3}$ one indeed obtains

$$\vec{K} \cdot \vec{R} = n\tilde{n} \underbrace{\vec{a}_1 \cdot \vec{b}_1}_{=2\pi} + m\tilde{m} \underbrace{\vec{a}_2 \cdot \vec{b}_2}_{=2\pi} + r\tilde{r} \underbrace{\vec{a}_3 \cdot \vec{b}_3}_{=2\pi} = 2\pi \cdot l, \tag{2.12}$$

since $\vec{a}_i \cdot \vec{b}_j = 2\pi\delta_{ij}$, as follows immediately from Equation (2.11). Note that Equation (2.12) implies that the reciprocal vectors $\vec{b}_{1,2,3}$ of a cubic real lattice with orthogonal basis vectors $\vec{a}_{1,2,3}$ point in the same direction as the basis vectors of the real lattice. This fact will be important to understand how the silicon band structure is modified due to quantization in a silicon inversion layer (see Chapter 4). Examples of how the reciprocal lattice and its basis vectors look like will be discussed in the next section in different dimensions.

2.3.2 The Brillouin Zone

In the reciprocal lattice, a Wigner–Seitz cell can be determined in the same way as described above. The resulting unit cell is called the first Brillouin zone, which plays a prominent role when displaying the band structure of a solid. In fact, the first Brillouin zone contains all k-values of waves that can be represented with the particular lattice structure under consideration, and hence the entire band structure. Due to its importance, the concept of the first Brillouin zone will be elaborated in more detail below covering 1D, 2D and finally the three-dimensional case.

In the one-dimensional case, the only possible Bravais lattice is a regular arrangement of lattice points with lattice constant a. As a result, the reciprocal lattice "vector" is simply $b = \frac{2\pi}{a}$. Constructing the Wigner–Seitz cell by intersecting the tie-lines between adjacent reciprocal lattice points exactly in the middle yields the first Brillouin zone to range from $-\frac{\pi}{a}$ to $\frac{\pi}{a}$.

As displayed in Figure 2.5, there are five Bravais lattices in the two-dimensional case. Using the fact that $\vec{a}_i \cdot \vec{b}_j = 2\pi\delta_{ij}$ it is straightforward to compute the reciprocal basis lattice vectors $\vec{b}_{1,2}$ from the basis vectors of the real lattice $\vec{a}_{1,2}$. As two representative examples, Figure 2.8 shows the real (left) and reciprocal (right) lattices of the rectangular and the body-centered rectangular Bravais lattices together with the appropriate basis vectors. The first Brillouin zones are shown on the right. A similar construction can be carried out for the remaining 2D Bravais lattices.

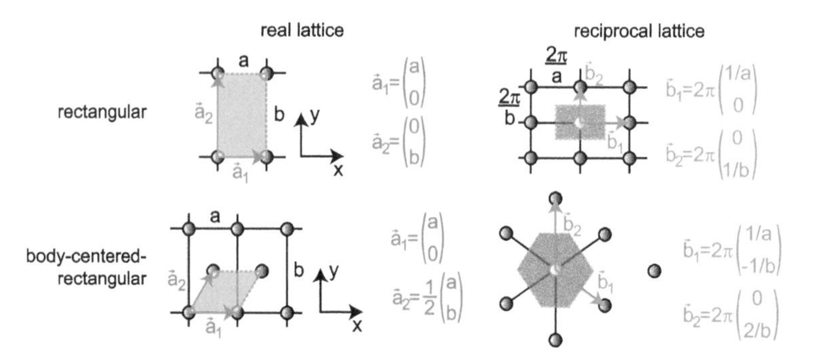

Figure 2.8: Real and reciprocal lattices for a 2D rectangular and a 2D body-centered rectangular lattice. The gray-shaded areas in the right panels are the first Brillouin zones of the two lattices.

In the 3D case, only the reciprocal lattice and the first Brillouin zone of the fcc lattice are provided here, because of their importance for elemental semiconductors such as silicon and germanium. When the reciprocal basis vectors $\vec{b}_{1,2,3}$ are constructed based on Equation (2.11), a body-centered cubic (bcc) lattice is obtained (as reciprocal lattice of a fcc lattice), which is illustrated in Figure 2.9. Constructing the Wigner–Seitz cell within the bcc reciprocal lattice finally yields the first Brillouin zone of the fcc lattice (i. e., of diamond, Si and Ge) as depicted in the right panel of Figure 2.9.

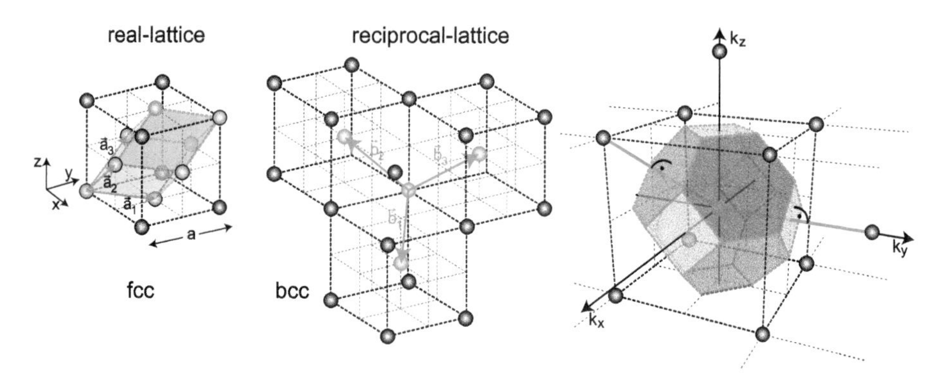

Figure 2.9: The reciprocal lattice of a fcc Bravais lattice spanned by the basis vectors $\vec{a}_{1,2,3}$ is a body-centered cubic lattice with reciprocal basis vectors $\vec{b}_{1,2,3}$.

2.4 Tight-Binding Calculation of Band Structures

In order to understand and compute the electronic properties of semiconductor devices, we need to know the band structure of the materials involved. To this end, the tight-binding method will be introduced in detail in the present section. The discussion starts with simplified and low-dimensional systems before a generalization will be discussed.

A recipe for tight-binding calculations will then be given that is employed to compute exemplarily the band structure of carbon nanotubes, 2D materials and silicon.

2.4.1 Discrete Schrödinger Equation

Before using the tight-binding formalism to compute the band structure of a solid, let us solve the Schrödinger equation on a discrete lattice with lattice constant a (illustrated in Figure 2.10(a)), i. e., we solve Schrödinger's equation using a finite difference scheme (cf. Section 6.1.1, in particular Equation (6.1)). To keep things as simple as possible, the calculation will be done in one-dimension but can easily be extended to higher dimensions (see Task 5). Using the difference quotient, it is straightforward to show that

$$\frac{d^2}{dx^2}\psi(x) \rightarrow \frac{\frac{\psi_{j+1}-\psi_j}{a} - \frac{\psi_j-\psi_{j-1}}{a}}{a} = \frac{\psi_{j+1} - 2\psi_j + \psi_{j-1}}{a^2}. \tag{2.13}$$

As a result, the Schrödinger equation can be rewritten in the following form:

$$-t(\psi_{j+1} - 2\psi_j + \psi_{j-1}) + V_j\psi_j = E\psi_j \tag{2.14}$$

with $t = \frac{\hbar^2}{2ma^2}$. As will become clear below, Equation (2.14) represents a tight-binding calculation of a 1D solid with nearest neighbor interaction if (i) the lattice points of the finite difference grid are interpreted as atomic positions of the solid, (ii) the wave functions ψ_j at each lattice point j are interpreted as atomic s-orbitals and (iii) t is replaced with the overlap integral (see below) of adjacent atomic orbitals.

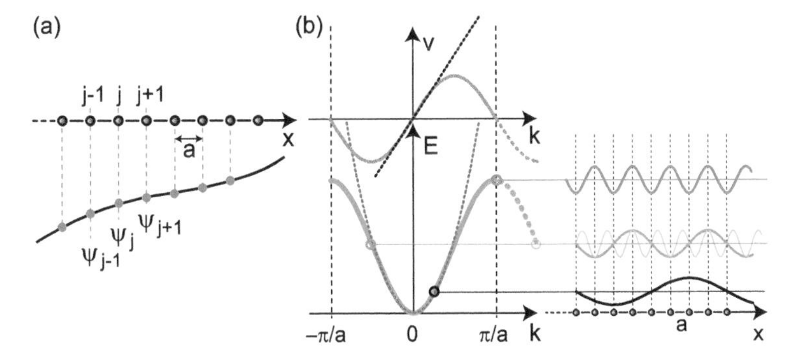

Figure 2.10: (a) Finite difference discretization of a 1D wave function $\psi(x)$ on a regular lattice $x = j \cdot a$ with lattice constant a. (b) Resulting dispersion relation (green) and group velocity (gray line in the top panel). The right panel shows the wave function for four different k-values. Due to the discreteness, the wave functions for $k > \frac{\pi}{a}$ are already accounted for in the first Brillouin zone. For instance, $k = \frac{3\pi}{2a}$ (light blue line) is represented by $k = -\frac{\pi}{2a}$ (solid dark blue line).

In the case of $V_j = V_0$, Equation (2.14) can be solved with a plane wave ansatz, i. e., by assuming that $\psi(x_j) = \psi_j = \Phi_0 \exp(ikx_j) = \Phi_0 \exp(ikja)$. As a result, $\psi_{j\pm1} = \Phi_0 \exp(ik(j \pm 1)a) = \psi_j \cdot \exp(\pm ika)$. Inserting this into Equation (2.14) yields

$$-t(e^{ika} - 2 + e^{-ika}) + V_0 = E. \tag{2.15}$$

Finally, with $e^{ika} + e^{-ika} = 2\cos(ka)$ the following dispersion relation $E(k)$ of a discrete finite difference lattice is obtained:

$$E(k) = 2t(1 - \cos(ka)) + V_0, \tag{2.16}$$

which is plotted in Figure 2.10(b) (solid green line) together with the quadratic dispersion relation of a free particle (dotted dark green line). When $a \to 0$, i. e., in the continuum limit $\cos(ka)$ can be approximated with a Taylor expansion and the quadratic dispersion relation of a free particle is recovered.

If we interpret the discretization on a finite difference grid as a one-dimensional crystal with lattice constant a, then the relevant k-states lie within the first Brillouin zone (cf. Section 2.3.2), i. e., within the k-interval $-\frac{\pi}{a} < k \leq \frac{\pi}{a}$. This fact becomes clear when looking more closely at Equation (2.16): for $k > \frac{\pi}{a}$ corresponding wave functions exhibit decreasing wavelengths λ as expected for increasing $k = \frac{2\pi}{\lambda}$ (see light blue line in Figure 2.10(b)). However, when the value of the respective wave function is projected onto the discrete lattice, these values are the same as the ones belonging to a wave function with a smaller k within the first Brillouin zone. This is illustrated in Figure 2.10(b) with the light and dark blue waves: While the light blue wave with $k > \frac{\pi}{a}$ shows a significantly smaller wavelength it has the same value at the lattice points (right panel), and hence the light and dark blue waves are not different states. The minimum wavelength is actually at the Brillouin zone boundary where $k = \frac{\pi}{a} \to \lambda = 2a$, meaning that the wave function becomes maximal/minimal at adjacent lattice points (red line in Figure 2.10(b)); more explanations are provided through the QR code #6.

The dispersion relation of a discrete lattice has a further interesting consequence for the group velocity $v(k) = \frac{1}{\hbar}\frac{\partial E(k)}{\partial k}$ of a particle: since the group velocity is proportional to $\sin(ka)$ as illustrated in the top panel of Figure 2.10(b) (gray line), the velocity does not increase linearly with k as in the continuum case (black dotted line in the top panel of Figure 2.10(b)) but first increases with k, slows down and eventually becomes zero when approaching the Brillouin-zone boundary (see QR code #6). The situation is similar to a spinning wheel observed under a stroboscope light source. When the frequency of rotation increases one first observes faster spinning. However, when the angular frequency of rotation approaches the frequency of the stroboscope, the wheel appears to be slowing down and eventually seems to be standing still.

In the following section, we will further consider discrete solids. However, while in the case discussed so far (finite difference lattice), the lattice was a mere consequence of the discretization of the Scrödinger equation (e. g., in order to be able to solve it numerically), the next sections will deal with discrete systems representing a periodic lattice of a (more or less) real solid.

Task 5.
2D discrete Schrödinger equation: Set up the 2D version of the Schrödinger equation with constant potential $V(x,y) = V_0$ using the finite difference scheme on a regular lattice with lattice constant a in x- and y-directions and solve it employing an appropriate discrete plane wave ansatz. Plot the resulting dispersion relation.

2.4.2 Linear Combination of Atomic Orbitals

Any wave function $\psi_l(\vec{r})$ can be written as a linear combination of a complete set of basis wave functions ϕ_n with unknown coefficients c_n^l, i. e., $\psi_l(\vec{r}) = \sum_n c_n^l \phi_n(\vec{r})$. Let us consider four (one-dimensional) atoms in a row each a distance a apart from its adjacent neighbors as depicted in Figure 2.11, left panels. If the coupling between nearest neighbors is not too strong, it appears reasonable to chose the atomic orbitals $\phi_n(x)$ (blue lines) of the individual atoms (black potential lines) as basis wave functions. For the time being, it is assumed that each atom has only a single orbital. The Hamiltonian of the four-atom system is then given by

$$\mathcal{H} = -\frac{\hbar^2}{2m}\frac{d^2}{dx^2} + V_1(x) + V_2(x) + V_3(x) + V_4(x). \tag{2.17}$$

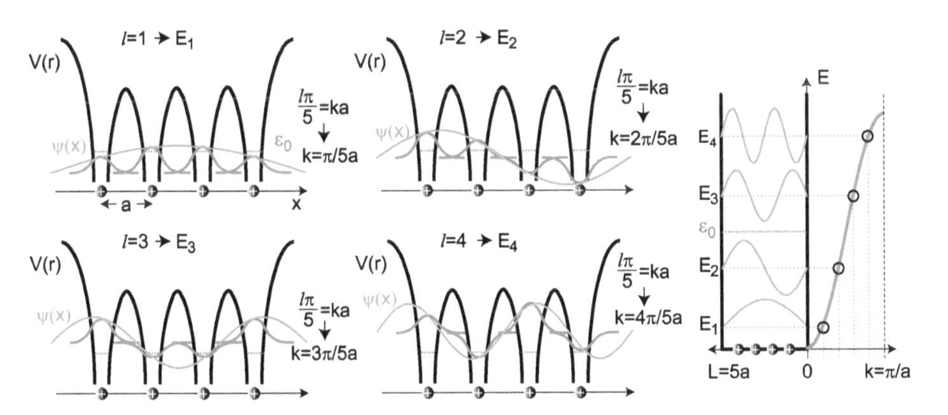

Figure 2.11: LCAO in a system consisting of four atomic Coulomb potentials (black lines). The overall wave function is made up of the four ground-state wave functions (blue lines) of the isolated atoms. The envelope (orange lines) is one of the four solutions shown in the right panel belonging to a PIB of length $L = 5a$ with four discrete energy eigenvalues due to quantization.

Using the linear combination of atomic orbitals (LCAO) as ansatz wave function, the Schrödinger equation becomes $\mathcal{H}(\sum_{n=1,\ldots,4} c_n^l \phi_n) = E(\sum_{n=1,\ldots,4} c_n^l \phi_n)$. To determine the unknown coefficients $c_1^l, \ldots, c_4^l$, the Schrödinger equation is multiplied from the left with the conjugate complex of the atomic orbital of site 1, i. e., $\phi_1^\star(x)$ and integrated over x. Next, we do the same with the other atomic orbitals $2, \ldots, 4$ and as a result obtain a set of four equations for the unknown coefficients $c_1^l, \ldots, c_4^l$. In these equations, several integrals of the form $\int dx \phi_m^\star(x)\mathcal{H}\phi_n(x)$ as well as $\int dx \phi_m^\star(x)\phi_n(x)$ will appear. To write this in a much more compact way, Dirac's "bra-ket"-notation is used in the following, in which we set $\int dx \phi_m^\star(x)\mathcal{H}\phi_n(x) = \langle m|\mathcal{H}|n\rangle$ and $\int dx \phi_m^\star(x)\phi_n(x) = \langle m|n\rangle$. The resulting four equations for the unknown coefficients can hence be written as

$$\begin{pmatrix} c_1^l\langle 1|\mathcal{H}|1\rangle + c_2^l\langle 1|\mathcal{H}|2\rangle + c_3^l\langle 1|\mathcal{H}|3\rangle + c_4^l\langle 1|\mathcal{H}|4\rangle \\ c_1^l\langle 2|\mathcal{H}|1\rangle + c_2^l\langle 2|\mathcal{H}|2\rangle + c_3^l\langle 2|\mathcal{H}|3\rangle + c_4^l\langle 2|\mathcal{H}|4\rangle \\ c_1^l\langle 3|\mathcal{H}|1\rangle + c_2^l\langle 3|\mathcal{H}|2\rangle + c_3^l\langle 3|\mathcal{H}|3\rangle + c_4^l\langle 3|\mathcal{H}|4\rangle \\ c_1^l\langle 4|\mathcal{H}|1\rangle + c_2^l\langle 4|\mathcal{H}|2\rangle + c_3^l\langle 4|\mathcal{H}|3\rangle + c_4^l\langle 4|\mathcal{H}|4\rangle \end{pmatrix} = E_l \begin{pmatrix} c_1^l\langle 1|1\rangle + c_2^l\langle 1|2\rangle + c_3^l\langle 1|3\rangle + c_4^l\langle 1|4\rangle \\ c_1^l\langle 2|1\rangle + c_2^l\langle 2|2\rangle + c_3^l\langle 2|3\rangle + c_4^l\langle 2|4\rangle \\ c_1^l\langle 3|1\rangle + c_2^l\langle 3|2\rangle + c_3^l\langle 3|3\rangle + c_4^l\langle 3|4\rangle \\ c_1^l\langle 4|1\rangle + c_2^l\langle 4|2\rangle + c_3^l\langle 4|3\rangle + c_4^l\langle 4|4\rangle \end{pmatrix}. \tag{2.18}$$

Since the wave functions $\phi_1, \ldots, \phi_4$ are the (normalized) eigenfunctions of each atom, the term $\langle n|\mathcal{H}|n\rangle \approx \epsilon_0\langle n|n\rangle = \epsilon_0$ yields the eigenenergy ϵ_0, which is the same for site $1, \ldots, 4$ since all atoms are considered to be equal (cf. Figure 2.11).[2] Furthermore, the wave function of each atom ϕ_n is regarded as being localized at its atomic sites and rapidly decays further away from the site n so that a nonnegligible overlap of wave functions (atomic orbitals) is only obtained between nearest neighbors. Since all atoms are considered to be the same and on a regular lattice the integrals $\langle n|\mathcal{H}|m\rangle$ all yield the same value when n and m are interchanged. We set $\langle m|\mathcal{H}|n\rangle = -V_{ss}$ if $m = n \pm 1$ and $\langle m|\mathcal{H}|n\rangle = 0$ if $|n - m| \geq 2$.[3]

Finally, with the same argument of wave functions being tightly bond to their host atom, the overlap integrals $\int dx \phi_m^\star(x)\phi_n(x) = \langle m|n\rangle$ are: $\langle n|n\rangle = 1$, $\langle m|n\rangle = \Delta$ if $m = n \pm 1$ and $\langle m|n\rangle = 0$ otherwise. Hence, Equation (2.18) can be written in matrix form as

2 Note that this is certainly an approximation since ϕ_n are not really eigenfunctions of the full Hamiltonian $\mathcal{H}$.

3 The change of nomenclature from $t \to V_{ss}$ (cf. Equation (2.16)) for the overlap integral acknowledges the fact that the considered overlap integral stems from two s-orbital-like wave functions (ground-state wave functions). Further below, the overlap between s- and p-like orbitals will also be taken into consideration and denoted accordingly.

$$\underbrace{\begin{pmatrix} \epsilon_0 & -V_{ss} & 0 & 0 \\ -V_{ss} & \epsilon_0 & -V_{ss} & 0 \\ 0 & -V_{ss} & \epsilon_0 & -V_{ss} \\ 0 & 0 & -V_{ss} & \epsilon_0 \end{pmatrix}}_{\mathsf{H}} \begin{pmatrix} c_1^l \\ c_2^l \\ c_3^l \\ c_4^l \end{pmatrix} = E_l \begin{pmatrix} 1 & \Delta & 0 & 0 \\ \Delta & 1 & \Delta & 0 \\ 0 & \Delta & 1 & \Delta \\ 0 & 0 & \Delta & 1 \end{pmatrix} \begin{pmatrix} c_1^l \\ c_2^l \\ c_3^l \\ c_4^l \end{pmatrix}. \tag{2.19}$$

This is a generalized eigenvalue problem since the matrix on the right side of the equation is not the unit matrix. However, since the overlap of the wave functions $\langle m|n\rangle$ even between nearest neighbors is small, Δ is a small number, which is frequently neglected such that Equation (2.19) becomes a simple eigenvalue problem and the energy eigenvalues E_l (i. e., the discrete dispersion relation) are obtained by solving $\det|\mathsf{H} - E_l\mathbf{1}| = 0$. In the present case, one obtains four eigenvalues $E_{l=1}, \dots, E_{l=4}$ and four eigenvectors $\vec{c}^{\,l=1}, \dots, \vec{c}^{\,l=4}$. Note that $l = 1, \dots, 4$ is an integer number indexing the eigenvalues and eigenvectors (and is of course related to the wave number k used above via $k_l a = l\frac{\pi}{5}$, see below). H is a symmetric Toeplitz-matrix, and thus, its eigenvalues can be computed analytically in the present case. These eigenvalues are explicitly given by $E_l = \epsilon_0 - 2V_{ss}\cos(\frac{l\pi}{4+1})$, which is similar to the dispersion relation Equation (2.16) obtained as a solution of the discretized Schrödinger equation if we set $\frac{l\pi}{4+1} = ka$. Two eigenvalues above and two below ϵ_0 are obtained, as shown in the right part of Figure 2.11. Moreover, the eigenvectors are given in the form $\vec{c}^{\,l} = (\sin(\frac{1\cdot\pi}{4+1}l), \dots, \sin(\frac{4\cdot\pi}{4+1}l))$.

Figure 2.11 shows the scenario discussed so far. The eigenvectors $\vec{c}^{\,l=1}, \dots, \vec{c}^{\,l=4}$ determine the amplitudes with which each $\phi_n(x)$ contributes and the resulting envelopes (orange lines in Figure 2.11) are equal to the eigenfunctions of a particle-in-the-box system depicted in the right panel of Figure 2.11. For instance, the eigenvalue in the case $l = 1$ is $E_1 = \epsilon_0 - 2V_{ss}\cos(\frac{\pi}{5})$ and the eigenvector is $\vec{c}^{\,1} = (\sin(\frac{\pi}{5}), \sin(\frac{2\pi}{5}), \sin(\frac{3\pi}{5}), \sin(\frac{4\pi}{5})) \approx (0.59, 0.95, 0.95, 0.59)$. Hence, the wave function ψ_l of a system of n coupled atoms (each providing a single orbital) is given by $\psi_l = \sum_n c_n^l \phi_n$ where the ϕ_n are the individual eigenfunctions of the atoms and the coefficients c_n^l are phase factors that determine the amplitude with which each ϕ_n contributes at its atomic site. This way of constructing the wave function from atomic orbitals and computing the energy eigenvalues can easily be generalized to facilitate the calculation of the dispersion relation of entire crystals with periodic lattice.

2.4.3 Periodic Potentials—Tight-Binding in 1D

The formalism of the preceding section will now be extended to describe large periodic crystals. First, a one-dimensional chain of N atoms ($N \to \infty$) with lattice constant a is considered where each atom n provides the same, single eigenstate $\phi_n(x)$. The Hamiltonian in this case is simply an extension of Equation (2.17) going from four to N atoms:

$\mathcal{H} = -\frac{\hbar^2}{2m}\frac{d^2}{dx^2} + \sum_n V(x - na)$ where $V(x)$ represents the Coulomb potential of one atom located at $x = 0$. The overall Coulomb potential at x is therefore given by summing the contributions from the different sites at positions na leading to $\sum_n V(x - na)$.

Based on the results of the preceding section, we can immediately write down the overall wave function ψ_k of the system: basically, it will be a sum of phase factors multiplied with the eigenstates of the individual atoms of the underlying crystal lattice. The phase factor (i. e., the eigenvectors $\vec{c}^{\,l}$ in the preceding section) is the envelope of the overall wave function. In Section 2.4.2, the envelope was proportional to sine-functions as appropriate for a (finite) PIB system; in the present case of a 1D infinite system, the appropriate phase factors will be plane waves, i. e., proportional to e^{ikx}. Here, $x = n \cdot a$ is used since the plane wave will only be evaluated at the positions of the discrete atoms. Hence,

$$\psi_k(x) = C\sum_{n=1}^{N} e^{ikna}\phi_n(x) = C\sum_{n=1}^{N} e^{ikna}\phi(x - na) \rightarrow C\sum_{n=1}^{N} e^{ikna}|n\rangle = |\psi_k\rangle \tag{2.20}$$

where C is a normalization constant that needs to be determined.[4] Next, the Schrödinger equation $\mathcal{H}|\psi_k\rangle = E_k|\psi_k\rangle$ must be solved. This is done in the same fashion as before by multiplying it from the left with $\langle m|$, which yields

$$C\sum_n e^{ikna}\langle m|\mathcal{H}|n\rangle = E_k C\sum_n e^{ikna}\langle m|n\rangle. \tag{2.21}$$

Note that in the present case of an infinitely large crystal we get $N \rightarrow \infty$ times the same Equation (2.21) for all $m = 1, \ldots, N$. In the next step, only nearest neighbor interaction will be considered; an extension to include next nearest neighbors will be discussed exemplarily in Task 6. As in the preceding section, the overlap integrals between adjacent atoms are neglected, i. e., $\langle m|n\rangle = \delta_{m,n}$. As a result, Equation (2.21) reduces to three summands on the left and a single one on the right-hand side:

$$\langle m|\mathcal{H}|m-1\rangle e^{ik(m-1)a} + \langle m|\mathcal{H}|m\rangle e^{ik(ma)} + \langle m|\mathcal{H}|m+1\rangle e^{ik(m+1)a} = E\langle m|m\rangle e^{ikma}. \tag{2.22}$$

Noting that $\langle m|m\rangle = 1$, $\langle m|\mathcal{H}|m\rangle = \epsilon_0$ (approximated with the eigenenergy of the isolated atom), denoting $\langle m|\mathcal{H}|m \pm 1\rangle = -V_{ss}$ and dividing the equation above by e^{ikma} we finally obtain

$$E(k) = -V_{ss}e^{-ika} + \epsilon_0 - V_{ss}e^{ika} = \epsilon_0 - 2V_{ss}\cos(ka). \tag{2.23}$$

This result has the exact same form as Equation (2.16) obtained in the case of the discretized Schrödinger equation. Thus, the same cosine-like dispersion relation is ob-

---

4 Normalization of the overall wave function has been dropped in the four-atom example in Section 2.4.2 to keep the discussion as simple as possible.

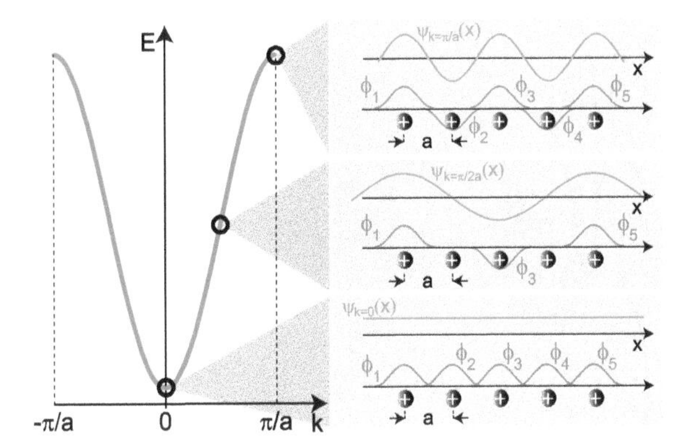

Figure 2.12: Dispersion relation (left) of a one-dimensional, infinite crystal with lattice constant a. The wave function $\psi(x)$ is the envelope of the summation of the atomic orbitals ϕ_n at lattice sites n with an appropriate phase factor.

tained that is depicted in Figure 2.12 together with the envelope wave functions (orange) consisting of a sum of orbitals of the individual sites.

Before discussing Equation (2.23) in more detail, let us determine the normalization constant. C can be computed by requiring $\langle\psi_k|\psi_k\rangle$ to be unity:

$$\langle\psi_k|\psi_k\rangle = |C|^2 \sum_{n,m} e^{-ikma+ikna}\langle m|n\rangle \stackrel{!}{=} 1. \tag{2.24}$$

With the approximation $\langle m|n\rangle = \delta_{m,n}$, the double sum in Equation (2.24) reduces to $\sum_n e^0 = N$. As a result, we obtain $C = \frac{1}{\sqrt{N}}$ and the wave function is given by

$$|\psi_k\rangle = \frac{1}{\sqrt{N}} \sum_{n=1}^{N} e^{ikna}|n\rangle. \tag{2.25}$$

Looking at the cosine band and its dependence on the lattice constant a, one can already extract a number of trends concerning the properties of a solid: in the case of a small a, the atoms are very close to each other leading to a strong overlap of the wave functions of adjacent atoms. As a result, a rather large band is obtained. On the contrary, if the atoms are moved away from each other the overlap is decreased and a shallow band is observed; in fact, in the extreme case of atoms being completely isolated from each other only the eigenenergy of the eigenfunction of the individual atom ϵ_0 is obtained giving rise to a completely flat band. This dependence of the band on the lattice constant is depicted in Figure 2.13. In Section 2.11.1, the concept of the density of states (DOS) will be discussed in detail. A small overlap integral leads to a shallow band, because the eigenenergies of the crystal do not differ much from ϵ_0. This results in a large density of states (i. e., many available states that can be occupied with electrons in a narrow energetic range) which is, for instance, important for magnetism. Therefore, one

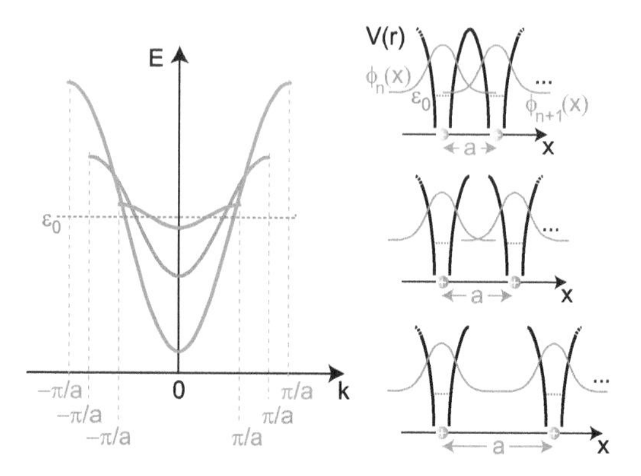

Figure 2.13: Dispersion relations of a one-dimensional lattice of atoms with a single basis function ϕ_n for three different lattice constants. The further away the atoms are from each other, the smaller will the overlap of adjacent wave functions be (right panels) leading to a decreasing energetic width of the resulting band; for simplicity, $V_{ss} \propto 1/a^2$ is assumed.

expects that tightly bond electrons (usually in energetically low lying atomic orbitals) lead to narrow bands with a high density of states.

Task 6.

1D tight-binding with next nearest neighbor interaction: Consider a one-dimensional solid consisting of a lattice with lattice constant a and with a hydrogen-like atom providing a single s-orbital. Based on Equation (2.25), compute the band structure of the solid taking nearest neighbor and next nearest neighbor interaction into account. Assume the corresponding overlap integrals to be $-V_{ss}^{n.n.}$ and $-V_{ss}^{nn.n.}$ with $|V_{ss}^{nn.n.}| < |V_{ss}^{n.n.}|$.

Up to now, only a singe s-like orbital per lattice site has been considered. The extension to two (or more) orbitals per lattice site is, however, rather straightforward. Using the orbitals as an orthonormal basis, the wave function of each atom is given by a linear combination of the orbitals that one wants to incorporate into the calculation. This is actually similar to the case discussed in Section 2.4.2. For instance, if one s- and a p_x-orbital are used the wave function becomes

$$\begin{aligned}\psi_k(x) &= \frac{1}{\sqrt{N}}\sum_n e^{ikna}(c_1\phi_s(x-na) + c_2\phi_{p_x}(x-na))\\ &\to \frac{1}{\sqrt{N}}\sum_n e^{ikna}(c_1|s,n\rangle + c_2|p_x,n\rangle)\end{aligned} \tag{2.26}$$

where the constants $c_{1,2}$ need to be determined. Note that a nomenclature was chosen for the ket vector that includes the index n for the lattice site and the type of orbital. Next, the Schrödinger equation is set up and multiplied from the left with $\langle s,m|$ and with $\langle p_x,m|$ to provide two equations that allow the computation of the constants $c_{1,2}$. To be specific, the following two equations are obtained where we use the fact that the basis functions $|s,n\rangle$ and $|p,n\rangle$ represent an orthonormal basis. Furthermore, only nearest neighbor interaction is considered:

$$\begin{aligned}
&(c_1 \underbrace{\langle s,m|\mathcal{H}|s,m+1\rangle}_{-V_{ss}} + c_2 \underbrace{\langle s,m|\mathcal{H}|p_x,m+1\rangle}_{-V_{sp}})e^{ika} + c_1 \underbrace{\langle s,m|\mathcal{H}|s,m\rangle}_{\epsilon_0} \\
&\quad + c_2 \underbrace{\langle s,m|\mathcal{H}|p_x,m\rangle}_{=0} + (c_1 \underbrace{\langle s,m|\mathcal{H}|s,m-1\rangle}_{-V_{ss}} + c_2 \underbrace{\langle s,m|\mathcal{H}|p_x,m-1\rangle}_{V_{sp}})e^{-ika} = c_1 E \\
&(c_1 \underbrace{\langle p_x,m|\mathcal{H}|s,m+1\rangle}_{V_{sp}} + c_2 \underbrace{\langle p_x,m|\mathcal{H}|p_x,m+1\rangle}_{V_{pp}})e^{ika} + c_1 \underbrace{\langle p_x,m|\mathcal{H}|s,m\rangle}_{=0} \\
&\quad + c_2 \underbrace{\langle p_x,m|\mathcal{H}|p_x,m\rangle}_{\epsilon_1} + (c_1 \underbrace{\langle p_x,m|\mathcal{H}|s,m-1\rangle}_{-V_{sp}} + c_2 \underbrace{\langle p_x,m|\mathcal{H}|p_x,m-1\rangle}_{V_{pp}})e^{-ika} = c_2 E
\end{aligned} \tag{2.27}$$

where again all $\langle s,m|s,m\pm 1\rangle$ and $\langle p_x,m|p_x,m\pm 1\rangle$ are neglected. The overlap integrals $\langle p_x,m|\mathcal{H}|p_x,m\pm 1\rangle$ and $\langle p_x,m|\mathcal{H}|s,m\pm 1\rangle$ depend on the relative orientation of the p_x-orbitals as illustrated in Figure 2.14(a) and (b). While the overlap integral between s-orbitals has a negative value (since the higher probability of finding electrons in between lattice sites reduces the energy), V_{pp} is positive since two p_x-like orbitals overlap with a phase of π (Figure 2.14(b), red and blue lines). The same is true for the overlap between an s-orbital at site m and a p_x-orbital at site $m+1$ (or a p_x-orbital at $m-1$ and an s-orbital at m). Equation (2.27) can be rewritten in matrix form and one obtains

$$\begin{aligned}
&\left[\begin{pmatrix} -V_{ss} & -V_{sp} \\ V_{sp} & V_{pp} \end{pmatrix} e^{ika} + \begin{pmatrix} \epsilon_0 & 0 \\ 0 & \epsilon_1 \end{pmatrix} + \begin{pmatrix} -V_{ss} & V_{sp} \\ -V_{sp} & V_{pp} \end{pmatrix} e^{-ika}\right] \begin{pmatrix} c_1 \\ c_2 \end{pmatrix} \\
&\quad = E \begin{pmatrix} 1 & 0 \\ 0 & 1 \end{pmatrix} \begin{pmatrix} c_1 \\ c_2 \end{pmatrix}.
\end{aligned} \tag{2.28}$$

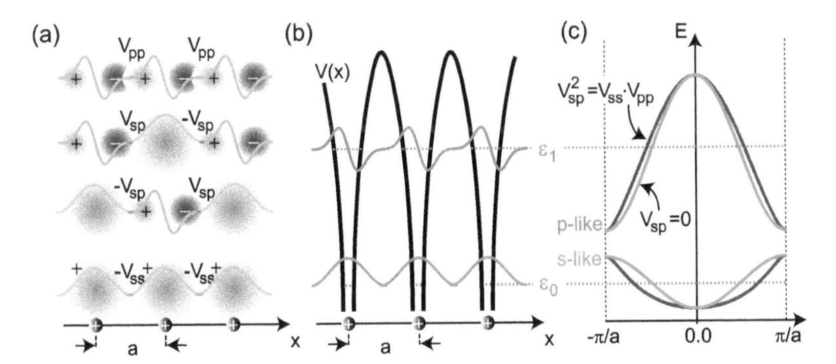

Figure 2.14: (a) Illustration of the s- and p_x-orbitals giving rise to the overlap integral used in Equation (2.28). (b) s- (blue lines) and p_x-like orbitals (red lines) with negative/positive overlap yielding an upward (s-like) and downward (p-like) opened cosine band (c). The two cosine bands (light green without s-p-coupling, dark green with $V_{sp}^2 = V_{ss}V_{pp}$) are centered around the energies ϵ_0 and ϵ_1.

This is a simple eigenvalue equation and the solution can be obtained by solving the characteristic polynomial:

$$\det \begin{vmatrix} -V_{ss}e^{ika} - V_{ss}e^{-ika} + \epsilon_0 - E & -2iV_{sp}\sin(ka) \\ 2iV_{sp}\sin(ka) & V_{pp}e^{ika} + V_{pp}e^{-ika} + \epsilon_1 - E \end{vmatrix} \stackrel{!}{=} 0. \tag{2.29}$$

The resulting quadratic equation for E yields the two bands (dark green lines in Figure 2.14(c))

$$E(k) = \frac{\Delta\epsilon^+ - 2(\Delta V^-)\cos(ka)}{2} \pm \sqrt{\left(\frac{\Delta\epsilon^- - 2(\Delta V^+)\cos(ka)}{2}\right)^2 + 4V_{sp}^2\sin^2(ka)}. \tag{2.30}$$

Here, $\Delta\epsilon^\pm = \epsilon_0 \pm \epsilon_1$ and $\Delta V^\pm = V_{ss} \pm V_{pp}$. If s- and p_x-orbitals are uncoupled, i. e., if $V_{sp} = 0$, two independent cosine bands are obtained that are plotted in Figure 2.14 (c) (light green line).

Task 7.
1D tight-binding with basis: Consider a one-dimensional solid consisting of a lattice with lattice constant a. There are two atoms A and B in the unit cell. Assume $\langle\phi_A, n|\mathcal{H}|\phi_A, n\rangle = \langle\phi_B, n|\mathcal{H}|\phi_B, n\rangle = 0$ and nearest neighbor interaction with overlap integrals $\langle\phi_A, n|\mathcal{H}|\phi_B, n\rangle = -V_{ss}$, $\langle\phi_A, n|\mathcal{H}|\phi_B, n-1\rangle = -\tilde{V}_{ss}$. Compute the band structure. Plot it in the first Brillouin zone for $V_{ss} = 3$ eV and $\tilde{V}_{ss} = 2$ eV. What happens if $V_{ss} = \tilde{V}_{ss}$?

2.4.4 Tight-Binding in 2D(3D)

In the last section, the tight-binding formalism was extended to the case of two orbitals per atom. As a next step toward a general tight-binding formalism, a 2D(3D) solid will be studied in this section.

Consider a regular 2D square lattice with $N \times N$ atoms (again $N \to \infty$) placed a distance a apart from each other such that two primitive lattice vectors $\vec{a}_1 = (a, 0)$ and $\vec{a}_2 = (0, a)$ can be defined as illustrated in Figure 2.15. To simplify the calculation, we will start again with a single s-orbital per atom. Based on the results of the preceding section, Equation (2.25) can be extended to a 2D system in a straightforward manner. The simple square lattice considered here is a Bravais lattice and the lattice vector $\vec{R}_{n,m}$ is given by $\vec{R}_{n,m} = n\vec{a}_1 + m\vec{a}_2 = (n \cdot a, m \cdot a)$; n and m are the indices of the atom positions along the x- and y-directions, respectively. As a result,

$$\psi(\vec{r}) = \frac{1}{\sqrt{N^2}} \sum_{n,m} e^{i\vec{k}\vec{R}_{n,m}} \phi_s(\vec{r} - \vec{R}_{n,m}) \tag{2.31}$$

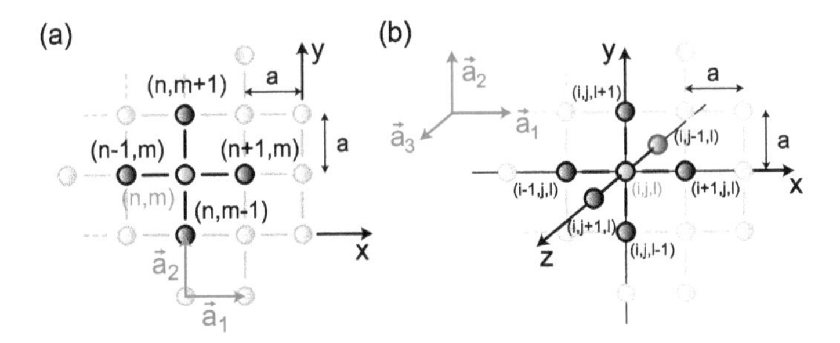

Figure 2.15: Schematic of a 2D(a) and a 3D(b) square lattice with lattice constant a. A central atom is connected to four or six nearest neighbors, respectively.

with $\vec{k} = (k_x, k_y)$. Note that the normalization constant is $\frac{1}{\sqrt{N^2}}$ as appropriate for a 2D system. Using Dirac's notation with $\phi(\vec{r} - \vec{R}_{n,m}) = \phi(x - na, y - ma) \to |n, m\rangle$ the Schrödinger equation becomes

$$\mathcal{H}\frac{1}{\sqrt{N^2}}\sum_{n,m} e^{i(k_x na + k_y ma)}|n, m\rangle = E(\vec{k})\frac{1}{\sqrt{N^2}}\sum_{n,m} e^{i(k_x na + k_y ma)}|n, m\rangle. \tag{2.32}$$

Next, Equation (2.32) will be multiplied with $\langle j, l|$ from the left (keep in mind that this implies an integration over $\vec{r}$). The resulting N^2 equations are all alike so that it is sufficient to compute only a single one. Again, we consider only nearest neighbor interaction, meaning that the double sum over n and m in Equation (2.32) yields nonzero contributions only if $j = n$ and $j = n \pm 1$ and $l = m$ and $l = m \pm 1$. Hence, the following equation is obtained:

$$\begin{aligned}
&\underbrace{\langle j, l|\mathcal{H}|j + 1, l\rangle}_{=-V_{ss}} e^{i(k_x(j+1)a + k_y la)} + \underbrace{\langle j, l|\mathcal{H}|j, l + 1\rangle}_{=-V_{ss}} e^{i(k_x ja + k_y(l+1)a)} \\
&\quad + \underbrace{\langle j, l|\mathcal{H}|j - 1, l\rangle}_{=-V_{ss}} e^{i(k_x(j-1)a + k_y la)} + \underbrace{\langle j, l|\mathcal{H}|j, l - 1\rangle}_{=-V_{ss}} e^{i(k_x ja + k_y(l-1)a)} \\
&\quad + \underbrace{\langle j, l|\mathcal{H}|j, l\rangle}_{=\epsilon_0} e^{i(k_x ja + k_y la)} = E(\vec{k})e^{i(k_x ja + k_y la)}
\end{aligned} \tag{2.33}$$

where it was assumed that the overlap between adjacent atoms is always $-V_{ss}$ as appropriate for a crystal with square lattice of the same atoms. Dividing Equation (2.33) by $e^{i(k_x ja + k_y la)}$ yields the dispersion relation given by

$$\begin{aligned}
E(k_x, k_y) &= -V_{ss}e^{ik_x a} - V_{ss}e^{ik_y a} - V_{ss}e^{-ik_x a} - V_{ss}e^{-ik_y a} + \epsilon_0 \\
&= \epsilon_0 - 2V_{ss}\cos(k_x a) - 2V_{ss}\cos(k_y a).
\end{aligned} \tag{2.34}$$

Figure 2.16 shows the dispersion relation (left panel) as a function of energy within the first Brillouin zone (see also the QR code #10). Since a 2D crystal with square lattice is

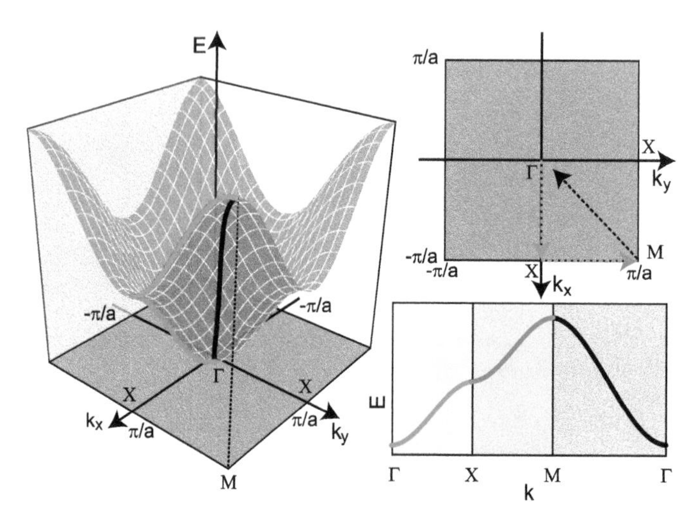

Figure 2.16: Left panel: Two-dimensional dispersion relation within the first Brillouin zone of a crystal with regular square lattice (depicted in Figure 2.15 (a)) and single *s*-orbital. The right panel shows the dispersion along directions of high symmetry.

considered the Brillouin zone is a square with $-\frac{\pi}{a} < k_x \leq \frac{\pi}{a}$ and $-\frac{\pi}{a} < k_y \leq \frac{\pi}{a}$. Often, the dispersion relation is plotted along high symmetry directions as depicted in the bottom right panel of Figure 2.16. Since the primitive unit cell of the 2D crystal consists only of a single atom with a single *s*-orbital, a single band is obtained.

The extension to 3D is straightforward based on Figure 2.15(b) resulting in $E(\vec{k}) = \epsilon_0 - 2V_{ss}(\cos(k_x a) + \cos(k_y a) + \cos(k_z a))$; for further details on the 2D/3D band structure, see QR code #10.

Task 8.
Tight-binding: 2D and two orbitals: Consider a 2D crystal on a square lattice with lattice constant a. The primitive unit cell contains a single atom. Each atom provides an *s*-like and a p_x-like orbital. Compute the band structure of the crystal assuming nearest neighbor interaction with overlap integrals $-V_{ss}$ for *s*-orbitals, V_{pp} for the overlap of p_x-orbitals along the *x*-direction and $-V_{pp\pi}$ for the overlap of the p_x-orbitals along the *y*-direction.

2.4.5 Overlap Integrals and Direction Cosines

In the preceding sections (see also Task 8), it became clear that we need to know the various overlap integrals between the orbitals (*s*- and p_x-orbitals so far) in order to carry out a tight-binding calculation. Moreover, in the two- or three-dimensional case, the crystalline orientation of the orbitals (*p*-, *d*-orbitals) will play a decisive role for the electronic properties of a solid.

Computing the overlap integrals can in principle be done numerically but is a nontrivial task and we are not going to do this in the present book. Instead, the overlap integrals can be considered as adjustable parameters to reproduce the band structure of the material under consideration, which is known as empirical tight-binding calculation. However, taking the relative orientation of the orbitals into account, the number of adjustable (fit-)parameters that are needed can be reduced to a minimum exploiting the symmetry of the crystal structure of the solid. In the following, the effect of differently oriented *p*-orbitals will be discussed in detail, since this will be used for the tight-binding calculation of graphene, carbon nanotubes and silicon; *d*-orbitals will be considered when dealing with transition metal dichalcogenides (Section 2.8.3).

Figure 2.17 shows the different overlap integrals when two atoms (blue and green) are bonded to each other via *s*- and *p*-orbitals; the respective wave functions are also plotted to illustrate the overlap. One can distinguish between an overlap of two *s*-orbitals, a *s*- and a *p*-orbital and two *p*-orbitals. For two *s*-orbitals, the orientation is irrelevant (due to spherical symmetry) and the value is denoted $-V_{ss}$. When *s*- and *p*-orbitals overlap, the orientation of the *p*-orbital with respect to the *s*-orbital needs to be taken into consideration: in Figure 2.17(b), a nonvanishing overlap with value denoted $-V_{sp\sigma}$ is obtained. On the other hand, if the *p*-orbital is oriented as shown in (e), zero overlap is obtained due to symmetry reasons (an equal positive and negative overlap that cancels out). The overlap of two *p*-orbitals yields two different configurations: a σ-overlap depicted in Figure 2.17(c) and a π-overlap shown in (d). Again, due to symmetry reasons, the overlap vanishes in the cases displayed in Figure 2.17(f) and (g).

With the overlap integrals shown in Figure 2.17, the general case of two arbitrarily oriented *s*-*p*- or *p*-*p*-orbitals can be constructed. The idea is now, to expand the *p*-orbital(s) as a linear combination of two *p*-orbitals $p_{\parallel}$ and $p_{\perp}$ that are colinear and

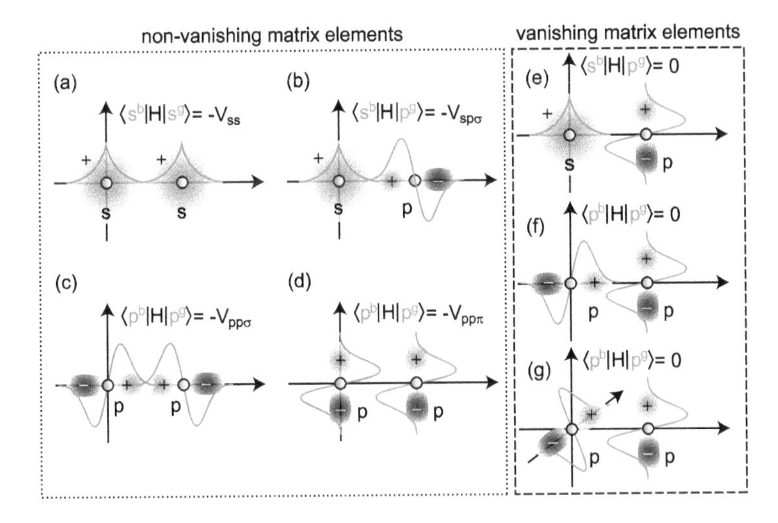

Figure 2.17: Nonvanishing and vanishing overlap integrals between *s*- and *p*-orbitals.

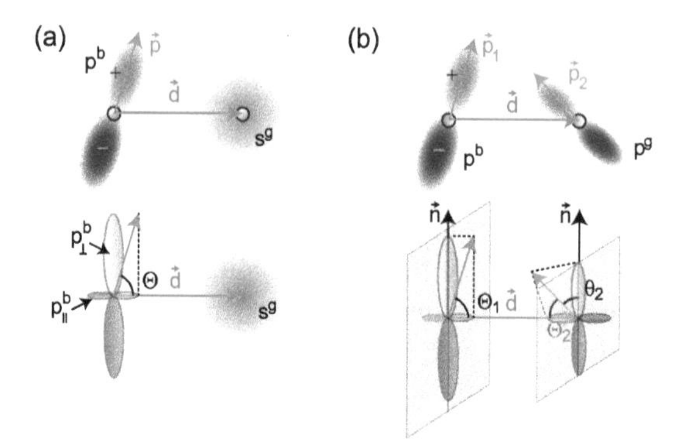

Figure 2.18: Relative orientation of an s- and a p-orbital (a) and two p-orbitals (b) with respect to each other yielding the so-called direction cosines. The superscripts b and g refer to the blue and green atom.

perpendicular to the bond vector $\vec{d}$ that connects the two atoms under consideration. Figure 2.18 shows this scenario in the case of an overlap between s- and p-orbitals (a) as well as two p-orbitals (b).

Let us start by considering the case illustrated in Figure 2.18(a). The vector $\vec{d}$ points from the blue to the green atom and the p-orbital is aligned along $\vec{p}$. $\vec{d}$ and $\vec{p}$ span a plane and we obtain two contributions of the p-orbital within this plane. The coefficients to expand the p-orbital as a linear combination of $p_\parallel$ and $p_\perp$ (with respect to $\vec{d}$) are simply given by $\cos(\Theta)$ (which is equal to the scalar product of the unit vectors along $\vec{p}$ and $\vec{d}$, i. e., $\frac{\vec{p}\cdot\vec{d}}{|\vec{p}||\vec{d}|} = \cos(\Theta)$) along the direction of $\vec{d}$ and $\sin(\Theta)$ along the direction perpendicular to $\vec{d}$ within the plane spanned by $\vec{d}$ and $\vec{p}$. As a result, one obtains

$$|p\rangle = \cos(\Theta)\cdot|p_\parallel\rangle + \sin(\Theta)\cdot|p_\perp\rangle. \tag{2.35}$$

Since the overlap between s and $p_\perp$ is zero for symmetry reasons (cf. Figure 2.17(e)) and the overlap between s and $p_\parallel$ yields $-V_{sp\sigma}$ the overlap integral in the general case leads to

$$\langle s|\mathcal{H}|p\rangle = \cos(\Theta)\underbrace{\langle s|\mathcal{H}|p_\parallel\rangle}_{=-V_{sp\sigma}} + \sin(\Theta)\underbrace{\langle s|\mathcal{H}|p_\perp\rangle}_{=0}. \tag{2.36}$$

In the case of two p-orbitals (see Figure 2.18(b)), one needs to compute the overlap integral $\langle p^b|\mathcal{H}|p^g\rangle$, and hence, both p^b and p^g need to be expanded in a linear combination of the p-orbitals in $p_\parallel$ and $p_\perp$. To do so, a coordinate system can be chosen that is aligned with respect to the plane spanned by the vectors $\vec{p}_b$ and $\vec{d}$ such that $|p^b\rangle = \cos(\Theta_1)\cdot|p_\parallel^b\rangle + \sin(\Theta_1)\cdot|p_\perp^b\rangle$ with $\cos(\Theta_1) = \frac{\vec{p}_1\vec{d}}{|\vec{p}_1||\vec{d}|}$ for the blue atom (similar to Equation (2.35)). In this coordinate system, the vector $\vec{p}_2$ of the green atom is projected onto $\vec{d}$ yielding the component for $|p_\parallel\rangle$ and onto $\vec{n}$ (see Figure 2.18(b)) providing the component $|p_\perp\rangle$. As a result, $|p^g\rangle = \cos(\Theta_2)\cdot|p_\parallel^g\rangle + \cos(\theta_2)\cdot|p_\perp^g\rangle$ with $\cos(\Theta_2) = \frac{-\vec{p}_2\vec{d}}{|\vec{p}_2||\vec{d}|}$ and $\cos(\theta_2) = \frac{\vec{p}_2\vec{n}}{|\vec{p}_2||\vec{n}|}$. Inserting this into the overlap integral yields

$$\langle p^b|\mathcal{H}|p^g\rangle = \cos(\Theta_1)\cos(\Theta_2)\underbrace{\langle p_\parallel^b|\mathcal{H}|p_\parallel^g\rangle}_{=-V_{pp\sigma}} + \sin(\Theta_1)\cos(\Theta_2)\underbrace{\langle p_\perp^b|\mathcal{H}|p_\parallel^g\rangle}_{=0}$$
$$+ \cos(\Theta_1)\cos(\theta_2)\underbrace{\langle p_\parallel^b|\mathcal{H}|p_\perp^g\rangle}_{=0} + \sin(\Theta_1)\cos(\theta_2)\underbrace{\langle p_\perp^b|\mathcal{H}|p_\perp^g\rangle}_{=-V_{pp\pi}}. \tag{2.37}$$

As a result, we can reduce the tight-binding computation to a few overlap integrals and take the orientation of the various orbitals with appropriate angular factors—called direction cosines—into account. The latter basically reflect the particular crystal structure of the solid at hand.

2.4.6 Recipe for General Tight-Binding Method

We are now in the position to generalize the tight-binding method and transfer this into a recipe. This recipe will be used in later sections to compute the band structure of various semiconductors (details are provided in the video accessible through the QR code #12). To be specific, the recipe will be written down explicitly for a two-dimensional crystal with body-centered cubic lattice shown in Figure 2.19, left panel. Here are the steps you need to go through in order to compute the band structure:

1. *Find the unit cell of the underlying crystal lattice and determine the associated primitive vectors.*
   Figure 2.19 shows a square lattice with primitive vectors $\vec{a}_1 = (a, 0)$ and $\vec{a}_2 = (0, a)$. The lattice has been shifted (gray tiles) to elucidate the fact that each unit cell contains two atoms A (blue) and B (green) (if $A = B$ the lattice could be reduced to a simple cubic with smaller lattice constant $(a/\sqrt{2})$).

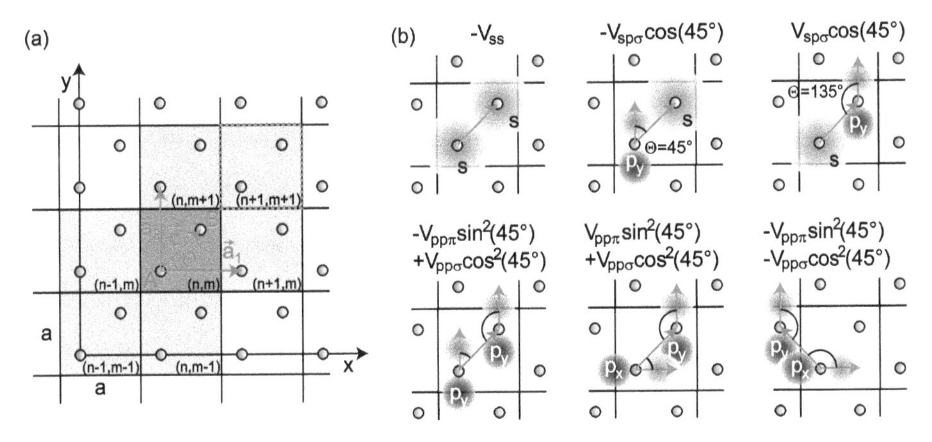

Figure 2.19: Two-dimensional "body-centered" cubic lattice with two different atoms A and B. The primitive lattice vectors are $\vec{a}_{1,2}$ and the vector $\vec{d}$ is at an angle of $\Theta = 45°$ with respect to the coordinate axes. The possible overlap integrals between the s- and p-orbitals are shown on the right.

2. *Choose an appropriate set of orbitals as basis wave functions.*
   For the 2D crystal, it is assumed that each atom A and B provides one s-, a p_x- and a p_y-orbital. Which orbitals to choose is not a priori known and it may be necessary to include more orbitals to describe the band structure (known, e. g., from experiments) appropriately. In the present case, this means that the basis contains $n_a = 2$ atoms with $n_o = 3$ orbitals and as such, solving the Schrödinger equation yields a matrix equation, which has dimension $N_b \times N_b$ with $N_b = n_a \times n_o = 6$. The (LCAO)-wave-function is given by $|\psi\rangle = \frac{1}{\sqrt{N^2}} \sum_{n,m} e^{i\vec{k}\vec{R}_{n,m}} (c_1|s^b, n, m\rangle + c_2|p_x^b, n, m\rangle + c_3|p_y^b, n, m\rangle + c_4|s^g, n, m\rangle + c_5|p_x^g, n, m\rangle + c_6|p_y^g, n, m\rangle)$ with $\vec{R}_{n,m} = n\vec{a}_1 + m\vec{a}_2$. The coefficients $c_{1..6}$ need to be determined by solving the (matrix) Schrödinger equation. Note that an additional phase factor to account for the relative position of the atoms within each unit cell with respect to the lattice (i. e., the basis atoms, giving rise to a factor of unity in the case of the blue atom and a factor of $e^{i\vec{k}\vec{d}}$ in the case of the green atom) is the same in each unit cell and is therefore integrated into the coefficients $c_{1..6}$.
3. *Determine the nearest neighbor unit cells.*
   Choose an arbitrary on-site unit cell (n, m) with lattice vector $\vec{R}_{n,m} = n\vec{a}_1 + m\vec{a}_2$ leading to this unit cell (dark gray in Figure 2.19(a)). Determine the number of neighboring unit cells $n_{\mathrm{n.\,n.}}$ with nearest neighbor interaction to atoms within the on-site unit cell. In the case of the 2D cubic crystal displayed in Figure 2.19, there are 8 unit cells around the on-site cell, of which the light gray shaded six contain atoms that are nearest neighbors of the atoms A or B in the on-site unit cell; only these unit cells need to be considered in the following.
4. *Set up the Schrödinger equation for the on-site unit cell.*
   Write down one empty on-site matrix $\mathsf{H}_{n,m}$ having dimensions $N_\mathrm{b} \times N_\mathrm{b}$ (6×6 in the present case). Write the bra-vectors of the basis orbitals of the on-site unit cell in a column left to the on-site matrix. Next, the ket vectors of same orbitals will be written in a row on top of the on-site matrix. As a result, one obtains

	$\vert s^b\rangle$	$\vert p_x^b\rangle$	$\vert p_y^b\rangle$	$\vert s^g\rangle$	$\vert p_x^g\rangle$	$\vert p_y^g\rangle$
$\langle s^b\vert$	ϵ_{s^b}	0	0	$-V_{ss}$	$\frac{V_{sp\sigma}}{\sqrt{2}}$	$\frac{V_{sp\sigma}}{\sqrt{2}}$
$\langle p_x^b\vert$	0	ϵ_{p^b}	0	$\frac{-V_{sp\sigma}}{\sqrt{2}}$	$\frac{-V_{pp\pi}+V_{pp\sigma}}{2}$	$\frac{V_{pp\pi}+V_{pp\sigma}}{2}$
$\langle p_y^b\vert$	0	0	ϵ_{p^b}	$-\frac{V_{sp\sigma}}{\sqrt{2}}$	$\frac{V_{pp\pi}+V_{pp\sigma}}{2}$	$\frac{-V_{pp\pi}+V_{pp\sigma}}{2}$
$\langle s^g\vert$	$-V_{ss}$	$-\frac{V_{sp\sigma}}{\sqrt{2}}$	$-\frac{V_{sp\sigma}}{\sqrt{2}}$	ϵ_{s^g}	0	0
$\langle p_x^g\vert$	$\frac{V_{sp\sigma}}{\sqrt{2}}$	$\frac{-V_{pp\pi}+V_{pp\sigma}}{2}$	$\frac{V_{pp\pi}+V_{pp\sigma}}{2}$	0	ϵ_{p^g}	0
$\langle p_y^g\vert$	$\frac{V_{sp\sigma}}{\sqrt{2}}$	$\frac{V_{pp\pi}+V_{pp\sigma}}{2}$	$\frac{-V_{pp\pi}+V_{pp\sigma}}{2}$	0	0	ϵ_{p^g}

Since the three s- and $p_{x,y}$-orbitals of each atom represent an orthonormal set of basis functions (for the isolated atom), two diagonal 3×3 matrix blocks are obtained with the eigenenergies $\epsilon_{s^{b,g},p^{b,g}}$ of the respective orbitals on the main diagonal; note that the eigenenergies of the p_x- and p_y-orbitals are degenerate. The overlap integrals are then given by $\langle s^b|\mathcal{H}|s^g\rangle = -V_{ss}$ (top left panel in Figure 2.19(b)), $\langle s^b|\mathcal{H}|p_x^g\rangle = \langle s^b|\mathcal{H}|p_y^g\rangle = -1/\sqrt{2}\cdot(-V_{sp\sigma})$ because $\cos(\Theta) = \cos(135°) = -\cos(45°) = -1/\sqrt{2}$ (top right panel in Figure 2.19(b)). Moreover, $\langle p_x^b|\mathcal{H}|s^g\rangle = \langle p_y^b|\mathcal{H}|s^g\rangle = -V_{sp\sigma}/\sqrt{2}$ (top middle panel in Figure 2.19(b)) and $\langle p_x^b|\mathcal{H}|p_x^g\rangle = \langle p_y^b|\mathcal{H}|p_y^g\rangle = -V_{pp\pi}\sin^2(45°) + V_{pp\sigma}\cos^2(45°) = \frac{-V_{pp\pi}+V_{pp\sigma}}{2}$ (bottom left panel in Figure 2.19(b)). Note that in the latter case the overlap integral due to π-bonding $-V_{pp\pi}$ needs to be multiplied with $\sin(\Theta_1)\sin(\Theta_2) = \sin^2(45°) = 1/2$ leaving the negative sign in front of $-V_{pp\pi}$. In contrast, $-V_{pp\sigma}$ is multiplied with $\cos(\Theta_1)\cos(\Theta_2) = \cos(45°\cos(135°)) = -1/2$. Finally, $\langle p_x^b|\mathcal{H}|p_y^g\rangle = \langle p_y^b|\mathcal{H}|p_x^g\rangle = \frac{V_{pp\pi}+V_{pp\sigma}}{2}$ (bottom middle panel in Figure 2.19(b)). The last overlap to be considered (will be used under 5.) is illustrated in Figure 2.19(b), bottom right. Here, the p_x^b-orbital of the on-site unit cell overlaps with a p_y^g-orbital of the unit cell $-\vec{a}_1$ away from the on-site cell. In this case, the prefactor in front of the $-V_{pp\sigma}$ is $\cos(135°)\cos(135°) = 1/2$, and hence $\langle p_x^b|\mathcal{H}|p_y^g\rangle_{-\vec{a}_1} = \frac{-V_{pp\sigma}-V_{pp\pi}}{2}$.

5. *Set up matrices for nearest neighbor coupling.*
   For each of the $n_{\text{n.n.}}$ nearest neighbor unit cells, write an empty coupling matrix, again with dimensions $N_\text{b}\times N_\text{b}$. Write the ket-vectors of the orbitals of the nearest neighbor unit cell under consideration on top of the matrix and a column of bra-vectors containing the orbitals of the on-site unit cell to the left of the matrix. Fill the matrix elements taking only nearest neighboring atoms and the orientation of the different orbitals with appropriate direction cosines into account. Next, multiply each nearest neighbor matrix with its appropriate phase factor. In the example considered here, the phase factors are given by $e^{\pm i\vec{k}\vec{a}_1}$, $e^{\pm i\vec{k}\vec{a}_2}$ and $e^{\pm i\vec{k}(\vec{a}_1+\vec{a}_2)}$. Exemplarily, for the red framed unit cell shown in Figure 2.19(a) (with $\vec{R}_{1,1} = \vec{a}_1+\vec{a}_2$) the phase factor is $e^{i\vec{k}\vec{R}_{1,1}} = e^{i\vec{k}(\vec{a}_1+\vec{a}_2)} = e^{i(k_x a+k_y a)}$. Thus, the coupling matrix multiplied with the phase factor is given by

	$\lvert s^b\rangle_{\vec{R}_{1,1}}$	$\lvert p_x^b\rangle_{\vec{R}_{1,1}}$	$\lvert p_y^b\rangle_{\vec{R}_{1,1}}$	$\lvert s^g\rangle_{\vec{R}_{1,1}}$	$\lvert p_x^g\rangle_{\vec{R}_{1,1}}$	$\lvert p_y^g\rangle_{\vec{R}_{1,1}}$
$\langle s^b\rvert$	0	0	0	0	0	0
$\langle p_x^b\rvert$	0	0	0	0	0	0
$\langle p_y^b\rvert$	0	0	0	0	0	0
$\langle s^g\rvert$	$-V_{ss}$	$\frac{V_{sp\sigma}}{\sqrt{2}}$	$\frac{V_{sp\sigma}}{\sqrt{2}}$	0	0	0
$\langle p_x^g\rvert$	$-\frac{V_{sp\sigma}}{\sqrt{2}}$	$\frac{-V_{pp\pi}+V_{pp\sigma}}{2}$	$\frac{V_{pp\pi}+V_{pp\sigma}}{2}$	0	0	0
$\langle p_y^g\rvert$	$-\frac{V_{sp\sigma}}{\sqrt{2}}$	$\frac{V_{pp\pi}+V_{pp\sigma}}{2}$	$\frac{-V_{pp\pi}+V_{pp\sigma}}{2}$	0	0	0

$\cdot\, e^{i(k_x a+k_y a)}$

$\mathsf{H}_{n+1,m+1}$

   where the matrix elements are filled in the same fashion as stated in 4.

6. *Set up and solve the secular equation*
   Sum all matrices, i. e., the on-site and all nearest neighbor coupling matrices (including the phase factors) yielding a single $N_\text{b}\times N_\text{b}$-matrix H_{tot}. Assume $\langle\ldots,n,m|\ldots,$

$l,r\rangle = \delta_{n,l}\delta_{m,r}$ as has been done above. We finally arrive at

$$\begin{aligned} \mathsf{H}_{\text{tot}} = \mathsf{H}_{n+1,m}e^{ik_xa} + \mathsf{H}_{n+1,m+1}e^{i(k_xa+k_ya)} + \mathsf{H}_{n,m+1}e^{ik_ya} + \mathsf{H}_{n-1,m}e^{-ik_xa} \\ + \mathsf{H}_{n-1,m-1}e^{-i(k_xa+k_ya)} + \mathsf{H}_{n,m-1}e^{-ik_ya} + \mathsf{H}_{n,m}. \end{aligned}$$

Note that the Schrödinger equation that needs to be solved can be written as $\mathsf{H}_{\text{tot}}\vec{c} = E\mathbf{1}\vec{c}$ where the vector $\vec{c}$ contains the coefficients $c_{1..6}$, and $\mathbf{1}$ and E are the unit matrix and the energy, respectively. Since we are only interested in the dispersion relation it is sufficient to solve $\det|\mathsf{H}_{\text{tot}} - E\mathbf{1}| = 0$. The result is the desired dispersion relation $E(\vec{k})$.

2.4.7 Complex Band Structure

In the calculations of the band structure so far, a purely real $\vec{k}$ was assumed since we were looking for solutions that yield waves $\propto e^{i\vec{k}\vec{r}}$ propagating through the periodic crystal. However, we can also find complex solutions with $\tilde{k} = k + i\kappa$. The imaginary part of $\tilde{k}$ leads to an exponential decay of the wave function according to $e^{i\tilde{k}x} \propto e^{i(i\kappa)x} = e^{-\kappa x}$. Therefore, the inverse of κ can be interpreted as a decay length of the wave function. At surfaces and interfaces, these complex solutions play a crucial role, because the connection of the complex solutions to, e. g., the wave function of a metal contact leads to an interface density of states within the band gap with a spatial extend $\sim 1/\kappa(E)$ (an example is shown in Figure 4.19(b) and Figure 4.20). Hence, the complex band structure facilitates the investigation of interfaces (for instance metal-semiconductor contacts, see Section 4.6), and tunneling (Section 9.1), etc. based on the computation of the bulk properties of a material, which is far less demanding than calculating a (realistic) surface/interface.

As an example for the complex band structure, let us discuss Equation (2.30). Without any coupling, i. e., $V_{sp} = 0$, the dispersion relation yields two independent cosine-bands due to the s- and the p_x-orbitals. Inserting $\tilde{k}$ yields $\cos(ka + i\kappa a) = \cos(ka)\cosh(\kappa a) - i\sin(ka)\sinh(\kappa a)$. It is important to note that Equation (2.30) yields only real energy values, if $\cos(ka + i\kappa a)$ is real, i. e., when $\sin(ka)\sinh(\kappa a) = 0$. This is the case at the Brillouin zone center and at the zone boundary where $\cos(i\kappa a) = \cosh(\kappa a)$ and $\cos(\pi + i\kappa a) = -\cosh(\kappa a)$, respectively. Only at these points complex $\tilde{k}$ can be found that yield real energies. At the Brillouin zone boundary (where the band gap is found in the present case), two independent bands with imaginary wavenumber are obtained as depicted with the thin black lines in the left panel of Figure 2.20. On the other hand, a coupling of s- and p_x-orbitals between two adjacent atoms yields s- and p_x-bands that are connected via the complex band structure. Figure 2.20, left, shows the band structure, which was computed for two different coupling strengths $|V_{\text{sp}}|$ based

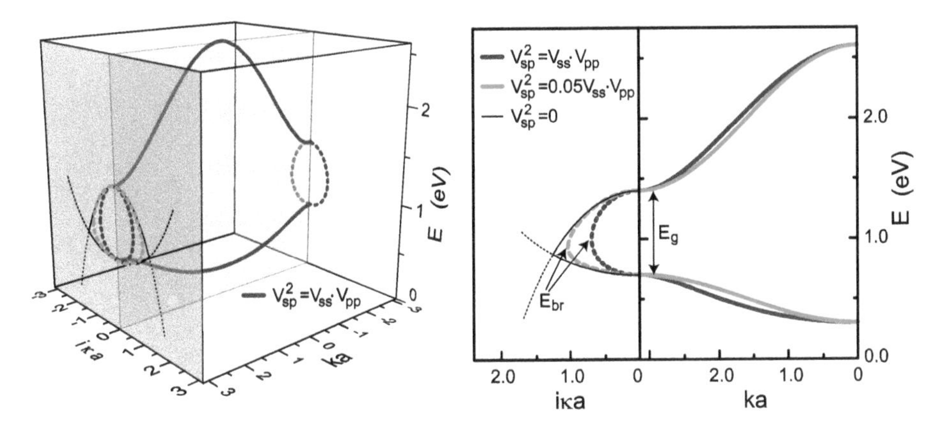

Figure 2.20: Left: Complex band structure of a 1D model with s- and p_x-orbitals (see Equation (2.30)). For nonzero $|V_{sp}|$, s- and p_x-bands are connected through a complex $\tilde{k} = k + i\kappa$. The right panel shows part of the same band structure for positive k and κ for different $|V_{sp}|$.

on Equation (2.30).[5] A stronger coupling leads to a smaller maximum value of κ within the band gap. This is important since it implies a larger decay length $\sim 1/\kappa$ of the wave function for energies within the band gap.

Close to the band edges the complex band structure can clearly be assigned to the conduction or to the valence band. This means, that the "character" of the bands changes from valence- to conduction-band like within the band gap. The energy of this transition is called the branching point E_{br} (see Figure 2.20, right panel), which is at the energy where $dE/d\kappa \to \infty$. E_{br} plays an important role for, e. g., metal-semiconductor contacts since it determines whether surface states carry a positive or negative charge.

In the simple case of Equation (2.30), it is straightforward to find the complex band structure by inserting $\tilde{k}$ and looking for real energies. In the general case, the complex band structure is usually found the other way around namely by fixing a certain real energy and then looking for the complex $\tilde{k}$ that yield this energy. Such an approach will be employed in Section 4.6 to discuss metal-semiconductor contacts based on a 1D model. The higher-dimensional case is more involved and the reader is referred to the available literature.

2.4.8 Band Structure at Surfaces

The tight-binding calculation discussed so far are based on a fully periodic crystal structure. However, in nanoelectronics devices, surfaces and interfaces play a very important

5 Note that the parts of the solution with larger κ are not plotted due to their much faster exponential decay.

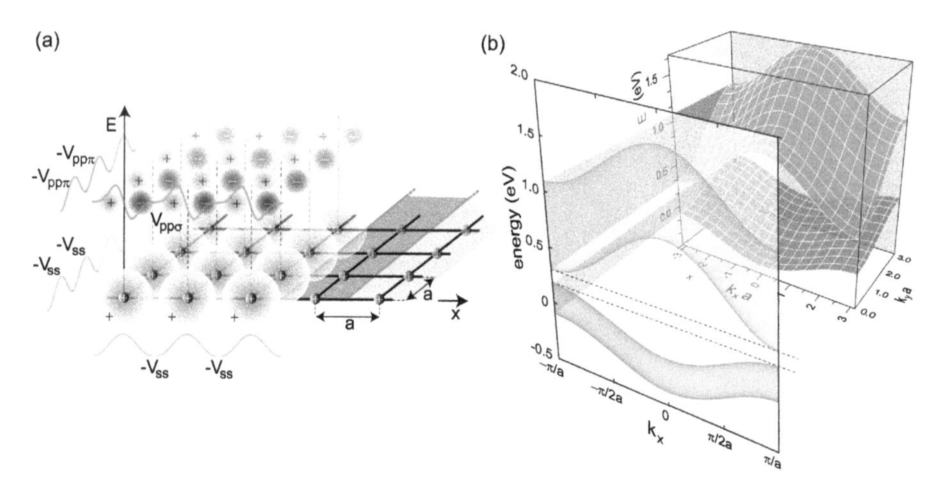

Figure 2.21: (a) Schematic of a 2D solid with s- and p_x-orbital with a surface that leads to a 1D system with semiinfinite unit cell (dark gray, red-framed). (b) Computed surface projected band structure; for simplicity $|V_{sp} = 0|$ is assumed. The intensity of the green color reflects the density of projected cosine bands. The 3D plot in the back shows the band structure of the full 2D system with surface at $y = 0$, and thus positive k_y.

role (e. g., in metal-semiconductor contacts, the MOS-interface, etc.). Therefore, it is necessary to understand, how the calculations so far change when a surface is involved. Let us start with a two-dimensional example and consider a square lattice with a single atom per site with an s- and a p_x-orbital (cf. Figure 2.21 and QR code #11). In order to compute the band structure, we can consider the system to be a one-dimensional solid consisting of semiinfinite unit cells (dark gray, red-framed in Figure 2.21) and use the recipe presented in Section 2.4.6. One on-site (left in Equation (2.38)) and two coupling matrices (right) multiplied with appropriate phase factors need to be set up. With the overlap integrals $-V_{ss}$, $-V_{pp\pi}$, $V_{pp\sigma}$ and V_{sp}, the following matrices are obtained:

$$\left(\begin{array}{cc|cc|cc} \epsilon_s & 0 & -V_{ss} & 0 & 0 & \cdots \\ 0 & \epsilon_p & 0 & -V_{pp\pi} & 0 & \cdots \\ \hline -V_{ss} & 0 & \epsilon_s & 0 & -V_{ss} & \ddots \\ 0 & -V_{pp\pi} & 0 & \epsilon_p & 0 & -V_{pp\pi} \\ \hline 0 & 0 & -V_{ss} & 0 & \epsilon_s & 0 \\ 0 & 0 & 0 & -V_{pp\pi} & 0 & \epsilon_p \\ \hline \vdots & \vdots & \vdots & \vdots & \ddots & \ddots \end{array}\right), \quad \left(\begin{array}{cc|cc|cc} -V_{ss} & \mp V_{sp} & 0 & 0 & 0 & \cdots \\ \pm V_{sp} & V_{pp\sigma} & 0 & 0 & 0 & \cdots \\ \hline 0 & 0 & -V_{ss} & \mp V_{sp} & 0 & \cdots \\ 0 & 0 & \pm V_{sp} & V_{pp\sigma} & 0 & \cdots \\ \hline 0 & 0 & 0 & 0 & -V_{ss} & \cdots \\ 0 & 0 & 0 & 0 & \pm V_{sp} & \cdots \\ \hline \vdots & \vdots & \vdots & \vdots & \ddots & \ddots \end{array}\right) e^{\pm i k_x a} \tag{2.38}$$

Due to the undisturbed periodicity along the x-direction the resulting Hamiltonian consists of 2×2 block matrices on the main diagonal that have the same form as Equation (2.28). These 2×2 block matrices are repeated infinitely many times and are coupled by the off-diagonal 2×2 block matrices that contain V_{ss} and $V_{pp\pi}$ (see left matrix in Equation (2.38)).

If the eigenvalue equation is set up for finite but rather large unit cells (known as the slab method), the quasicontinuous dispersion displayed in Figure 2.21(b) is obtained. This is called the surface projected band structure which can be constructed by projecting all states from the band structure plotted as a function of k_x and positive k_y in the background of (b) onto a plane (similar to the projection of the 1D subbands in Figure 2.3).

2.5 Effective Mass Approximation

With a tight-binding calculation, the band structure of a crystalline solid can be computed. However, knowledge of the electronic states in the vicinity of the bottom of the conduction and the top of the valence band is often sufficient to assess the electronic behavior of devices. This facilitates an incredible simplification: instead of dealing with the entire band structure, the vicinity around the bottom of the conduction and top of the valence bands can be Taylor-expanded up to second order. As a result, conduction and valence bands can be replaced with a quadratic dispersion relation that resembles the dispersion of a free particle. The only difference is that the curvature of the approximate dispersion relation may be different from that of a free particle. The different curvature can be described by the so-called effective mass. Hence, electrons in a solid can be regarded as free particles that exhibit an effective mass $m^\star$ different from the electron mass m_0. This "effective-mass approximation" allows disregarding the impact of the crystal (lattice structure, constituents, etc.) on the electronic states and thus on the electronic behavior of devices. One cannot overestimate the benefits and impact of this tremendous simplification.

According to classical mechanics, a force F (due to, e. g., an electric field) changes the momentum of a charge carrier as $F = \frac{dp}{dt} = \frac{d}{dt}(\hbar k)$ where the de Broglie relationship has been used. Noting that $F = m\frac{dv}{dt}$ it is clear that

$$m\frac{dv}{dt} = \hbar\frac{dk}{dt} = \hbar\frac{dk}{dv}\frac{dv}{dt}, \tag{2.39}$$

and thus, $m = \hbar\frac{dk}{dv}$. Since the group velocity is $v(k) = \frac{1}{\hbar}\frac{dE(k)}{dk}$ (see Section 2.1.1) the well-known expression for the effective mass is obtained:

$$m^\star = \left(\frac{1}{\hbar}\frac{dv}{dk}\right)^{-1} = \left(\frac{1}{\hbar^2}\frac{d^2}{dk^2}E(k)\right)^{-1} \tag{2.40}$$

This means that the curvature of the band determines the effective mass: a strong increase of $E(k)$ with increasing k results in a strong curvature, and hence in a small effective mass. On the other hand, a rather flat band yields a small curvature and, therefore, a large effective mass $m^\star$. Thus, the effective mass strongly depends on the crystal structure, on the direction within a crystal and also on what orbital the band we are calculating an effective mass from belonged to. Energetically low lying orbitals that are tightly bond to their nucleus have a small overlap with adjacent atoms and result in a large effective mass and vice versa (see also Figure 2.13).[6] This is the reason why the valence band usually has a larger effective mass compared to the conduction band. Using Equation (2.23), the effective mass around the bottom of the conduction band (i. e., $k \approx 0$) is $m^\star = \frac{\hbar^2}{2V_{ss}a^2}$, which seems to suggest that $m^\star$ increases for smaller a. However, keep in mind that the value of the overlap integral $|V_{ss}|$ increases stronger than a^2 decreases due to the strong spatial dependence of the overlap of the wave functions of two adjacent atoms. As a result, the effective mass decreases if a decreases due to a significantly larger overlap. Hence, exerting compressive strain on a solid may lead to larger overlap integrals that yield a band structure with small effective mass.

If the entire k-interval of the cosine-band equation (2.16) is considered, i. e., $k = -\frac{\pi}{a}, \ldots, \frac{\pi}{a}$ a rather peculiar behavior of $m^\star$ as a function of energy is observed: Around the band minimum the effective mass is constant but strongly increases and in fact diverges to $+\infty$ when $k \to \frac{\pi}{2a}$. For larger k-values $m^\star$ takes on negative values, approaching a constant as $k \to \frac{\pi}{a}$. This behavior can be understood when looking at the velocity $v(k) = \frac{1}{\hbar}\frac{dE(k)}{dk}$: increasing k implies an increase of energy. But although E increases, the velocity approaches a constant value at $k = \frac{\pi}{2a}$, which can only occur if the effective mass increases. Furthermore, for even higher energies the velocity drops, meaning that carriers are decelerated, which requires a negative effective mass. This situation is schematically shown in Figure 2.22.

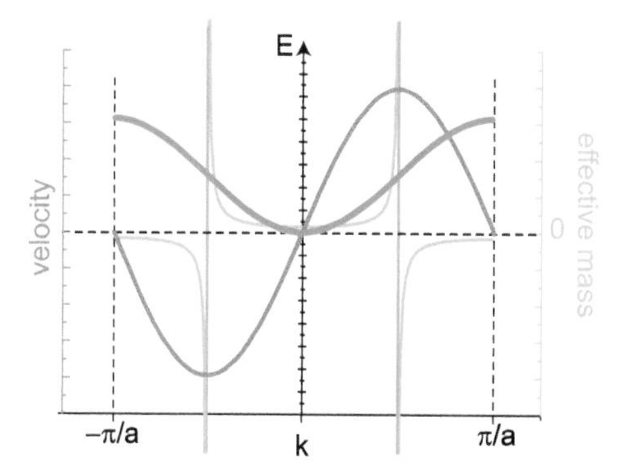

Figure 2.22: Cosine band (green) of a 1D lattice within the first Brillouin zone. The dark blue line shows $v = \frac{1}{\hbar}\frac{\partial E}{\partial k}$ and the light blue line displays the effective mass. $m^\star$ diverges in the middle of the Brillouin zone and approaches a constant where $E(k)$ becomes flat.

6 Moreover, electrons in inner orbitals screen the Coulomb potential of the nucleus further contributing to the difference between tightly bond inner orbitals and orbitals with higher energies.

The fact that the effective mass increases and eventually diverges (when the dispersion becomes linear) is due to the deviation of the band structure from the parabolic behavior. When using Newton's 2nd law in Equation (2.39) the implicit assumption of the mass being independent of time (and hence energy) has been made; the time derivative is applied only to the group velocity. As a result, the effective mass extracted above is strictly valid only in the energy range where the dispersion is parabolic, i. e., where the effective mass is constant. If, however, the band structure is nonparabolic (for instance, in the case of graphene (see Section 2.8.1) that exhibits a linear dispersion relation), the definition of effective mass given in Equation (2.40) is not valid anymore [237]. Indeed it has been shown [237, 15] that the momentum effective mass is appropriate in this case: with $p = m_p^\star v = \hbar k$ where v is the group velocity (see Equation (2.5)) we have

$$m_p^\star = \hbar \frac{k}{v} = \hbar^2 k \left(\frac{\partial E}{\partial k} \right)^{-1} = \frac{\hbar^2}{2} \frac{\partial}{\partial E}(k^2). \tag{2.41}$$

Note that this expression yields the same effective mass as Equation (2.40) in the case of a quadratic dispersion relation. However, in the case of a nonparabolic band structure, the expression above is appropriate and will be used to extend the effective mass approximation also to such instances.

Task 9.
Effective mass from tight-binding band structure: Consider the 1D solid displayed in Figure 2.14(a), (b) with the band structure shown in Figure 2.14(c). Here, the s-like and the p-like bands can be interpreted as valence and conduction bands, respectively. Compute the effective masses that are obtained for the valence and conduction band using $\langle m, s|\mathcal{H}|m \pm 1, s\rangle = -V_{ss}$, $\langle m, p|\mathcal{H}|m \pm 1, p\rangle = V_{pp}$, $\langle m, p|\mathcal{H}|m, s\rangle = 0$ and $V_{sp} = 0$ with $2|V_{ss}| = |V_{pp}|$. Why is the valence band effective mass larger than the effective mass in the conduction band?

2.5.1 Effective Mass Tensor

Expression (2.40) that relates the effective mass to the curvature of the band structure is in the present form valid only in one dimension. In a general three-dimensional solid, the expression can be extended and yields an effective mass tensor. In order to derive this effective mass tensor, the acceleration $\vec{a} = \frac{d}{dt}\vec{v}$ is rewritten as $\vec{a} = \frac{d}{dt}(\frac{1}{\hbar}\nabla_k E(\vec{k}))$. Next, the derivatives $\frac{d}{dt}$ and ∇_k are interchanged yielding $\vec{a} = \frac{1}{\hbar}\nabla_k \frac{dE(\vec{k})}{dt}$. Since the derivative with respect to t is a total derivative, one obtains $\frac{dE}{dt} = \frac{\partial E}{\partial k_x}\frac{dk_x}{dt} + \frac{\partial E}{\partial k_y}\frac{dk_y}{dt} + \frac{\partial E}{\partial k_z}\frac{dk_z}{dt}$, which is equal to the scalar product $\nabla_k E(\vec{k}) \cdot \frac{d\vec{k}}{dt}$. Noting that the force is $\vec{F} = \hbar\frac{d\vec{k}}{dt}$ the acceleration becomes $\vec{a} = \frac{1}{\hbar^2}\nabla_k(\nabla_k E(\vec{k}) \cdot \vec{F})$. The latter expression is equal to $\vec{a} = \frac{1}{\hbar^2}\nabla_k(\frac{\partial E}{\partial k_x}F_x + \frac{\partial E}{\partial k_y}F_y + \frac{\partial E}{\partial k_z}F_z)$. Moreover, from classical mechanics $\vec{a} = \frac{1}{m^\star}\vec{F}$, and thus one can identify the effective mass tensor from

$$\vec{a} = \frac{1}{\hbar^2}\begin{pmatrix} \frac{\partial^2 E}{\partial k_x^2}F_x + \frac{\partial^2 E}{\partial k_x \partial k_y}F_y + \frac{\partial^2 E}{\partial k_x \partial k_z}F_z \\ \frac{\partial^2 E}{\partial k_y \partial k_x}F_x + \frac{\partial^2 E}{\partial k_y^2}F_y + \frac{\partial^2 E}{\partial k_y \partial k_z}F_z \\ \frac{\partial^2 E}{\partial k_z \partial k_x}F_x + \frac{\partial^2 E}{\partial k_z \partial k_y}F_y + \frac{\partial^2 E}{\partial k_z^2}F_z \end{pmatrix} = \underbrace{\frac{1}{\hbar^2}\begin{pmatrix} \frac{\partial^2 E}{\partial k_x^2} & \frac{\partial^2 E}{\partial k_x \partial k_y} & \frac{\partial^2 E}{\partial k_x \partial k_z} \\ \frac{\partial^2 E}{\partial k_y \partial k_x} & \frac{\partial^2 E}{\partial k_y^2} & \frac{\partial^2 E}{\partial k_y \partial k_z} \\ \frac{\partial^2 E}{\partial k_z \partial k_x} & \frac{\partial^2 E}{\partial k_z \partial k_y} & \frac{\partial^2 E}{\partial k_z^2} \end{pmatrix}}_{(m^\star)^{-1}} \vec{F}. \tag{2.42}$$

If the band structure is such that it can be approximated by an ellipsoid whose principal axes are aligned with respect to $k_{x,y,z}$ the off-diagonal elements, e. g., $\frac{\partial^2 E}{\partial k_x \partial k_y}$, vanish and the main diagonal elements of Equation (2.42) carry the inverse of the effective masses along the k_x-, k_y- and k_z-directions. The most prominent example of an ellipsoidal band structure leading to direction-dependent effective masses is silicon, which will be discussed in detail in Section 2.6.3.

2.5.2 Energy-Dependent Effective Mass

The effective mass given by Equation (2.40) is only valid in the case of a quadratic dispersion relation. However, the band structure is often nonparabolic and as a result an energy-dependent effective mass has to be introduced in order to preserve the simplicity of the effective mass approximation. Nonparabolicity can be taken into consideration with a dispersion relation of the form

$$\frac{\hbar^2 k^2}{2m_c^\star} \approx E(1 + \alpha E) \tag{2.43}$$

where the subscript "c" has been added to $m_c^\star$ for clarity since this is the effective mass at the (conduction) band minimum (i. e., it is the effective mass according to Equation (2.40) without nonparabolicity). α is the nonparabolicity factor where $\alpha \approx 1/E_g$. Solving Equation (2.43) for k^2 and using Equation (2.41) for the momentum effective mass, the following energy-dependent effective mass is obtained:

$$m^\star(E) = m_c^\star(1 + 2\alpha E). \tag{2.44}$$

In Section 2.4.7, the complex band structure has been calculated with tight-binding showing how an overlap between s-and p_x-orbital connects the conduction to the valence bands through a complex wave number. Figure 2.23 (thin black lines) shows that the complex band structure within the band gap cannot be described solely with two constant effective masses for the conduction and valence bands since this leads to a substantially stronger decay of the wave function within the band gap, and thus, may lead to an underestimated band-to-band tunneling rate (cf. Chapter 9). In order to preserve the simplicity of the effective mass approximation yet properly describe the complex band

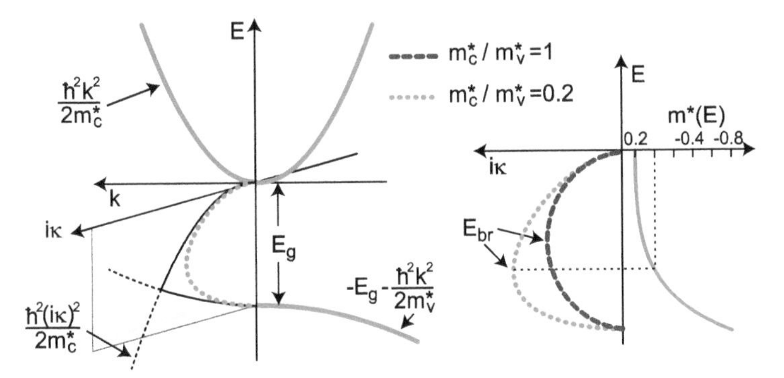

Figure 2.23: Left: Flietner's dispersion relation in the case of $m_c^\star = 0.2m_0$ and $m_v^\star = 0.8m_0$ with complex band structure within the band gap (green dotted line, $E_g = 1$ eV). The right panel shows a comparison of the complex band structure for $m_c^\star/m_v^\star = 1$ (dashed dark green) and $m_c^\star/m_v^\star = 0.2$ (green dotted) [74, 101] and the momentum effective mass computed with Equation (2.41).

structure within the band gap, Flietner's dispersion relation[7] may be used [74, 101]:

$$\frac{\hbar^2 k^2}{2m_c^\star} = E\left(1 + \frac{E}{E_g}\right)\left(1 + \left(1 - \sqrt{\frac{m_c^\star}{m_v^\star}}\right)\frac{E}{E_g}\right)^{-2} \tag{2.45}$$

which is shown in Figure 2.23 (see the figure caption for details). The complex band structure (green dotted line) shows the same appearance as has already been discussed in Section 2.4.7 with a significantly different behavior compared to simply continuing the band structure within the band gap based on the constant effective masses $m_{c,v}^\star$ (thin black lines). The right panel shows the complex band structure for different ratios $m_c^\star/m_v^\star$. In the case $m_c^\star/m_v^\star = 1$, the resulting complex band structure is circular (compare with the complex band structure of carbon nanotubes in Section 2.9) with the branching point $E_{\rm br}$ at $E_g/2$. If $m_v^\star > m_c^\star$, $E_{\rm br}$ is moved to lower energies again reflecting the discussion in Section 2.4.7.

If we again use the definition for the momentum effective mass Equation (2.41), Flietner's dispersion Equation (2.45) can be solved for $k^2 \to -\kappa^2$. Computing the derivative results in the energy-dependent effective mass within the band gap given by $m_p^\star(E) = m_c^\star(1 + (1 + \sqrt{\frac{m_c^\star}{m_v^\star}})\frac{E}{E_g})/(1 + (1 - \sqrt{\frac{m_c^\star}{m_v^\star}})\frac{E}{E_g})^3$, which incorporates the impact of different (constant) effective masses in the conduction and valence bands $m_c^\star$ and $m_v^\star$, respectively. $m_p^\star(E)$ approaches $m_c^\star$ at the conduction band (i. e., $E \to 0$), is equal to zero at $E_{\rm br}$ and approaches $-m_v^\star$ at the valence band edge (see right panel, Figure 2.23). Since it is more

7 Note that in Equation (2.45), $E = 0$ and $E = E_g$ are the conduction and valence band edges, respectively. To compute the complex band structure (depicted in Figure 2.23), $k^2 = (i\kappa)^2$ needs to be inserted in Equation (2.45) yielding an extra negative sign.

convenient to work with positive effective masses, the absolute value of $m_p^\star(E)$ is used keeping in mind that at E_{br}, $m_p^\star(E)$ changes from electron- to hole-character; for more details see QR code #14.

Finally, one can also define an (energy-dependent) energy effective mass $m_E^\star$ [127, 51], which follows directly from the dispersion. For instance, solving Equation (2.43) for $k = \sqrt{\frac{2m_c^\star E(1+\alpha E)}{\hbar^2}}$ the energy effective mass can be read off this expression by setting it equal to $\sqrt{\frac{2m_E^\star E}{\hbar^2}}$ resulting in $m_E^\star = m_c^\star(1 + \alpha E)$. The energy effective mass will be used in Chapter 6 in order to obtain a proper description of the density of states within the band gap based on Flietner's dispersion relation in a single-band description.

While an energy-dependent effective mass can be used to take nonparabolic effects and the complex band structure into account it is important to note that it renders the Hamiltonian to be non-Hermitian. Nevertheless, in the following chapters many of the simulation examples and results presented are based upon the use of an energy-dependent effective mass, which turned out to provide excellent results.

2.6 Bulk Materials

Bulk materials for nanoelectronics include group IV elemental semiconductors and their heterostructures, III–V as well as II–VI compound semiconductors. Group IV elemental semiconductors, i. e., diamond, silicon germanium and gray tin (α-Sn) and III-arsenides (GaAs, InAs, etc.) and III-antimonides crystallize in the diamond structure. Due to its importance and ubiquitous use, silicon will be discussed exemplarily in detail regarding crystal structure, band structure, etc.

2.6.1 Silicon—Crystal Structure and Material Properties

Silicon is arguably the most important and the most studied semiconductor and crystallizes in the diamond lattice due to the four sp^3 hybrid orbitals. These orbitals are formed since it is energetically favorable to put silicon's four valence electrons into orbitals that provide maximum spatial distance between the electrons to lower the electron-electron interaction. Thus, the sp^3 orbitals are tetrahedral, leading to a diamond lattice structure.

The diamond structure consists of two face-centered cubic lattices, shifted by $(\frac{a}{4}, \frac{a}{4}, \frac{a}{4})$ as depicted in Figure 2.24. Here, a is the lattice constant of the underlying cubic fcc lattice. The diamond lattice is therefore no Bravais lattice but can be described by a fcc Bravais lattice with basis containing two atoms where, the vector $\vec{d} = (\frac{a}{4}, \frac{a}{4}, \frac{a}{4})$ (see Figure 2.24) leads from the first to the second atom within the unit cell.

Figure 2.25(a) shows the Wigner–Seitz cell of a fcc lattice within the (non-primitive) cubic unit cell of the fcc lattice as was already shown in Figure 2.7. The additional atoms required for the diamond lattice are represented as green spheres. Note that in the case

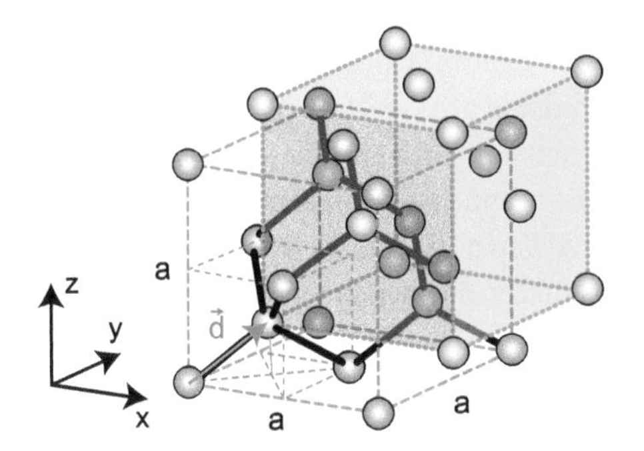

Figure 2.24: Two interwoven fcc lattices constitute the diamond lattice. sp^3-hybridization yields each Si atom to be tetrahedrally surrounded by four adjacent atoms as illustrated in the red framed box.

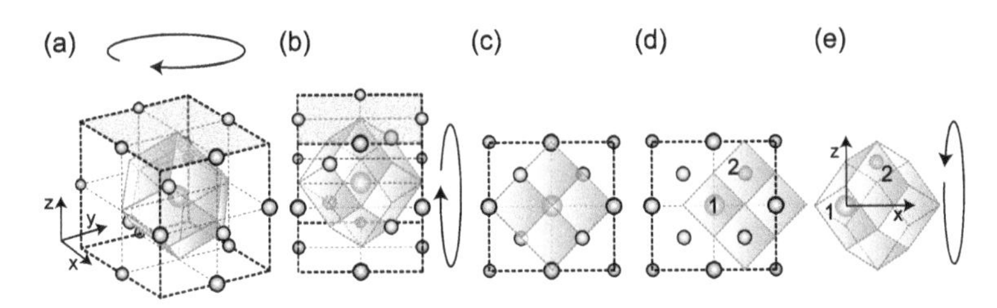

Figure 2.25: Wigner–Seitz cell of a fcc lattice (a). If the cell is turned and tilted appropriately, it becomes clear that the Wigner–Seitz cell of Si contains two atoms (b)–(d). Shifting the Wigner–Seitz cell slightly yields an unambiguous definition of the unit cell (e).

of, e. g., GaAs, the blue and green spheres would embody gallium and arsenic atoms, respectively; for elemental semiconductors such as silicon all atoms are the same. Obviously, there is a central atom (large blue) within the Wigner–Seitz cell and four green atoms are at the respective corners shown in Figure 2.25(a). Each green atom is shared by four Wigner–Seitz cells such that overall two ($1 + 4 \cdot \frac{1}{4}$) atoms are in each cell. If the cubic cell in (a) is rotated around the z-axis as illustrated in (b), and subsequently rotated around the x-axis (c) and if in addition, the Wigner–Seitz cell is shifted by $\frac{a}{4}$ along the x-direction, the two atoms denoted "1" and "2" in Figure 2.25(d) are fully within the Wigner–Seitz cell. Using the shifted Wigner–Seitz cell, it is clear that only two atoms are within each unit cell of the silicon crystal structure. This shifted cell will be used to construct a tight-binding calculation of the band structure in the succeeding section since it allows to immediately see where nearest neighbor interaction needs to be taken into account. Before we turn our attention to the band structure calculation, a few words about crystallographic planes/surfaces of the silicon crystal are appropriate.

For the work-horse of integrated circuits—metal-oxide-semiconductor field-effect transistors (MOSFETs)—surfaces and interfaces of silicon play a decisive role: in MOSFETs, inversion charge is induced at the interfaces between the silicon substrate and the gate dielectric (see Chapters 4 and 5). Moreover, even the behavior of some wet-chemical etchants (see Section 3.6.4 for details) is determined by crystallographic surfaces. The most important surfaces/interfaces of silicon are the (100), (110) and (111) crystallographic planes displayed in Figure 2.26. Silicon atoms at a surface are not bonded to four next neighboring atoms anymore, and thus provide so-called dangling bonds that are localized states with energies within the band gap of silicon. As will be discussed in detail in Section 4.5, these gap states are important for the switching of MOSFETs.

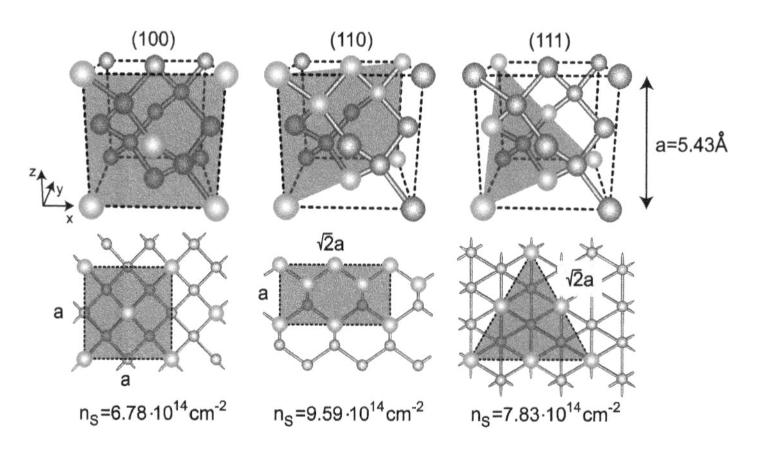

Figure 2.26: Cubic lattice of silicon in real space. The (100), (110) and (111) crystallographic planes (gray-shaded area) together with the surface atom density n_S are given.

The surface density of dangling bonds is determined by the surface density of atoms n_S and the number of bonds per atom. In the three examples displayed in Figure 2.26, the surface atoms are shown in blue and their surface density can easily be computed by counting the number of surface atoms in one cubic cell and dividing this number by the gray area. The (100) surface contains 2 atoms ($4 \cdot \frac{1}{4}$ due to the corner atoms and the central fcc-atom) per area a^2 resulting in $n_S = 6.78 \cdot 10^{14}\ \mathrm{cm}^{-2}$. Because each of the two atoms has two unsatisfied bonds out of plane, the total bond density is $1.356 \cdot 10^{15}\ \mathrm{cm}^{-2}$. The (110) surface has 4 atoms per $\sqrt{2} \cdot a^2$, and hence $n_S = 9.59 \cdot 10^{14}\ \mathrm{cm}^{-2}$; in contrast to the (100) surface, here only a single bond per atom is unsaturated giving rise to a corresponding bond density of $9.59 \cdot 10^{14}\ \mathrm{cm}^{-2}$. Finally, 2 atoms ($3 \cdot \frac{1}{6}$ due to the corner atoms plus $3 \cdot \frac{1}{2}$ from the side edges of the triangular area) are on the (111) surface within the gray triangle with an area of $\frac{\sqrt{3}}{2}a^2$ that have a single free bond. Thus, the surface atom and bond density of the (111) plane are equal and given by $n_S = 7.83 \cdot 10^{14}\ \mathrm{cm}^{-2}$.

2.6.2 Tight-Binding Calculation of Silicon

According to the recipe for band structure calculation given in Section 2.4.6, we need to first determine the unit cell, the number of atoms within the unit cell and which orbitals we have to choose to describe the band structure. The number of atoms within the unit cell has already been determined in the previous section to be two ($n_a = 2$). In the following, the Wigner–Seitz cell displayed in Figure 2.25(e) is used as unit cell. Furthermore, in order to describe the band structure properly it has been shown that in the case of silicon at least the 3s, 3$p_{x,y,z}$ and the 4$s^\star$ orbitals (i. e., $n_o = 5$) need to be included [261]. Consequently, $N_b = n_a \cdot n_o$, and thus 10×10 matrices have to be set up. Moreover, each unit cell has 12 neighboring cells that are displayed in Figure 2.27. However, from Figure 2.27 it is obvious that not all 12 cells need to be incorporated since we only consider interaction between nearest neighboring atoms (i. e., between blue and green atoms with a distance of $|\vec{d}|$) and some of the neighboring unit cells do not contain nearest neighboring atoms. To be specific, Figure 2.27 shows the 12 neighboring cells that are located at $\vec{a}_1 = (a/2, 0, -a/2)$ and $\vec{a}_2 = (-a/2, 0, a/2)$ (a), $\vec{a}_3 = (a/2, -a/2, 0)$ and $\vec{a}_4 = (-a/2, a/2, 0)$ (b), $\vec{a}_5 = (0, -a/2, a/2)$ and $\vec{a}_6 = (0, a/2, -a/2)$ (c), $\vec{a}_7 = (a/2, 0, a/2)$ and $\vec{a}_8 = (-a/2, 0, -a/2)$ (d), $\vec{a}_9 = (-a/2, -a/2, 0)$ and $\vec{a}_{10} = (a/2, a/2, 0)$ (e), $\vec{a}_{11} = (0, a/2, a/2)$ and $\vec{a}_{12} = (0, -a/2, -a/2)$ (f) provided that the origin of the coordinate axes is at the central blue atom and again, a is the lattice constant of the underlying fcc Bravais lattice.

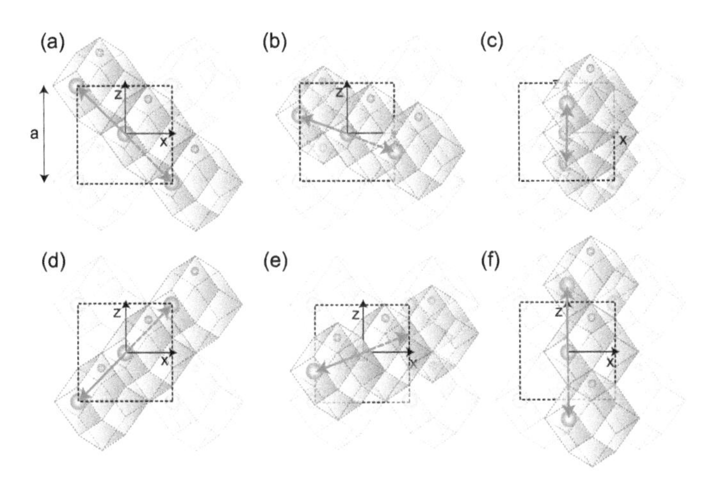

Figure 2.27: 12 nearest neighboring cells of the silicon Wigner–Seitz cell. The six nearest neighbor cells displayed in panels (a)–(c) do not contain any nearest neighbor atoms with respect to the central cell. Panels (d)–(f) show the six relevant nearest neighbor unit cells.

Vectors $\vec{a}_{1,\dots,6}$ lead to neighboring cells that do not contain a nearest neighbor atom, and hence do not contribute (as obvious from Figure 2.27(a)–(c)). As a result, seven 10×10 matrices need to be set-up, namely the on-site matrix and six nearest neighbor coupling matrices that are multiplied with appropriate phase factors. These matrices are explicitly given through the QR code #15 where details of the tight-binding calculation are provided. For improved readability, only the final result $\mathrm{H}_{\mathrm{tot}}$ is stated here:

$$\begin{pmatrix} \epsilon_{s^b} & 0 & 0 & 0 & 0 & -V_{ss}g_0 & V_{sp}g_1 & V_{sp}g_2 & V_{sp}g_3 & 0 \\ 0 & \epsilon_{p_x^b} & 0 & 0 & 0 & -V_{sp}g_1 & -V_{xx}g_0 & V_{xy}g_3 & V_{xy}g_2 & -V_{s^\star p}g_1 \\ 0 & 0 & \epsilon_{p_y^b} & 0 & 0 & -V_{sp}g_2 & V_{xy}g_3 & -V_{xx}g_0 & V_{xy}g_1 & -V_{s^\star p}g_2 \\ 0 & 0 & 0 & \epsilon_{p_z^b} & 0 & -V_{sp}g_3 & V_{xy}g_2 & V_{xy}g_1 & -V_{xx}g_0 & -V_{s^\star p}g_3 \\ 0 & 0 & 0 & 0 & \epsilon_{s^{b\star}} & 0 & V_{s^\star p}g_1 & V_{s^\star p}g_2 & V_{s^\star p}g_3 & -V_{s^\star s^\star}g_0 \\ -V_{ss}g_0^\star & -V_{sp}g_1^\star & -V_{sp}g_2^\star & -V_{sp}g_3^\star & 0 & \epsilon_{s^g} & 0 & 0 & 0 & 0 \\ V_{sp}g_1^\star & -V_{xx}g_0^\star & V_{xy}g_3^\star & V_{xy}g_2^\star & V_{s^\star p}g_1^\star & 0 & \epsilon_{p_x^g} & 0 & 0 & 0 \\ V_{sp}g_2^\star & V_{xy}g_3^\star & -V_{xx}g_0^\star & V_{xy}g_1^\star & V_{s^\star p}g_2^\star & 0 & 0 & \epsilon_{p_y^g} & 0 & 0 \\ V_{sp}g_3^\star & V_{xy}g_2^\star & V_{xy}g_1^\star & -V_{xx}g_0^\star & V_{s^\star p}g_3^\star & 0 & 0 & 0 & \epsilon_{p_z^g} & 0 \\ 0 & -V_{s^\star p}g_1^\star & -V_{s^\star p}g_2^\star & -V_{s^\star p}g_3^\star & -V_{s^\star s^\star}g_0^\star & 0 & 0 & 0 & 0 & \epsilon_{s^{g\star}} \end{pmatrix}. \tag{2.46}$$

The factors $g_{0,\dots,3}$ are given by $g_0 = 1 + e^{i\vec{k}\vec{a}_7} + e^{i\vec{k}\vec{a}_9} + e^{i\vec{k}\vec{a}_{11}}$, $g_1 = 1 - e^{i\vec{k}\vec{a}_7} - e^{i\vec{k}\vec{a}_9} + e^{i\vec{k}\vec{a}_{11}}$, $g_2 = 1 + e^{i\vec{k}\vec{a}_7} - e^{i\vec{k}\vec{a}_9} - e^{i\vec{k}\vec{a}_{11}}$, $g_3 = 1 - e^{i\vec{k}\vec{a}_7} + e^{i\vec{k}\vec{a}_9} - e^{i\vec{k}\vec{a}_{11}}$. The different signs in front of the phase factors are due to the orientation of the orbitals with respect to each other, as has been discussed in Section 2.4.5. The vectors $\vec{a}_{7,9,11}$ are given by $\vec{a}_7 = \frac{a}{2}(1,0,1)$, $\vec{a}_9 = \frac{a}{2}(1,1,0)$ and $\vec{a}_{11} = \frac{a}{2}(0,1,1)$ (cf. Figure 2.27); vectors $\vec{a}_{8,10,12}$ are taken into account via the conjugate complex of the factors $g_{0,\dots,3}$.[8]

The matrix Equation (2.46) is not only valid for silicon. Provided appropriate values for the matrix elements are inserted it can be used for all semiconductors with diamond crystal structure. Empirical parameters for the different semiconductors can, for instance, be found in [261]. For convenience, the parameters for C, Si and Ge are reprinted in Table 2.1. In this respect, it is important to note that the tight-binding parameters stated allow reproducing only parts of the band structure correctly. The reason is that the $sp^3s^\star$-model is insufficient to describe all details simultaneously (see [134, 35] for more information). In order to obtain a proper description of the conduction band the tight-binding parameters provided by Klimeck et al. for the $sp^3s^\star$-model can be adopted [134] that are reprinted in the third line of Table 2.1. The final step of the calculation is to solve the secular equation det $|\mathsf{H}_{\text{tot}} - E\mathbf{1}|$, which results in the dispersion relation $E(\vec{k})$.

2.6.3 Band Structure of Silicon

Using the tight-binding Hamiltonian H_{tot} stated in Equation (2.46) within the $sp^3s^\star$-model, the band structure is calculated with the parameters given in Table 2.1. Since silicon is basically a fcc lattice with basis, in reciprocal space one obtains a body-centered cubic lattice (cf. Figure 2.9). The Wigner–Seitz cell in reciprocal space, i. e., the first Brillouin zone is thus constructed as has been discussed in Section 2.3.2 and is

8 Note that in contrast to the literature, the matrix elements here have an additional minus sign so that the overlap of, e. g., two s-orbitals yields $-V_{ss}$ as has been used in earlier sections.

Table 2.1: Tight-binding parameters of the sp^3s^*-model for diamond, silicon and germanium reprinted from [261]. The third line states alternative parameters for silicon that reproduce the conduction band correctly (taken from [134]).

	$\epsilon_s^{b,g}$	$\epsilon_{p_{x,y,z}}^{b,g}$	ϵ_{s^*}	V_{ss}	V_{sp}	V_{xx}	V_{xy}	V_{s^*p}
C	−4.545	3.84	11.37	22.725	15.22	−3.84	11.67	8.211
Si	−4.2	1.715	6.685	8.3	5.7292	−1.715	4.575	5.3749
Si(N.N, E_c)	−3.659	1.679	3.876	7.971	8.875	−1.696	23.32	5.412
Ge	−5.88	1.61	6.39	6.78	5.4649	−1.61	4.9	5.2191

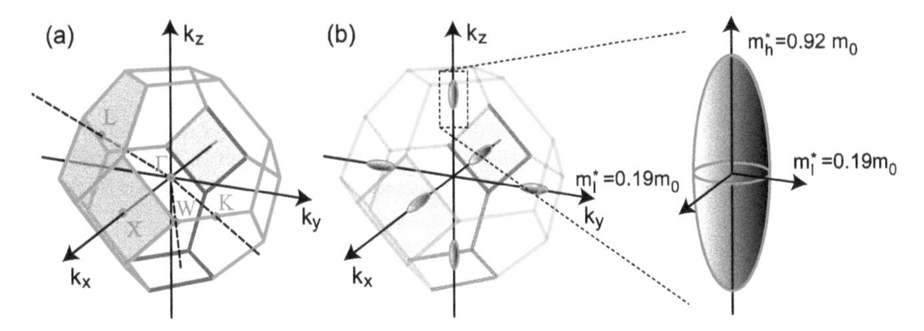

Figure 2.28: (a) Brillouin zone of silicon with high symmetry directions. (b) Constant energy surface for an energy slightly above the conduction band bottom. The right panel shows a close-up of a constant energy ellipsoid. A light effective mass m_l^* and a heavy effective mass m_h^* determine the conduction band.

displayed in Figure 2.28(a) with the high symmetry directions highlighted. Due to the three-dimensionality, the full band structure cannot be displayed anymore. Therefore, the band structure is plotted either only along the high symmetry directions (as in Figure 2.16, bottom right panel), or only a constant energy surface is shown. The latter is depicted exemplarily in Figure 2.28(b) in the case of the silicon conduction band revealing the well-known sixfold degeneracy. One observes rotationally symmetric ellipsoids along the $k_{x,y,z}$ axes; the video accessible through QR code #16 illustrates the connection between the typical representations of the silicon band structure further.

A plot of $E(\vec{k})$ along the Γ–L- and Γ–X-directions is shown in Figure 2.29, left panel. The band gap E_g is clearly visible; the close-up shows the conduction band in more detail so that the ellipsoidal form of the constant energy surfaces becomes apparent. A surface plot of the conduction band within the first Brillouin zone is displayed in the right panel, where again the ellipsoidal conduction band is visible. A light effective mass with $m_l^* = 0.19m_0$ and a heavy mass $m_h^* = 0.92m_0$ can be extracted from the band structure.

The consequences of this rather peculiar conduction band on the carrier density in bulk silicon and in particular in the case of an inversion layer in a metal-oxide-semiconductor field-effect transistor will be discussed in more detail in Section 2.11.2 and in Section 4.5, respectively.

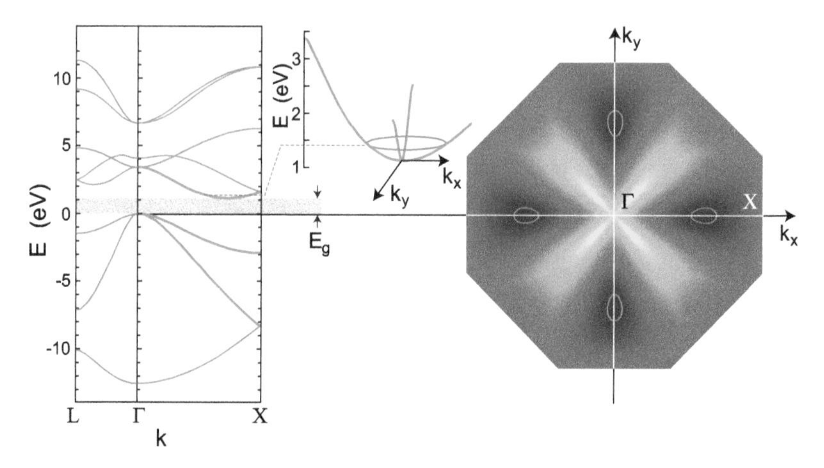

Figure 2.29: Left panel: Part of the Si band structure computed with the sp^3s^*-model based on the tight-binding parameters given in Table 2.1 [261]. The inset and the right panel show details of the conduction band as computed with the alternative parameters stated in the third line of Table 2.1 (taken from [134]).

2.7 The Surface of Bulk Crystals

As an example for the band structure of a bulk crystal, we consider the (100) surface of silicon shown in Figure 2.30(a). The topmost layer of atoms is arranged in a simple, body-centered square lattice with a surface unit cell tilted by 45° with respect to the fcc cubic cell and an area $\frac{a}{\sqrt{2}} \times \frac{a}{\sqrt{2}}$ where a is the lattice constant of the fcc cubic cell of silicon. Using this square lattice, semiinfinite unit cells can be defined that are arranged on the square lattice along x- and y-directions. This also means that we can in principle plot the band structure $E(\vec{k}_{\|})$ as a function of the component $\vec{k}_{\|}$ parallel to the (100) surface. By stretching the fcc cubic cell as shown in Figure 2.30(b), the unit cell for the

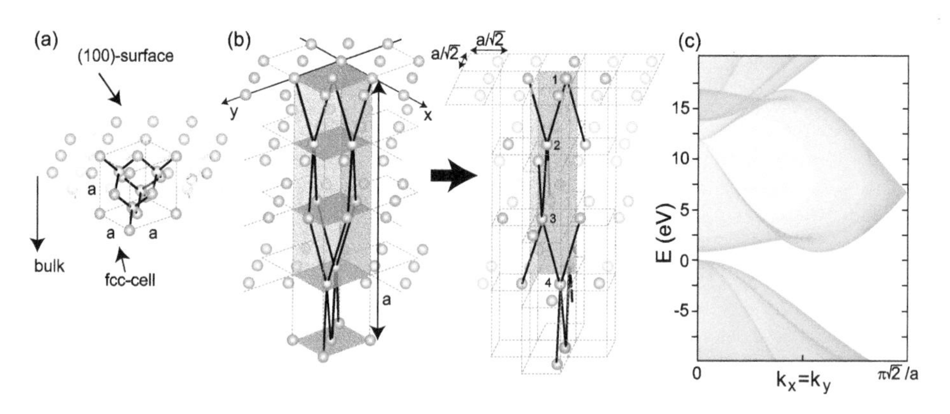

Figure 2.30: (a) Si (100) surface with fcc cubic cell of the diamond lattice. (b) The topmost layer of atoms at the (100) surface is arranged in a simple, body-centered square lattice of the bulk fcc lattice, which leads to a new simple cubic unit cell titled by 45°. (c) Projected surface band structure along the (110) direction.

computation of the surface band structure can be identified containing a repetitive arrangment of the four atoms highlighted in the figure (red circle). Eventually, from the right panel in (b), the coupling to nearest neighboring atoms can be deduced. The tight binding calculation then implies setting up one on-site and four coupling matrices. The coupling matrices have to be multiplied with appropriate phase factors. The detailed calculation and the result are provided through the QR code #17 leading to the surface projected band structure (see Section 2.4.8) displayed as an example in Figure 2.30(c). Note, that in the present case, the same sp^3s^*-model with the overlap integrals stated in Table 2.1 has been used. Furthermore, it is important to mention that for simplicity an unreconstructed Si-(100) surface has been assumed.

2.8 Two-Dimensional Materials

The first demonstration of a monolayer of graphene based on simple exfoliation by Novoselov and Geim [203, 82] initiated the new field of 2D materials research and culminated in the 2010 Nobel price in physics. Recently, 2D materials have attracted a strongly increasing attention fueled by the synthesis and exploration of many new 2D material classes based, e. g., on transition metal dichalcogenides (TMDCs) and on monoatomic allotropes of silicon (silicene), germanium (germanene) and tin (stanene) as well as black phosphorous and borophene.

In the present section, the electronic properties of selected 2D materials will be discussed. First, graphene will be considered since it allows a straightforward calculation of the band structure using the tight-binding formalism explained in the preceding chapters. Moreover, TMDCs will be studied as an example of novel 2D materials with promising electronic properties with respect to nanoelectronics devices.

2.8.1 Monolayer Graphene

Graphene is a special material with some very unusual properties that have inspired scientists to use the material in various directions including opto-electronic devices, sensors, etc. To understand why the material is unusual, let us compute the band structure employing the tight-binding recipe detailed in Section 2.4.6.

1. Graphene has a honeycomb lattice structure, as shown in Figure 2.31, and a suitable unit cell is shown as the dark gray diamond containing two carbon atoms (colored blue and green). Next, we can determine the lattice vectors $\vec{a}_{1,2}$ from the right panel of Figure 2.31. Using elementary geometrical considerations, one obtains $\vec{a}_1 = (a\sqrt{3}/2, a/2)$ and $\vec{a}_2 = (a\sqrt{3}/2, -a/2)$.
2. The honeycomb lattice is a result of the sp^2-hybridization of the s- and the three $p_{x,y,z}$-orbitals of carbon yielding three sp^2-hybrid orbitals in-plane that form 120° angles with respect to each other. These three orbitals form σ-bonds with adjacent

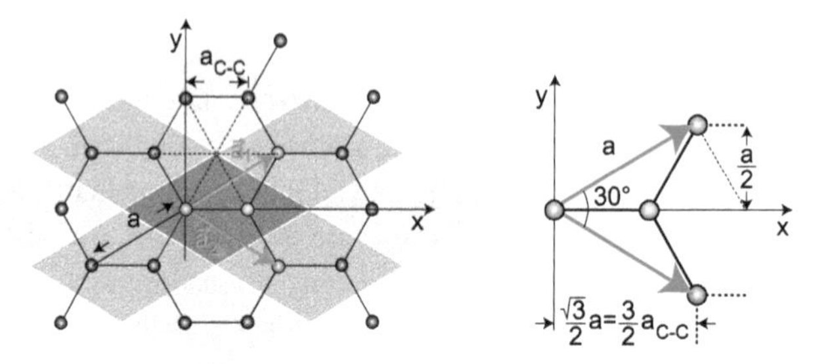

Figure 2.31: Graphene lattice with diamond-shaped central unit cell (dark gray) and nearest neighboring cells (light gray). Right panel: construction of the two unit vectors $\vec{a}_1$ and $\vec{a}_2$.

carbon atoms that are responsible for the unique mechanical strength of graphene. The remaining p_z-orbital is oriented perpendicular to the plane of the sp^2-orbitals forming π-bonds with the adjacent neighbors (illustrated in Figure 2.17(d)). It is this p_z-orbital that is responsible for the electronic behavior of graphene around the Fermi level, and thus, we can restrict our calculation of the band structure to the p_z orbitals. As a result, with two atoms (blue and green in Figure 2.31) per unit cell each providing a single p_z-orbital we have to solve simple 2×2 matrix equations.

3. If only nearest neighbor interaction is considered, one needs to take into account the contributions from the four gray diamonds displayed in Figure 2.31; note that the remaining diamonds do not have any nearest neighbor atoms and are thus neglected here. This means that five 2×2 matrices need to be set up.
4. Since we are dealing only with p_z-orbitals of carbon atoms, $\langle p_z^g|\mathcal{H}|p_z^g\rangle = \langle p_z^b|\mathcal{H}|p_z^b\rangle = \epsilon_p$ and because the p_z-orbitals are rotationally symmetric with respect to the graphene plane, only a single overlap integral $\langle p_z^b|\mathcal{H}|p_z^g\rangle = -V_{pp\pi}$ for nearest neighbor atoms needs to be considered. As a result, the on-site matrix is given by

$$\begin{pmatrix} \epsilon_p & -V_{pp\pi} \\ -V_{pp\pi} & \epsilon_p \end{pmatrix}.$$

5. Next, write down four empty 2×2-matrices with the ket vectors of the p_z-orbitals of the two carbon atoms of a nearest neighbor unit cell on top and the bra vectors of the on-site cell in a column in front of the matrix. Multiply the matrices with the appropriate phase factors. For the upper left neighboring unit cell, which is located at $-\vec{a}_2$ with respect to the on-site unit cell, this yields

$$\begin{pmatrix} \langle p_z^b|\mathcal{H}|p_z^b\rangle_{-\vec{a}_2} = 0 & \langle p_z^b|\mathcal{H}|p_z^g\rangle_{-\vec{a}_2} = -V_{pp\pi} \\ \langle p_z^g|\mathcal{H}|p_z^b\rangle_{-\vec{a}_2} = 0 & \langle p_z^g|\mathcal{H}|p_z^g\rangle_{-\vec{a}_2} = 0 \end{pmatrix} e^{-i(k_x \frac{a\sqrt{3}}{2} - k_y \frac{a}{2})}$$

where $|p_z^{b,g}\rangle_{-\vec{a}_2}$ are meant to be the ket vectors of the p_z-orbitals of the blue and green atoms of the nearest neighbor unit cell that is $-\vec{a}_2$ away from the on-site cell. Note that only the blue atom in the on-site cell has a nonzero overlap with the green atom in this particular nearest neighbor cell. The phase factor is simply $e^{-i\vec{k}\vec{a}_2} = e^{-i(k_x \frac{a\sqrt{3}}{2} - k_y \frac{a}{2})}$. Similarly, the remaining three 2 × 2 coupling matrices are derived, which are given by

$$\begin{pmatrix} 0 & 0 \\ -V_{pp\pi} & 0 \end{pmatrix} e^{i\vec{k}\vec{a}_1}, \quad \begin{pmatrix} 0 & 0 \\ -V_{pp\pi} & 0 \end{pmatrix} e^{i\vec{k}\vec{a}_2}, \quad \begin{pmatrix} 0 & -V_{pp\pi} \\ 0 & 0 \end{pmatrix} e^{-i\vec{k}\vec{a}_1}.$$

Finally, the overall matrix H_{tot} is obtained by adding all five matrices leading to

$$\mathsf{H}_{tot} = \begin{pmatrix} \epsilon_p & -V_{pp\pi}(1 + e^{-i\vec{k}\vec{a}_1} + e^{-i\vec{k}\vec{a}_2}) \\ -V_{pp\pi}(1 + e^{i\vec{k}\vec{a}_1} + e^{i\vec{k}\vec{a}_2}) & \epsilon_p \end{pmatrix} \tag{2.47}$$

6. Solving $\det|\mathsf{H}_{tot} - E\mathbf{1}| = 0$ yields the final band structure

$$\begin{aligned} E(\vec{k}) &= \epsilon_p \pm \sqrt{V_{pp\pi}(1 + 2e^{ik_x\sqrt{3}/2}\cos(k_y a/2))V_{pp\pi}(1 + 2e^{-ik_x\sqrt{3}/2}\cos(k_y a/2))} \\ &= \epsilon_p \pm V_{pp\pi}\sqrt{1 + 4\cos(k_x a\sqrt{3}/2)\cos(k_y a/2) + 4\cos^2(k_y a/2)} \end{aligned} \tag{2.48}$$

where $e^{i\vec{k}\vec{a}_1} + e^{i\vec{k}\vec{a}_2} = 2e^{ik_x a\sqrt{3}/2} \cdot \cos(k_y a/2)$ and $e^{-i\vec{k}\vec{a}_1} + e^{-i\vec{k}\vec{a}_2} = 2e^{-ik_x a\sqrt{3}/2} \cdot \cos(k_y a/2)$ has been used. Equation (2.48) represents a conduction and a valence band which are completely symmetric. At $T = 0$ K, all states of the valence band would be filled completely while the conduction band would be empty. Hence, the Fermi energy $E_f = \epsilon_p$. For simplicity, we set $E_f = 0$ in the following.

The band structure of a monolayer graphene is depicted in Figure 2.32(a). The first Brillouin zone is a hexagon with six K-points. Each K-point belongs to three adjacent Bril-

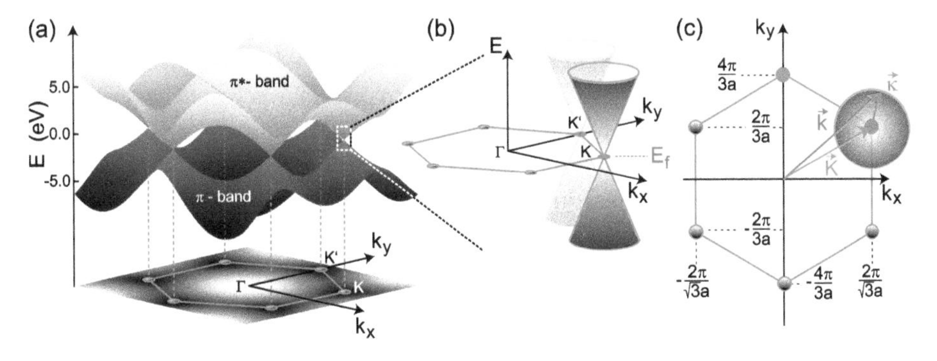

Figure 2.32: (a) Dispersion relation of a graphene monolayer calculated with one p_z-orbital per carbon atom. (b) Close-up of the band structure around the Fermi energy. A linear dispersion relation is obtained.

louin zones and, therefore, only two are independent (denoted K and K' in the figure). A close-up of the band structure around the Fermi energy (i. e., $E = 0$) is shown in (b). The most prominent features of the band structure are (i) the lack of a band gap and (ii) the linear dispersion relation around the Fermi energy. In fact, at E_f the conduction band cone and the valence band cone touch each other such that graphene has no band gap. Moreover, the linear behavior of the dispersion relation is obtained when $E(\vec{k})$ is Taylor-expanded around K (or K'). To this end, the reduced vector $\vec{\kappa}$ (here, this is *not* meant to be the imaginary part of $\vec{k}$) is defined with $\vec{\kappa} = \vec{k} - \vec{K}$ as illustrated in Figure 2.32(c). Since the band structure at each K-point is the same and $E(\vec{K}) = 0$, it is sufficient to expand the band structure around a single K point. Looking at Equation (2.47) it is obvious that the off-diagonal elements of the matrix are the conjugate complex of each other. Let us call the upper right element h and expand it around $\vec{K}$, which yields

$$\begin{aligned} h(\vec{k}) &\approx \underbrace{h|_{\vec{k}=\vec{K}}}_{=0} + \frac{\partial h}{\partial k_x}(k_x - K_x) + \frac{\partial h}{\partial k_y}(k_y - K_y) \\ &= -V_{pp\pi} ia \frac{\sqrt{3}}{2} \underbrace{(k_x - K_x)}_{=\kappa_x} + V_{pp\pi} a \frac{\sqrt{3}}{2} \underbrace{(k_y - K_y)}_{=\kappa_y}. \end{aligned} \tag{2.49}$$

With this expansion, the band structure is simply given by $E(\vec{k}) = \pm\sqrt{|h|^2}$ and as a result, the dispersion around $\vec{K}$ as a function of the reduced vector $\vec{\kappa}$ is

$$E(\vec{\kappa}) \approx \pm V_{pp\pi} a \frac{\sqrt{3}}{2} \sqrt{\kappa_x^2 + \kappa_y^2} = \pm V_{pp\pi} a \frac{\sqrt{3}}{2} |\vec{\kappa}|. \tag{2.50}$$

Indeed, Equation (2.50) shows the expected linear dispersion relation that can be seen in Figure 2.32(a) and (b).

2.8.2 Bilayer Graphene

A central result of the preceding section was that a monolayer graphene does not exhibit a band gap. As will be discussed in Section 10.1, this yields field-effect transistor devices with very high off-state leakage currents. As an alternative, bilayer graphene can be used to create a band-gap. In order to understand how and why a band gap can be opened in bilayer graphene, a tight-binding calculation will be carried out.

Consider a bilayer graphene in A-B-stacking (also known as Bernal stacking) as illustrated in Figure 2.33, left panel. Here, the blue carbon atom (b1) of layer 1 sits exactly on top of the green atom of layer 2 (g2). These two atoms are coupled to each other via van der Waals interaction. The unit cell has indeed not changed and is still the same diamond as in the case of monolayer graphene (illustrated with the gray diamond in Figure 2.33). In contrast to the monolayer, the unit cell now stretches over both layers and as a result, the unit cell contains four atoms with p_z-orbitals (note the four corner

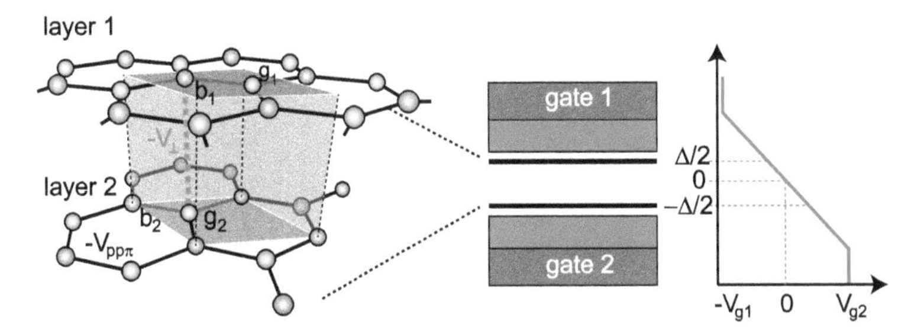

Figure 2.33: Bilayer graphene (left) with *A*-*B*-stacking. A projection of the unit cell of layer 1 is illustrated with the gray diamond in layer 2. The interlayer van der Waals coupling is described by the coupling parameter $V_\perp \approx 0.3$ eV. The right panel shows the bilayer in a double-gate configuration with opposite voltages $\pm V_g$ applied at the gate electrodes resulting in a potential of $\pm\Delta/2$ in the two layers.

atoms in the layer 2 diamond). Hence, a total of four p_z orbitals need to be taken into consideration. However, the bilayer graphene exhibits the same four nearest neighbor unit cells as in the case of the monolayer.

The coupling between layer 1 and 2 can be restricted to the on-site unit cell (between atoms b1 and g2) since this coupling stems from van der Waals interaction and as such drops off even faster than the interaction due to wave function overlap within each layer. As a result, when setting up the matrix $\mathsf{H}_{\mathrm{tot}}$, which is a 4×4 matrix, there will be two 2×2-blocks on the main diagonal, which are simply given by Equation (2.47). Since van der Waals coupling needs to be accounted for only within the on-site cell, one arrives at

$$\begin{array}{c} \langle p_z^{b1}| \\ \langle p_z^{g1}| \\ \langle p_z^{b2}| \\ \langle p_z^{g2}| \end{array} \underbrace{\overset{\begin{array}{cccc} |p_z^{b1}\rangle & |p_z^{g1}\rangle & |p_z^{b2}\rangle & |p_z^{g2}\rangle \end{array}}{\begin{pmatrix} 0 & h & 0 & -V_\perp \\ h^\star & 0 & 0 & 0 \\ 0 & 0 & 0 & h \\ -V_\perp & 0 & h^\star & 0 \end{pmatrix}}}_{=\mathsf{H}_{\mathrm{tot}}} \tag{2.51}$$

where the Dirac point has been set to zero and the overlap integral due to van der Waals interaction is $\langle p_z^{b1}|\mathcal{H}|p_z^{g2}\rangle = -V_\perp$. Note that the linear approximation around the K-points has been used with $h = -V_{pp\pi} ia\frac{\sqrt{3}}{2}\kappa_x + V_{pp\pi} a\frac{\sqrt{3}}{2}\kappa_y$.

Next, the bilayer graphene is considered to be embedded into a dual-gate structure with a top-gate (gate 1) and a bottom gate (gate 2) as illustrated in Figure 2.33, right panel. For simplicity, the gates are taken to be symmetric with equal gate dielectrics exhibiting the same thickness. In addition, equal but opposite gate voltages $V_{g1} = -V_g$ and $V_{g2} = V_g$ are applied. Consequently, the Dirac points of the two layers are shifted in energy. In the symmetric case considered here, the Dirac points are shifted to $\pm\frac{\Delta}{2}$ due to the perpendicular electric field (cf. Figure 2.33, right panel). Therefore, the Hamiltonian is

$$\mathsf{H}_{\text{tot}} = \begin{pmatrix} \frac{\Delta}{2} & h & 0 & -V_\perp \\ h^\star & \frac{\Delta}{2} & 0 & 0 \\ 0 & 0 & -\frac{\Delta}{2} & h \\ -V_\perp & 0 & h^\star & -\frac{\Delta}{2} \end{pmatrix}. \tag{2.52}$$

Solving $\det|\mathsf{H}_{\text{tot}} - E\mathbf{1}| \stackrel{!}{=} 0$ eventually yields the band structure. In the present case, the secular equation can be solved analytically and $E(\kappa_x, \kappa_y)$ is explicitly given by

$$E(\kappa_x, \kappa_y) = \pm\sqrt{V_{pp\pi}^2 a^2 \frac{3}{4}|\vec{\kappa}|^2 + \frac{V_\perp^2}{2} + \frac{\Delta^2}{4} \pm \sqrt{\frac{V_\perp^4}{4} + V_{pp\pi}^2 a^2 \frac{3}{4}(\Delta^2 + V_\perp^2)|\vec{\kappa}|^2}}. \tag{2.53}$$

While the approximation used to arrive at Equation (2.53) (van der Waals interaction only between g1 and b2) neglects some details of bilayer graphene a closed expression for the band structure is obtained that captures the effects relevant for the opening of a band gap. Figure 2.34(a) shows the resulting band structure $E(|\vec{\kappa}|)$ for $\Delta = 0$ (black lines) and $\Delta = 0.26$ eV (red lines) realized by applying appropriate gate voltages (cf. Figure 2.33, left panel). Note that the straight lines (denoted $E_{2,3}$) belong to the solution with the negative sign under the square root and the dotted (E_1 and E_4) to the positive sign as can easily be deduced by looking at the band structure in the case of $\kappa = 0$.

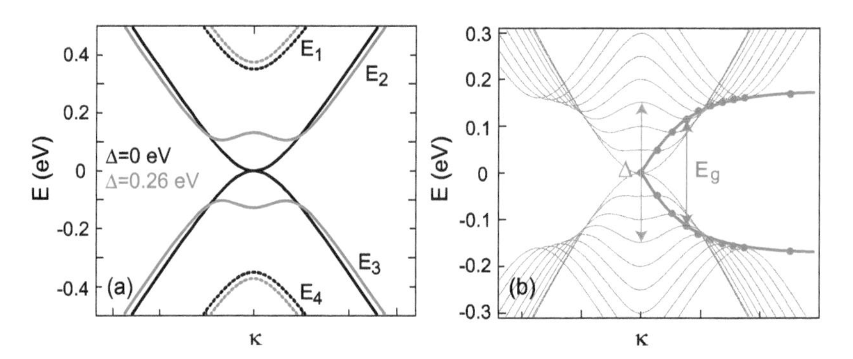

Figure 2.34: (a) Band structure $E(|\vec{\kappa}|)$ of the bilayer graphene in the case $\Delta = 0$ (black) and $\Delta = 0.26$ eV (red). (b) Due to the Mexican hat-shaped band structure $E_g \leq \Delta$ and the induced band gap approaches a limit determined by $E_g \stackrel{\Delta\to\infty}{\longrightarrow} V_\perp$.

Since the band structure exhibits the shape of a Mexican hat, the minimum band gap is not at $\kappa = 0$. Computing the minimum E_{min} with $\frac{\partial E(\kappa)}{\partial \kappa} = 0$ yields $E_g/2$ and one eventually obtains

$$E_g = \frac{\Delta V_\perp}{\sqrt{V_\perp^2 + \Delta^2}} \tag{2.54}$$

which clearly shows that a gate-tunable band gap can be generated in the case of bilayer graphene. For small Δ, i. e., $\Delta < V_\perp$, the band gap depends linearly on Δ with $E_g \approx \Delta$. However, for large Δ the achievable band gap saturates at the value of $V_\perp$. Figure 2.34(b) shows the behavior of $E_g(\Delta)$; here, the straight gray lines show $E_{3,4}$ for increasing Δ and the red line illustrates the behavior of E_g as a function of Δ showing a saturation. Since $V_\perp$ is the overlap integral due to the van der Waals interaction its value $V_\perp = 0.3\,\text{eV}$ is rather small compared to the overlap integral between the covalently bonded carbon atoms within each graphene layer. Thus, the achievable band gaps are rather small, too. This is consistent with experimentally found values of the band gap in bilayer graphene discussed in Section 10.1.3. For logic transistor operation, the band gap would be too small in order to provide the desired low off-state leakage currents. Nevertheless, the possibility of a gate-tunable band gap whose magnitude depends on the interaction, and hence on the distance between the two graphene layers opens a number of interesting applications (see section 11.3.2, for instance).

Task 10.
Tight-binding of bilayer graphene: Consider bilayer graphene with Bernal stacking and carry out the full tight-binding calculation with nearest neighbor interaction and van der Waals coupling between the atoms b1 and g2 as illustrated in Figure 2.33. Compute the band structure explicitly.

2.8.3 Transition Metal Dichalcogenides

A major drawback of monolayer graphene for the realization of nanoelectronics devices is the lack of a band gap. Even in the case of bilayer graphene, the maximum band gap is rather small. Fortunately, there are various alternative 2D materials with different properties ranging from metallic to semiconducting, to insulating and even superconducting. One of the most widely studied class of 2D materials are transition metal dichalcogenides (TMDCs) consisting of a transition metal M (blue) and a chalcogen X (green) in an MX_2-configuration with upper and lower layer of chalcogen atoms centered around a layer of transition metal atoms as depicted in Figure 2.35.

Monolayer TMDCs can be existent in the 2H- or 1T-configurations shown in Figure 2.35. The difference between the two configurations is the way the chalogen atoms are arranged: in the 2H-configuration, the chalcogen in the upper layer sits exactly on top of the one in the lower layer (see the close-up in Figure 2.35(a)) whereas in the 1T-configuration the two chalcogen atoms lie on top of a straight line that crosses the central transition metal atom (cf. the close-up in Figure 2.35(b)). In the case of the 2H-configuration,[9] TMDCs exhibit a graphene-like honeycomb lattice as depicted in (a) when observed from the top; in the 1T configuration the lattice resembles a hexagonal

9 In the case of a single layer, the 2H is also called 1H phase.

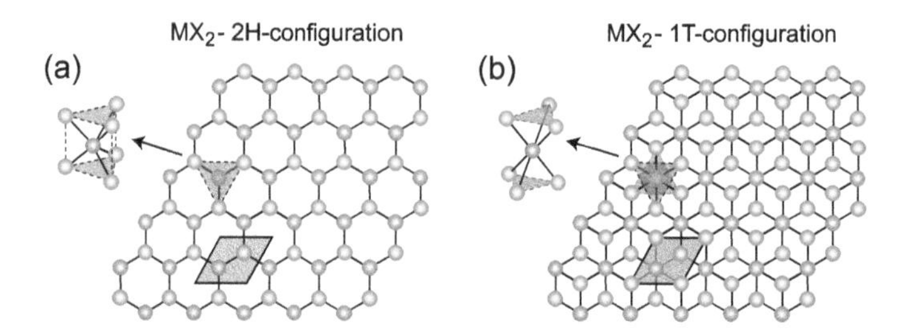

Figure 2.35: Top view of the 2H(1H) (a) and 1T (b) configurations of a transition metal dichalcogenide. The close-ups depict the arrangement of the two chalcogen layers around the central transition metal atom. The gray diamonds show the primitive unit cell that contains two chalcogen and one transition metal atom.

lattice (note, however, the difference from a hexagonal lattice). In both cases, the primitive unit cell is as illustrated with the gray diamond that contains two chalcogen and one transition metal.

The following discussion will concentrate on the 2H-configuration. As already mentioned, the honeycomb lattice of the 2H-configuration has the same diamond-shaped unit cell as in the case of graphene with the difference that the unit cell now contains three instead of two atoms. Primitive vectors $\vec{a}_1 = (a, 0)$ and $\vec{a}_2 = (a/2, \sqrt{3}a/2)$ can be defined (Figure 2.37, bottom left) and a tight-binding calculation can be carried out using the recipe detailed in Section 2.4.6.

The transition metal makes the incorporation of d-orbitals into the basis for the tight-binding calculation necessary. Since in the transition metals the five d-orbitals (d_{z^2}, $d_{x^2-y^2}$, d_{xy}, d_{yz} and d_{xz}) are most relevant for the bonding and since the p-orbitals are the important orbitals in the case of the chalcogen we expect in total a number of 11 orbitals (five d- and 2×3 p-orbitals of the two chalcogen atoms) to be incorporated into the computation. However, it has been shown in [174] that the complexity of the tight-binding calculation can be strongly reduced by noting that the three orbitals d_{z^2}, $d_{x^2-y^2}$ and d_{xy} play the major role and mainly determine the conduction and valence bands. In this so-called three-band approximation, the chalcogen atoms are neglected which implies that the lattice changes from a honeycomb lattice to a hexagonal one as illustrated in the left panel of Figure 2.37. As a result, in the three-band approximation there is only a single transition metal in the unit cell whose bonding is described by three d-orbitals. Each transition metal has six nearest neighbor atoms so that we need to set up seven 3×3 matrices in order to solve for the band structure.

The presence of the d-orbitals adds in complexity and requires somewhat more consideration in order to understand how the matrix elements in the tight-binding calculation need to be computed. As an example, Figure 2.36 shows a top view of the hexagonal transition metal lattice. In the central on-site unit cell, a d_{z^2} is assumed surrounded by d_{xy} (a) and $d_{x^2-y^2}$-orbitals (b). Taking the bond vector $\vec{d}$ (equal to $\vec{a}_2$ in Figure 2.36) connecting two adjacent atoms as reference, the d-orbitals are rotated around the z-axis to yield the right panels of (a) and (b). Using the direction cosines for d-orbitals (cf. Sec-

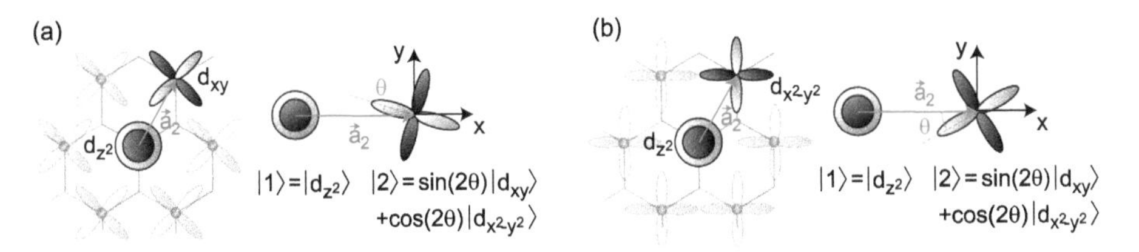

Figure 2.36: Hexagonal lattice within the three-band model for tight-binding calculations of TMDCs. (a) shows the overlap between a d_{z^2}-orbital on the on-site unit cell with d_{xy}-orbitals on the nearest neighbor atoms. (b) shows the same in the case of a $d_{x^2-y^2}$-orbital.

tion 2.4.5) allows reducing the overlap between nearest neighbors to the minimal set of overlap integrals [235] and one finally arrives at (see QR code #19 for details)

$$\mathrm{H} = \begin{pmatrix} h_0 & h_1 & h_2 \\ h_1^\star & h_{11} & h_{12} \\ h_2^\star & h_{12}^\star & h_{22} \end{pmatrix} \tag{2.55}$$

where the different entries are given by $h_0 = \epsilon_1 - 2V_0(\cos 2\alpha + 2\cos\alpha\cos\beta)$, $h_{11} = \epsilon_2 + 2V_{11}\cos 2\alpha + (V_{11} + 3V_{22})\cos\alpha\cos\beta$, $h_{22} = \epsilon_2 + 2V_{22}\cos 2\alpha + (3V_{11} + V_{22})\cos\alpha\cos\beta$, $h_1 = -2\sqrt{3}V_2\sin\alpha\sin\beta + 2iV_1(\sin 2\alpha + \sin\alpha\cos\beta)$, $h_2 = 2V_2(\cos 2\alpha - \cos\alpha\cos\beta) + 2\sqrt{3}iV_1\cos\alpha\sin\beta$ and $h_{12} = \sqrt{3}(V_{22} - V_{11})\sin\alpha\sin\beta + 4iV_{12}\sin\alpha(\cos\alpha - \cos\beta)$. Here, the parameters α and β are $\alpha = \frac{1}{2}k_x a$ and $\beta = \frac{\sqrt{3}}{2}k_y a$. Appropriate parameters for three typical TMDCs are given in Table 2.3 [174]. Figure 2.37, right panel, shows the result of a tight-binding calculation using the three band model of a monolayer of MoS_2 (green curves) and WSe_2 (red curves). Obviously, both TMDCs exhibit a band gap at the K-points consistent with the values given in Table 2.2.

Table 2.2: Typical material properties of common TMDCs.

	MoS_2	WSe_2	$MoSe_2$	WS_2	WTe_2
E_g (eV)	1.79 (ML)	1.61 (ML)	1.49 (ML)	2.0 (ML)	0.71 (ML)
	1.2 (bulk)	1.2 (bulk)	1.0 (bulk)	1.3 (bulk)	
$m_c^\star$	0.46	0.34	0.55	0.3	0.31
$m_v^\star$	0.56	0.44	0.64	0.44	0.41

Table 2.3: Tight-binding parameters for the three-band model for different TMDCs [174].

	a (Å)	ϵ_1 (eV)	ϵ_2 (eV)	V_0 (eV)	V_1 (eV)	V_2 (eV)	V_{11} (eV)	V_{12} (eV)	V_{22} (eV)
MoS_2	3.19	1.046	2.104	0.184	0.401	0.507	0.218	0.338	0.057
WS_2	3.191	1.13	2.275	0.206	0.567	0.536	0.286	0.384	−0.061
WSe_2	3.325	0.943	2.179	0.207	0.457	0.486	0.263	0.329	0.034

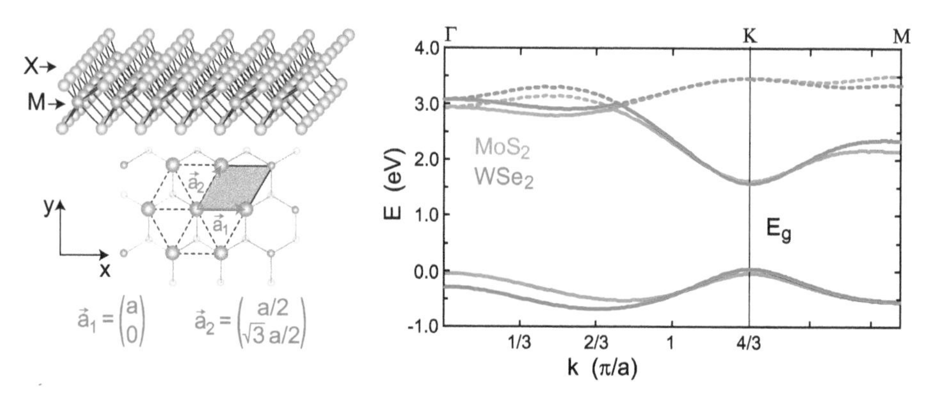

Figure 2.37: Left: Schematic of the 2H-configuration of MX_2. The bottom panel shows the primitive lattice vectors together with the change of the honeycomb to a hexagonal lattice in the three-band approximation [174]. Right: Band structure ($\Gamma \rightarrow K \rightarrow M$) of MoS_2 (green curves) and WSe_2 (red curves) computed using the three-band model based on the parameters given in Table 2.3.

2.8.4 Other Two-Dimensional Materials

Apart from the 2D materials mentioned so far, there are many more 2D materials. Worth mentioning is certainly hexagonal boron-nitride (h-BN). h-BN has the same, single-atom thin honeycomb lattice structure as graphene; the only difference is that alternating boron and nitrogen atoms sit on the lattice sites [266]. Hexagonal boron nitride is an insulator with a band gap of $E_g = 5.9$ eV and can thus serve as a gate dielectric in an all-2D field-effect transistor (see Section 10.3). Moreover, h-BN is a perfect substrate providing very smooth surfaces and, therefore, reduced disorder due to potential fluctuations in graphene [81]. In fact, very high carrier mobilities in graphene encapsulated in h-BN were observed and recently employed for ultrasensitive Hall sensors [61]. In Section 11.3.2, the use of h-BN for cryogenic 2D material MOSFETs will be explored.

A very interesting 2D material is also black phosphorous (BP). What makes this material so interesting is the fact that it exhibits a band gap that can be tuned over a wide range with the number of layers that are stacked on top of each other. The reason for this dependence is the relatively strong interaction between adjacent layers. Apart from this, silicene, germanene, stanene and borophene have recently attracted a great deal of attention. A vast amount of literature is available on each of the materials and the reader is encouraged to explore these materials further.

2.8.5 Graphene Nanoribbons

It was explained in Section 2.2 how quantum confinement allows reducing the dimensions of a system and how quantization effectively increases the band gap of a semiconducting nanostructure. The same reasoning can be applied to graphene in order to realize a band gap that may be suitable for field-effect transistor devices.

Cutting a graphene sheet into a so-called nanoribbon allows exploiting vertical quantization. Let us compute the band gap that can in principle be realized with this method and determine what size of nanoribbons would be needed in order to achieve an appropriate E_g. To this end, consider the nanoribbon displayed in Figure 2.38. Due to the particular form of the nanoribbon along its edges reminiscent of an armchair, the displayed type of nanoribbon is called armchair graphene nanoribbon (AGNR). A graphene nanoribbon can be specified by a vector $\vec{C} = n\vec{a}_1 + m\vec{a}_2$ along the width of the nanoribbon that is a linear combination of the primitive lattice vectors $\vec{a}_1$ and $\vec{a}_2$ multiplied with integer numbers n and m; the nanoribbon is, however, only specified up to ± one dimer, which becomes evident from Figure 2.38(a): when adding the dimer "7"–"8" to the smallest nanoribbon, the vector $\vec{C}$ does not change although a somewhat larger unit cell is obtained (gray instead of the dark gray area). Since the vector $\vec{C}$ is aligned along the width of the nanoribbon, the axis of the nanoribbon lies perpendicular to $\vec{C}$. An AGNR always has the indices $(n, 0)$, i. e., $\vec{C}$ is simply a multiple of $\vec{a}_1$ (cf. Figure 2.38(a)).

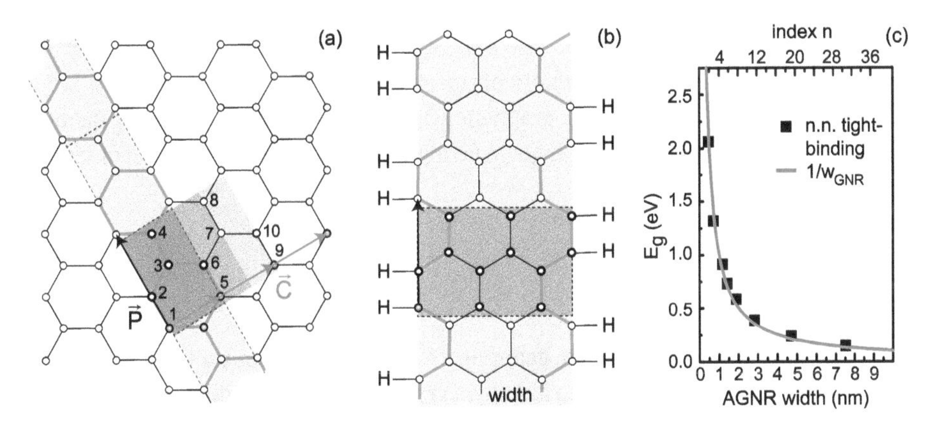

Figure 2.38: (a) Armchair graphene nanoribbon with different unit cells and numbering of the atoms involved. The σ-bonds of the broken sp^2 hybrid orbitals at the armchair edges are saturated with hydrogen atoms in order to avoid additional, localized states within the band gap (b). (c) Band gap E_g of AGNRs as a function of the width w_{GNR} (lower x-axis and index n used to set up $\vec{C} = (n, 0)$ on the upper x-axis). The red line is a $1/w_{GNR}$-fit.

At the edges of a nanoribbon, there is an unsaturated σ-bond of one of the sp^2-orbitals, which can be terminated with a hydrogen atom as illustrated in Figure 2.38(b). As such, a nanoribbon represents a one-dimensional crystal with a unit cell that may contain a relatively large amount of carbon atoms (at least compared to the unit cells of the materials we considered so far). The periodicity of this one-dimensional crystal is given by the vector $\vec{P}$ (see Figure 2.38(a)) whose length $|\vec{P}|$ represents the new lattice constant of the one-dimensional crystal. Consequently, the first Brilluoin zone of the nanoribbon extends from $-\frac{\pi}{|\vec{P}|}, \ldots, \frac{\pi}{|\vec{P}|}$. In the particular case of an AGNR, $\vec{P}$ is always given by $\vec{P} = \vec{a}_1 - 2\vec{a}_2$ and has a length $|\vec{P}| = a\sqrt{3}$ (with a being the lattice constant of the underlying graphene lattice).

AGNRs allow setting up a simple relation between the width of a nanoribbon and the respective band gap. Using the nearest neighbor tight-binding calculation (for details, see the QR code #20), the band structure can be computed that allows extracting E_g. Figure 2.38(a) and (b) show AGNRs with different widths. The smallest AGNR consists of six carbon atoms (denoted "1", ..., "6") yielding the dark gray unit cell.[10] Depending on the number of dimers added to the nanoribbon (e. g., atoms "7"–"8" and "9"–"10"), the AGNR is either metallic or semiconducting. In fact, if the number of dimers is $2 + 3n$ with n being an integer, the band gap vanishes (see QR code #20 for details). For all other AGNR, a semiconductor is obtained with a direct band gap at the Brillouin-zone center. As displayed in Figure 2.38(c), very small widths w_{GNR} are required to obtain reasonably high band gaps suitable for room-temperature electronics.[11] However, manufacturing nanoribbons in the range of $w_{GNR} \approx$ 1–2 nm is a very difficult task to do. Moreover, one has to deal with edge states, i. e., unsaturated dangling carbon bonds that lead to additional localized states. Furthermore, line-edge roughness due to a lithography and etch process is transferred into a fluctuating potential along the nanoribbon, and hence yields substantial variability of the electronic properties (see Section 10.1.2 for experimental data). As a result, graphene nanoribbons are not necessarily the material of choice when thinking of a high yield, highly reproducible fabrication of highly integrated digital circuits.

2.9 Carbon Nanotubes

Exploiting quantum confinement to generate a band gap is certainly a viable approach but turned out to be very difficult to handle in graphene because widths of the nanoribbons in the 1–2 nm regime are required. This extremely small width renders the edges to be very important. In fact, a missing dimer would lead to a locally strongly varying size of the band gap. Moreover, missing carbon atoms and the unsaturated σ-bond of the sp^2 orbitals lead to electronic states that strongly deteriorate the usefulness of the nanoribbon (cf. Figure 10.8).

---

10 Note that the unit cell has been shifted slightly in order to facilitate an easy identification of the atoms that belong to the unit cell.

11 The reason for the band gap in a AGNR discussed here is quantum confinement. There are also other mechanisms that may lead to the formation of a band gap in a graphene nanoribbon. Further discussion of this is, however, beyond the scope of the present book and the reader is referred to [236] for aspects related to this topic.

A very elegant way to avoid all these issues is to roll-up the graphene nanoribbon to a carbon nanotube (CNT), a hollow, cylindrical object displayed in Figure 2.40(b). To enable rolling up the graphene sheet to a nanotube, it is necessary that the atoms on both sides of the graphene edge are equivalent, i. e., they “merge” to a single atom when the nanotube is rolled up (this is of course not an engineered top-down approach but occurs during the growth of the nanotubes). Similar to graphene nanoribbons, a carbon nanotube can be denoted by the vector $\vec{C} = n\vec{a}_1 + m\vec{a}_2$, which in a CNT is the circumference vector. Contrary to graphene nanoribbons where $\vec{C}$ specifies the ribbon only up to one dimer, in the case of a carbon nanotube the circumference vector $\vec{C}$ uniquely specifies the CNT because a nanotube can only be obtained if carbon atoms of the same sublattice (blue or green atoms in Figure 2.31) are connected to each other. The vector $\vec{C} = n\vec{a}_1 + m\vec{a}_2$ only leads to atoms of the same sublattice and as a result, nanotubes can be defined unambiguously with the indices (n, m). $\vec{C}$ is explicitly given by

$$\vec{C} = n \cdot \vec{a}_1 + m \cdot \vec{a}_2 = \frac{a}{2}\begin{pmatrix} \sqrt{3}(n+m) \\ n-m \end{pmatrix}. \tag{2.56}$$

With the circumference vector $\vec{C}$, the diameter d_{CNT} of the nanotube is given by $d_{\mathrm{CNT}} = \frac{|\vec{C}|}{\pi} = \frac{a}{\pi}\sqrt{n^2 + m^2 + nm}$.

There are basically three different types of nanotubes that are depicted in Figure 2.39: armchair, zig-zag and chiral nanotubes. Since there is no “edge” in a nanotube anymore, the nomenclature refers to the appearance along the circumference (in contrast to a GNR): Armchair CNTs are (n, n)-tubes since the circumference resembles an armchair (see green line in Figure 2.39). Zig-zag nanotubes are $(n, 0)$-tubes and their circumference shows the zig-zag pattern they are named after. Finally, chiral nanotubes are general (n, m) nanotubes (Figure 2.39, right panel). Exemplarily, Figure 2.40(a) shows a graphene lattice with circumference vector $\vec{C}$ that would result in a (2, 1) nanotube.[12]

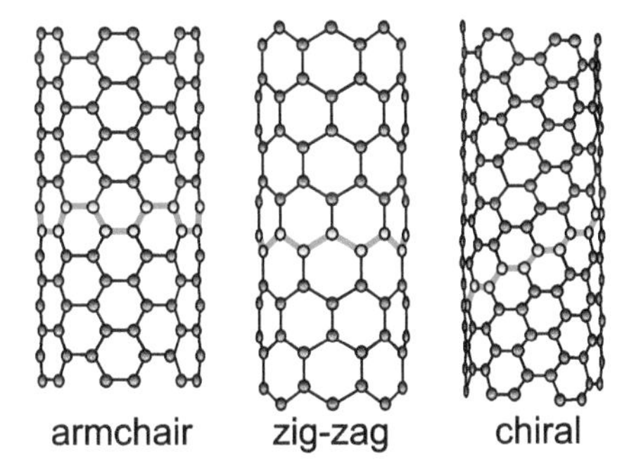

Figure 2.39: Three different types of carbon nanotubes: armchair ((n, n)-tubes), zig-zag ($(n, 0)$-tubes) and chiral ((n, m)-tubes).

12 Note that such a nanotube does not exist, the indices were chosen for convenience.

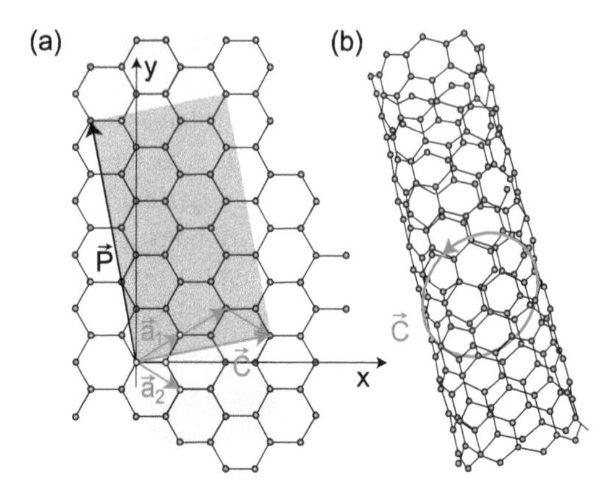

Figure 2.40: (a) The circumference vector $\vec{C} = n\vec{a}_1 + m\vec{a}_2$ determines the (n, m)-CNT under consideration. (b) $\vec{C}$ is perpendicular to the axis of the CNT. $\vec{P}$ and $\vec{C}$ span a new unit cell.

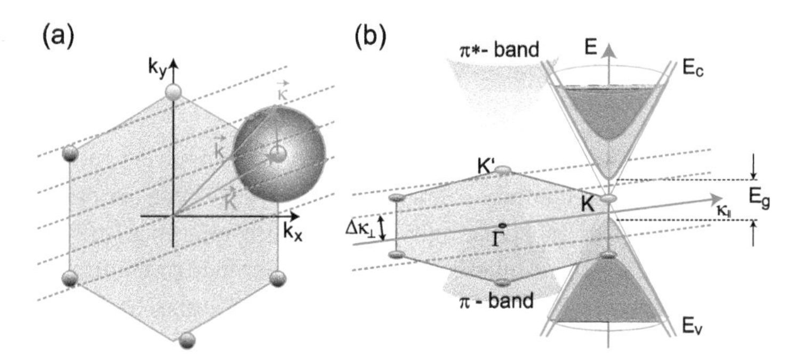

Figure 2.41: (a) Hexagonal Brillouin zone of the graphene lattice. The $\vec{k}$-vector can be written as a sum of $\vec{K}$ and the reduced vector $\vec{\kappa}$ with origin at the respective K-point. (b) Rolling up a graphene sheet to a nanotube enforces periodic boundary conditions. As a result, allowed states lie on the red dashed lines. Cuts between the graphene cones and $E - \vec{\kappa}$-planes oriented along the red dashed lines give rise to one-dimensional subbands (green lines). If the boundary condition is such that there are no states at K, a semi-conducting tube with band gap E_g is obtained.

The circumference vector $\vec{C}$ is aligned perpendicular to the axis of the resulting nanotube and perfect periodic boundary conditions apply to the electronic wave function along $\vec{C}$. To compute the resulting band structure, the linearized dispersion relation of the underlying graphene lattice, Equation (2.50), can be used. Here, it is sufficient to concentrate on a single K-point as illustrated in Figure 2.41. For the wave function of the electrons that move along the nanotube, we can make a plane wave ansatz $\psi(\vec{r}) \propto e^{i\vec{k}\vec{r}}$. The periodic boundary conditions impose the following quantization condition on the wave function: $\psi(\vec{r} + \vec{C}) \stackrel{!}{=} \psi(\vec{r})$ and since $\psi(\vec{r} + \vec{C}) \propto e^{i\vec{k}(\vec{r}+\vec{C})} = e^{i\vec{k}\vec{r}} e^{i\vec{k}\vec{C}}$ it is required that $e^{i\vec{k}\vec{C}} = 1$, which is fulfilled when $\vec{k}\vec{C} = \tilde{n} \cdot 2\pi$ (with $\tilde{n}$ being an integer number).

Next, a coordinate system is used, which is more appropriate for the nanotube under consideration with orthonormal unit vectors along the nanotube axis $\vec{e}_\parallel$ and along the circumference with $\vec{e}_\perp = \frac{\vec{C}}{|\vec{C}|}$. Within the new coordinate system, the reduced vector $\vec{\kappa} = \vec{k} - \vec{K} = \kappa_\parallel \vec{e}_\parallel + \kappa_\perp \vec{e}_\perp$, and thus, the dispersion relation Equation (2.50) becomes $E(\kappa_\parallel, \kappa_\perp) = \pm V_{pp\pi} a \frac{\sqrt{3}}{2} \sqrt{\kappa_\parallel^2 + \kappa_\perp^2}$. The periodic boundary conditions will now result in a quantized $\kappa_\perp$ whose value can be computed with

$$\vec{k}\vec{C} = \vec{K}\vec{C} + \vec{\kappa}\vec{C} = \vec{K}\vec{C} + \kappa_\perp |\vec{C}| \overset{!}{=} \tilde{n} \cdot 2\pi. \tag{2.57}$$

Here, the fact is used that $\vec{e}_\parallel \vec{C} = 0$. Finally, since $\vec{C} = n\vec{a}_1 + m\vec{a}_2$ and using exemplarily the K-point shown in Figure 2.41(a) with $\vec{K} = 2\pi/a(1/\sqrt{3}, 1/3)$, this results in

$$\vec{K}\vec{C} = \begin{pmatrix} 2\pi/\sqrt{3}a \\ 2\pi/3a \end{pmatrix} \begin{pmatrix} \sqrt{3}a/2(n+m) \\ a/2(n-m) \end{pmatrix} = \pi(n+m) + \frac{\pi}{3}(n-m) = 2\pi \frac{2n+m}{3}. \tag{2.58}$$

Thus, Equation (2.57) can be written as

$$\tilde{n} \cdot 2\pi = 2\pi \frac{2n+m}{3} + \kappa_\perp |\vec{C}|. \tag{2.59}$$

Solving this equation for $\kappa_\perp$ and insertion into $E(\kappa_\parallel, \kappa_\perp)$ leads to the one-dimensional dispersion relation of a carbon nanotube given by

$$E(\kappa_\parallel, \tilde{n}) = \pm V_{pp\pi} a \frac{\sqrt{3}}{2} \sqrt{\kappa_\parallel^2 + \left(2 \frac{\tilde{n} - \frac{2n+m}{3}}{d_{\text{CNT}}}\right)^2} \tag{2.60}$$

where $d_{\text{CNT}} = |\vec{C}|/\pi$ is the diameter of the nanotube. Obviously, there are only states at the Fermi level, i. e., at $E = 0$ for $\kappa_\parallel = 0$, if $\tilde{n} - \frac{2n+m}{3} = 0$. This means that, whenever $\frac{2n+m}{3}$ is an integer, the CNT is metallic because the quantization condition due to the periodic boundary conditions crosses the K-points. Note that this condition is equivalent to the requirement that $n - m$ is an integer multiple of 3, usually found in the literature. In all other cases, a semiconducting nanotube is obtained that can be used for device applications. In this case, the band gap E_g can be determined by setting $\kappa_\parallel = 0$ resulting in

$$E_g = 2 \times V_{pp\pi} a \frac{\sqrt{3}}{2} \frac{2}{d_{\text{CNT}}} \left| \tilde{n} - \frac{2n+m}{3} \right| \tag{2.61}$$

where $\tilde{n}$ needs to be chosen in order to make $|\tilde{n} - \frac{2n+m}{3}|$ as small as possible. Hence, the band gap depends inversely proportional on the diameter of the nanotube similar to the $1/w_{\text{GNR}}$ dependence of the band gap of AGNRs.

Equation (2.60) can be used to compute an effective mass of the CNT. To this end, we can first use a Taylor expansion up to second order around the conduction band (valence band) minimum (maximum) and then employ the definition Equation (2.40), which yields

$$E(\kappa_\parallel, \tilde{n}) \approx \underbrace{V_{pp\pi} a \frac{\sqrt{3}}{d_{\mathrm{CNT}}}\left(\tilde{n} - \frac{2n+m}{3}\right)}_{=E_g/2} + \underbrace{V_{pp\pi} a \frac{\sqrt{3}}{8} \frac{d_{\mathrm{CNT}}}{\tilde{n} - \frac{2n+m}{3}}}_{=\frac{\hbar^2}{2m^\star}} \kappa_\parallel^2. \tag{2.62}$$

The equation can immediately be compared with a quadratic dispersion relation, i. e., $E(k) = \frac{\hbar^2 \kappa_\parallel^2}{2m^\star} + V_0$ (where $V_0 = E_g/2$) to identify the effective mass as

$$m_{\mathrm{CNT}}^\star = \frac{4\hbar^2(\tilde{n} - \frac{2n+m}{3})}{V_{pp\pi} a d_{\mathrm{CNT}} \sqrt{3}}. \tag{2.63}$$

Figure 2.42 shows the (first) conduction and valence bands of a semiconducting carbon nanotube (green lines) together with the effective mass approximation (thin black lines). As a result, for not too large energies, a CNT can be considered as a "normal" one-dimensional semiconductor with a symmetric band structure described by a certain energy gap E_g and equal effective masses $m_c^\star = m_v^\star$. However, as has been discussed in Section 2.5, we should use the momentum effective mass Equation (2.41) in the case of a nonparabolic dispersion leading to

$$m_p^\star(E) = \hbar^2 \kappa_\parallel \left(\frac{\partial E(\kappa_\parallel, \tilde{n})}{\partial \kappa_\parallel}\right)^{-1} = \frac{2\hbar^2}{V_{pp\pi} a \sqrt{3}} \sqrt{\kappa_\parallel^2 + \left(2\frac{\tilde{n} - \frac{2n+m}{3}}{d_{\mathrm{CNT}}}\right)^2} = \frac{4\hbar^2}{V_{pp\pi}^2 a^2 3} E(\kappa_\parallel, \tilde{n}), \tag{2.64}$$

which results in the same $m_{\mathrm{CNT}}^\star$ given in Equation (2.63) if $\kappa_\parallel \to 0$. Noting that $E(\kappa_\parallel = 0) = E_g/2$ we can replace $E(\kappa_\parallel)$ with $E + E_g/2$ resulting in an energy-dependent effective

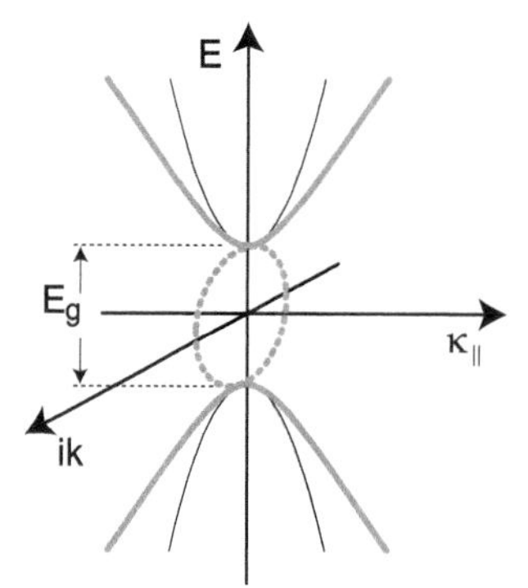

Figure 2.42: Conduction/valence bands of a semiconducting CNT. The thin black lines show the effective mass approximation. Within the band gap, allowed κ-values are purely imaginary and lie on the green dashed circle.

mass given by

$$m_p^\star(E) = \frac{4\hbar^2(\tilde{n} - \frac{2n+m}{3})}{V_{pp\pi} a d_{\mathrm{CNT}} \sqrt{3}} \left(1 + \frac{2E}{E_g}\right) = m_{\mathrm{CNT}}^\star \left(1 + \frac{2E}{E_g}\right). \tag{2.65}$$

This has the same form as Equation (2.44) with $\alpha = 1/E_g$ as already mentioned above.

Equation (2.60) does also have purely imaginary solutions within the band gap with $\kappa_\parallel \to i\kappa$. These solutions yield a complex band structure of the CNT connecting the conduction with the valence band. Indeed, inserting $i\kappa$ yields the equation of a circle as illustrated in Figure 2.42 (green dashed line). Note that this is consistent with an energy-dependent effective mass given by $m_p^\star(E) = m_{\mathrm{CNT}}^\star(1 + \frac{2E}{E_g})$ to describe the complex band structure within the band gap as discussed in Section 2.5. Indeed, inserting $m_p^\star(E)$ into a quadratic dispersion relation yields the same solution as obtained based on Equation (2.60) with purely complex $i\kappa$.

As an alternative approach to computing the band structure by imposing periodic boundary conditions on the graphene dispersion, a carbon nanotube can also be considered as a one-dimensional crystal (as was done in the case of a graphene nanoribbon) and one can use the recipe for tight-binding calculations given above in order to determine the band structure. To this end, consider the (2, 1) nanotube depicted in Figure 2.40(a) that serves as a simple model system here to demonstrate the approach.

Task 11.
Periodicity of CNTs: In general, a carbon nanotube is uniquely specified by the circumference vector $\vec{C} = n\vec{a}_1 + m\vec{a}_2$, which wraps around the nanotube circumference and is aligned perpendicular to the axis of the nanotube. Find a general expression for the vector $\vec{P}$ as a function of n and m and determine the size of the one-dimensional Brillouin zone in k-space of the (n, m) nanotubes.

As in the case of the graphene nanoribbon, a periodicity vector $\vec{P}$ can be found that is perpendicular to $\vec{C}$ (see Task 11 for details), i. e., along the axis of the new one-dimensional nanotube crystal. In the case of the (2, 1) nanotube displayed in Figure 2.40(a), $\vec{P} = \tilde{j}\vec{a}_1 + \tilde{m}\vec{a}_2$ with $\tilde{j} = 4$ and $\tilde{m} = -5$. As a result, the unit cell spanned by $\vec{C}$ and $\vec{P}$ contains 28 atoms (dark gray area in Figure 2.40(a)). The unit cell must consist of an even number of atoms giving rise to an even number of bands. Filling those bands with electrons yields half of the bands filled (at $T = 0$ K), which are the valence bands. The other, empty half represents the conduction bands.

As already mentioned above, the unit cell of the (2, 1)-CNT consists of 28 atoms each providing one p_z-orbital. Hence, we obtain 14 valence and 14 conduction bands and just as in the case of a graphene nanoribbon, the Brillouin zone extends from $-\frac{\pi}{|\vec{P}|}, \ldots, \frac{\pi}{|\vec{P}|}$, i. e., $|\vec{P}|$ is the new lattice constant of the (2, 1)-carbon nanotube. Figure 2.43(a) shows exemplarily the conduction and valence bands of a few different (n, m) nanotubes. As

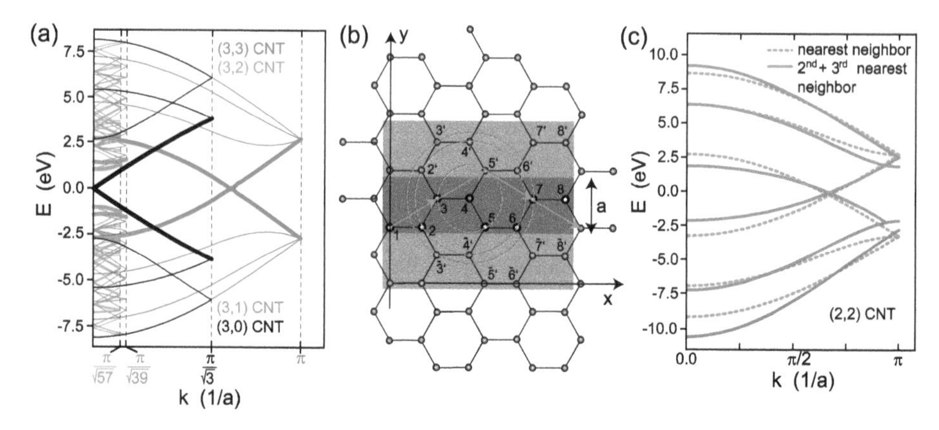

Figure 2.43: Conduction and valence bands of selected carbon nanotubes. Metallic tubes with states at $E_f = 0$ eV are obtained if $\frac{2n+m}{3}$ is an integer. (b) Graphene lattice with the on-site and nearest neighboring unit cells (gray shaded) of a (2, 2) carbon nanotube. The dashed red circles show where the nearest, second and third nearest neighbor atoms around the atom 4 are located. (c) Band structure of a (2, 2) carbon nanotube computed with tight-binding using nearest neighbor interaction only (green dashed lines) and including second and third nearest neighbor interaction (blue lines).

expected, (n, n) nanotubes and CNTs where $\frac{2n+m}{3}$ is an integer yield metallic behavior with electronic states at the Fermi level ($E = 0$).

Up to now, tight-binding was only considered with nearest neighbor interaction. This frequently yields unsatisfying results or can only reproduce the band structure within certain energy ranges (cf. Section 2.6.2). To improve the calculation, either more orbitals need to be taken into consideration (as, for instance, in the case of silicon by including $4s^*$-orbitals) or the interaction with second and third nearest neighbors has to be incorporated. As an example, how such a calculation would have to be carried out, a (2, 2) carbon nanotube is considered. Again, a CNT with rather low indices is chosen, so that the nanotube and the tight-binding matrices can be stated explicitly; an extension to nanotubes with higher indices is straightforward. The appropriate parameters for the tight-binding overlap integrals are taken from [219].

Figure 2.43(b) shows a part of the graphene lattice in the case of the (2,2) nanotube. The on-site unit cell with 8 atoms is dark gray shaded; atoms in the upper unit cell are denoted with a prime and atoms in the lower unit cell with a minus sign on top of the number (and additional prime). As an example, the red atom number 4 has been selected and the nearest, second nearest and third nearest neighbors around this atom have been indicated with the red dashed circles. Each carbon atom has three nearest, six second nearest and three third nearest neighbor atoms. In order to set-up the overall matrix H_{tot}, the recipe introduced above can be applied. The only difference is that there will be more matrix elements in H due to the various overlap integrals that take the coupling with the nearest, second and third nearest neighbors into account. Details of the calculation are accessible through the QR code #22. Figure 2.43(c) shows the band

structure of the (2, 2) nanotube with nearest neighbor interaction only (green dashed lines) and with second and third nearest neighbor interaction. Differences in the band structure can be observed around the Fermi level as well as at $k = 0$ and $k = \pi/a$.

2.10 The Fermi Distribution Function

In this section, a brief derivation of the Fermi distribution function is given. The reason for this is the importance of this distribution function for the calculation of the carrier density and the current (for a deeper discussion, see Chapters 2 and 13 in [133]). Moreover, the derivation is necessary to understand the slightly different Fermi distribution function for localized (such as impurity, donor and acceptor states) and delocalized states.

The Fermi distribution function is derived from statistical physics and represents the most probable distribution function that obeys the laws of thermodynamics as well as quantum physics. To start with, let us first consider a set of N distinguishable particles. Particles with this property can be labeled, which means that one can keep track of them either because they all have different physical properties or one is able to localize them unambiguously at any time. Suppose the particles are randomly arranged on a linear chain and each particle configuration represents a thermodynamic microstate of the linear chain. Combinatorics tells us that one obtains $N!$ different permutations, representing $N!$ different microstates. As an example, consider three particles (A, B, C) that one can arrange in $3! = 6$ different permutations, i. e., (A, B, C), (B, A, C), (A, C, B), (B, C, A), (C, B, A) and (C, A, B). If some fraction n_1 of the particles is identical, say $A = B$, then we would overcount the number of microstates and, therefore, the number of microstates has to be reduced to $\frac{N!}{n_1!}$. In the example above, we get $\frac{3!}{2!} = 3$ different states, namely (A, A, B), (A, B, A) and (B, A, A). Suppose there is another subset n_2 of identical particles, then the number of microstates would be $\frac{N!}{n_1!n_2!}$ and so on for further subsets of identical particles.

Next, consider a system that contains many electronic states that are closely spaced in terms of the energetic difference between the states. We can group them in equidistant energy intervals indexed with i where the energetic difference between adjacent groups $E_{i+1} - E_i \rightarrow dE$ is considered to be very small. The number of states in each group is denoted g_i; note that g_i/dE will become the density of states introduced in the coming sections. This means that in each group g_i plays the role of N according to our consideration above. Hence, one obtains $g_i!$ microstates that can be filled with electrons. However, since electrons are identical particles and cannot be distinguished the number of microstates has to be reduced. In addition to being indistinguishable, electrons are Fermions meaning that one state can only be occupied by a single electron. This means that, if there are less electrons than available states, the remaining states are empty. As a result, we can think of the g_i as states being filled with n_i electrons (leading to an occupied state) and $(g_i - n_i)$ holes (i. e., a state that is not occupied). Since both electrons and holes are indistinguishable, the number of microstates becomes $\frac{g_i!}{n_i!(g_i-n_i)!}$ (i. e., $N! \to g_i!$, $n_1! \to n_i!$ and $n_2! \to (g_i - n_i)!$). Including all different energy intervals i, the overall number of microstates Ω is

$$\Omega = \prod_i \frac{g_i!}{n_i!(g_i - n_i)!}. \tag{2.66}$$

With the knowledge of Ω, we are now in a position to compute the entropy S based on its statistical definition, i. e., $S = k_B \ln \Omega$ [111]. Exploiting that $\ln(a \cdot b) = \ln a + \ln b$ and using Sterling's approximation $\ln n! \approx n \ln n - n$, we obtain

$$\frac{S}{k_B} \approx \sum_i g_i \ln(g_i) - g_i - (n_i \ln(n_i) - n_i + (g_i - n_i)\ln(g_i - n_i) - (g_i - n_i)). \tag{2.67}$$

In order to get the most probable distribution, we need $dS = 0$ under the additional constraints that the total number of particles $N_{\text{tot}} = \sum_i n_i$ is constant and that the inner energy U is conserved; the latter means that $U = \sum_i n_i E_i$ is constant. The two constraints are best taken into consideration with Lagrange multipliers, so that one needs to find the n_i that makes the function F stationary:

$$F = \underbrace{\sum_i (g_i \ln(g_i) - n_i \ln(n_i) - (g_i - n_i)\ln(g_i - n_i))}_{=S/k_B} - \alpha\Big(\sum_i n_i - N_{\text{tot}}\Big) - \beta\Big(\sum_i E_i n_i - U\Big). \tag{2.68}$$

Now, $dF \overset{!}{=} 0$ is fulfilled if $\frac{\partial F}{\partial n_i} = 0$ for all i. With Equation (2.68), it is then easy to show that

$$\frac{\partial F}{\partial n_i} = -\Big(\ln(n_i) + n_i \frac{1}{n_i}\Big) - \Big(-\ln(g_i - n_i) - (g_i - n_i)\frac{1}{g_i - n_i}\Big) - \alpha - \beta E_i \overset{!}{=} 0. \tag{2.69}$$

Since the second terms in the two parentheses cancel each other, Equation (2.69) becomes $0 = \ln(\frac{g_i - n_i}{n_i}) - \alpha - \beta E_i$. The probability that a state E_i is occupied is given by $\frac{n_i}{g_i}$. So, rewriting the equation above as $\ln(\frac{g_i}{n_i} - 1) = \alpha + \beta E_i$ yields $\frac{n_i}{g_i} = \frac{1}{e^{\alpha + \beta E_i} + 1}$, which is the desired Fermi distribution function $f(E_i)$. What remains to be done is the determination of the Lagrange multipliers α and β. They can be obtained by acknowledging the fact that at large energy E_i, the computed distribution function needs to behave like a Boltzmann distribution, i. e., $f(E_i) \to \exp(-\frac{E - E_f}{k_B T})$. As a result, one obtains $\beta = \frac{1}{k_B T}$ and $\alpha = -\beta E_f$ where E_f is the Fermi energy.[13] Eventually (from now on the index i is dropped and we consider the energy as being continuous),

13 This is not fully correct since the chemical potential μ_c should have appeared here. However, setting $\mu_c = E_f$ yields appropriate results in the cases considered in this book.

$$f(E) = \frac{1}{1 + e^{\frac{E-E_f}{k_B T}}}, \tag{2.70}$$

which is the well-known expression for the Fermi distribution function.

The top panel of Figure 2.44 shows $f(E)$ for different temperatures T. When $T \to 0$ K, $f(E)$ becomes a step function meaning that all states up to the Fermi energy are filled and all states with energies larger than E_f are empty. At finite temperatures, the Fermi distribution deviates from the step function behavior only in a narrow energy range of approximately $4 \times k_B T$ around E_f. This makes perfect sense, since electrons in low lying (energy) states will not obtain enough energy to be excited into an energy range where empty states are available.

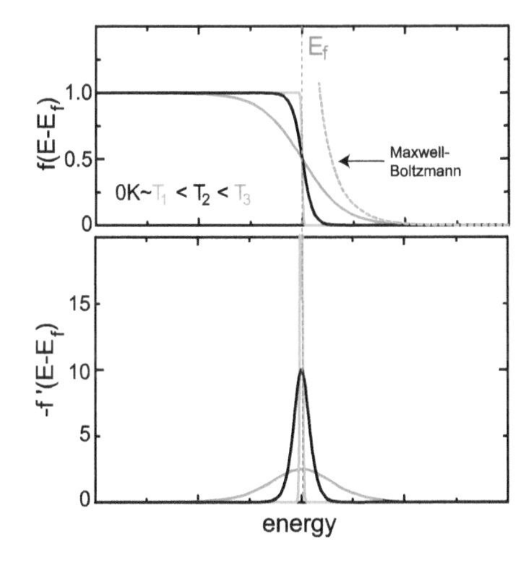

Figure 2.44: The top panel shows $f(E)$ for different temperatures. The orange dashed line is the Boltzmann approximation, which is appropriate if $E - E_f \gg k_B T$. The lower panel shows the derivative of $f(E)$ with respect to E.

For sufficiently high energies where $E - E_f \gg k_B T$, the exponential term dominates in the denominator, i. e., $e^{\frac{E-E_f}{k_B T}} \gg 1$ and the Fermi distribution function evolves into the Maxwell–Boltzmann distribution as illustrated in the top panel of Figure 2.44. The lower panel of Figure 2.44 shows the (negative) derivative of the Fermi distribution function. It is clear that at $T = 0$ K, i. e., when $f(E) = \Theta(E_f - E)$, the derivative becomes the delta function $\delta(E - E_f)$. Even at finite temperatures, the derivative strongly peaks at $E = E_f$ such that in many cases the approximation $-\frac{\partial f(E)}{\partial E} \approx \delta(E - E_f)$ is applicable.

2.10.1 The Fermi Distribution for Holes

The Fermi distribution for holes—called $f_h(E)$ in the following—can be derived easily from the result obtained for $f(E)$ by noting that the probability of finding an empty state at a certain energy is just $1 - f_e(E)$ where the Fermi distribution was now denoted $f_e(E)$ to distinguish it from $f_h(E)$. Hence, the Fermi distribution for holes is

$$f_h(E) = 1 - \frac{1}{1 + e^{\frac{E-E_f}{k_B T}}} = \frac{1 + e^{\frac{E-E_f}{k_B T}}}{1 + e^{\frac{E-E_f}{k_B T}}} - \frac{1}{1 + e^{\frac{E-E_f}{k_B T}}}$$
$$= \frac{e^{\frac{E-E_f}{k_B T}}}{1 + e^{\frac{E-E_f}{k_B T}}} = \frac{1}{1 + e^{\frac{E_f-E}{k_B T}}}. \tag{2.71}$$

This means that one obtains either of the two distribution functions by simply reversing the sign of the exponent as shown in Figure 2.45. Choosing one of the distributions over the other is thus simply a matter of convenience and depends on the situation and device under consideration. Indeed, states that are not occupied with electrons can be interpreted as being occupied with a hole. Both the electron and the hole picture can be used if the appropriate distribution functions are employed. In Chapter 9, so-called band-to-band tunnel transistors will be discussed. Since in these devices carriers are transferred from the conduction to the valence band (and vice versa), it is important to use *either* the electron- *or* the hole-picture in both the conduction and valence bands.

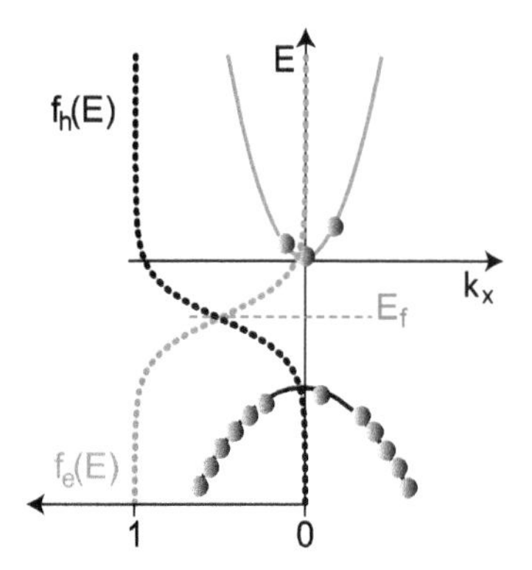

Figure 2.45: Fermi distribution of electrons (green) and holes (black). An occupied electronic state is an empty hole state and, therefore, $f_h = 1 - f_e$.

2.10.2 Fermi Distribution Function for Dopants

If the semiconductor under consideration is doped (see Section 4.1 for details on doping), a frequently made approximation is to consider all dopants as being fully ionized. This yields a constant background of immobile charge that is responsible, e. g., for the band bending within *p*-*n*-junctions or within the depletion zone in a metal-oxide-semiconductor capacitor (more on this in Chapter 4). However, when incorporated into a host semiconductor lattice, the dopant atom can be considered as a hydrogen atom within a dielectric matrix with a ground-state energy E_d below the conduction band (i. e., E_d is the ionization energy of the dopant); Section 4.3.1 provides further details on this hydrogen model of dopants. The state E_d can be occupied or empty depending on the temperature and the position of the Fermi level E_f with respect to E_d. Naively, one might use the Fermi distribution function (2.70) setting $E = E_d$ to account for the occupation of the dopant state. However, looking into a textbook on the subject one finds in the case of donor levels that the Fermi distribution is

$$f_{\text{donor}}(E_d) = \frac{1}{1 + \frac{1}{2} e^{\frac{E_d - E_f}{k_B T}}}. \tag{2.72}$$

The factor 1/2 in front of the exponential term is said to be due to the fact that the donor state E_d can be occupied in two ways, namely with an electron with spin up or one with spin down. At first sight, this argument appears to be the same as is used to account for spin degeneracy in the conduction or valence band: within the conduction/valence bands, each state can also be occupied with a spin-up or -down electron. This basically doubles the amount of available states and can be accounted for by doubling g_i in Equation (2.66) and all following equations of the derivation. Obviously, this will not change the derivation and results in a probability that a state E_i is occupied of $\frac{n_i}{2g_i}$. Effectively, this means that spin degeneracy is taken into account by simply multiplying the Fermi distribution with a factor of two. The same is true if there is more than a single band available as is the case in the valence band. Ultimately, any degeneracy in the bands can be taken care of by multiplying the Fermi distribution function with the appropriate degeneracy factor. So, the question is, why does the spin degeneracy in the case of donors not yield a simple factor of two but instead a factor of $\frac{1}{2}$ in front of the exponential term? Where is the difference compared to conduction/valence band electrons?

The difference has to do with the counting of available (micro)states leading to a different way how spin degeneracy needs to be incorporated. As was mentioned above, particles can be distinguished when they can be labeled. This is either the case when the particles have different physical properties or if you can track them down (i. e., follow their trajectories) at any time. Electrons are certainly identical particles with identical physical properties and within the conduction or valence band the electrons/holes are described by plane waves with a well-defined momentum $\hbar k$ (up to a crystal momentum). Heisenberg's uncertainty relation tells us that in this case they are delocalized making it impossible to track them down. Electrons/holes in the conduction/valence bands are indistinguishable and as a result, if we take spin degeneracy into consideration, the number of available microstates in Equation (2.66) simply doubles yielding a factor of two in front of the Fermi distribution function (alternatively, the density of states can be considered to be twice as large to account for spin degeneracy) as already mentioned.

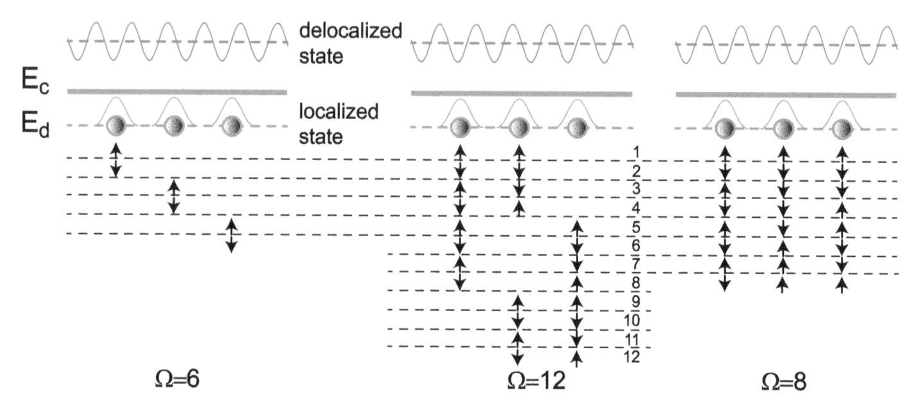

Figure 2.46: Illustration of localized versus delocalized states. The $N = 3$ localized states due to donors at E_d are occupied with $n = 1$ (left), $n = 2$ (center) and $n = 3$ (right) electrons giving rise to $\Omega = 6, 12, 8$ microstates because of the possibility to occupy with either spin up or down.

In the case of dopants, this is fundamentally different. The reason is that dopants are localized meaning that their occupation leads to distinguishable particles. As a result, the number of available microstates will be substantially larger compared to the case discussed above. If each dopant state E_d can be occupied by either a spin-up or a spin-down electron, the number of microstates increases by a factor of 2^n where n is the number of electrons. Thus, if there are N dopant atoms each providing a state at E_d then the number of microstates is given by $\frac{N!2^n}{n!(N-n)!}$. To give an example, in Figure 2.46 $N = 3$ dopants are displayed and the left panel shows the number of microstates due to a varying occupation of the three states at E_d with a single electron (arrows indicate spin up and spin down, respectively). Obviously, six microstates are obtained and indeed the same number results from the calculation: $\frac{3!2}{1!(3-1)!} = 6$. With two electrons accommodating the three dopants atoms (energy levels E_d), $\frac{3!2^2}{2!(3-2)!} = 12$, and thus the 12 microstates are obtained that are explicitly shown in the center panel of Figure 2.46. Finally, with $n = 3$ electrons, $\frac{3!2^3}{3!(3-3)!} = 8$ microstates are obtained as depicted in the right panel of Figure 2.46. Overall, the number of microstates therefore is

$$\Omega = \prod_i \frac{N! \cdot 2^{n_i}}{n_i!(N-n_i)!}. \tag{2.73}$$

Using again the Stirling approximation as above, the entropy S looks basically the same as given in Equation (2.67) if (i) g_i is replaced with N and (ii) an additional term $\ln(2^{n_i}) = n_i \ln(2)$ is added due to the extra factor 2^{n_i} (cf. Equation (2.73)) because of the localized energy levels that can be occupied with spin-up or spin-down electrons. Computing again $dF = 0$ with the same constraints as above taken into consideration yields

$$\ln\left(\frac{N-n_i}{n_i}\right) + \ln(2) - \alpha - \beta E_d \stackrel{!}{=} 0. \tag{2.74}$$

Solving for $\frac{n_i}{N}$, one finally arrives at the Fermi distribution function of donors stated in Equation (2.72).

Looking at the derivation of the Fermi distribution function of an electronic state at E_d that can be occupied with spin up or down electrons, it is obvious that the difference from the derivation of the Fermi distribution function of electrons in the conduction band is the factor 2^{n_i} in Equation (2.73), which in general can be written as $g_d^{n_i}$ with g_d being the degeneracy factor. As a result, the generalized Fermi distribution for a donor or—more generally speaking—for donor-like localized states (such as defects) with energy E_d and a degeneracy factor g_d reads

$$f_{\text{donor}}(E_d) = \frac{1}{1 + \frac{1}{g_d}\exp(\frac{E_d - E_f}{k_B T})} \xrightarrow{E_d \gg E_f} g_d e^{-\frac{E_d - E_f}{k_B T}} . \tag{2.75}$$

To complete this section, let us briefly also look at the Fermi distribution for a localized acceptor(-like) state E_a occupied with holes. Since the valence band is twice degenerate with the two spin directions, $g_d = 4$ and hence the Fermi distribution function of holes in the valence band is again modified with the factor $1/g_d$ resulting in

$$f_{\text{acceptor}}(E_a) = \frac{1}{1 + \frac{1}{g_d}\exp(\frac{E_f - E_a}{k_B T})} \xrightarrow{E_f \gg E_a} g_d e^{-\frac{E_f - E_a}{k_B T}} . \tag{2.76}$$

In both Equations, (2.75) and (2.76), the Boltzmann tail is as expected $\propto g_d$.

2.11 Density of States and Carrier Density

Equipped with the knowledge about band structure calculations, effective mass approximation and Fermi distribution functions we are now in a position to compute the carrier density in a material. At first, a material with an isotropic effective mass $m^\star$ is assumed. Exemplarily, the electron density in the conduction band of a semiconductor will be calculated.

To compute the carrier density, we consider a volume $L_x \cdot L_y \cdot L_z$ with $L_{x,y,z} \to \infty$ and constant potential. In this case, the exact lengths are irrelevant and we can set $L_x = L_y = L_z \equiv L$. The wave function within the volume L^3 is $\psi_{\vec{k}}(\vec{r})$. For a certain $\vec{k}$-state, the carrier density $n(\vec{r})$ is given by the probability density of finding a particle at a position $\vec{r}$ multiplied with the probability that this particular $\vec{k}$-state is occupied; the former is given by the absolute square of the wave function $\psi_{\vec{k}}(\vec{r})$ and the latter by the Fermi distribution function $f(\vec{k})$. Finally, a summation over all $\vec{k}$-states yields the carrier density according to

$$n(\vec{r}) = \sum_{\vec{k}} |\psi_{\vec{k}}(\vec{r})|^2 \cdot f(\vec{k}) \tag{2.77}$$

2.11.1 Density of States

In Equation (2.77), the Fermi distribution was written as $f(\vec{k})$ in order to make it clear that f represents the probability that a certain state $\vec{k}$ is occupied. However, as we know from Equation (2.70), f depends explicitly on energy E and since the relation between E and $\vec{k}$ is given by the dispersion relation we have $f(\vec{k}) = f(E(\vec{k}))$. Accordingly, Equation (2.77) can be rewritten in the following form:

$$n(\vec{r}) = \sum_{\vec{k}} |\psi_{\vec{k}}(\vec{r})|^2 \int dE\delta(E - E(\vec{k}))f(E) \tag{2.78}$$

where $\delta(E - E(\vec{k}))$ is a delta function. Next, the energy integral can be moved in front of the summation and one obtains

$$n(\vec{r}) = \int dE \underbrace{\sum_{\vec{k}} |\psi_{\vec{k}}(\vec{r})|^2 \delta(E - E(\vec{k}))}_{D(E,\vec{r})} f(E). \tag{2.79}$$

This means that the calculation of the carrier density can be cast into an energy integration over the so-called local density of states (lDOS) $D(E, \vec{r})$ multiplied with the Fermi distribution function $f(E)$. The delta function in the lDOS picks the eigenenergies of the system under consideration (given by the dispersion relation) and the absolute square of the eigenfunction $\psi_{\vec{k}}(\vec{r})$ provides the probability density of finding a particle at $\vec{r}$ in state $\vec{k}$. Therefore, a full solution of the Schrödinger equation is required to compute the lDOS. However, in many cases, we deal with quasiinfinite systems with a constant potential V_0 such that the wave functions $\psi_{\vec{k}}(\vec{r})$ can be approximated with plane waves yielding $|\psi_{\vec{k}}(\vec{r})|^2 = \frac{1}{L_x L_y L_z} e^{i\vec{k}\vec{r}} e^{-i\vec{k}\vec{r}} = \frac{1}{L_x L_y L_z}$ and the dependence on space drops out of the equation. In this case, the lDOS reduces to the density of states (DOS) given by

$$D(E) = \frac{1}{L_x L_y L_z} \sum_{\vec{k}} \delta(E - E(\vec{k})). \tag{2.80}$$

Hence, knowledge of the dispersion relation $E(\vec{k})$ is sufficient to compute the density of states in large systems with (quasi)constant potential. It is clear that the density of states is proportional to a sum over delta functions (cf. Equation (2.80)) since the delta functions peak whenever the energy E is equal to an eigenenergy $E(\vec{k})$. Integrating a single delta peak $\delta(E - E(\vec{k}))$ over energy yields the correct carrier density, namely $\frac{1}{L_x L_y L_z}$ since there is only one state at $E(\vec{k})$ that is filled with one electron (provided the Fermi level is well above the energy level) in a volume of $L_x L_y L_z$; if there are more states, the carrier density increases accordingly.

To carry out the summation in Equation (2.77), we need to know the discrete $\vec{k}$-values. From our calculation of the particle-in-the-box system with fixed boundary conditions, we know that confinement results in quantized values for the respective k-states and we found $k_n = n\frac{\pi}{L}$. However, in this case only k-values with $k > 0$ are allowed. For our calculation, things would be substantially easier if all k-values (positive, negative and $k = 0$) were allowed. Therefore, we will use so-called Born–von Karman or periodic boundary conditions in the following. When dealing with carbon nanotubes in Section 2.9, periodic boundary conditions have already been used. But while in CNTs, periodic boundary conditions must be used due to the physical situation, here the considered system is very large such that the specific boundary conditions used will not have any impact on physical quantities inside the system; periodic boundary conditions are merely used to simplify our calculations. In the next step, the sum over $\vec{k}$-states needs to be transferred into an integral as detailed in the info-box below.

How to compute the sum over k-states? In calculations of the charge carrier density and/or the current in nanoscale devices, one frequently needs to compute a sum over all k-states. Summations are always troublesome because they require in most cases a numerical computation. However, if the system we consider is large, the k-values are quasicontinuous and it is then possible to transfer the summation over discrete k-values into an integral. To do so, let us consider a one-dimensional conductor of length L that stretches out from $-L/2$ to $L/2$; for simplicity, the potential is set to zero throughout. With Born–von Karmann or periodic boundary conditions, the wave function (plane wave) is required to be the same (apart from a phase factor) if $x \to x + L$. This means that $\psi(x+L) \stackrel{!}{=} \psi(x) \to e^{ik(x+L)} \stackrel{!}{=} e^{ikx}$, which is true if $e^{ikL} = 1$. As a result, $k_n = n \cdot \frac{2\pi}{L}$ and the difference between adjacent k-values is $k_{n+1} - k_n = \Delta k = \frac{2\pi}{L}$. In turn, this means $\frac{\Delta k}{2\pi/L} = 1$. We obtain as a general recipe of how to transfer the typical sums over k-states ihto an integral the following: insert a factor of $1 = \frac{\Delta k}{2\pi/L}$ into the sum, then we can write $\Delta k \to dk$ since $\Delta k = \frac{2\pi}{L}$ becomes very small with increasing system size L, and replace the sum with an integral. In summary, this means

$$\frac{1}{L}\sum_k f(k) = \frac{1}{L}\sum_k \frac{\Delta k}{2\pi/L} f(k) \stackrel{\Delta k \to dk}{\longrightarrow} \frac{1}{2\pi}\int dk f(k). \tag{2.81}$$

This relation will be used frequently throughout the book. An extension to higher dimensions is obvious. For instance, $\frac{\Delta k}{2\pi/L} \to \frac{\Delta k_x}{2\pi/L_x}\frac{\Delta k_y}{2\pi/L_y}\frac{\Delta k_z}{2\pi/L_z}$.

Using Equation (2.81), the DOS can be written as

$$D(E) = \frac{1}{(2\pi)^d}\int d\vec{k}\,\delta(E - E(\vec{k})) \tag{2.82}$$

where $d = 1, 2, 3$ is the dimension. In the case of a quadratic dispersion relation with isotropic effective mass m^*, the dispersion relation $E(\vec{k}) = \frac{\hbar^2\vec{k}^2}{2m^*}$ (here, $V_0 = 0$ for simplicity) depends only on the magnitude $|\vec{k}|$, and thus, a one-dimensional integral over dk is obtained for all $d = 1, 2, 3$. For instance, for $d = 3$ the triple integration in Equation (2.82) $d\vec{k} \to 4\pi k^2 dk$. Now, changing variables to $\epsilon = \frac{\hbar^2 k^2}{2m^*}$ yields $dk = \frac{m^*}{\hbar^2 k} d\epsilon$ and $k = \sqrt{\frac{2m^*\epsilon}{\hbar^2}}$. Eventually, in three dimensions

$$D_{3D}(E) = \frac{1}{(2\pi)^3} 4\pi \int d\epsilon \frac{m^*}{\hbar^2}\sqrt{\frac{2m^*\epsilon}{\hbar^2}}\,\delta(E - \epsilon) = \frac{m^*}{2\pi^2\hbar^3}\sqrt{2m^*E} \tag{2.83}$$

Similarly, the DOS can be computed in 2D. Here, $d\vec{k} \to 2\pi k dk$ and with $dk = \frac{m^\star}{\hbar^2 k} d\epsilon$ and using Equation (2.82) it is clear that the density of states is given by

$$D_{2D}(E) = \frac{1}{4\pi^2} \int d\epsilon\, 2\pi k \frac{m^\star}{\hbar^2 k} \delta(E - \epsilon) = \frac{m^\star}{2\pi\hbar^2}, \tag{2.84}$$

which is constant. Finally, in the 1D case the computation of the DOS is straight-forward since in the integral over k (cf. Equation (2.81)) one simply replaces $dk = \frac{m^\star}{\hbar^2 k} d\epsilon = \frac{m^\star}{\hbar^2} \sqrt{\frac{\hbar^2}{2m^\star \epsilon}} d\epsilon$, and hence (there is an additional factor of 2 since k can assume positive and negative values for the same ϵ)

$$D_{1D}(E) = \frac{1}{2\pi} \int d\epsilon\, 2 \frac{m^\star}{\hbar^2} \sqrt{\frac{\hbar^2}{2m^\star \epsilon}}\, \delta(E - \epsilon) = \frac{2}{h} \sqrt{\frac{m^\star}{2E}}. \tag{2.85}$$

Note that all three expressions for the DOS in 1D, 2D and 3D are stated without spin degeneracy. As has been discussed in Section 2.10.2, spin degeneracy can be easily incorporated by multiplying the DOS with a factor of two.

In most textbooks, the DOS is derived in an alternative way that is reproduced here for completeness. Assume again an isotropic effective mass and a quadratic dispersion relation $E(\vec{k}) = \frac{\hbar^2 \vec{k}^2}{2m^\star}$ (again, $V_0 = 0$ for simplicity). We now ask the question how many additional states dZ one obtains if k is increased to $k + dk$. Looking at Figure 2.47(a), in 3D, additional states lie within the volume between the two spheres $\frac{4\pi}{3}(k + dk)^3 - \frac{4\pi}{3}k^3 \approx \frac{4\pi}{3} 3k^2 dk$ where all factors $\propto dk^2$ and $\propto dk^3$ have been neglected. A single k-state occupies the volume $\frac{2\pi}{L_x}\frac{2\pi}{L_y}\frac{2\pi}{L_z} = \frac{8\pi^3}{V}$ such that the number of additional states becomes

$$dZ = \frac{4\pi k^2 dk}{8\pi^3 / V}. \tag{2.86}$$

The DOS is now the number of additional states per change of energy and volume. Therefore, we replace $k = \sqrt{2m^\star E/\hbar^2}$ and $dk = \frac{m^\star}{\hbar^2 k} dE = \frac{1}{\hbar}\sqrt{\frac{m^\star}{2E}}$ in the equation above and obtain the same expression as in Equation (2.83):

$$D_{3D}(E) = \frac{1}{V}\frac{dZ}{dE} = \frac{m^\star}{2\pi^2\hbar^3}\sqrt{2m^\star E}. \tag{2.87}$$

With spin degeneracy, the expression is multiplied with a factor of two.

The DOS in a two-dimensional system can be obtained in a similar fashion: Again, we ask how many additional states dZ one obtains in the area in-between k and $k + dk$ (cf. Figure 2.47(b)). In the 2D case, this area is given by $\pi(k + dk)^2 - \pi k^2 \approx 2\pi k dk$ where again contributions $\propto dk^2$ have been neglected. Since the area one k-state occupies is $\frac{2\pi}{L_x}\frac{2\pi}{L_y} = \frac{4\pi^2}{A}$, $dZ = A 2\pi k dk/(4\pi^2)$. The DOS is then given as the number of additional states per energy interval and area. Since $k dk = \frac{m^\star}{\hbar^2} dE$, the result is of course the same as Equation (2.84).

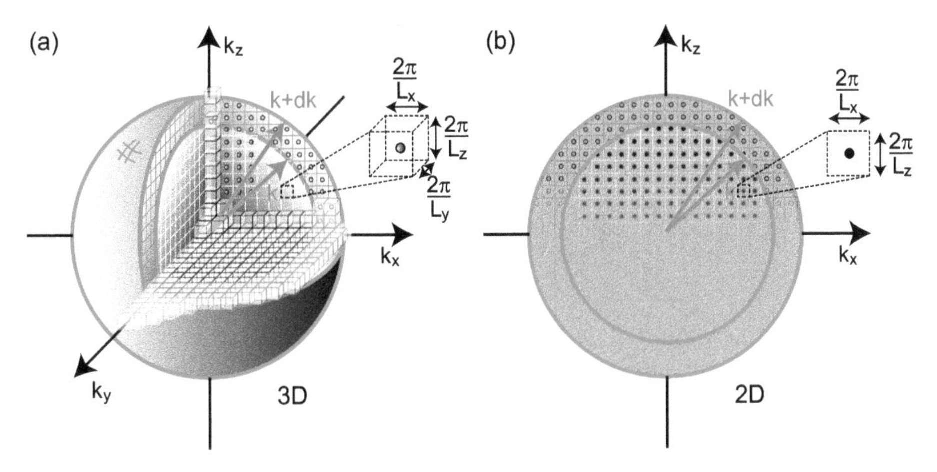

Figure 2.47: Constant energy surface in the case of an isotropic effective mass in a 3D (a) and a 2D system (b). Increasing k by dk yields a shell in-between the two constant energy surfaces with additional k-states (marked red) that can be occupied with carriers. Each state needs a volume/area within the k-space given by $(\frac{2\pi}{L})^d$ with d being the dimension.

Table 2.4 summarizes the density of states expressions in all three dimensions. Note that in contrast to the expressions above a factor of two due to spin degeneracy has been included.

Table 2.4: Density of states in various dimensions. Spin degeneracy has been taken into account.

	1D	2D	3D
DOS	$D(E) = \frac{2}{h}\sqrt{\frac{2m^\star}{E}}$	$D(E) = \frac{m^\star}{\pi\hbar^2}$	$D(E) = \frac{m^\star}{\pi^2\hbar^3}\sqrt{2m^\star E}$
	$D(E) \propto E^{-1/2}$	$D(E) \propto E^0$	$D(E) \propto E^{1/2}$

2.11.2 Density of States in the Case of Anisotropic Effective Masses

So far, we only considered the case of an isotropic effective mass giving rise to circular/spherical constant energy surfaces. However, as was shown in Section 2.6.3, the conduction band of silicon is determined by six rotational ellipsoids with one heavy and two light effective masses along the k_x-, k_y- and k_z-directions. In more general terms, the dispersion relation (the potential energy is neglected for simplicity) is given by

$$E(k_x, k_y, k_z) = \frac{\hbar^2 k_x^2}{2m_x^\star} + \frac{\hbar^2 k_y^2}{2m_y^\star} + \frac{\hbar^2 k_z^2}{2m_z^\star}, \tag{2.88}$$

which can be rewritten as

$$E(k_x, k_y, k_z) = \frac{\hbar^2}{2m^\star_{\text{DOS}}} \Big[\underbrace{k_x^2 \frac{m^\star_{\text{DOS}}}{m^\star_x}}_{(k'_x)^2} + \underbrace{k_y^2 \frac{m^\star_{\text{DOS}}}{m^\star_y}}_{(k'_y)^2} + \underbrace{k_z^2 \frac{m^\star_{\text{DOS}}}{m^\star_z}}_{(k'_z)^2} \Big]. \tag{2.89}$$

With the shift of variables,

$$dk_{x,y,z} = dk'_{x,y,z} \sqrt{\frac{m^\star_{x,y,z}}{m^\star_{\text{DOS}}}} \tag{2.90}$$

the ellipsoidal dispersion relation appears as if it exhibited an isotropic effective mass. In turn, this allows using one of the approaches outlined above to determine the DOS (see Task 12 for details). The so-called "density of states effective mass" $m^\star_{\text{DOS}} = (m^\star_x \cdot m^\star_y \cdot m^\star_y)^{1/3}$ is an "average" of the three effective masses in the three spatial directions. In the case of silicon, two of the effective masses ($m^\star_{x,y,z}$) are equal to the light effective mass $m^\star_l$ and the remaining one is $m^\star_h$. As a result, with $m^\star_{\text{DOS}}$ the density of states reads

$$D(E)_{3D} = \frac{(m^\star_{\text{DOS}})^{3/2}}{2\pi^2\hbar^3} \sqrt{2E} \xrightarrow{\text{Si}} \frac{\sqrt{(m^\star_l)^2 m^\star_h}}{2\pi^2\hbar^3} \sqrt{2E}. \tag{2.91}$$

Multiplying this with a factor of six due to the six ellipsoids one finally arrives at the DOS of the conduction band in bulk silicon.

Task 12.
Density-of-states effective mass: Consider a two-dimensional system that exhibits two different effective masses with $m^\star_x \neq m^\star_z$. Compute the density of states of the two-dimensional system. What is the density-of-state effective mass $m^\star_{\text{DOS}}$?

2.11.3 Fermi Integrals

In the case of large systems with constant potential, a constant carrier density is obtained according to

$$n = \int dE\, D_{1D/2D/3D}(E) \cdot f(E) \tag{2.92}$$

in one, two or three dimensions. Even this simplified expression can only be solved analytically in very few cases; all other instances need either an appropriate approximation or a numerical solution.

In a three-dimensional semiconductor that is not degenerately doped, the Fermi level lies within the band gap. In addition, if E_f is energetically sufficiently separated from the bottom of the conduction(valence) band $E_c(E_v)$ the Fermi distribution can be approximated with the Boltzmann distribution. In the following, the results for the carrier concentration will be discussed exemplarily for electrons in the conduction band;

similar expressions can be derived for holes in the valence band in the same fashion. Using the Boltzmann approximation, the carrier density is

$$n \approx \int_{E_c}^{\infty} dE \frac{m^\star}{2\pi^2\hbar^3} \sqrt{2m^\star(E - E_c)} e^{-\frac{E-E_f}{k_B T}} \tag{2.93}$$

where the lower boundary of the integration is the conduction band bottom since energies below E_c lie within the band gap where no states (band tails are neglected here) are available. The integral can be rewritten in the form of the definite integral $\int_0^\infty dx \sqrt{x} \exp(-x) = \sqrt{\pi}/2$. As a result, one obtains

$$n \approx \underbrace{2\left(\frac{2\pi m^\star k_B T}{h^2}\right)^{\frac{3}{2}}}_{=\mathrm{DOS}_{\mathrm{eff}}=N_c} \cdot \underbrace{\exp\left(-\frac{E_c - E_f}{k_B T}\right)}_{\text{Boltzmann-distribution}} . \tag{2.94}$$

This result can be interpreted as a product of an effective density of states $\mathrm{DOS}_{\mathrm{eff}}$ for the single level at E_c and the Boltzmann distribution provides the probability that the level E_c is occupied.

In a two-dimensional system with constant density of states, Equation (2.92) basically yields an integration of the Fermi distribution function, which can be computed analytically resulting in

$$\begin{aligned} n &= \frac{m^\star}{\pi\hbar^2} \int_{E_c}^{\infty} dE \frac{1}{1 + \exp(\frac{E-E_f}{k_B T})} = -\frac{m^\star}{\pi\hbar^2} k_B T \ln(1 + e^{\frac{E_f - E}{k_B T}}) \Big|_{E_c}^{\infty} \\ &= \frac{m^\star}{\pi\hbar^2} k_B T \ln(1 + e^{\frac{E_f - E_c}{k_B T}}) \end{aligned} \tag{2.95}$$

The integral of the Fermi distribution function is called “supply” function. Again, the carrier density can be interpreted as a product of a density of states and a distribution (i. e., supply) function as in Equation (2.94).

Fully quantized systems (in all three spatial dimensions) as displayed, e. g., in Figure 2.4 exhibit a density of states where separate, discrete energy levels E^{rln} are obtained (cf. Section 2.2.3). As a result, the density of states is a sum of delta functions where the multiplicity (i. e., the degeneracy) of each level needs to be taken into consideration with an appropriate factor g_{rln}. For the carrier density (which is equal to the number of carriers in the present case), one gets

$$n = \int dE \sum_{r,l,n} g_{rln} \delta(E - E^{rln}) f(E - E_f) = \sum_{r,l,n} \frac{g_{rln}}{1 + \exp(\frac{E^{rln} - E_f}{k_B T})} \tag{2.96}$$

2.12 Density of States Beyond the Parabolic Dispersion

In the preceding sections, the density of states has been derived analytically based on a quadratic approximation of the dispersion relation where the underlying solid is described by (direction-dependent) effective masses. However, sometimes an effective mass approximation is not appropriate. In the present section, the density of states is computed and discussed in several of such cases.

2.12.1 Density of States of an Arbitrary Band Structure

In order to compute the density of states directly based on a band structure $E(\vec{k})$ – for instance calculated with the tight-binding method—Equation (2.82) can be used where the summation needs to be carried out over the first Brillouin zone only. The question is how to do the integration over the δ-function with (numerically) computed dispersions? Two suitable methods to do so are the smearing and the tetrahedron method [251, 250]. While the latter, in particular with the so-called Blöchl corrections [29], is more accurate (see, for instance, [251, 250]), the smearing method is straightforward to implement and yields decent results.

At the end of Section 2.10 (see also the lower panel of Figure 2.44), it was shown that the negative of the derivative of the Fermi distribution yields the δ-function for $T \to 0$. As a result, we can write Equation (2.82) as

$$D(E) = \frac{1}{V} \sum_{\vec{k}} \delta(E - E(\vec{k})) \approx \frac{1}{(2\pi)^d} \sum_{n} \int_{1^{st}BZ} d\vec{k} \frac{e^{\frac{E-E_n(\vec{k})}{\sigma}}}{\sigma(1 + e^{\frac{E-E_n(\vec{k})}{\sigma}})^2} \tag{2.97}$$

where $\sigma \to 0$ is the smearing factor (similar to temperature), which should be as small as possible. A band index n has been introduced to account for multiple bands present in the Brillouin zone. Examples of the DOS computed with the smearing method for a number of dispersion relations calculated in the preceding sections are accessible through QR code #24.

2.12.2 Disorder and Band-Tailing

In the preceding sections, the band structure and density of states were computed based on a strictly periodic, perfect crystal. However, in reality, disorder of the crystal (due to defects, charged Coulomb centers, electron-phonon interaction, etc.) leads to a phenomenon called band-tailing. As a result, the band edges are blurred with a strongly decreasing but substantial, nonzero density of states within the band gap. While band-tailing is often less important for the functionality of regular field-effect transistors at room temperature (and consequently neglected), it plays an important role in band-to-band tunneling transistors (cf. Chapter 9.1.1) and is decisive for cryogenic field-effect transistors (see Chapter 11).

In order to illustrate the effect of disorder on the density of states, the 1D model discussed in Section 2.4.3 with one s-and a p_x-orbital (see Figure 2.14) is used; disorder can simply be taken into consideration by randomly varying the overlap integrals V_{ss} and V_{pp} (note that V_{sp} is neglected in the following) and the on-site energies ϵ_0 and ϵ_1 at each lattice site (see Figure 2.48(a)). To compute the band structure, a 1D linear chain with more than 3000 sites and periodic boundary conditions is assumed and the eigenvalues of the correspondingly large matrix are calculated numerically. In order to include disorder, 500 different, random configurations of varying overlap integrals and varying on-site energies are computed and added to obtain an appropriate average of the resulting band structure. Finally, the smearing method (see preceding section) is used to compute the DOS. Figure 2.48(b) shows the resulting band structure (left panel) and the DOS (right panel) in the case of weak (green and red lines) and strong (light green and orange lines) disorder. The right inset depicts a close-up of the lower edge of the p-band. Obviously, strong disorder yields pronounced band-tails (orange line) that are also known as Urbach tails [254]. In Chapter 11, it will be discussed how band-tailing affects the performance of cryogenic field-effect transistors and how it can be mitigated or even circumvented.

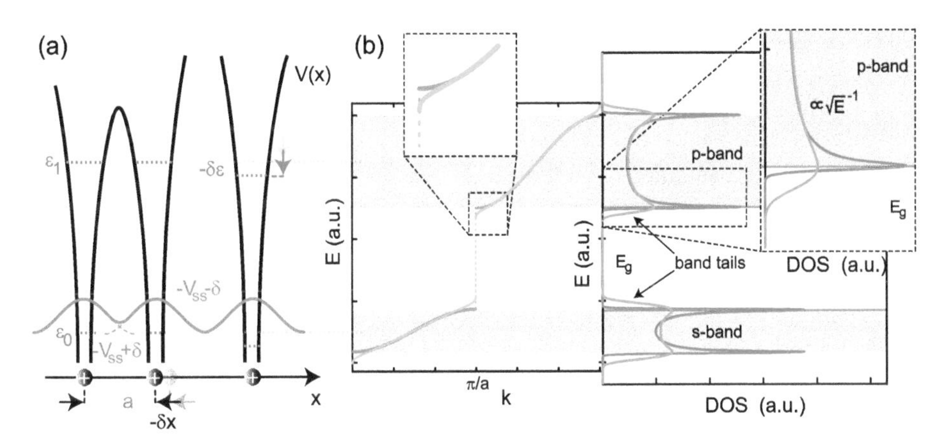

Figure 2.48: (a) Schematic of a 1D crystal with disorder leading to varying overlap integrals (e. g., $-V_{ss} \pm \delta$) and on-site energies $\epsilon_{0,1} \pm \delta\epsilon$. (b) Band structure (left panel) of a 1D solid with weak (green) and strong disorder (light green). The corresponding DOS (red for low disorder and orange for strong disorder) is displayed in the right panel. The insets show close-ups of the band structure (left) and the resulting DOS (right).

2.12.3 Density of States at Surfaces

Up to now, the density of states was computed in 1D, 2D and 3D solids. However, in practical implementations of nanoelectronics devices, the surface of a bulk material plays a very prominent role (see, for instance, Section 4.6 on metal-semiconductor contacts). In order to complement our considerations so far, the surface DOS (sDOS) will be computed. To this end, let us assume that a bulk material is cut into two halves along the $z = 0$ plane. Since the surface basically breaks the symmetry in the direction perpendicular to it, the sDOS will become a function of z (at least close to the surface) and, as a result, we need to compute a local DOS. This can in principle be done using Equation (2.79) if the wavefunction $\psi(\vec{r})$ is known. For instance, if an effective mass approximation is adopted, the semiinfinite bulk material can be considered as a slab of length L with $L \to \infty$. As a result, a separation ansatz with plane waves along x- and y-directions and the PIB solutions ($\propto \sin(n\pi/L \cdot z)$, Equation (2.6)) can be used to construct ψ. However, in order to provide a more generally valid framework, the sDOS is computed based on a Green's function ansatz. Details and derivations of the approach are given in Section 6.2.1 (see Task 24, p. 289); here, only the necessary basics are provided.

The defining equation for the retarded Green's function is

$$(E - \mathcal{H} + i\eta)G^r(x, x', E) = \delta(x - x') \tag{2.98}$$

where $i\eta$ is an infinitesimally small imaginary part. The relevance for our considerations here is that the imaginary part of the $G^r(x, x' = x, E)$ is the local DOS, and thus we can calculate the surface DOS. If a discrete lattice is considered, this equation turns into a matrix equation showing that G^r can be computed by inverting the matrix $(E - \mathcal{H} + i\eta)$. The determining matrix equation has infinite dimensions and we therefore need a way how to compute it. As an example of how to do this, let us consider the 2D semiinfinite crystal displayed in Figure 2.49(a) (same as in Section 2.4.8). In addition, let us assume that the potential energy is $V_0 = 0$ throughout the semiinfinite system and $V_{sp} = 0$ for simplicity. One can then subdivide the semiinfinite crystal into a surface layer and a semiinfinite rest. Following the formalism detailed in Section 6.2.3 and using Equation (2.38), the surface Green's function describing the semiinfinite crystal can be computed by solving the following (Dyson) equation:

$$\begin{pmatrix} G_{11}^r & G_{12}^r \\ G_{21}^r & G_{22}^r \end{pmatrix} = \left[\begin{pmatrix} \tilde{E} & 0 \\ 0 & \tilde{E} \end{pmatrix} - \begin{pmatrix} \epsilon_s - 2V_{ss}\cos(k_x a) & 0 \\ 0 & \epsilon_p + 2V_{pp\sigma}\cos(k_x a) \end{pmatrix} - \begin{pmatrix} V_{ss}^2 G_{11}^r & V_{ss}V_{pp\pi}G_{12}^r \\ V_{ss}V_{pp\pi}G_{21}^r & V_{pp\pi}^2 G_{22}^r \end{pmatrix} \right]^{-1} \tag{2.99}$$

where $\tilde{E} = E + i\eta$. In order to determine the local DOS of the first few layers of the crystal shown in Figure 2.49(a), the equation of the Green's function for the desired number of layers is set up and the semiinfinite crystal is accounted for with a self-energy based on the solution of Equation (2.99). As an example, Figure 2.49(b) shows the DOS for the first four layers of the crystal shown in (a); for details on the calculation, see QR code #25.

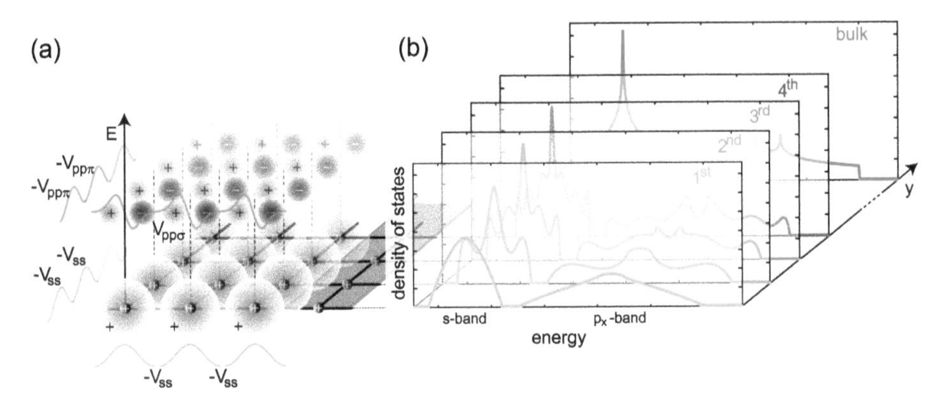

Figure 2.49: Surface DOS (b) of the first four layers of a 2D semiinfinite crystal with cubic lattice and (cf. Figure 2.21) one s- and p_x-orbital per unit cell (a). The red curve denoted with "bulk" in (b) is the DOS of an infinite 2D crystal.

2.12.4 Density of States of Graphene

In Section 2.11.1, the density of states in various dimensions has been derived based on a quadratic dispersion relation. This means that the band structure of the underlying semiconductor has been considered as being described with an appropriate effective mass, which is certainly a good approximation for rather conventional semiconductors such as silicon, III–V semiconductors and alike. In the case of graphene, it was shown in Section 2.8.1 that the band structure around the Fermi energy can be approximated with two cones whose tips touch at the Dirac point (i. e., at E_f). Therefore, we need to recalculate the DOS in the case of graphene.

In the conduction band, the dispersion is given by $E(\vec{\kappa}) = V_{pp\pi}a\frac{\sqrt{3}}{2}|\vec{\kappa}|$. Due to the rotational symmetry (E only depends on the magnitude $|\vec{\kappa}|$), the density of states can easily be computed using one of the approaches discussed in Section 2.11.1. Introducing cylindrical coordinates, one obtains $d\vec{\kappa} = 2\pi\kappa d\kappa$, and thus the density of states in graphene $D_{\mathrm{Gr}}(E)$ is (the factor $2 \cdot 2$ is due spin degeneracy spin and the two K-points)

$$D_{\mathrm{Gr}}(E) = 2 \cdot 2\frac{1}{(2\pi)^2}\int d\vec{\kappa}\delta(E - E(|\vec{\kappa}|)) = 4\frac{2\pi}{(2\pi)^2}\int d\kappa\kappa\delta(E - E(|\vec{\kappa}|)). \tag{2.100}$$

Next, changing variables to $\epsilon = E(|\vec{\kappa}|)$ yields $d\kappa = \frac{d\epsilon}{V_{pp\pi}a\sqrt{3}/2}$ and with $\kappa = \epsilon\frac{1}{V_{pp\pi}a\sqrt{3}/2}$ one obtains

$$D_{\mathrm{Gr}}(E) = \frac{4}{2\pi}\int d\epsilon\frac{\epsilon}{V_{pp\pi}^2 a^2 3/4}\delta(E - \epsilon) = \frac{8}{3\pi V_{pp\pi}^2 a^2}|E|. \tag{2.101}$$

This result is completely different from the DOS in a conventional 2D semiconductor with quadratic dispersion relation, which is constant. In the case of graphene, however, the particular crystalline structure leads to a linear dispersion, and hence to a density

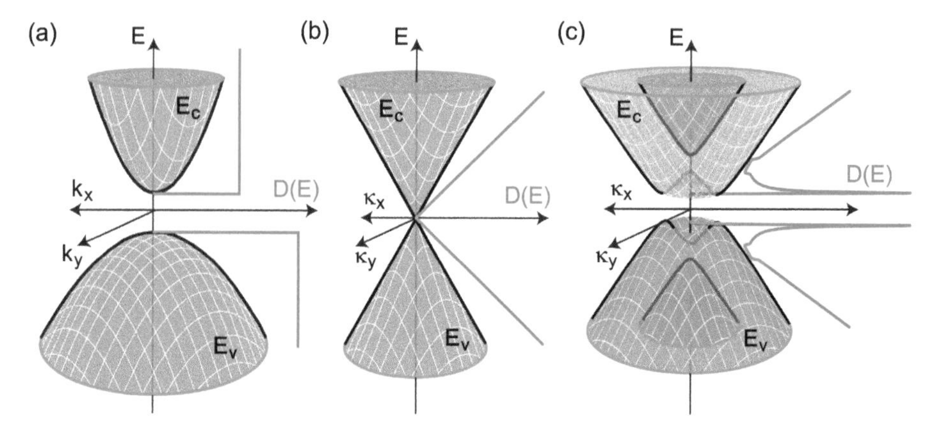

Figure 2.50: Dispersion relation and density of states of a 2D semiconductor with quadratic dispersion relation (a), monolayer graphene (b) and bilayer graphene (c).

of states that increases linearly with energy for $E > 0$ and $E < 0$. In particular, the DOS vanishes at the Fermi energy, which is why graphene can be called a zero-gap semiconductor. Figure 2.50 shows a comparison between the band structure and the DOS (red lines) in a conventional 2D semiconductor with quadratic dispersion relation (a), monolayer (b) and bilayer graphene (c) exhibiting a distinctly different energy-dependence with van Hove singularities in the case of bilayer graphene [122]. While the DOS in (a) and (b) can be computed analytically (in (a) with a larger valence band effective mass compared to the conduction band, typically found in semiconductors) as has been done above, the DOS of bilayer graphene has been calculated with the smearing method based on Equation (2.53).

2.12.5 Density of States of Carbon Nanotubes

The density of states of a carbon nanotube also shows an unusual behavior that needs to be discussed. We have to distinguish between two cases: first, when looking at the band structure of a carbon nanotube (see Equation (2.60)) it is obvious that if $\frac{2n+m}{3}$ is an integer value the band structure of the lowest (i. e., closest to the Fermi energy) subband will be given by

$$E(\kappa_{\|}) = \pm V_{pp\pi} a \frac{\sqrt{3}}{2} |\kappa_{\|}|, \tag{2.102}$$

so it basically represents one "slice" of the graphene band structure that contains the Dirac point. However, in contrast to graphene, the density of states $D_{\mathrm{CNT}}(E)$ of such a carbon nanotube does not depend on energy (in a certain energy range, see below). Computing the DOS as above and changing variables to $\epsilon = V_{pp\pi} a \sqrt{3}/2\kappa_{\|}$ yields

$$D_{\mathrm{CNT}}(E) = \frac{2}{2\pi}\int d\kappa_{\|}\delta(E - E(\kappa_{\|})) = \frac{4}{2\pi V_{pp\pi}a\sqrt{3}}\int d\epsilon\delta(E-\epsilon) = \frac{2}{\pi V_{pp\pi}a\sqrt{3}}, \tag{2.103}$$

which has to be multiplied by a factor of four to account for spin degeneracy and the two independent K-points of the underlying graphene band structure. Surprisingly, the DOS is constant although the band structure is linear as in the case of graphene. The difference stems from the dimensionality of the considered system. As a result of a finite DOS at the Fermi energy, carbon nanotubes where $\frac{2n+m}{3}$ is an integer value are metallic.

Semiconducting CNTs are obtained if $\frac{2n+m}{3}$ is not an integer value. In this case, $E(\kappa_{\|}) \neq 0$ when $\kappa_{\|} = 0$. Close to the conduction band bottom (valence band top) the band structure can be approximated with a quadratic dispersion relation (cf. Equation (2.63)) and as a result, a 1D density of states would be obtained. However, for not too large energies the dispersion of a CNT differs from a quadratic band structure and becomes linear. Therefore, the DOS in a carbon nanotube is expected to behave as $1/\sqrt{E}$ close to the bottom of a 1D subband eventually becoming constant for higher energies. Indeed, using the full dispersion relation, given in Equation (2.60), allows computing the full density of states of CNTs. For a single one-dimensional subband, the following expression is obtained (multiplication with four is again required):

$$D_{\mathrm{CNT}}(E) = \frac{4E}{\pi V_{pp\pi}a\sqrt{3}\sqrt{4E^2 - 4V_{pp\pi}^2 a^2 3(\frac{\tilde{n}-\frac{2n+m}{3}}{d_{\mathrm{CNT}}})^2}} \tag{2.104}$$

Figure 2.51 (left panel) shows the band structure of a (10, 8) nanotube exhibiting several conduction and valence bands. Since $(2 \cdot 10 + 8)/3$ is not an integer, the nanotube is semiconducting and a band gap of $E_g = 0.697$ eV is obtained from Equation (2.61); remember that in order to determine the band gap the smallest value of $\tilde{n} - \frac{2n+m}{3}$ (here $\tilde{n} = 9$) needs to be found and inserted into Equation (2.61). The panel on the right shows

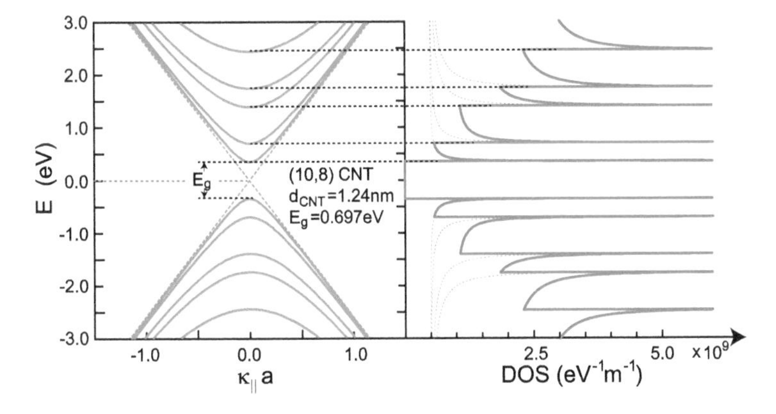

Figure 2.51: Band structure of a semiconducting (10, 8) carbon nanotube exhibiting multiple 1D subbands (left panel). Density of states in a (10, 8) nanotube. The DOS diverges proportional to $1/\sqrt{E}$ at the bottom of each subband as expected in a 1D system. No states are available within the band gap E_g.

the density of states of the (10, 8) nanotube. As discussed above, the DOS diverges whenever a new 1D subband sets in and subsequently drops $\propto 1/\sqrt{E}$ with increasing E (decreasing in the valence band). Because the dispersion becomes linear at higher energies the DOS in a CNT approaches a constant for each subband (illustrated by the thin dotted lines) such that the overall DOS increases as displayed in Figure 2.51.

2.13 Current Flow

Up to now, the discussion has been restricted to equilibrium situations. However, electronic devices are usually operated in nonequilibrium due to the current flow. We therefore consider current transport in the final section of this chapter.

2.13.1 Quasiclassical Approach

Let us begin by considering an electron as a classical particle with effective mass m^* that is accelerated due to an electric field (the effect of a magnetic field will not be discussed here). Moreover, the particle will be scattered with a total mean time between two scattering events denoted by τ. If an electric field $\vec{\mathcal{E}}$ is applied, the following equation of motion is obtained:

$$m^* \frac{d\vec{v}}{dt} - \frac{m^* \vec{v}}{\tau} = -e\vec{\mathcal{E}}. \tag{2.105}$$

In the stationary case, there is no acceleration, i. e., $\frac{d\vec{v}}{dt} = 0$ and the resulting velocity is called drift velocity v_d, which is proportional to the electric field, i. e., $\vec{v}_d = \frac{e\tau}{m^*}\vec{\mathcal{E}}$. The factor in front of $\mathcal{E}$ is called the carrier mobility $\mu = \frac{e\tau}{m^*}$. The current density $\vec{j}$ is proportional to the carrier density n as well as to the carrier velocity according to

$$\vec{j} = e \cdot n \cdot \vec{v}. \tag{2.106}$$

Inserting the expression for the drift velocity into the Equation (2.106) yields

$$\vec{j} = \underbrace{\frac{e^2 n\tau}{m^*}}_{=\sigma} \vec{\mathcal{E}}. \tag{2.107}$$

Because Ohm's law relates the current density to the electric field via the conductivity, σ is given as shown in the equation above. Equation (2.107) is valid in an isotropic semiconductor, where the acceleration of carriers due to an electric field points in the direction of the electric field. As a result, the scalar conductivity $\sigma = \frac{e^2 n\tau}{m^*} = en\mu$ is obtained. The expression for the conductivity can be extended in a straightforward manner to account for multiple scattering mechanism as well as multiple channels that contribute to the conductivity.

Different scattering mechanisms impacting electronic transport in semiconductors such as impurity scattering, electron-phonon scattering, surface roughness or alloy scattering can each be described by a mean-time τ between two successive scattering events. If the scattering mechanisms are uncorrelated, their impact can be combined into a single, total scattering time τ_{tot}, which is given by Matthiessen's rule as

$$\frac{1}{\tau_{\text{tot}}} = \frac{1}{\tau_{\text{imp}}} + \frac{1}{\tau_{\text{e-ph}}} + \frac{1}{\tau_{\text{s-rough}}} + \cdots \tag{2.108}$$

This is obvious since the resistivity $\rho \propto 1/\sigma$, and thus ρ is inversely proportional to the scattering time τ. In the case of uncorrelated scattering mechanism, their impact to the total resistivity is added as stated in Equation (2.108). Furthermore, if several parallel "channels" for conduction are present, the conductivities associated with these channels have to be added. For instance, in the case of silicon, there are six equivalent valleys in the conduction band that contribute. When computing the carrier density, the different effective masses in the silicon conduction band and the sixfold degeneracy have been "merged" into the density of states effective mass $m^\star_{\text{DOS}}$ (cf. Section 2.11.2). In a similar fashion, current flow can be described by an effective mass for transport $m^\star_{\text{trans}}$, which however is completely different compared to $m^\star_{\text{DOS}}$. If we consider transport in the x-direction, for instance, then there will be two valleys contributing with the heavy(longitudinal) mass and four with the light(transversal) mass. Divided by the number of valleys, we obtain $m^\star_{\text{trans}}$ as

$$\frac{1}{m^\star_{\text{trans}}} = \frac{1}{6}\left(\frac{2}{m^\star_l} + \frac{4}{m^\star_t}\right) = \frac{1}{3}\left(\frac{1}{m^\star_l} + \frac{1}{m^\star_t} + \frac{1}{m^\star_t}\right). \tag{2.109}$$

Finally, the simultaneous conductivity of electrons and holes through a semiconductor leads to

$$\sigma = e(n\mu_e + p\mu_h) \tag{2.110}$$

where $\mu_{e/h} = \frac{e\tau}{m^\star_{\text{trans,e/h}}}$ denotes the electron/hole mobility with $m^\star_{\text{trans,e/h}}$ being the transport effective mass of electrons and holes, and n/p are the carrier densities of electrons and holes, respectively. It is important to note that Equation (2.110) can only be used if electron and hole transport are completely separated from each other. In Chapter 9, band-to-band tunneling in transistors will be discussed where carriers from the conduction band are injected into the valence band (and vice versa). In this case, electron and hole transport are interlinked and it is thus important to describe the current transport exclusively either with electrons or holes.

The discussion so far led to an expression for the conductivity with a scattering time τ that is independent of energy, and thus has the same value for all carriers. In order to generalize this result, we need to know how the charge carriers behave when external forces act on them. The charge carriers are described by the distribution function $f(\vec{x},\vec{k},t)$, which depends on the location $\vec{x}$, on the wave-vector $\vec{k}$ and time t; in the absence of any external forces, $f(\vec{x},\vec{k},t)$ is simply the Fermi distribution function $f(E_f)$. The idea is now that the total change of $f(\vec{x},\vec{k},t)$ (e. g., due to external forces) with time is equal to the change of f due to scattering. Explicitly, this means

$$\frac{df(\vec{x},\vec{k},t)}{dt} = \left(\frac{\partial f(\vec{x},\vec{k},t)}{\partial t}\right)_{\text{scat}}$$

$$\rightarrow \frac{\partial f}{\partial t} + \underbrace{\nabla_{\vec{r}} f(\vec{x},\vec{k},t)\frac{\partial \vec{r}}{\partial t}}_{\nabla f \vec{v}_{\vec{k}}} + \underbrace{\frac{\partial f}{\partial E}}_{\approx\frac{\partial f(E_f)}{\partial E}}\underbrace{\frac{1}{\hbar}\frac{\partial E}{\partial \vec{k}}}_{=\vec{v}_{\vec{k}}}\underbrace{\hbar\frac{\partial \vec{k}}{\partial t}}_{=\frac{\partial \vec{p}}{\partial t}=\vec{F}} = \left(\frac{\partial f}{\partial t}\right)_{\text{scat}} \tag{2.111}$$

Equation (2.111) is called the Boltzmann transport equation (see, e. g., [285]). The first term in the equation relates to the explicit time dependence of f, the second term is the diffusion term due to a gradient in the distribution function and the third term is due to an external force $\vec{F}$. In the following, an electric field is considered with $\vec{F} = -e\vec{\mathcal{E}}$. A stationary scenario is considered where $\partial f/\partial t = 0$ and it is assumed that the diffusion term can be neglected. Moreover, the third term due to the external force is supposed to be small enough so that the distribution function can be linearized around the equilibrium Fermi distribution function, i. e., $f(\vec{r},\vec{k},t) \approx f(E_f) + f_1(\vec{r},\vec{k},t)$. As a next step, the so-called relaxation time approximation is applied in order to obtain an approximate expression for the scattering term in Equation (2.111). If we assume that the distribution function $f(\vec{r},\vec{k},t)$ will relax exponentially toward the equilibrium distribution $f(E_f)$ with a time constant τ, then $f(\vec{r},\vec{k},t) - f(E_f) = f_1(\vec{r},\vec{k},t) \propto \exp(-t/\tau)$. As a result,

$$\left(\frac{\partial f}{\partial t}\right)_{\text{scat}} = \frac{\partial f_1}{\partial t} = -\frac{1}{\tau_{\vec{k}}} f_1(\vec{r},\vec{k},t). \tag{2.112}$$

Combining Equations (2.111) and (2.112) yields

$$f_1(\vec{r},\vec{k},t) \approx \left(-\frac{\partial f(E_f)}{\partial E}\right)\tau\vec{v}_{\vec{k}} e\vec{\mathcal{E}}. \tag{2.113}$$

Next, Equation (2.106) can be readily extended to account for situations where the carrier velocity is not constant but depends on the wave-vector $\vec{k}$. Together with Equation (2.113), one obtains

$$\vec{j} = e\int d\vec{k} f(\vec{r},\vec{k},t)\vec{v}_{\vec{k}} = \underbrace{e\int d\vec{k} f(E_f)\vec{v}_{\vec{k}}}_{=0} + e\int d\vec{k} f_1(\vec{r},\vec{k},t)\vec{v}_{\vec{k}}$$

$$= e\int d\vec{k}\left(-\frac{\partial f(E_f)}{\partial E}\right)e\tau_{\vec{k}}\vec{v}_{\vec{k}} e\vec{\mathcal{E}}\vec{v}_{\vec{k}}. \tag{2.114}$$

Explicitly writing down the equation and bringing it into the form $\vec{j} = \sigma_{ij}\vec{\mathcal{E}}$ allows extracting the conductivity tensor σ_{ij} from

$$\vec{j} = \underbrace{e \int d\vec{k} e \tau_{\vec{k}} \left(-\frac{\partial f(E_f)}{\partial E}\right) \begin{pmatrix} v_{\vec{k}}^x v_{\vec{k}}^x & v_{\vec{k}}^x v_{\vec{k}}^y & v_{\vec{k}}^x v_{\vec{k}}^z \\ v_{\vec{k}}^y v_{\vec{k}}^x & v_{\vec{k}}^y v_{\vec{k}}^y & v_{\vec{k}}^y v_{\vec{k}}^z \\ v_{\vec{k}}^z v_{\vec{k}}^x & v_{\vec{k}}^z v_{\vec{k}}^y & v_{\vec{k}}^z v_{\vec{k}}^z \end{pmatrix}}_{\sigma_{ij}} \vec{\mathcal{E}}. \tag{2.115}$$

In an isotropic semiconductor, the conductivity tensor is diagonal with the same term on all three diagonal entries. Although silicon has a cigar-shaped dispersion relation, it is isotropic because this dispersion relation is on all three $k_{x,y,z}$-axes and as a result, on average, an isotropic behavior is obtained. If a single, isolated conduction band valley of silicon would be considered, the directions of the electric field and current transport would be different.

2.13.2 Landauer Formalism

Today's nanoscale devices have become so small that the assumption of ballistic transport, i. e., scattering-free transport is in many cases justified. In the present section, we will therefore investigate how current transport can be described in the limit of ballistic transport. For simplicity, a one-dimensional system such as a nanowire of length L (where L is considered to be very large) is considered. Quantization is such that only the first 1D subband is occupied with carriers (cf. Section 2.2.2 and in particular Figure 2.3).

In the following, we will assume the nanowire to be on a constant potential Φ_0 where the quantization energy of the first subband (cf. Section 2.2.2) has been incorporated into Φ_0. Note that the nomenclature has been changed for the potential from $V_0 \to \Phi_0$. The reason for this is to avoid confusion with the applied voltage V at the drain contact that separates the Fermi levels in source and drain. With constant potential, the wave function $\phi_k(x)$ will be a simple plane wave, i. e., $\phi_k(x) = \frac{1}{\sqrt{L}}e^{ikx}$ where the prefactor ensures that the wave function is normalized. At the two ends of the nanowire, a source and a drain contact are attached that are considered to be in equilibrium and whose carrier distribution is given by two Fermi distribution functions with Fermi levels E_f^s and E_f^d as illustrated in Figure 2.52.

From the dispersion relation, the group velocity of the carriers can be computed to be $v_k = \frac{1}{\hbar}\frac{\partial E(k)}{\partial k}$. This implies a negative group velocity for carriers on the left branch of the dispersion relation (blue carriers in Figure 2.52) and a positive on the right branch (red carriers in Figure 2.52). In equilibrium, carriers occupy states equally on both branches of the dispersion relation up to the Fermi energy $E_f^s = E_f^d$, and thus the two current components, i. e., left-to-right and right-to-left, cancel out each other exactly yielding zero net current.

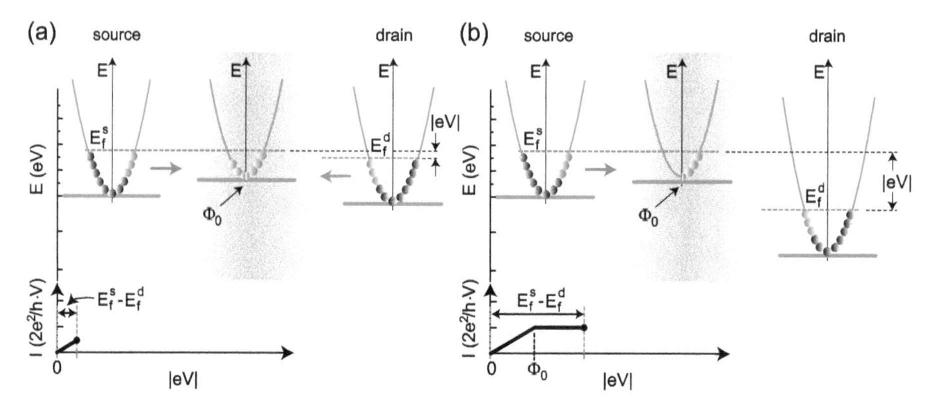

Figure 2.52: Occupation of the dispersion relation in a 1D nanowire at $T = 0\,K$ due to carrier injection from equilibrium Fermi distributions in a left source and right drain contact. (a) shows the near equilibrium and (b) the nonequilibrium case where the two Fermi levels have been separated by applying a bias voltage V. The lower panels show the respective current flows. Current saturation is obtained in (b) once V is large enough to avoid carrier injection from drain (i. e. when $E_f^d \le \Phi_0$).

In the limit of ballistic transport, i. e., in the absence of back-scattering, one can associate the carriers on the right branch of the dispersion relation with the Fermi distribution function in the source contact and the carriers on the left branch with the Fermi distribution in the drain contact of the considered 1D nanowire structure. If the Fermi levels of both contacts $E_f^{s,d}$ are separated by the applied drain-source voltage V, a net current will flow through the 1D nanowire due to carriers in the dispersion whose velocity component is not compensated by a component with equal magnitude but opposite direction; this situation is illustrated in Figure 2.52(a) for small V and in (b) for a large V. The current (density) is given by

$$j = e \sum_{k \ge 0} |\phi_k(x)|^2 f_s(k) v_k - e \sum_{k<0} |\phi_k(x)|^2 f_d(k) |v_k| \tag{2.116}$$

where $f_{s,d}$ are the Fermi distribution functions in the left and right contacts, respectively. Equation (2.116) can be understood as a generalization of $j = env$: $|\phi_k(x)|$ is the probability density of finding a carrier in state k at x and $f(k)$ is the probability that the state k is occupied. Since the carriers keep the distribution function of the contact they were injected from, $f_s(k)$ applies for $k > 0$ on the right branch of the dispersion relation and $f_d(k)$ needs to be used for negative k on the left branch. Because on the left branch $\frac{1}{\hbar}\frac{\partial E(k)}{\partial k} = -|v(k)|$, the sum of the two current contributions in Equation (2.116) can be written as the difference between left-to-right and right-to-left current contributions, which is illustrated in Figure 2.52. Next, since L is considered to be very large yielding plane waves as appropriate wave functions, the absolute square of the wave function is simply $|\phi_k(x)|^2 = \frac{1}{L}$. The sum over k is transferred into an integration as has been done above (see the info-box in Section 2.11.1). Finally, let us use the dispersion relation to compute $v(k) = \frac{1}{\hbar}\frac{\partial E(k)}{\partial k} = \frac{\hbar k}{m^*} \to v(E) = \sqrt{\frac{2(E-\Phi_0)}{m^*}}$ and to change from integrating over k to an integration over energy, i. e., $dk = dE\frac{m^*}{\hbar^2 k} = dE\frac{1}{\hbar}\sqrt{\frac{m^*}{2(E-\Phi_0)}}$. Note that $\frac{1}{\hbar}\sqrt{\frac{m^*}{2(E-\Phi_0)}} = 2\pi\frac{D_{1D}}{2}$ with $D_{1D}/2$ being half of the 1D DOS (cf. Equation (2.85)). This is reasonable because the net current is the difference between left-to-right and right-to-left flowing carriers each occupying either the right or left part of the dispersion leading to half of the full DOS. Since the current density in 1D equals the current I, we obtain

$$I = 2e\frac{L}{2\pi}\left(\int dk\frac{1}{\hbar}\frac{\partial E}{\partial k}\frac{1}{L}f_s(E(k)) - \int dk\frac{1}{\hbar}\frac{\partial E}{\partial k}\frac{1}{L}f_d(E(k))\right)$$
$$= 2e\int_{\Phi_0}^{\infty} dE\,\underbrace{\frac{D_{1D(E)}}{2}v(E)}_{1/h}(f_s - f_d) = \frac{2e}{h}\int_{\Phi_0}^{\infty} dE(f_s - f_d) \tag{2.117}$$

where appropriate boundaries for the integration over the energy have been introduced; the additional factor of 2 accounts for spin degeneracy. The same result can also be obtained by noting that in the first part of Equation (2.117) $dk\frac{1}{\hbar}\frac{\partial E}{\partial k} = \frac{1}{\hbar}dE$. However, this way the understanding is obscured that $\frac{1}{2}D_{1D}(E)\cdot v(E) = \frac{1}{h}$, which makes a lot of sense because it means that when carriers are accelerated (for instance, at the drain end of a MOSFET), the current does not increase because the increase of carrier velocity is compensated by a reduced carrier density (density of states). Ultimately, this ensures current continuity, which is required in a MOSFET if gate leakage and impact ionization can be neglected (see Section 5.2.3 for more discussion).

Task 13.
Current in 1D: Based on Equation (2.117) compute a closed expression for the current through a one-dimensional nanowire device at finite temperature T and at a fixed Φ_0 that depends explicitly on the drain-source voltage V (see Figure 2.52).

In the limit of $T = 0\,\mathrm{K}$, the Fermi distribution functions become step-functions, and hence the integration over energy yields simply $\int dE(f_s - f_d) \to \int_{E_f^d}^{E_f^s} dE$. The difference between the left and right Fermi energies is maintained by applying the voltage $|eV| = E_f^s - E_f^d$, and thus $I = \frac{2e^2}{h}V$, which results in a finite conductance of $\frac{I}{V} = \frac{2e^2}{h}$ (including spin degeneracy). This is an interesting result: Although we consider completely ballistic transport, a finite conductance is obtained. The reason for the rather high resistance of $\frac{h}{2e^2} \approx 12.9\,\mathrm{k}\Omega$ is the way the current is determined: In the present case, there are two contacts attached to the nanowire under consideration and the voltage is applied in-between the two contacts. The contacts are large in order to ensure that they are in equilibrium, and thus the applied voltage ensures that carriers are injected into the nanowire according to the left/right Fermi distributions leading to a net current flow I. If we were able to drive the same current through the nanowire but hook up voltage probes at the edges of the nanowire without disturbing the distribution of carriers, one would obtain a zero voltage drop between the voltage probes as expected in the case of ballistic transport. However, such a measurement can usually not be carried out: Even if ideal voltage probes were attached to the nanowire that carry zero net current, there would be a finite coupling between the nanowire and the probes. As a result, the probes would mix left and right moving particle populations, and thus yield a finite resistance showing up in a voltage drop. Such a scenario will be used in Section 6.3.1 to mimic scattering with so-called Buettiker probes.

Let us go back to Equation (2.117) to clarify where the finite resistance comes from. Assume, that instead of having only a single 1D subband that can carry current, there would be M 1D subbands available for current flow. In the ballistic case, they can all be considered as being independent, and thus their current contribution can simply be added up. Therefore, the total current becomes

$$I = \frac{2e^2}{h} M \cdot V. \tag{2.118}$$

If we increase the size of the nanowire that we consider here, then the energetic difference between 1D subbands will decrease leading to an increase of M as an increasing number of 1D subbands will be occupied with carriers contributing to the current. As a result, the conductance increases and the resistance decreases according to $\propto 1/M$. This means that in a large ballistic system where a very large number M of 1D subbands contributes to the current, the resistance indeed approaches zero as expected for scattering-free electronic transport. In turn, this means that the finite resistance associated with ballistic current transport through a single 1D subband can be interpreted as a contact resistance: a part of the applied voltage drops at the left contact interface and the other part across the right contact interface.

2.13.3 Multimode Transport

It was shown above that one subband or mode can carry a current (at $T = 0\,\mathrm{K}$) of $\frac{2e^2}{h}V$. If the subband is degenerate, this value has to be multiplied by the associated degeneracy factor d_g. For instance, in a nanowire with quadratic cross-section and isotropic effective mass it was shown in Section 2.2.2 that the second subband is twice degenerate and would carry a current of $2\frac{2e^2}{h}V$ accordingly. In the case of M subbands, their contributions have to be summed. The question, however, is how many modes in a device contribute to the current and how this appears in the electrical device characteristics?

Figure 2.53 shows the occupation of multiple subbands (again $T = 0\,\mathrm{K}$) for four different energies of the potential Φ_0 within the nanowire. From (a)–(b), the current increases linearly and only the first subband contributes. When Φ_0 is moved below E_f^d, the current stays constant (c). Although the carriers occupy states at higher energies their density drops leading to the constant current shown in the lower panel of Figure 2.53(c). In (d), the contribution of a second subband is displayed also showing a linear increase

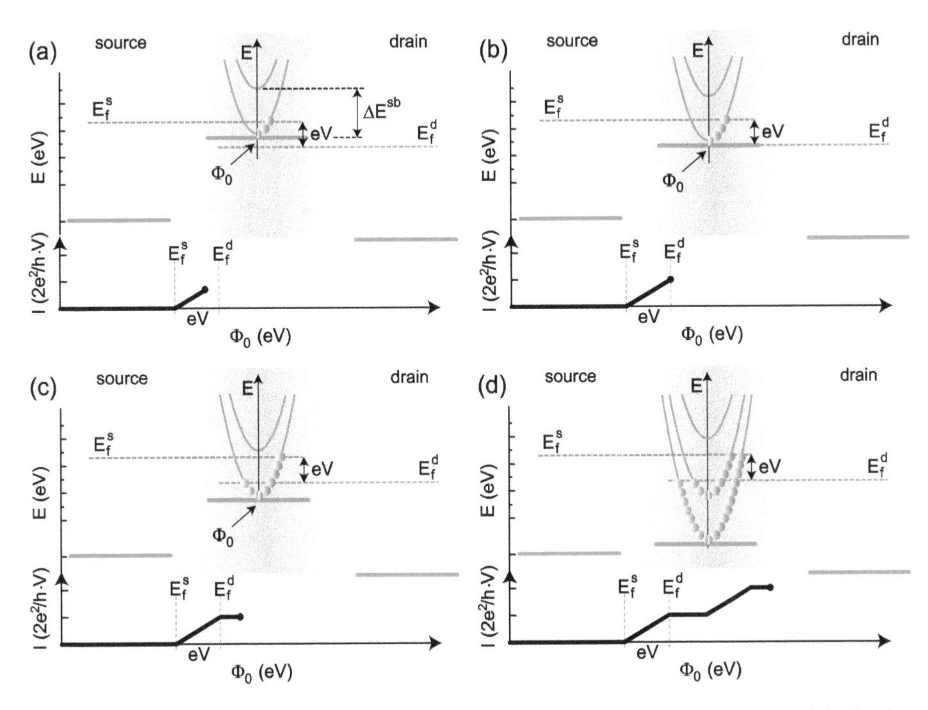

Figure 2.53: Occupation of the 1D dispersion relations in a nanowire at $T = 0$ K due to carrier injection from equilibrium Fermi distributions in a left source and a right drain contact. (a) shows the carrier injection into the first 1D subband. If the potential Φ_0 of the nanowire is moved towards lower energies, two and more subbands will be occupied as displayed in (b)-(d).

and current saturation once the subband bottom is moved below E_f^d. Therefore, the current increases stepwise and if the current of the Mth subband saturates it is simply given by $\frac{2e^2}{h}V{\cdot}M$. The stepwise current increase can only be observed if the separation between subbands ΔE^{sb} is larger than the applied voltage V.

Task 14.
Multimode transport: Consider a one-dimensional nanowire device at a temperature $T = 0$ K and $T = 300$ K. The nanowire has a quadratic cross-section $d \times d$ and was etched out of a bulk semiconductor that exhibits an isotropic effective mass. Suppose that carrier confinement yields a quantization energy of 0.1 eV for the first subband. In addition, a small bias voltage $V = 0.01$ V has been applied at drain and source is connected to ground; E_f^s is at an energy of 0.05 eV. Furthermore, assume that a gate electrode has been connected to the nanowire that has perfect gate control. This means that changing the gate voltage from V_g^1 to V_g^2 with $V_g^2 > V_g^1$ leads to moving the potential in the nanowire, Φ_0, energetically to $\Phi_0 - e(V_g^2 - V_g^1)$; it is assumed that $\Phi_0 = 0$ eV at $V_g = 0$ V. Compute the current I through the nanowire for gate voltages ranging from 0 V, ..., 1.1 V and plot I in the case of both temperatures.

So far, the discussion of multimode transport has been carried out assuming a temperature of $T = 0$ K. In this case, a linear current increase as displayed in Figure 2.53 is ob-

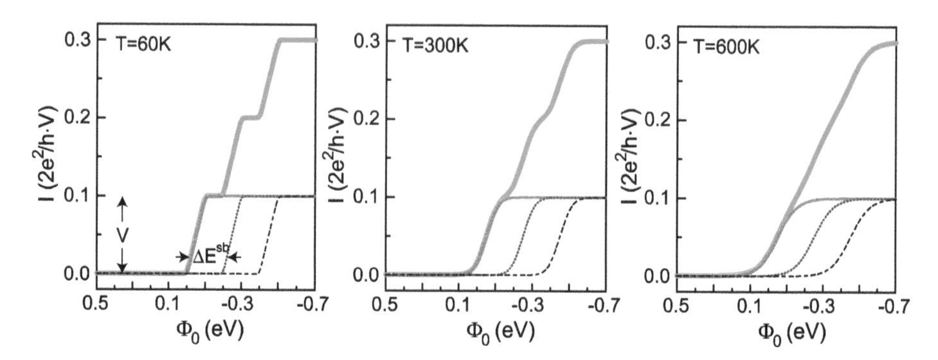

Figure 2.54: Current normalized to $2e^2/h$ with a bias voltage of 0.1 V for different temperatures. If the thermal broadening is sufficiently increased such that ~$4 \times k_B T$ is of the same order as the energetic subband spacing ΔE^{sb} the contribution of the different modes overlap and cannot be distinguished anymore. The same is true if $|eV| \approx \Delta E^{sb}$.

tained. At finite temperature, however, the current steps will be broadened. In fact, if the thermal broadening $4 \times k_B T$ is of the same order as the energetic separation ΔE^{sb} between the subbands, the $(n+1)$th subband will already contribute to the current before the current in the nth subbands saturates. Figure 2.54 shows this scenario for three different temperatures. In the present case, $\Delta E^{sb} = 0.2$ eV was assumed. Indeed, for high temperatures the step-like increase of current cannot be observed anymore (in Figure 2.54, right panel, $4 \times k_B T \approx 0.2$ eV) although up to three subbands contribute. The same is true if $|eV| \approx \Delta E^{sb}$: in this case, the current increases in the nth mode up to the point when the $(n+1)$th mode sets in.

Exercises

Exercises together with solutions are accessible via the QR code.

3 Semiconductor Fabrication

Apart from the fundamental properties of the materials that nanoscale devices are made of, their behavior is intimately connected to the technology used to fabricate them. Therefore, an introduction to the most important semiconductor fabrication techniques is necessary to complement the physics background provided in the preceding chapter. A thorough treatment of the topic certainly goes well beyond the scope of this book. There are quite a few textbooks available that cover semiconductor device fabrication with the required depth and breadth (the reader is referred to [215, 248]). Thus, the focus here is on fabrication techniques that have been used to realize the experimental devices discussed in later chapters. These are mostly techniques that students come across during their work in the clean room. For selected topics that I felt would be useful and worth mentioning, more details are provided.

The sequence of presenting the material is oriented along an actual fabrication of a device starting from wafer cleaning, lithography processes, wet and dry etching, chemical and physical vapor deposition, etc. A large part of the material deals with the processing of silicon. Seemingly old-fashioned, silicon is still the basis for most experimental research devices. Moreover, silicon has recently regained substantial attention as material for spin qubits (as isotopically purified ^{28}Si) and for conventional devices operating at cryogenic temperatures (see Chapter 11) to control qubits.

3.1 Wafer Cleaning

The cleaning of samples plays a crucial role during the fabrication of semiconductor devices. Appropriate cleaning can remove organic residuals, particles and metal ions from the wafer surface and is necessary to obtain proper device functionality, yield and reproducibility of the processes. Cleaning is mandatory prior to high temperature process steps such as the growth of a gate dielectric in order to prevent dielectric breakdown, drift in device characteristics and to prevent a reduction of carrier lifetime within the substrate as a consequence of diffusing contaminants. On the other hand, cleaning often involves the removal of contamination by oxidizing the top surface layer of a substrate resulting in a loss of material. In addition, it may also add to increasing the surface roughness of substrates, which deteriorates the carrier mobility in metal-oxide-semiconductor devices. As a result, samples should be processed in a way that reduces the number of cleaning steps to a minimum. In the present section, different cleaning methods and solutions will be discussed briefly; for more details on the topic, the reader is referred to the available literature (cf. [220], for instance).

3.1.1 Solvents

Solvents such as acetone, n-methyl-2-pyrrolidone (NMP) or dimethylsulfoxide (DMSO) are being used to remove photoresist remainders or other organic contaminants from

https://doi.org/10.1515/9783111054421-003

the sample surface prior to, e. g., photoresist coating. If persistent organic layers such as baked photoresist must be removed, cooking the samples in acetone may be effective. However, acetone has a very low flashpoint (–18 °C) and a very low boiling temperature of 56 °C, so great care has to be taken when doing so. An alternative may be NMP; due to its significantly higher boiling temperature of 203 °C, samples can be left in hot NMP for hours. However, in comparison to acetone, NMP has a substantially lower self-ignition temperature (265 °C compared to 465 °C in the case of acetone) and in addition, NMP is toxic to reproduction and a substance of very high concern[1] and has therefore been banned from usage in many labs. A nontoxic alternative is DMSO that has flash and ignition points (88 °C and 270 °C) comparable to NMP and also a rather high boiling temperature (189 °C).

Since acetone leaves residues on the sample surface,[2] samples are usually rinsed in isopropyl alcohol (IPA) after acetone; in some labs, a rinse in methanol is inserted known as AMI clean (acetone, methanol, IPA clean). IPA can be removed from the wafer surface without residues either by blowing dry the sample with a nitrogen gun or by retracting the sample from boiling IPA. The latter is also used for ultraclean wafer drying processes that will be briefly discussed below (see, e. g., [87, 167, 190]). Because IPA is hygroscopic, the wafer surface will be dehydrated after blow-dry with nitrogen similar to a dehydration on a hot plate. Due to its lower surface tension than water, hot IPA is an option for drying nano-electromechanical structures.

3.1.2 Piranha Cleaning Solution

Piranha or "sulfuric acid hydrogen peroxide mixture" (SPM) is a mixture of H_2SO_4 and H_2O_2 (i. e., a strong acid and oxidant) with ratios ranging from 2:1 to 8:1 (typical is a ratio of 3:1–4:1) and temperatures in the range of 90 °C–130 °C. The corrosiveness of Piranha can be adjusted by changing the mixing ratio: the higher the H_2O_2-fraction, the stronger the oxidizing characteristic and the more violently the reaction. Since after the mixing of the two chemicals the temperature rises to approximately 140 °C due to the strong exothermal reaction, the solution needs to cool down to the desired process temperature. A too high process temperature and mixing ratios smaller than ~3:1 will lead to increased surface roughness.

Piranha is intended to remove organic and metallic contaminants and is usually used, if substantial amounts of (organic) contaminations have to be removed from a wafer. In fact, even entire photoresist layers can be removed from substrates and such wet chemical stripping processes are investigated as an alternative to oxygen plasma in order to avoid plasma-induced damages of the substrate. However, if a large amount of

---

1 https://echa.europa.eu/de/substance-information/-/substanceinfo/100.011.662

2 When working with hot acetone, drying of the sample surface must be prevented.

organic material such as a photoresist layer is to be removed, the common immersion into a Piranha bath leads to the formation of a residue layer that often cannot be removed anymore. To circumvent this, Piranha spray processes have been developed with temperatures up to 200 °C. The high temperature is necessary in order to increase the concentration of HO^* and HSO_4^* radicals in the SPM that are necessary to remove heavily carbonized photoresist as present on samples after high-dose ion implantation [46].

Wafers treated with Piranha are known to exhibit an increased area density of particles and sulfur contaminations on the wafer surface such that a dip in diluted hydrofluoric acid (HF) should always be carried out after the cleaning.

3.1.3 RCA Clean

The best-known cleaning solutions are the so-called RCA-clean method and its modifications [129]. RCA-clean consists of two cleaning solutions called standard clean 1 (SC1), RCA1 or due to its constituents ammonium hydroxide (NH_3(aq)) hydrogen peroxide mixture (APM), and standard clean 2 (SC2), RCA2 or hydrochloric acid (HCl) hydrogen peroxide mixture (HPM). Prior to a RCA-clean, a Piranha clean is frequently carried out if severe contaminations are present on the sample surface (see the preceding section) [128]. Furthermore, HF etch steps were added to the original RCA-clean in order to remove the SiO_2 chemically grown during the cleaning.

Various cleaning sequences have been tested in the literature, yielding different efficiencies. For instance, it was found that the sequence SPM/SC1/HF/SC2 provides the best cleaning efficiency for removing metals. However, the hydrophobic surface after the HF-dip is prone to be recontaminated with organics and particles from the environment, which will not be removed by the following SC2. Therefore, an HF-dip is usually only done prior to SC1. The sequence SPM/HF/SC1/SC2 was found to be most effective in particle removal (see review [128] and the references therein). Several modifications of the original recipe resulting in the use of strongly diluted cleaning solutions have been suggested and studied (see, e. g., [192]) that showed an equal cleaning effectiveness. Table 3.1 gives an overview over the cleaning solutions with different mixing ratios. After the cleaning, wafers need to be rinsed and dried, which is a critical process step requiring extra care to avoid recontamination.

Table 3.1: Wet chemical cleaning solutions. Ratios are stated by % vol. based on aqueous NH_3(29 %), H_2O_2(30 %), HCl(37 %) and H_2SO_4(≥96 %).

Cleaning solution	Content	Mixing ratio	Temperature
piranha	H_2SO_4:H_2O_2	2:1–4:1	~80 °C
SC1	NH_3(aq):H_2O_2:H_2O	1:1:5–1:5:20	60–80 °C
SC2	HCl:H_2O_2:H_2O	1:1:5–1:1:20	60–80 °C

3.1.3.1 Standard Clean 1

Standard Clean 1 or APM consists of NH_3(aq):H_2O_2:H_2O in a wide range of different mixing ratios (1:1:5 to 1:1:200) used at temperatures from 80 °C down to room temperature. While the original solution had a ratio of 1:1:5 and a temperature of 80 °C, today ratios of 1:1:20 and temperatures of 50–60 °C are commonly used. Reasons for the dilution include saving resources, a reduction of waste and the fact that APM etches silicon. While the etch rate is indeed very small, it nevertheless increases the surface roughness, deteriorating the carrier mobility at MOS interfaces.

SC1 removes organic contaminants and some metals as well as particles from the wafer surface. The removal of organics occurs via oxidation and dissolution while metals are removed due to complex formation of the ammonia with metals. The removal of particles relies on the fact that SC1 etches silicon and that most surfaces immersed in a solution with a pH $>$ 10 build up a negative surface charge. Thus, once a particle is detached from the surface due to the etching a positively charged electrostatic double layer builds up around the particle that prevents redeposition of the particles due to electrostatic repulsion (see Figure 3.1(a)). Figure 3.1(b) shows the ζ-potential[3] of different material surfaces as a function of the pH value of the solution the respective particles are immersed in (values are extracted from [262]). Obviously, in solutions with pH $>$ 9 the surfaces of the different materials (including the wafer surface and common particle contaminants) exhibit the same polarity leading to the electrostatic repulsion mentioned above. Therefore, alkaline solutions such as SC1 are effective in terms of particle removal [117]. However, keeping the pH-value below approximately 3, and hence

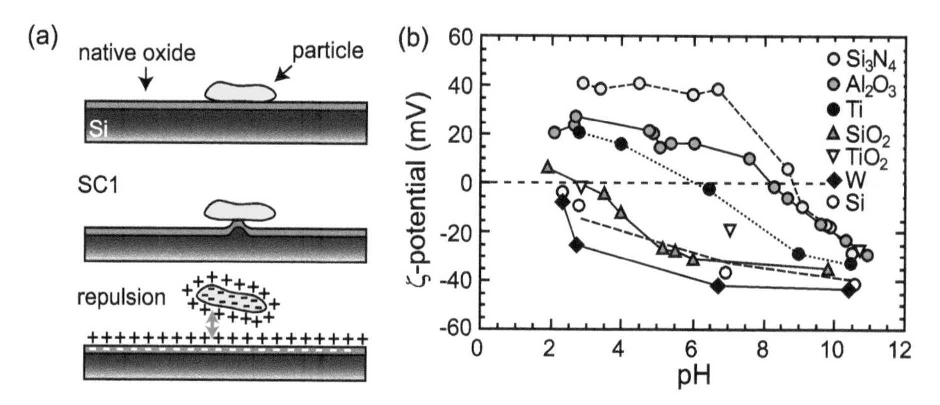

Figure 3.1: (a) Particle removal due to underetching and electrostatic repulsion. (b) ζ-potential at the surface of different materials as a function of pH-value (extracted from [262]).

3 At the interface between the solid and the cleaning solution, the negatively charged surface attracts positively charged ions that form the so-called Stern layer. A second layer of ions is attracted forming the double layer. The ζ-potential is the potential between the surfaces of the solid and the double layer (see, e. g., [152]).

ensuring that particles and surfaces exhibit the same polarity of the ζ-potential (cf. Figure 3.1(b)) acidic solutions are also suited for removing particles; the latter is used in the so-called IMEC clean [192]. Finally, particle removal can be made more effective by applying megasonic.

In order to achieve an appropriately efficient particle removal, it was shown that approximately 2–3 nm of silicon need to be etched during the cleaning step [105]. The etching of silicon occurs with a very small rate due to the following reaction: Silicon (hydrogen passivated Si-H) is predominantly oxidized due to H_2O_2, which yields a chemical oxide on top of the surface. This oxide is etched due to the presence of OH^--groups according to the reaction: $OH^- + Si_2O \rightarrow SiO_3H^-$. The latter reaction is very slow and basically determines the rate of the etching process. The etch rate also depends on temperature. Reported values of the etch rate are 0.1 nm/min at 60 °C and 0.37 nm/min at 85 °C [239]. Due its dependence on the OH^--concentration, the etch rate can be reduced controllably by diluting the APM (up to ratios of 1:1:200). Consequently, dilution of SC1 is meaningful, since a too high pH-value does not necessarily increase the particle removal due to the electrostatic repulsion. But if the concentration is rather high, increased loss of material due to etching and a higher surface roughness are expected.

3.1.3.2 Standard Clean 2

Standard clean 2 (SC2) is a mixture consisting of HCl, H_2O_2 and H_2O and is used to remove metal contaminations. In particular, alkali metal ions, Al^{+3}, Fe^{+3} and Mg^{+2} are removed that form insoluble hydroxides in alkaline ammonia solutions (such as SC1). Similar to SC1, diluted solutions at temperatures ~60 °C are in use (see Table 3.1) for the same reasons mentioned above. However, rather concentrated solutions are also employed. For instance, a modified SC2 can be used after anisotropic etching of silicon with KOH (see Section 3.6.4). Potassium is known to diffuse through gate oxides and it also deteriorates the carrier lifetime. Thus, it is detrimental for most devices. Therefore, removing the K-ions is of utmost importance prior to continuing fabrication in clean process tools. While silicon can be etched anisotropically with tetramethylammonium-hydroxide (TMAH), which is CMOS-compatible, TMAH has somewhat different etch characteristics from KOH and has recently been found to be a very hazardous chemical (see info-box in Section 3.6.4 for further details). Using SC2 with a mixture of 1:1:1 allows removing the K-ions effectively. The high concentration of HCl yields a self-heating etch solution that does not need external heating (in contrast to the usual SC2).

3.1.3.3 Drying Procedures and Surface Conditioning

Drying of the substrates is an integral and very important part of each cleaning procedure. In particular, if the last step of the cleaning procedure is an HF-dip in order to remove the chemically grown oxide, the wafers become hydrophobic and the Si dangling bonds at the surface are saturated with hydrogen atoms. However, a hydrophobic surface is prone to becoming recontaminated. Furthermore, if the HF-dip is followed by a DI-water rinse and blow-dry, an incomplete hydrogen passivation is obtained with OH-groups being attached to the surface leading to a reoxidation of the surface after approximately 20 min. With an appropriate drying, on the other hand, some studies claim that reoxidation can be avoided up to several days. An extensive review on wet chemical surface passivation with cleaning solutions can be found in [91].

3.2 Oxidation of Silicon

When exposed to oxygen silicon readily oxidizes and if done properly forms stoichiometric silicon dioxide SiO_2. Si/SiO_2 is an extraordinary material combination in that it is the native oxide of silicon that exhibits excellent chemical, mechanical and thermal stability and provides superb electrical insulation properties. Even more important, SiO_2 yields an orders of magnitude reduction of electronic states at the Si/SiO_2 interface. The latter is key to the realization of metal-oxide-semiconductor field-effect transistors, which will be discussed in detail in Sections 4.5 and 5.2.2. In addition, for the fabrication of the silicon/SiO_2 material system, etch processes with very high selectivities of the two materials with respect to each other are available. Hence, SiO_2 plays a prominent role for the fabrication of sophisticated semiconductor structures serving not only as insulator but also as etch mask, sacrificial layer, etc.

Figure 3.2(a) displays a high-resolution transmission electron microscopy image[4] of the silicon/SiO_2 interface, clearly showing the border between the crystalline silicon structure and the amorphous SiO_2. In Figure 3.2(b), a schematic close-up of the interface is shown. Most of the bonds of the silicon surface atoms are involved in the oxide formation. However, some bonds (marked with a red line) can be unsaturated providing electrically active defects. At a silicon (100) surface, one distinguishes between P_{b0}

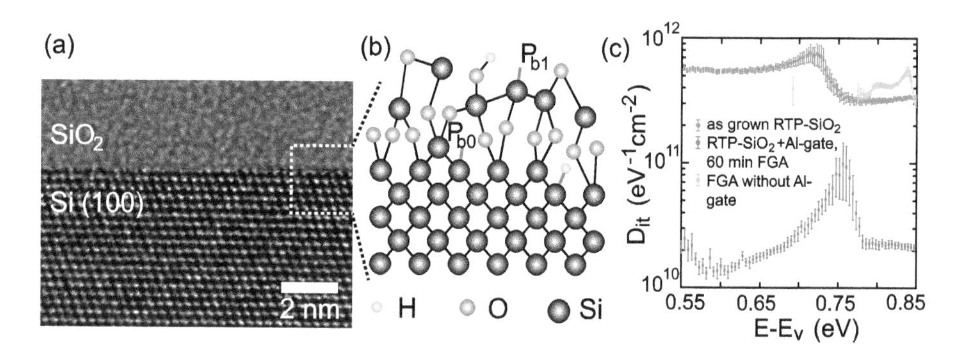

Figure 3.2: (a) TEM image of the Si/SiO_2 interface. (b) Illustration of the interface layer showing P_{b0} and P_{b1} defects. (c) Strong reduction of D_{it} due to FGA in the presence of aluminum.

4 A. Meledin, Central Facility for Electron Microscopy (GFE), RWTH Aachen University.

and P_{b1} defects;[5] in the former case, a Si atom is back-bonded to three other Si atoms whereas in the latter case, the Si atom is bonded to two Si and one oxygen atom (see Figure 3.2(b)). These so-called dangling bonds are amphoteric in nature, i. e., they provide energy levels located in the lower half of the band gap that are donor-like as well as acceptor-like states within the upper half of E_g.

Fortunately, the dangling bonds can be passivated with hydrogen (depicted in Figure 3.2(b)) that allows reducing the density of interface states D_{it} within the band gap substantially. Hydrogen is usually provided by a forming gas anneal (FGA), a mixture of nitrogen and hydrogen (~9:1), at 400–450 °C for approximately 20–30 min. Since forming gas consists of molecular hydrogen, it needs to be split into atomic hydrogen in order to be effective. In this respect, the presence of aluminum (for instance, as part of the gate electrode) is important because it acts catalytically yielding the required atomic hydrogen to saturate the dangling bonds as shown in Figure 3.2(c) [234]; without the presence of aluminum, hydrogen should be provided in atomic form, for instance with a hydrogen plasma [234]. With proper oxidation in combination with forming gas anneal, the density of interface states can be reduced ~10^5-fold compared to the density of unsaturated bonds at a bare Si surface. In the following section, the process of thermal oxidation of silicon is briefly discussed.

3.2.1 Thermal Oxidation

The oxidation of silicon is a straightforward process and requires only a quartz furnace that can be heated to elevated temperatures (up to approximately 1200 °C). If a sample (after carrying out a standard cleaning) is mounted into the furnace and heated up, the provided oxygen diffuses through the (possibly already) existing SiO_2 layer and further oxide grows at the Si/SiO_2 interface. One distinguishes between wet thermal and dry thermal oxidation. In the case of wet thermal oxidation, water is introduced into the furnace by driving oxygen as carrier gas through a so-called bubbler filled with deionized water. Alternatively, H_2/O_2 is burned in a controlled way at a torch connected to the quartz furnace. Dry oxidation, on the other hand, only requires the injection of oxygen gas into the furnace.

It is important to note that the oxidation of silicon is a growth and not a deposition process meaning that silicon from the oxidized sample is consumed to form the SiO_2. This has two important implications. First, oxide growth always occurs at the silicon/SiO_2-interface and as such it is expected that the growth rate drops with increasing SiO_2 thickness. This fact is often mentioned as the reason why a wet thermal

5 On a Si (111) surface, only the P_b defect exists, which is similar to P_{b0}.

oxidation leads to a larger oxide growth rate:[6] the water molecule is simply smaller than O_2 and can thus diffuse faster through the existing oxide. Second, the consumption of silicon during oxidation needs to be taken into account, particularly when working with Si nanostructures. Considering the fact that the ratio of the molecular density of SiO_2 and the atomic density of Si[7] is inversely proportional to the ratio of the consumed Si and grown SiO_2 thicknesses, one obtains

$$\frac{N_{\text{ox}}}{N_{\text{si}}} = \frac{2.3 \cdot 10^{22}\ \text{molecules/cm}^3}{5 \cdot 10^{22}\ \text{atoms/cm}^3} = \frac{d_{\text{si}}}{d_{\text{ox}}} \rightarrow d_{\text{si}} = 0.46 \cdot d_{\text{ox}}. \tag{3.1}$$

As a rule of thumb, oxidation consumes roughly half the thickness of the grown SiO_2.

The oxidation of silicon can be described with the Deal–Grove model [62]. To derive an expression for the SiO_2 growth based on it, a silicon sample in an oxygen atmosphere is considered as depicted in Figure 3.3(a). The interface between the gas phase with concentration c_g and the boundary layer with an oxygen concentration at the surface of c_{sf} is illustrated with the red dashed line. The difference in concentration drives a flux $j_g = h_g(c_g - c_{\text{sf}})$ toward the sample surface where h_g is the gas transport coefficient. Furthermore, the oxygen concentration at the Si/SiO_2 interface is denoted with c_{ox}, and hence there will be a flux of oxygen $j_{\text{diff}} = D_{\text{diff}}^{\text{ox}} \frac{\partial c}{\partial x} \approx D_{\text{diff}}^{\text{ox}} \frac{c_{\text{sf}} - c_{\text{ox}}}{d_{\text{ox}}}$ diffusing through the oxide with $D_{\text{diff}}^{\text{ox}}$ being the diffusion coefficient. Finally, a flux for the oxidation reaction can be defined as $j_{\text{ox}} = k_{\text{ox}} \cdot c_{\text{ox}}$ where k_{ox} is the rate of surface reaction. The latter flux can also be written as $j_{\text{ox}} = N \frac{\partial d_{\text{ox}}}{\partial t}$ where N is the concentration of oxidant molecules (H_2O or O_2) per unit volume of grown oxide ($N = 2.2 \cdot 10^{22}\ \text{cm}^{-3}$ in the case of dry oxidation).

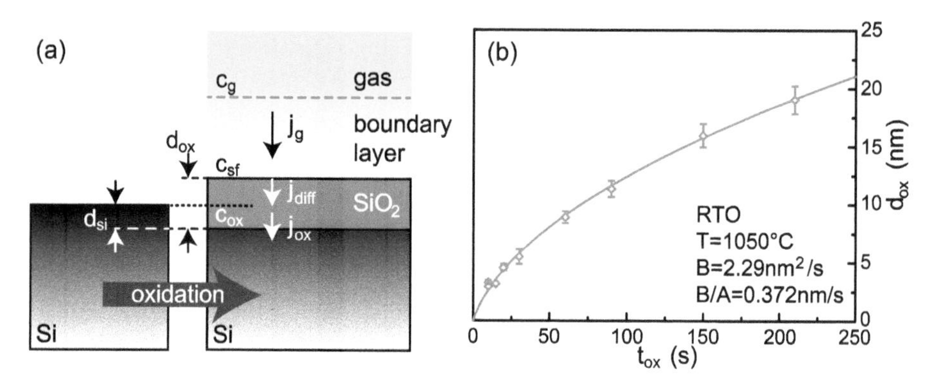

Figure 3.3: (a) Deal–Grove model of silicon oxidation. (b) SiO_2 thickness versus oxidation time grown at 1050 °C in a RTO furnace. The straight line is a Deal–Grove model fit.

6 It has been shown recently that oxygen indeed diffuses in its molecular form through defect-free SiO_2 without oxygen exchange with the network [180].

7 $N_{\text{ox}} = N_A \cdot \frac{\rho_{\text{ox}}}{m_{\text{ox}}}$ and $N_{\text{si}} = N_A \cdot \frac{\rho_{\text{si}}}{m_{\text{si}}}$ where N_A is Avogadro's constant, $\rho_{\text{ox,si}}$ are the densities of SiO_2 and Si and $m_{\text{ox,si}}$ are their molecular weights.

Since the continuity equation holds, all fluxes must be equal, i. e., $j_g = j_{\mathrm{diff}} = j_{\mathrm{ox}} = N\frac{\partial d_{\mathrm{ox}}}{\partial t} \equiv j$. Solving $j = h_g(c_g - c_{\mathrm{sf}})$ for c_{sf} and inserting this into $j = D_{\mathrm{diff}}^{\mathrm{ox}} \frac{c_{\mathrm{sf}} - c_{\mathrm{ox}}}{d_{\mathrm{ox}}}$ allows solving for c_{ox}. Finally, this is inserted into $j = k_{\mathrm{ox}} \cdot c_{\mathrm{ox}}$ and the resulting equation is solved for j, which in turn is equal to $N\frac{\partial d_{\mathrm{ox}}}{\partial t}$. This leads to the following first-order differential equation:

$$N\frac{\partial d_{\mathrm{ox}}}{\partial t} = \frac{c_g}{\frac{1}{k_{\mathrm{ox}}} + \frac{1}{h_g} + \frac{d_{\mathrm{ox}}}{D_{\mathrm{diff}}^{\mathrm{ox}}}}, \tag{3.2}$$

which—rewritten as $\frac{\partial d_{\mathrm{ox}}}{\partial t} = \frac{B}{A + 2d_{\mathrm{ox}}}$—can immediately be integrated leading to

$$d_{\mathrm{ox}}(t) = \frac{A}{2}\left(\sqrt{1 + \frac{4B}{A^2}(t + \tau)} - 1\right). \tag{3.3}$$

The factor τ takes an existing oxide thickness (e. g., the native oxide) into consideration. For rather short oxidation times, Equation (3.3) can be approximated with $d_{\mathrm{ox}} \approx \frac{B}{A}(t+\tau)$ and for long times $d_{\mathrm{ox}} \approx \sqrt{B(t+\tau)}$. B and B/A have been found to be of Arrhenius type with $B = C_1 \exp(-E_1/k_B T)$ and $B/A = C_2 \exp(-E_2/k_B T)$. Table 3.2 specifies values for the parameters $C_{1,2}$ and the activation energies $E_{1,2}$ in the case of wet and dry oxidation for (100) and (111) silicon surfaces. Note that $E_1^{\mathrm{dry}} > E_1^{\mathrm{wet}}$ is consistent with the fact that the different diffusion behavior of O_2 and H_2O through SiO_2 is responsible for the difference in the oxidation rate. Moreover, the oxidation of the {111} silicon crystallographic planes is higher compared to the {100} plane due to the larger density of Si atoms on {111} planes (cf. Figure 2.26).

For oxide thicknesses below approximately 35 nm, it is known that the Deal–Grove model underestimates the oxidation; in the case of thin oxides, there is a rapid initial oxidation phase, which is not well understood. However, phenomenological extensions of the Deal–Grove model allow for the description of thin SiO_2 (see, e. g., [187]). Indeed, Figure 3.3(b) shows d_{ox} of a thin SiO_2 grown with rapid thermal oxidation (RTO) at $T = 1050\,°\mathrm{C}$ as a function of oxidation time. With the appropriate parameters B and B/A (significantly larger than the values provided in Table 3.2), excellent qualitative agreement with the Deal–Grove model is obtained.

Table 3.2: Parameters for dry and wet thermal oxidation according to the Deal–Grove model.

	B((100) Si)	B/A((100) Si)	B((111) Si)	B/A((111) Si)
dry	$C_1 = 7.72 \cdot 10^2\ \frac{\mu m^2}{h}$	$C_2 = 3.71 \cdot 10^6\ \frac{\mu m}{h}$	$C_1 = 7.72 \cdot 10^2\ \frac{\mu m^2}{h}$	$C_2 = 6.23 \cdot 10^6\ \frac{\mu m}{h}$
	$E_1 = 1.23$ eV	$E_2 = 2.00$ eV	$E_1 = 1.23$ eV	$E_2 = 2.00$ eV
wet	$C_1 = 3.86 \cdot 10^2\ \frac{\mu m^2}{h}$	$C_2 = 0.97 \cdot 10^8\ \frac{\mu m}{h}$	$C_1 = 3.86 \cdot 10^2\ \frac{\mu m^2}{h}$	$C_2 = 1.63 \cdot 10^8\ \frac{\mu m}{h}$
	$E_1 = 0.78$ eV	$E_2 = 2.05$ eV	$E_1 = 0.78$ eV	$E_2 = 2.05$ eV

3.2.2 Local and Geometry-Dependent Oxidation of Silicon

As already mentioned above, the rate of oxidation depends on temperature, oxidation method (wet/dry), crystallographic orientation and on the doping level (not discussed here). In addition, the oxidation rate also depends on the geometry of the structure that is oxidized. Due to compressive strain, the oxidation is substantially lower in corners. The effect is more pronounced in concave than in convex corners and stronger at low oxidation temperatures.

The oxidation can also be done locally by covering parts of the silicon that should not be oxidize with an appropriate diffusion barrier layer for which silicon nitride (Si_3N_4) has proven most suitable. In the so-called local-oxidation-of-silicon (LOCOS) process, a nitride layer is patterned, and subsequently the sample is oxidized. Combining a geometry dependence with local oxidation of silicon allows realizing Si nanostructures that can be used for nanoelectronics devices and are difficult to realize with other processes.

The main panels of Figure 3.4 show two cross-sections of thermally oxidized V-grooves realized in bulk-Si with anisotropic silicon etching (see Section 3.6.4). In (a), a rather thick nitride (~100 nm), and in (b), an ultrathin silicon nitride (~3 nm) was used. In both cases, it is apparent that oxygen diffusion through the layer has been suppressed. In (a), the diffusion at the interface between the silicon and the nitride is suppressed, too. This is expected since the nitride was deposited onto an HF-dipped silicon wafer such that the nitride is in intimate contact with the silicon (i. e., without a so-called pad oxide). In contrast, in the case of (b), oxygen can diffuse from the (111)-flanks because the nitride is ultrathin, and hence mechanically flexible leading to the formation of the

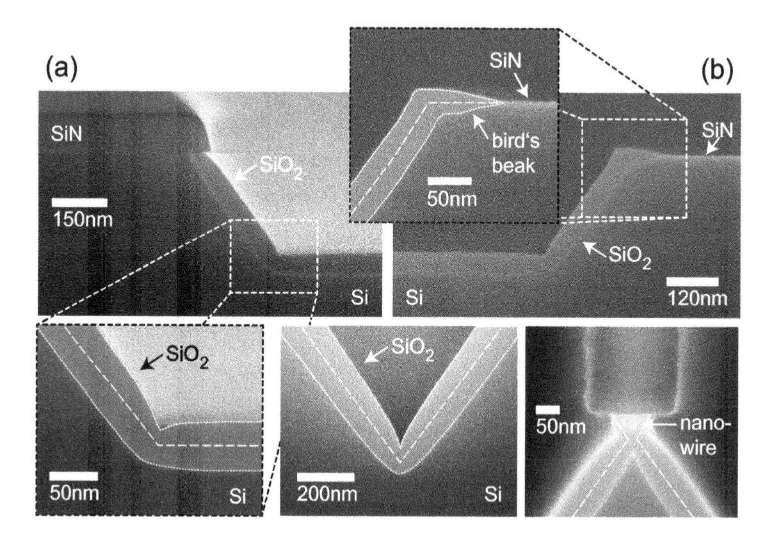

Figure 3.4: Local oxidation of a TMAH-etched V-groove with a thick SiN (a) and an ultrathin SiN (b). The inset of (b) shows the so-called bird's beak in the case of thin SiN. The lower panels show cross-sections after oxidation; the dashed lines indicate the original silicon structure.

so-called bird's beak (see inset of Figure 3.4(b)). The lower panels of Figure 3.4 show how compressive strain in convex corners leads to a reduction of the oxidation thickness. Finally, the combination of an anisotropically etched Si structure with LOCOS shows the formation of a nanowire with ~20 nm diameter (lower right panel).

3.2.3 Chemical Oxidation of Silicon

Extremely thin silicon dioxide layers can also be produced by oxidizing silicon wet chemically. As was discussed in Section 3.1, piranha, SC1 or SC2 can in principle be used to oxidize silicon resulting in a chemical oxide with a thickness of approximately 1–2 nm. However, the preferred way to oxidize silicon chemically is the use of nitric acid (called NAOS=nitric acid oxidation of silicon). The reason for the strong oxidizing capability of nitric acid even at temperatures as low as room temperature is the decomposition reaction

$$2HNO_3 \rightarrow 2NO + H_2O + 3O, \tag{3.4}$$

which provides atomic oxygen for the oxidation of silicon. NAOS can be carried out with different concentrations of HNO_3, at different temperatures, and in addition, oxidation is possible either by immersing the sample into the nitric acid [151] or using acid vapor [132]. Interestingly, NAOS can be described very well (apart from the thinnest oxides below ~0.8 nm) with a Deal–Grove-like model, that was originally used to describe the thermal growth of silicon nitride [274]. In this model, the grown oxide thickness d_{ox}^{NAOS} is assumed to be much larger than the diffusion length l_{diff} of oxidant through the grown oxide. Thus, d_{ox}^{NAOS} depends logarithmically on the oxidation time according to [274]

$$d_{ox}^{NAOS} = l_{diff} \ln\left(\frac{B \cdot t}{l_{diff}^2 + A/2l_{diff}}\right) \tag{3.5}$$

and not $\propto \sqrt{B \cdot t}$ as in Section 3.2.1.

Figure 3.5(a) shows data for oxidation with HNO_3-vapor (blue) and by immersion into HNO_3 (green) extracted from [151, 132]. The grown oxide thickness can be well described with the logarithmic time-dependence mentioned above as shown with the dashed lines. The different behavior of the curves can be mostly explained with the exponential temperature dependence of the factor B according to $B = C\exp(-E_A/k_BT)$. To fit the experimental data, $\frac{C}{l_{diff}^2 + A/2l_{diff}}$ and l_{diff} have been used as fitting parameters. In order to reproduced the oxide growth at different temperatures, it is sufficient to use the temperature dependence due to the exponential factor and assume only a weak temperature dependence of l_{diff}. E_A is again the activation energy related to the diffusion of the oxidant through the already grown oxide. In the present case, $E_A = 0.14$ eV, i. e., much lower compared to dry or wet thermal oxidation, which can be understood

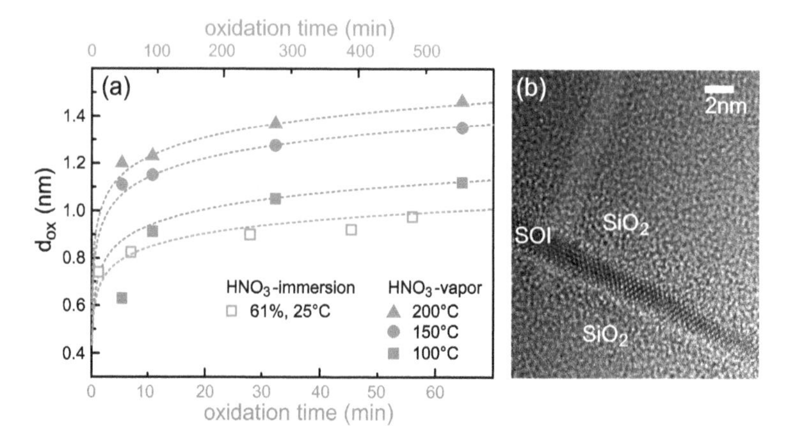

Figure 3.5: (a) SiO_2-thickness as a function of oxidation time using nitric-acid oxidation of silicon. The data was extracted from [151, 132]. (b) Ultrathin-body silicon-on-insulator thinned down with NAOS and HF-stripping [153].

because the oxidant is atomic oxygen compared to molecular oxygen in the dry thermal oxidation and water in the wet thermal oxidation case.

The important point here is that the logarithmic behavior of the oxidation yields basically a self-limiting oxide growth, and hence ultrathin d_{ox} become feasible that can be much better controlled compared to thermal oxidation. Therefore, NAOS is an excellent method for very controlled thinning of silicon structures. An example is displayed in Figure 3.5(b), showing silicon-on-insulator that has been thinned down to ~1.7 nm using multiple NAOS and HF-stripping steps. Moreover, the silicon dioxides grown with NAOS were shown to exhibit excellent insulating properties with even less leakage than their thermally grown counterparts with equal d_{ox} [132, 131].

3.3 Rapid Thermal Nitridation

Ultrathin silicon nitride layers can be grown using rapid thermal nitridation (RTN) in a NH_3/Ar-atmosphere. As mentioned above, the thickness d_{SiN} of the grown nitride can be described by Equation (3.5) with exponential dependence on process temperature (cf. Figure 3.6(a)) and logarithmical dependence on growth time (Figure 3.6(b)). The thickness of experimentally grown SiN layers is shown in Figure 3.6 where spectroscopic ellipsometry has been used to measure d_{SiN} [72, 223]. The exponential temperature dependence (a) together with the logarithmic time-dependence (b) provide excellent process control over d_{SiN}. As a result, even sub-1 nm SiN layers can be grown reproducibly.

It was already mentioned above that silicon nitride is a very effective diffusion barrier and is therefore used for the LOCOS process and also to encapsulate chips to prevent them from being exposed to atmosphere. Interestingly, even sub-1 nm thin SiN allows suppressing the oxidation of silicon during the deposition of a high-k gate dielectric.

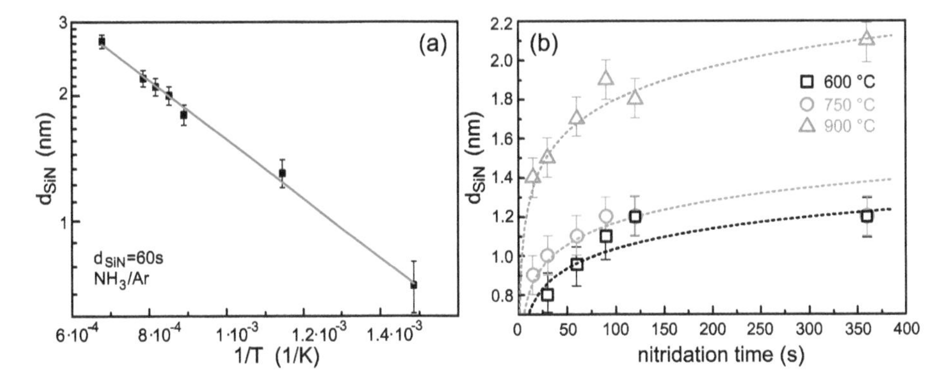

Figure 3.6: (a) Growth of SiN as a function of temperature (for constant time) using rapid thermal nitridation in Ar/NH_3 atmosphere. The thickness of the various SiN layers is measured by spectroscopic ellipsometry. (b) Measured SiN thickness (symbols) as a function of time for three different RTN growth temperatures. The lines are fits using Equation (3.5).

This property can, for instance, be used to engineer the silicon gate dielectric interface to improve the performance of cryogenic MOSFETs as will be discussed in Chapter 11.

3.4 Wafer Bonding

The fabricating of a device basically consists of a suitable sequence of deposition, lithography and etching process steps in order to achieve a certain functionality. A major difficulty in this respect is the combination of crystalline and amorphous materials. While it is usually not a problem to deposit an amorphous material on top of a crystalline one, the other way around is not possible or only in a polycrystalline form (such as the poly-Si gate electrodes). However, quite often it is desirable to have a crystalline layer of material on an amorphous layer. The most prominent example of this is arguably silicon-on-insulator (SOI) substrates. While different methods to realize SOI are reported in literature, the highest quality of such substrates is achieved using direct (also called fusion) wafer bonding.

During wafer bonding, two materials with very smooth surfaces are brought into contact with each other. The surface needs to be preprocessed; for instance, in the case of Si one distinguishes between hydrophilic (HL) and hydrophobic (HB) bonding depicted in Figure 3.7(a). When carrying out HL bonding, there is a SiO_2 layer on both wafers, and consequently, water is chemisorbed on the surface. In HB bonding, the oxide layers are removed with HF and both surfaces are hydrogen passivated.

Bonding is usually done in a tool that allows to separate the two wafers and keep them at a rather small distance from each other. Then, under vacuum, the center of one wafer is pressed against the second wafer. Subsequently, van der Waals forces (HL and HB bonding) and polymerization of silonal due to chemisorbed water at the wafer sur-

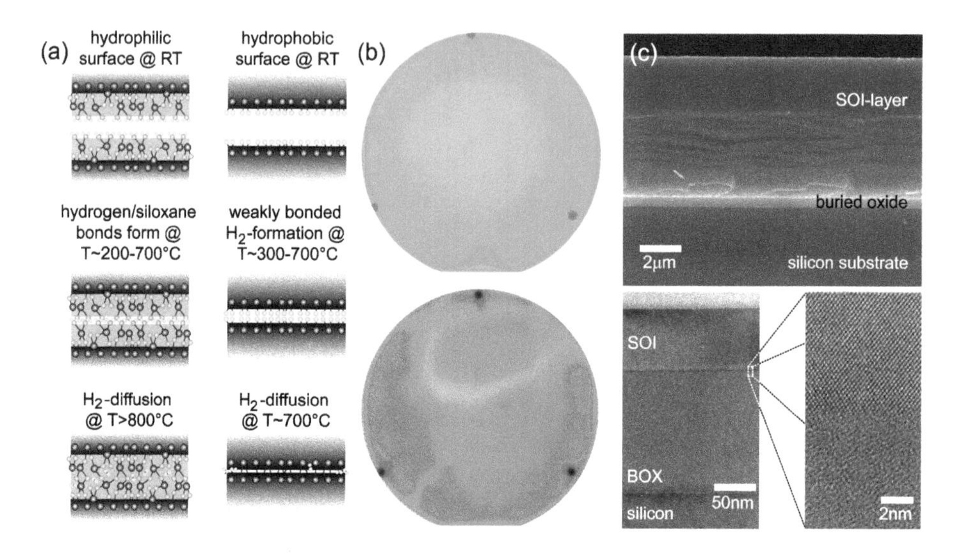

Figure 3.7: (a) Hydrophilic (left) and hydrophobic (right) wafer bonding. (b) Results of hydrophilic bonding of two Si 4″ wafers imaged with an infrared camera. The top image shows proper bonding, the lower a failed bonding process. (c) Cross-section electron micrograph of homemade SOI (top) and transmission electron microscopy image of commercially available SOI (bottom).

face (HL bonding) lead to a spontaneous bonding that starts from the center of the wafer and extends toward the edges. Retracting the separators results in a bonding wave that yields a void-free (if properly done) bonding. The two wafers are pressed against each other and while exerting a substantial pressure the bonded pair of wafers is annealed at temperatures up to ~200–300 °C. Finally, in order to strengthen the bond, the wafers are annealed at elevated temperatures.

Figure 3.7(b) shows infrared images of results after HL bonding of two 4″ Si wafers. The top image shows proper, void-free bonding (the three dark blue spots at the wafer edges are the wafer supports of the measurement apparatus) apart from an area around the main flat whereas the lower image displays a failed bonding with voids and reduced contact area. A cross-sectional electron micrograph of a homemade SOI wafer is shown in the top panel of Figure 3.7(c); the lower panel displays a transmission electron microscopy image of a commercial SOI wafer showing the excellent quality of the SiO_2/Si (i. e., amorphous to crystalline) interface.

Apart from a void-free bonding, the bonding strength is the decisive parameter to quantify the bond process. The bond strength is strongly increased with an anneal at elevated temperatures in HL and HB bonding. Hydrophilic bonding has the advantage that it provides a larger bonding strength below annealing temperatures of ~500 °C. In addition, the hydrophilic surface is less prone to be contaminated by particles (see Section 3.1.3). The drawback, however, is that in order to reach the full bonding strength annealing temperatures >800 °C are required; HB bonding on the other hand already reaches the full bonding strength at approximately 700 °C [216]. A very interesting way to facilitate low temperature bonding is a preprocessing with an oxygen plasma activation, which was shown to yield the full bonding strength already at a temperature as low as 105 °C [44].

3.5 Lithography

Lithography is one of the most important and most sophisticated process steps, needed to print patterns into a resist that will subsequently be transferred into the substrate or into a layer deposited on the substrate. Interestingly, for more than two decades industry had been printing features substantially smaller than the wavelength of the used light source (only recently the change to extreme UV lithography succeeded). This has been made possible by the so-called projection lithography equipped with a number of resolution-enhancing techniques including off-axis illumination, phase-shift masks, optical proximity correction, immersion lithography, and double-patterning. Since the present book is intended as a text-book for (under)graduate students, projection lithography is not dealt with here but the book concentrates on techniques that are relevant for students' daily work, i. e., contact lithography with mask aligners, optical lithography with laser scanners and electron-beam lithography. However, recently industrial resolution enhancement techniques, in particular phase-shift masks and optical proximity correction [269, 260], have also been adopted for contact lithography with mask aligners and will therefore be briefly mentioned. Finally, the so-called spacer patterning, which is independent of the actual lithography process, will be discussed.

3.5.1 Resist Coating

Photoresist coating is done by spinning the resist onto the sample. Prior to this, the sample needs to be dehydrated by baking it on a hot plate for several minutes. In the case of SiO_2 surfaces, an adhesion promoter such hexamethyldisilacane (HMDS) should be used in order to turn the hydrophilic SiO_2 surface hydrophobic. Afterwards, the sample is mounted onto a vacuum chuck. After prespinning at low spinning speed, the sample is accelerated to the final spinning speed ω in order to evenly distribute the resist across the sample. In the central part of the sample, a uniform resist thickness d_{pr} is obtained whose thickness is proportional to $1/\sqrt{\omega}$. Finally, a soft-bake of the samples is carried out in order to dry the resist, drive out solvents and enhance the contrast (cf. Section 3.5.2.2). Temperature and time depend on the specific resist in use; either a hot plate or a convection oven can be used.

At the edges of the sample, in particular at sharp edges and even more in corners, the resist thickness can be significantly larger. This so-called edge bead should be re-

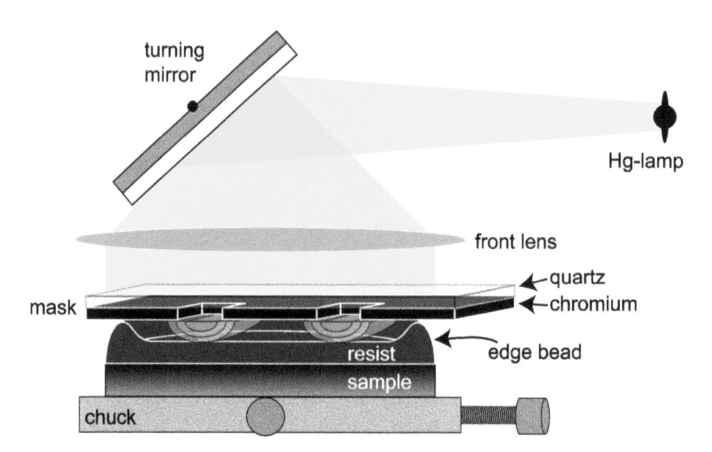

Figure 3.8: Schematic of a mask aligner. A resist coated sample (with edge bead) is brought into contact with the mask. A turning mirror or an array of light emitting diodes "switches" the light on and off resulting in very precise exposure times.

moved when doing contact lithography. If not removed, the resolution of the lithography process will be significantly deteriorated because the edge bead results in a several μm large separation between mask and the central region of the sample (see Figure 3.8), and hence in a blurry mask image within the resist due to diffraction. Removing the edge bead can either be done with an additional lithography process with high exposure dose and short development time or simply with a q-tip soaked in acetone. A successful removal can be easily observed with Newton interference rings across the entire sample when contact between sample and mask is made.

3.5.2 Optical Lithography

In a university clean room environment, optical lithography is usually carried out either as contact lithography with a mask aligner or contactless with a laser scanner. A mask aligner allows adjusting a resist-coated sample to be aligned with respect to a static mask that consists of a quartz glass plate with chromium patterns on it as illustrated in Figure 3.8. After alignment, sample and mask are brought into firm contact facilitating the printing of a one-to-one image of the chromium mask pattern into the resist by exposing the mask/sample sandwich to UV light with a certain dose. The resolution limit of the lithography pattern is determined by the diffraction of the UV light, as illustrated in Figure 3.9(a); here, Huygen's principle was used to compute the cross-section of the dose pattern shown in the lower panel. One clearly observes an exposure of the resist underneath the mask pattern due to diffraction. Minimum feature sizes of a typical contact lithography with mercury vapor lamps (wavelength $\lambda_{\text{g-line}} = 436\,\text{nm}$ or $\lambda_{\text{i-line}} = 365\,\text{nm}$) are therefore 0.5–1 μm. To minimize diffraction, it is very important

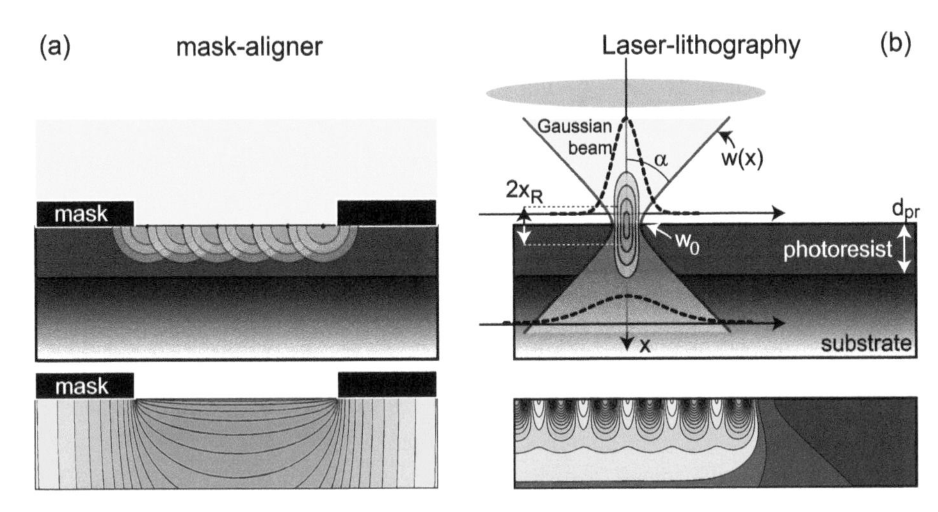

Figure 3.9: Resolution limit of optical lithography due to diffraction at the mask patterns (a); Huygen's principle was used to compute the cross-sectional dose pattern (interference has been neglected in the present case) shown in the lower panel. In the case of laser lithography illustrated in (b), the minimum beam waist of the Gaussian beam and the optical proximity effect associated to it limit the resolution. The cross-section of the dose profile depicted in the lower panel was computed as the overlap of six line exposures.

that the sample surface is in intimate contact with the mask. Hence, edge bead removal after the photoresist coating is essential for optimal pattern transfer. The direct contact between mask and sample requires frequent cleaning of the mask to remove debris and contamination.

While contact lithography is a parallel, and hence quick process, there is no flexibility since the required mask cannot be modified anymore. Laser lithography, on the other hand, is a maskless, and thus contactless, flexible yet slow lithography method since it is a serial process. The result of a laser lithography process is directly related to the shape of the laser beam. Here, a Gaussian beam with a beam waist (i. e., the diameter of the focal spot) w_0 and beam edge $w(x) = w_0\sqrt{1+(\frac{x}{x_R})^2}$ (cf. Figure 3.9(b)) with Rayleigh length $x_R = \frac{\pi w_0^2 n}{\lambda}$ is assumed. The resolution of laser lithography is then limited by w_0 and the optical proximity effect (see below). To determine the resolution, let us first compute the sine of the angle α of the beam divergence, which is the numerical aperture $N.A. = \sin(\alpha)$, given by $N.A. = \frac{w(x)}{\sqrt{x^2+w(x)^2}}$ (with refractive index $n = 1$). Next, the beam edge for large x becomes $w(x) \approx \frac{w_0 x}{x_R}$. Inserting this into the expression for $N.A.$, one obtains $N.A. \approx \frac{w_0}{\sqrt{x_R^2+w_0^2}} \to \frac{\lambda}{\pi w_0}$ since $\frac{x_R^2}{w_0^2} \gg 1$. The last expression means that $w_0 \approx \frac{\lambda}{\pi N.A.}$. Therefore, the resolution of laser lithography can be improved by choosing a laser source with small wavelength λ and focusing the beam with a lens with large numerical aperture. However, care has to be taken when the numerical aperture is large:

the Rayleigh length multiplied with a factor of 2 can be interpreted as the depth of focus DOF, and thus, DOF $= 2x_R \propto \frac{\lambda}{(N.A.)^2}$; this is illustrated in Figure 3.9(b) where the lines around the center of the beam represent locations of constant intensity. As a result, increasing $N.A.$ requires a strong reduction of the resist thickness to maintain $d_{\mathrm{pr}} \leq x_R$ (note that in Figure 3.9(b) the resist thickness would be too large).

The resolution of laser lithography is also determined by the so-called optical proximity effect. Due to the Gaussian beam shape, a substantial undeliberate exposure of the photoresist with UV light outside the constant dose contours is obtained as depicted in Figure 3.9(b). Exposing at several spots in close proximity (six in the case illustrated in (b)), this undeliberate dose adds up resulting in the dose profile shown in the lower panel of Figure 3.9(b). Eventually, the added dose may exceed the threshold dose needed for the developer to attack the photoresist. The undeliberate exposure needs to be taken into consideration when small feature sizes with high pattern fidelity are printed. This can either be done iteratively by adjusting the dose after inspection of the lithography result in a scanning electron microscope or by adjusting the dose distribution within the computer generated file of the mask employing, e. g., a (non-negative) least-square optimization as is done in Section 3.5.3.1.

3.5.2.1 Positive- and Negative-Tone Lithography Processes

A lithography step is used in order to prepare a resist structure that in turn is used to pattern the substrate (or parts of it). As an example, let us assume we want to prepare a metallic contact pad onto a substrate. In principle, this can be accomplished in two different ways: first, a so-called positive process can be carried out as illustrated in Figure 3.10. In this case, the metal thin film is deposited first onto the substrate followed by

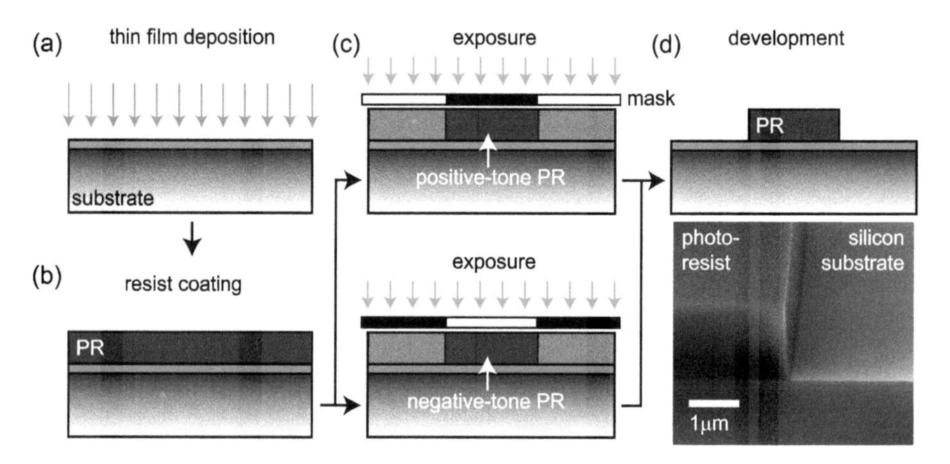

Figure 3.10: Pattern transfer after thin film deposition (a) and resist coating (b) with a positive (top panel of (c)) and negative (bottom panel of (c)) lithography process. The exposed (positive) or unexposed (negative) areas are dissolved during the development. The resist pattern can be used as mask to etch the thin film.

coating with a positive-tone resist; subsequently, a lithography is carried out where the specific dose depends on the chosen resist. In the case of a positive-tone resist, the areas where the mask is transparent and UV light hits the photoresist are washed away in an alkaline developer solution. The remaining resist structure can then be used as an etch mask to pattern the metallic film either with wet chemical or dry etching. Afterwards, the resist mask is removed, which completes the pattern transfer. As a second alternative, a very similar process but with so-called negative-tone resist can be done. In this case, the areas that are exposed initially remain on the substrate whereas the area that was not exposed to UV light is dissolved in the developer. In order to obtain the same metal pattern as in Figure 3.10, a mask with inverted mask pattern is needed.

3.5.2.2 Resist Development and Resist Contrast

A very important figure of merit for a resist is its contrast. While in the case of an excellent lithographic pattern transfer even a resist with rather low contrast yields good results, a resist with high contrast can compensate for a blurry lithography image thereby extending the resolution to smaller pattern dimensions. To elaborate further on this, let us consider a positive-tone resist with thickness d_{pr} in the following. During exposure, the long-chain polymers of the photo-active component of the resist will be disintegrated such that they can be dissolved in an alkaline solution (developer). Let us define an exposure dose D_0, below which the resist basically stays fully intact. The remaining resist thickness $d_{\text{pr}}^{\text{remain}}$ after development can be expressed as [279]

$$d_{\text{pr}}^{\text{remain}} = \frac{d_{\text{pr}}}{1+\eta}\left[e^{-(D-D_0)/A} + \eta e^{-(D-D_0)/B}\right] \tag{3.6}$$

where D is the deposited dose and η, A, B are fitting parameters.[8] Moreover, one can define an exposure dose D_c (dose to clean) where $d_{\text{pr}}^{\text{remain}} \to 0$. Let us assume that for a resist of ~1 µm thickness the dose to clean is reached when $d_{\text{pr}}^{\text{remain}} < 1\,\text{nm}$. In this case, a contrast γ of the photoresist can be extracted to be

$$\gamma = \frac{0.999}{\log(D_c) - \log(D_0)} \approx \left[\frac{\log(D_c)}{\log(D_0)}\right]^{-1}. \tag{3.7}$$

The closer the levels D_0 and D_c are the larger γ, i. e., the better the contrast of the resist. In the limit of $D_c = D_0$, the contrast becomes infinite. Typically, $\gamma \approx 2$ is considered a good contrast; for PMMA (cf. Section 3.5.3.2) $\gamma \approx 5$–10.

Figure 3.11 shows two blurred exposure dose curves $D(x)$. Unintended exposure occurs due to diffraction and optical proximity effect as discussed in Section 3.5.2. If a photoresist is used with a low contrast of $\gamma = 1.17$, the blurred dose profile such as the

8 Note that in [279] Equation (3.6) is stated for a negative resist.

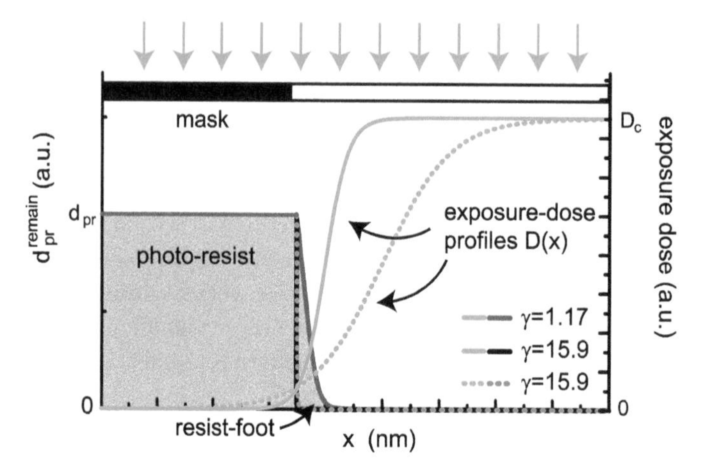

Figure 3.11: Resist profile calculated using Equation (3.6) with the dose profiles $D(x)$ shown with the yellow straight and dotted lines. Resists with two different contrasts $\gamma = 1.17$ and $\gamma = 15.9$ are assumed. A low contrast leads to a resist foot whereas a high contrast yields excellent pattern fidelity even if the dose profile is strongly broadened (dotted yellow line).

straight yellow line yields a resist profile as shown in the bottom left with a pronounced resist foot (dark brown line) and a resist side wall inclining an angle smaller than 90°. However, if the contrast could be increased to, e. g., $\gamma = 15.9$ the resist flank becomes very steep and the foot has almost vanished (black line). The excellent result remains even if the broadening of the exposure dose profile is increased (dotted lines). Therefore, care has to be taken when doing an optical lithography process to obtain the highest possible contrast. Factors that diminish contrast are, for instance, a too low humidity of the clean room air and the use of concentrated developer.

3.5.2.3 Image Reversal Process

A third alternative for pattern transfer with lithography is the so-called image reversal process. While positive/negative processes yield reliable and reproducible results, they require the availability of a highly selective etch mechanism to transfer the photoresist pattern into the thin (metal/dielectric) film without attacking the substrate underneath. Often, however, an etch process with suitable selectivity does not exist, or is not available in the lab. In this case, a so-called image reversal and lift-off process can be used to pattern a thin film. An image reversal process requires the use of a suitable photoresist (for instance, AZ5214E).

The process is illustrated in Figure 3.12 and starts with a short initial exposure step. This step should be substantially shorter than in the case of a positive process: the aim here is to obtain a gradient of the UV dose in the direction perpendicular to the substrate surface. After the exposure, a so-called reverse bake on a hot plate is carried out that activates a cross-linking agent within the exposed volume of the photoresist making it

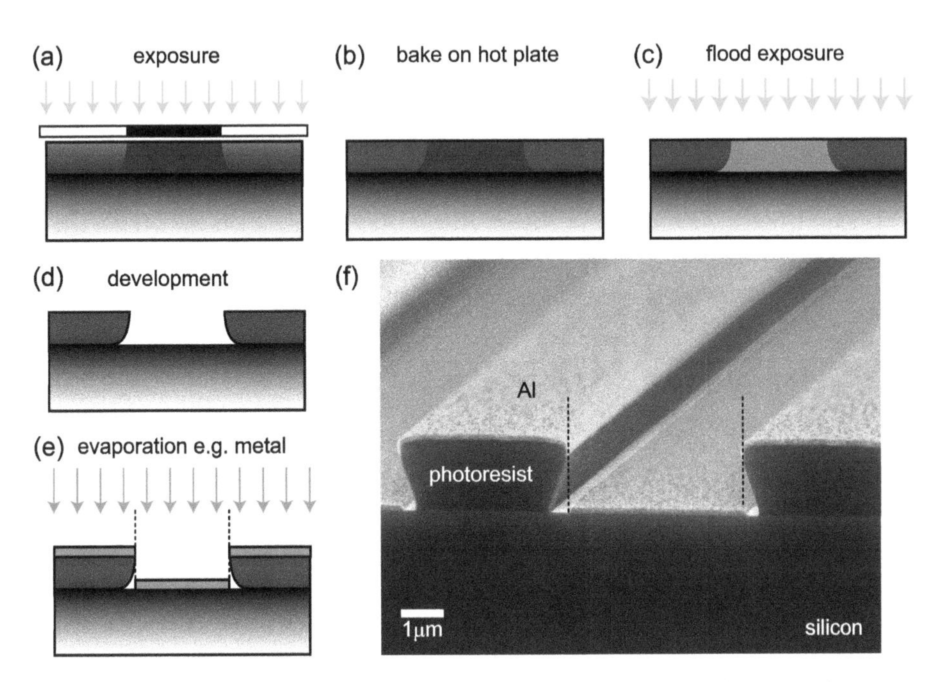

Figure 3.12: Schematic process flow ((a)–(e)) of an image reversal process that leads to inverted resist flanks ideally suited for a lift-off process. (f) Electron micrograph of a sample with resist exhibiting inverted flanks. A metal film is deposited on top with electron beam evaporation.

almost insoluble in the developer. Finally, a flood exposure is done so that the areas shadowed by the mask pattern during the first lithography step will be exposed, and thus dissolved in the subsequent development.

An image reversal process yields a resist pattern with inverse resist flanks (illustrated in Figure 3.12(d), see also (f)) in contrast to a positive process where the resist flanks are usually at best vertical and exhibit a more or less pronounced resist foot (cf. Figure 3.11); see QR code #29 for the progress toward inverse resist flanks during the resist development. If a nonconformal deposition process such as electron-beam evaporation (Section 3.8.1) is used, where the evaporated metal hits the sample at ~90 ° with respect to the surface and if the thickness of the deposited material (usually a metal) is substantially below the thickness of the resist, the deposited metal film will be discontinuous across the resist step (cf. Figure 3.12(e) and (f)). As a result, when immersed in a solvent such as acetone, the photoresist and with it the superficial metal on top of the photoresist will be washed away leaving the (approximately) same metal pattern as with a positive process with the same mask but without the necessity of etching the metal.

In order to obtain a resist pattern suitable for a lift-off, the dose of the initial exposure and the time/temperature of the reverse bake are decisive for the final resist cross-section. Figure 3.13 shows resist cross-sections obtained after varying the time of

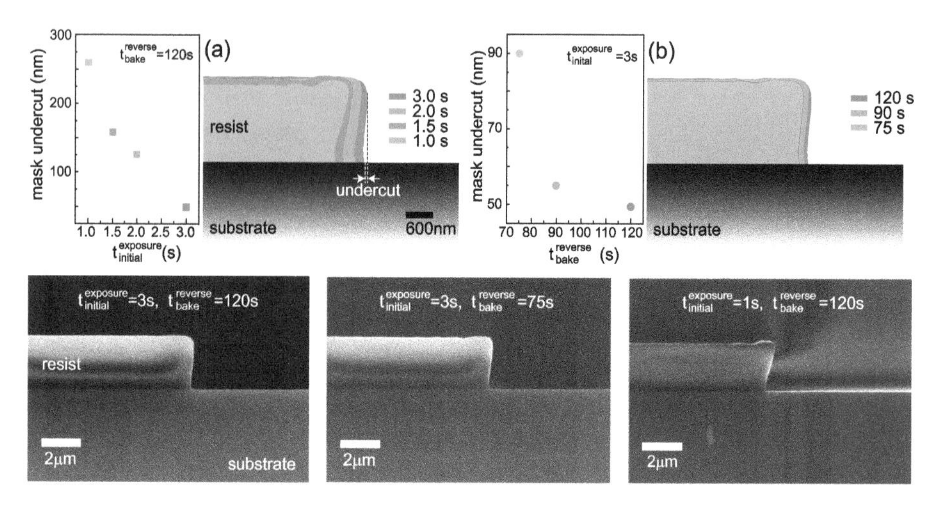

Figure 3.13: Inverted resist flanks realized with an image reversal process with different process parameters. (a) Dependence of the mask undercut on the time of the initial exposure with a reverse bake of 120 s. (b) Mask undercut as a function of reverse bake time for an initial exposure time of 3 s. The electron micrographs show exemplary resist cross-sections (on a silicon substrate) for the conditions stated in the figures.

the initial exposure (a) and the time for the reverse bake (b) (all other parameters are given in the figure). Note that the colored resist cross-sections are one-to-one copies of cross-sectional electron micrographs (as shown in the lower panels); the schematic presentation has been chosen for clarity. From the resist cross-sections, the mask undercut can be extracted. Obviously, reducing the initial exposure dose and the time of the reverse bake yield resist flanks better suited for lift-off. But one has to keep in mind that both measures also lead to a reduction of the resist thickness and, more importantly, to a lateral reduction of the resist mask (leading to larger metal patterns). However, if the exact size of the metal pattern is not important—as is often the case when the metal merely serves as a contact lead—the resist pattern can be tuned in an image reversal process to enable a proper lift-off process.

3.5.2.4 Interference Phenomena During Optical Lithography

In Section 3.5.2, the exposure dose profile within the photoresist due to diffraction and the proximity effect were discussed (see the lower panels of Figure 3.9). However, an important issue, namely interference phenomena, has been neglected so far. When highly reflecting substrates are being used, the formation of a standing wave pattern is very likely. In particular, the monochromatic light in laser lithography leads to rather pronounced standing wave patterns. When a photoresist with high contrast is used, the standing wave patterns can be seen as ripples in the resist side walls after the development. Figure 3.14, left panel, shows such a resist flank after laser lithography and development (the standing wave pattern is illustrated with the dark yellow lines) with

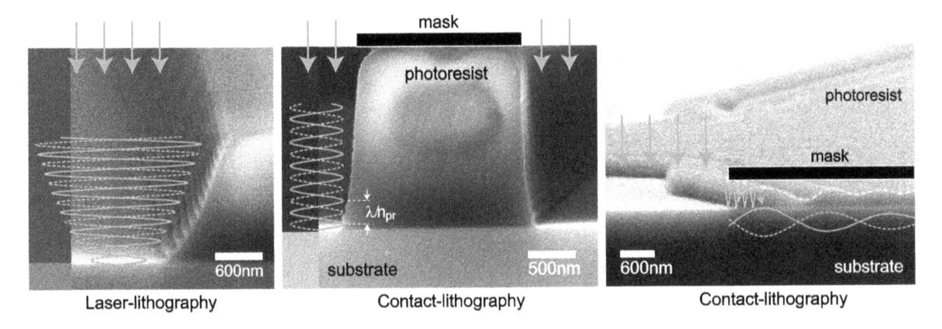

Figure 3.14: Electron micrographs of resist patterns after exposure and development. Standing wave patterns can lead to wavy resist patterns at the side walls and even along a pattern that was covered by the mask.

strongly pronounced ripples. Moreover, the limited depth of focus and a rather thick resist lead to the nonvertical resist flank.

Standing wave patterns in the photoresist also appear during lithography with a mask aligner that are usually equipped with less monochromatic mercury vapor lamps. The center panel of Figure 3.14 shows that at the substrate-resist interface the typical standing wave pattern appears. Moreover, if rather thin resists are used it may also occur that a horizontal mode, guided in-between substrate and mask is generated as illustrated with the dark yellow zig-zag pattern in the right panel of Figure 3.14. A horizontally guided mode can lead to an exposure of the photoresist underneath the chromium pattern. Again, interference leads to a wavy resist pattern shown in the figure (the white dashed line is a guide to the eye).

To diminish standing wave patterns within the resist either an antireflection coating (for instance, as a backside antireflective coating on the wafer surface) and/or a post-exposure bake prior to the development can be employed. The latter leads to a diffusion of the photoactive component of the photoresist that got exposed into regions with low exposure such that the ripple pattern is substantially reduced.

3.5.2.5 Resolution Enhancement

In order to increase the resolution of (optical) projection lithography, the industry implemented a number of measures already mentioned above. Some of them, namely phase-shift masks and optical proximity correction (OPC), have been adopted in contact lithography [269, 260]. Both resolution enhancement techniques exploit the coherence properties of the light source used for lithography.

Constructive interference of light waves diffracted from patterns on the mask (see Figure 3.15, left) yields an intensity, i. e., a lithography pattern, where the structures cannot be resolved anymore if the patterns on the mask are too close to each other (red line). In a phase-shift mask, indents are etched into the mask such that the light waves diffracted at adjacent mask patterns acquire a 180° phase shift. The resulting destructive

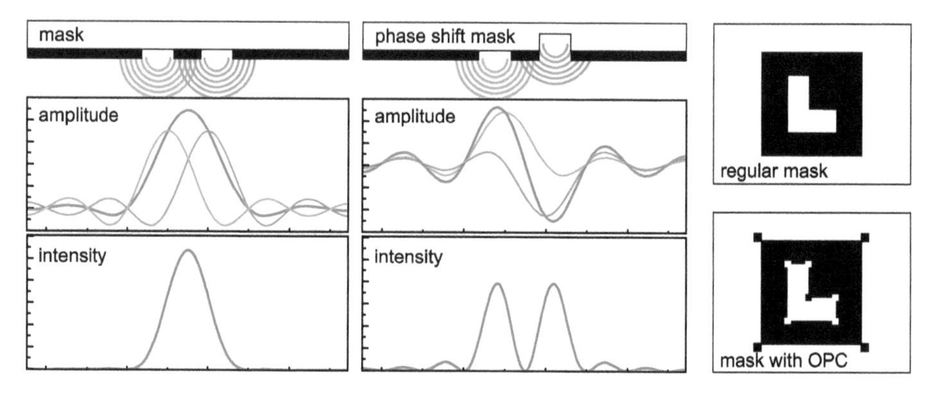

Figure 3.15: Left: Diffracted light from adjacent patterns leads to constructive interference (red line) and as a result, the two patterns cannot be resolved. Center: Indents in a phase-shift mask lead to a 180° phase shift between light diffracted from adjacent mask patterns resulting in an intensity minimum due to destructive interference. Right: Added/removed features on a mask with OPC result in diffraction patterns that better reproduce the desired structures.

interference leads to a minimum intensity in between the structures such that patterns, not resolved with a conventional mask, are clearly separated within the resist.

In the case of OPC, features are added at convex patterns (such as the corner of the square structure shown in Figure 3.15, right) and removed in concave corners as illustrated in the OPC mask in the figure. Diffraction of light at the mask including the additional/removed features will then lead to an improved image printed into the photoresist that better reproduces the initially intended pattern.

Both resolution enhancement techniques can certainly be combined. Phase-shift masks are more expensive than a mask with OPC. On the other hand, while the additional/removed features on an OPC mask can in principle be added easily, OPC usually requires solving a high-dimensional optimization problem to provide the optimum additional/removed features. However, in the case of contact lithography the limited coherence of the light is beneficial in that it limits the lateral size of patterns on the mask that contribute to the diffraction pattern at a particular spot on the sample. Therefore, with some experience, appropriate OPC can be implemented at least in a rudimentary way without sophisticated computations.

3.5.2.6 Resist Residues

Resist residues after the development of photoresist is an important issue. Figure 3.16 (a) shows an electron micrograph cross-section of a developed photoresist mask and one can clearly observe significant resist residues at the edge of the photoresist. During an etch process, these residues would be partially transferred into the substrate resulting in substantial line-edge roughness. Usually, resist residues are removed with a so-called DESCUM process. To this end, the sample is exposed to a short oxygen plasma preferably in a barrel reactor (cf. Section 3.7.5), but it can also be carried out in, e. g., an ICP-RIE tool

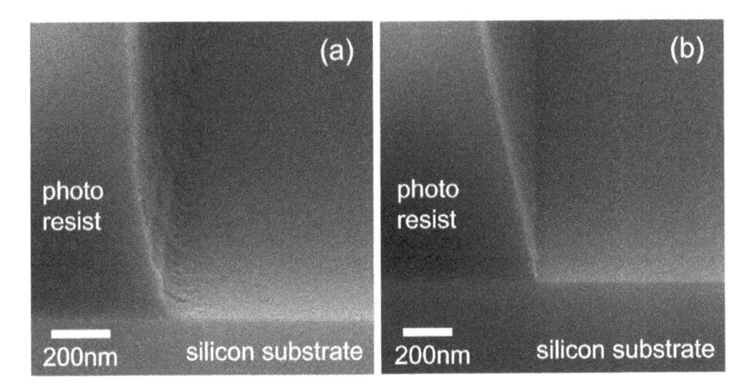

Figure 3.16: Electron micrograph cross-sections of a photoresist pattern on a silicon substrate before (a) and after (b) a DESCUM process. An oxygen plasma in a barrel reactor was used to remove the residues at the resist edge that remained after the development.

with lowest possible RF power. After a DESCUM process, the line-edge roughness will be strongly improved as is shown in Figure 3.16(b).

Resist residues are not only an issue at the edge of the resist but may also play a role for the contact resistance between metal and semiconductor in particular when dealing with semiconductors such as graphene where the use of an oxygen plasma is impossible. For instance, in [38] it was shown that a 3–4 nm thick photoresist residue leads to increased contact resistances that are higher compared to contacts fabricated based on electron-beam lithography and PMMA as resist.

3.5.2.7 Resist Hard Bake

After the development a hard bake is often carried out in order to drive out any remains of solvents in the resist, making it more resistant against an attack during an etch process. In addition, a hard bake improves the adhesion of the resist to the substrate, which is particularly important during wet chemical etching. However, a hard bake may lead to a reflow of the photoresist depending on the temperature and the time of the hard bake, which is usually an unwanted effect. Figure 3.17 shows cross-section electron micrographs of photoresist patterns after a hard bake for 1 min at 100 °C, 105 °C, 110 °C and 120 °C. Apparently, a small difference in temperature can have a rather strong impact on the resulting resist profile.

A rather rounded resist profile as obtained with a hard bake at 120 °C could be problematic during reactive ion etching (RIE): Ion bombardment or a low chemical selectivity of the RIE process may result in severe mask erosion and as such the pattern would be widened during the etch leading to nonvertical etch flanks in the substrate underneath. Extreme examples of mask erosion and its impact on the etch result are shown in Figure 3.36 where very shallow etch flanks are deliberately generated by mixing a certain amount of oxygen to the etch gas to promote mask erosion due to resist ashing.

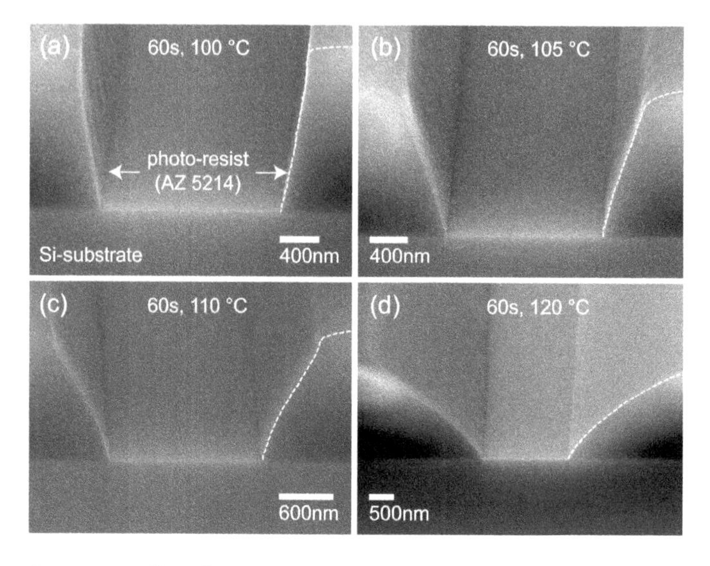

Figure 3.17: Reflow of resist patterns during a resist hard bake for 60 s at different temperatures. The white dashed lines are guides to the eye.

3.5.3 Electron-Beam Lithography

While optical lithography is the method of choice for industrial semiconductor fabrication because of its capability for parallel processing, nanoscale structures are usually realized with electron-beam lithography in research labs. An electron-beam lithography tool is basically a scanning electron microscope (SEM) equipped with an additional beam blank to switch the beam on and off as illustrated in Figure 3.18, left panel. The beam deflection coils, the beam blank and the sample stage are controlled with a single piece of software that allows moving the beam across the sample exposing each part of the sample with a predefined area dose of electrons accelerated into the keV-range.

In the nonrelativistic case (for acceleration voltages ≲100 kV), $eV_{\text{acc}} = \frac{1}{2}m_0 v^2$ where V_{acc} is the acceleration voltage, m_0 the electron mass and v its velocity. With de Broglie's relation $\lambda = \frac{h}{p}$, one obtains

$$\lambda = \sqrt{\frac{h^2}{2m_0 eV_{\text{acc}}}}. \tag{3.8}$$

Typical acceleration voltages are 10–30 kV meaning that extremely small wavelengths of $\lambda < 10$ pm are obtained. With state-of-the-art SEM-columns, a beam spot with a diameter of a few nanometers can be realized enabling in principle the printing of very small structures. However, the real resolution limit of electron-beam lithography is to a large extent determined by the interaction of the electrons with the substrate. Back-scattering of electrons from the substrate and the generation of secondary electrons lead to an ef-

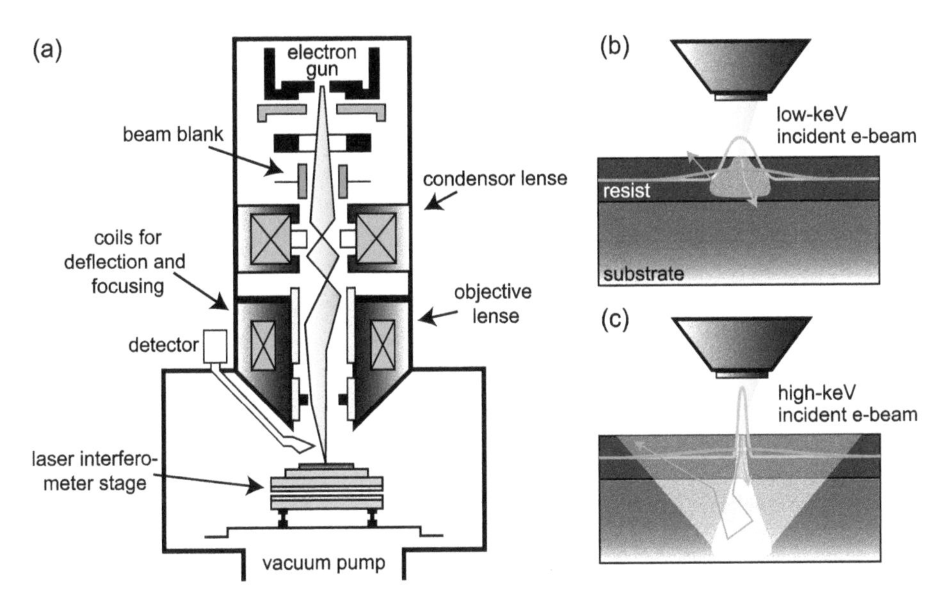

Figure 3.18: Left: Schematic of a typical electron-beam lithography tool. The most important addition compared to a scanning electron microscope is the beam blank that allows switching on and off the beam while maneuvering across the sample. Right panel: Scattering of the primary beam (blue) and the generation of secondary electrons (green) can be described with a double Gaussian beam that leads to an effective exposure of the resist.

fective deposition of electron dose within the resist in an area with a rather large radius around the spot of initial exposure. This proximity effect (cf. Section 3.5.2) will be further elaborated on in the next section.

The pattern transfer obtained with electron-beam lithography depends on the resist type, the electron dose injected into the resist as well as on the resist coating and solvent drive-out and finally the development. Therefore, details along an entire electron-beam lithography process chain will be discussed in the following sections.

3.5.3.1 Proximity Effect and How to Avoid It

As already mentioned above, the resolution of electron-beam lithography is to a large extent determined by the interaction of the electrons with the sample (substrate and resist). When injected into the sample, electrons are scattered depending on their energy and this gives rise to a broadening of the beam. Moreover, electrons may get backscattered within the substrate and leave the sample (and thus lead to an exposure of the resist) at distances rather far away from the location of the incident beam. Panels (b) and (c) of Figure 3.18 illustrate this scenario for a low and high electron energy. This leads to a certain charge background that adds to the overall exposure dose. As a result, each spot on the sample may be exposed directly by incident electrons plus electrons that stem from neighboring exposure spots as well as from the local environment leading to a substantially different printed image within the resist compared to the intended mask pattern. This phenomenon is again called proximity effect and is similar to the optical proximity effect discussed above.

To describe the impact of the proximity effect and find out how to avoid it, proper knowledge of the interaction of the beam with the sample is essential. Most accurate results are obtained with a Monte Carlo simulation. From such simulations, a function $f_{MC}(r)$ can be extracted that provides the distribution of electrons in the resist as a function of the radius r around the initial electron-beam exposure spot. The function $f_{MC}(r)$ can then be used to compute the overall electron dose in certain areas on a sample obtained during a lithography process. A less accurate method that still captures the essentials of the proximity effect is to describe the electron distribution function $f_{DG}(r)$ with a double Gaussian [233]:

$$f_{DG}(r) = \frac{1}{(1+\eta)\pi}\left[\frac{1}{\alpha^2}\exp\left(-\left(\frac{r}{\alpha}\right)^2\right) + \frac{\eta}{\beta^2}\exp\left(-\left(\frac{r}{\beta}\right)^2\right)\right], \tag{3.9}$$

where r is again the radius measured from the center of the beam. α, β and η are fit parameters taken from [206] that are reproduced for convenience in Table 3.3. Using $f_{DG}(r)$, a calculation of the area dose during electron-beam lithography based on a certain mask can be implemented in a straightforward manner.

Table 3.3: Double-Gaussian electron-beam parameters [206].

Beam energy (keV)	**α (μm)**	**β (μm)**	**η**
5	1.33	[0.18]	[0.74]
10	0.39	[0.60]	[0.74]
20	0.12	2.0	0.74
50	0.024	9.5	0.74
100	0.007	31.2	0.74

Figure 3.19(a), left panel, shows a structure consisting of two back-to-back L-patterns (gray area), which is intended to be printed subject to a distribution of exposure spots on a regular quadratic lattice with 4 nm distance between adjacent spots. The patterns have a lateral extension of 15 nm and are 20 nm apart from each other. Here, a 100 keV exposure of a silicon substrate covered with 200 nm PMMA is considered (see Table 3.3 for the used parameters).

In (a), all spots obtain a dose factor of 1, i. e., no proximity correction is applied in this case, leading to the exposure pattern shown in the center panel of Figure 3.19(a). Obviously, the pattern significantly deviates from the intended one with underexposed convex and overexposed concave corners leading to a blurred image. The right panel shows the resist pattern after development (an arbitrary threshold dose for development was assumed here).

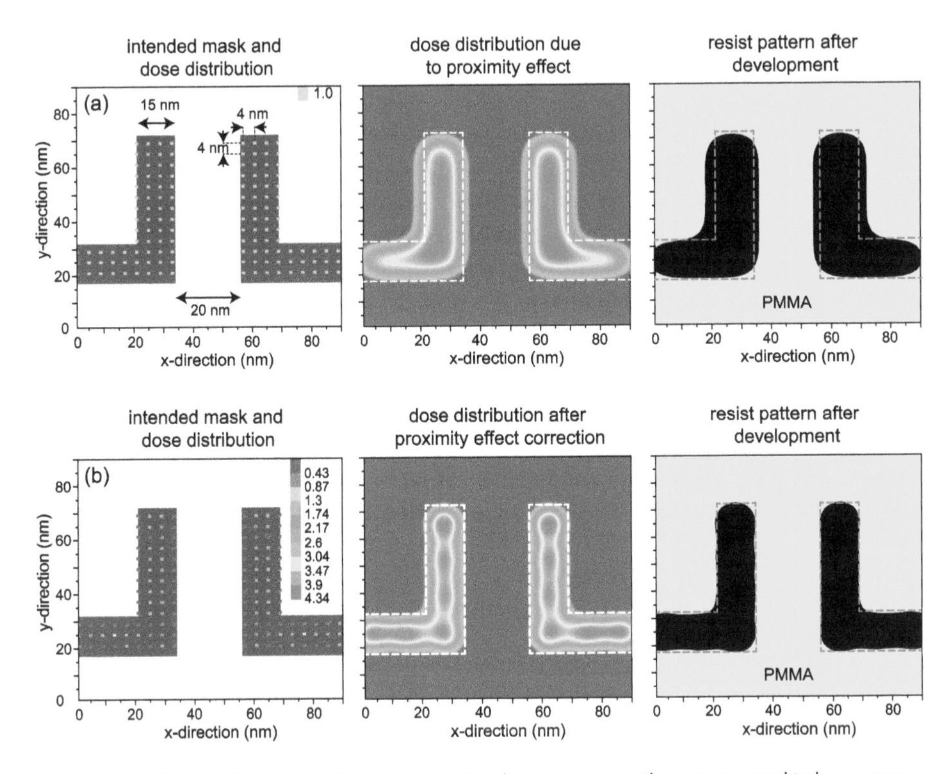

Figure 3.19: Left: Intended mask pattern (gray area) and exposure spots that are assumed to be on a regular 4 nm × 4 nm grid. Simulated dose distribution (center panels) of e-beam lithography with 100 keV electrons based on a double Gaussian distribution. The substrate consists of 200 nm PMMA on silicon. Right: resist pattern after development. (a) Uniform initial dose and (b) optimized dose distribution determined with a nonnegative least-square calculation.

A correction of the proximity effect involves decreasing the exposure dose in areas with neighboring structures and increasing the dose in sections that are rather isolated. In principle, this can be done in an iterative process by carrying out a number of exposures and comparing the obtained results after development with the intended mask pattern (see QR code #30 and try yourself). Alternatively, an optimization algorithm can be used to adjust the exposure dose at each spot in order to realize the best match with the intended mask. Figure 3.19(b), left panel, shows the dose each spot obtains after a proximity correction based on the double Gaussian exposure function Equation (3.9) using a nonnegative least-square optimization scheme that minimizes the sum of the squares of the difference between actual and intended dose pattern; the appropriate dose factors are displayed in the left panel, too. With proximity correction substantially better dose distribution (center panel) and printing results (right panel) are obtained (using the same threshold for full resist development as in (a)). Adapting the dose distribution to the Gaussian beam, the same least-square algorithm can also be used to suppress the proximity effect in laser lithography.

3.5.3.2 E-Beam Resists—PMMA

There are a number of electron beam resists with different characteristics commercially available. The most prominent and mostly used e-beam resist is polymethylmethacrylate (PMMA). PMMA is relatively easy to handle, allows a high resolution and can be used as a positive- as well as negative-tone resist depending on the exposure dose.

After resist coating of a sample, the prebake of PMMA (driving the solvents out of the PMMA) is usually done at temperatures of approximately 165–180 °C for 5–10 min either on a hot plate or in a convection oven. The prebake may have an impact on the line-edge roughness of the printed patterns. The reason for this is that during the prebake aggregations of the polymer can form that exhibit a different solubility in the developer compared to the remainder of the resist. Interestingly, if the prebake is carried out at a rather high temperature of 250 °C for a very short time of 15 s the polymers of the PMMA cannot form aggregations and as a result, a reduced LER has been observed [16]. However, the prebake at such a high temperature needs to be very short since otherwise the PMMA will degrade.

3.5.3.3 PMMA Development

After exposure, the PMMA needs to be developed. Several different solutions can be used to do so. The most common developer is a 1:3 mixture of methylisobutylketone (MIBK) and isopropyl alcohol (IPA). Samples are immersed in the solution for ~30 s and subsequently rinsed for 30 s in pure IPA and blown dry with nitrogen. In particular, when cooled to −15 °C, optimum development characteristics with highest contrast have been found [55].

Alternatively, a 7:3 mixture of IPA and DI water in combination with ultrasonic agitation has proven to yield a superb development [281]. In [204], a 4:1 ethanol/DI water mixture is used, which also leads to excellent results. Instead of rinsing the sample in IPA or pure ethanol, the best pattern fidelity is obtained when simply blowing dry the samples with nitrogen immediately after the development [204].

3.5.3.4 E-Beam Resist Residues

Similar to photoresist, it is important to be aware of resist residues after the development when working with PMMA (and other e-beam resists). These residues are not only found in electron-beam exposed and developed areas but residues also remain on the substrate after simply removing the resist with a solvent such as acetone. The resist residues have a thickness of approximately 1–2 nm [185], and thus may play an important role in the formation of contacts when using a lift-off process.

Numerous publications on resist residues and PMMA removal strategies can be found in literature. A recent study shows that after almost all common treatments, residues can still be found on the substrate [185]. The most effective treatment removing all residues is an oxygen plasma. However, it is often impossible to carry out such a DESCUM process, in particular when 2D materials such as graphene are to be contacted.

With sufficiently high electron dose in the contact areas, the thickness of the residues can be reduced to approximately 0.5 nm. Nevertheless, one needs to keep in mind that even such a thin layer may have an impact on the contact formation and the interaction between semiconductor and metal. Laser cleaning has been successfully employed to locally remove residues [120] and annealing the samples after the lift-off has been found to be rather effective to obtain low contact resistances. Alternatively, a sacrificial interlayer, e. g., a very thin deposited SiO_2, can be used that allows carrying out a DESCUM process with an oxygen plasma. Then immediately before the metal deposition, the sacrificial layer is removed (e. g., with HF in the case of SiO_2). When contacting silicon, this procedure is done automatically, since the native oxide (=sacrificial layer) needs to be removed in order to contact the silicon.

3.5.3.5 Electron Beam Lithography on Insulating Substrates

It was mentioned above that the resolution of electron-beam lithography is to a large extent determined by the interaction of the injected electrons with the substrate. In particular, if an insulating substrate is used, the injected electrons can lead to a strong charging effect that deteriorates the lithography result. A sufficiently large negative charge-up deflects the electron beam such that patterns may even be printed at a different location, as was shown in [121]. To avoid this, several strategies can be used. For instance, a thin conducting layer connected with ground deposited on top of the resist can be employed to reduce excessive charging of the substrate.

An alternative route is to avoid charging in the first place. This can be accomplished by noting that a substrate subject to irradiation with electrons can be charged positively or negatively depending on the energy of the electrons. At very low energies ($\lesssim$100 eV), injected electrons are captured in the insulating substrate leading to a net negative charge. At higher electron energies, incident electrons knock out secondary electrons from the substrate; in this mid-range of energies, both electrons can leave the sample resulting in a net positive charge. Finally, at even higher energies electrons penetrate deeply into the substrate such that they cannot leave it anymore giving rise to a net negative charge. Therefore, two cross-over energies exist where the effects cancel each other leaving the substrate uncharged [121]. Choosing the acceleration energy of the electrons to be equal to a cross-over energy avoids substrate charging effects altogether. The cross-over energy can be found experimentally by taking an SEM image of a certain area of the insulating substrate. Then zooming out, the originally observed area either appears dark, whitish or cannot be observed.[9] In the first case, (dark) the area is positively charged, in the second case it is negatively charged. If it cannot be observed, electrons are injected into the substrate at the cross-over energy, which should be used for the electron-beam lithography.

9 This is a well-known phenomenon when imaging with a scanning electron microscope.

3.5.4 Spacer Patterning

The final patterning technique discussed here is the so-called spacer patterning. The basic idea is to replace the critical part, i. e., the lithographic definition of very narrow, lateral patterns, with a fabrication process that can be controlled with highest precision, namely thin film deposition (which will be presented in detail in Section 3.8). Indeed, using atomic layer deposition (ALD), the thickness of the deposited layer can be controlled down to plus or minus a single monolayer.

For spacer patterning, a sacrificial layer (amorphous Si in Figure 3.20) is deposited onto a substrate and patterned with regular lithography and etching yielding vertical side walls (see Figure 3.20(a)). Then conformal deposition with ALD, low pressure chemical vapor deposition (CVD) or remote plasma-enhanced CVD is used to realize a layer with the same thickness on horizontal and vertical areas (b). Subsequently, anisotropic reactive ion etching (see Section 3.7.1) removes the deposited layer from horizontal areas. But due to the effectively thicker film (with respect to the direction of the anisotropic etching) on the vertical areas a so-called spacer remains (c). Finally, selectively removing the sacrificial layer leaves the spacer as a mask pattern on the substrate (d) whose lateral size is determined by the deposition and not a lithography process. The spacer can now be used as etch mask to transfer the pattern into the substrate underneath. Furthermore, the process can be carried out repeatedly using the spacer itself as the sacrificial structure for a subsequent spacer patterning process. With such a double spacer pro-

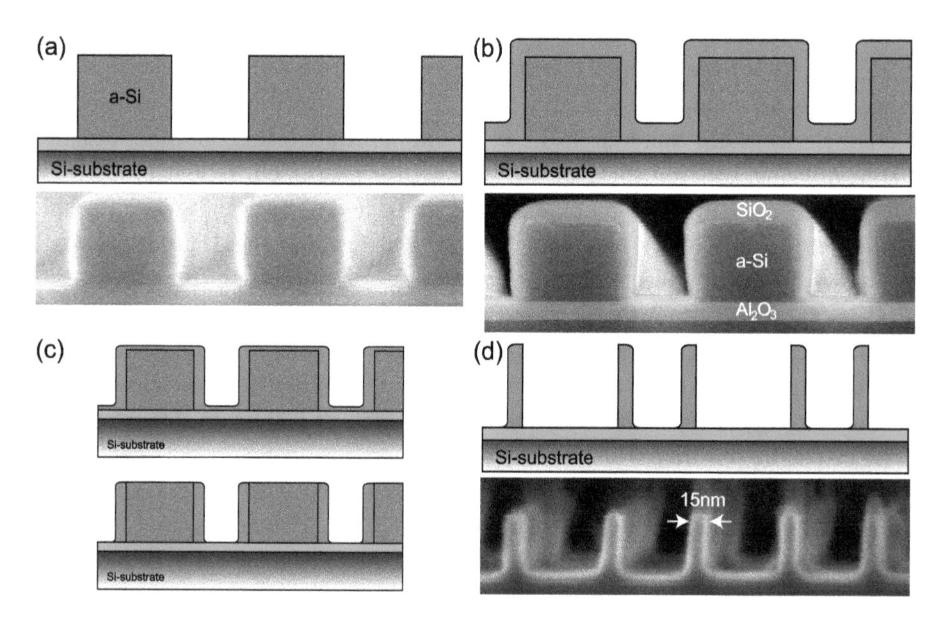

Figure 3.20: Schematic illustration and electron micrographs of a spacer patterning process based on an amorphous Si sacrificial layer and (ICP)PE-CVD-grown SiO_2. After anisotropic reactive ion etching of the SiO_2 and selective removal of the sacrificial layer 15 nm, SiO_2 patterns are obtained.

cess, it is possible to strongly increase the density of mask patterns based on an original lithography pattern that can be substantially larger since each spacer process doubles the number of features.

Apart from enabling the generation of lithography-independent nanoscale structures, spacer patterning has another benefit, which however requires a bit more explanation. Suppose a line pattern is printed using, e. g., optical lithography and etching. There will always be a certain amount of line-edge roughness (LER) at each of the two sides of the line pattern. If we assume that the LER on both sides are completely uncorrelated, there will be a line-width-roughness (LWR) in addition to LER with LWR = $\sqrt{2}$·LER. The top panels of Figure 3.21 illustrate the difference between LER and LWR; here, a sacrificial mask pattern is depicted. Creating a spacer (light blue) and using this as the mask for pattern transfer, allows reducing LWR to almost zero provided a thickness d_{dep} of the deposited spacer layer is chosen that is sufficiently thin (see electron micrographs in Figure 3.21). In this respect, sufficiently thin means the following: the LER (of a lithography process) as a function of the spatial coordinate along the edge of the mask can be Fourier analyzed [107, 92]. Plotting the obtained spectrum as a function of (spatial) wavelength, LWR approaches zero if the deposited spacer thickness d_{dep} is smaller than the wavelength(s) of the dominant frequency (range) since the spacer reproduces the edge of the sacrificial mask on both sides of the spacer. In other words, the LER on both sides of the spacer are fully correlated (see Figure 3.21, lower panel).

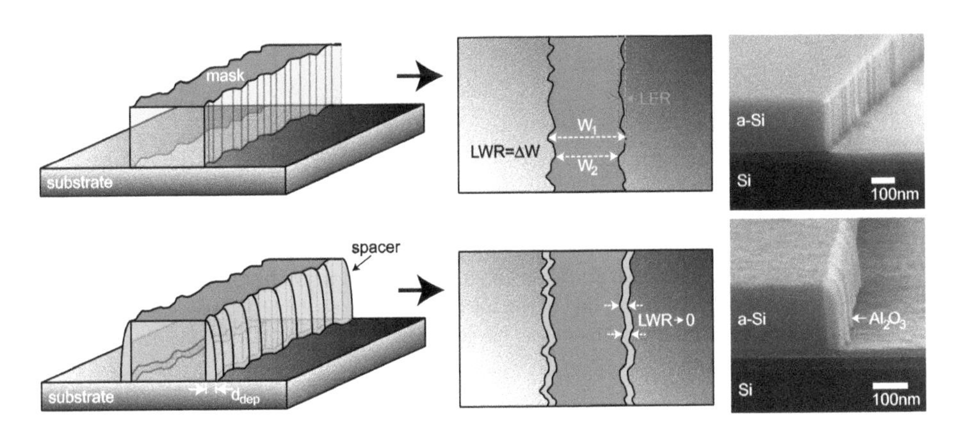

Figure 3.21: Line-edge roughness (LER) and line-width roughness (LWR) of a mask pattern on top of a substrate (top panel). Creating an appropriately thin spacer that will be used as the actual mask for pattern transfer allows reducing LWR to zero (lower panels). Right panels: electron micrographs of a sacrificial amorphous Si pattern (top) with Al_2O_3 spacer (bottom).

If, on the other hand, the thickness of the deposited spacer layer is larger than those wavelengths of the LER that are of the same order and smaller than the critical dimensions of the patterned structure,[10] the conformal deposition yields a smoothing of the LER on one side of the spacer. In order to illustrate this smoothing, a Monte-Carlo simulation of a fully conformal deposition process (light blue areas in Figure 3.22) that could be realized with ALD has been carried out. Figure 3.22 shows a bird's eye view of the edges of two sacrificial patterns (dark blue) with two different dominant wavelengths (top and bottom panels). From left to right, the number of (ALD) deposited layers increases as stated in the figure. One can clearly see that a deposition of 15 layers (left) yields a d_{dep} thin enough that in both cases LER is unchanged but LWR tends to zero. If the number of deposited layers increases, LER at the upper edge decreases. Moreover, in the case of the smaller wavelength (lower panels), a smaller deposition thickness d_{dep} is sufficient to remove most of the LER at the resulting edge.

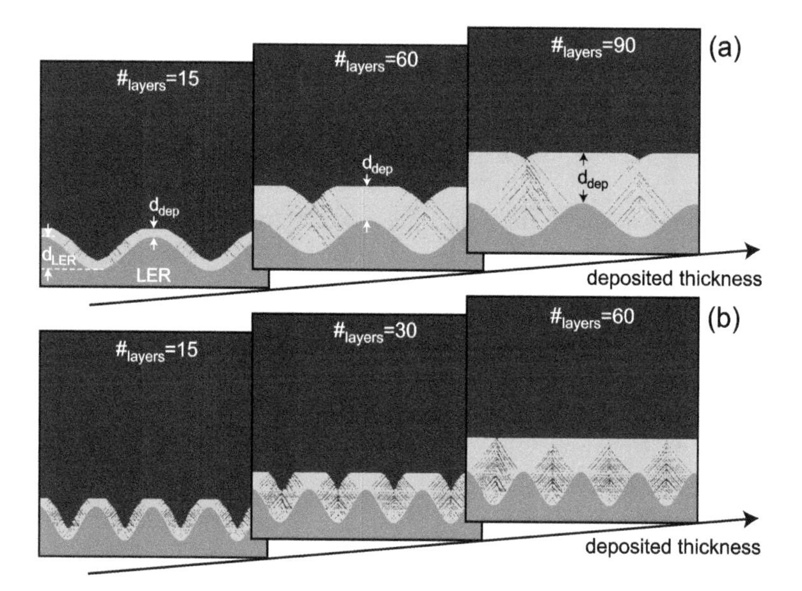

Figure 3.22: Monte Carlo simulation of a fully conformal atomic layer deposition process. The panels show a bird's eye view of the edge of patterns that exhibit LER with two different dominant wavelengths at three different numbers of deposited layers. The smaller the dominant wavelength of LER, the thinner can d_{dep} be to yield a reduction of LER.

In order to exploit the reduced LER, a second spacer layer needs to be deposited while the sacrificial pattern and the first spacer have to be removed. In the latter case, the second spacer can be thin, yielding a suppression of LWR as well as LER. Thus, spacer patterning allows realizing very small and dense features with suppressed LER as well as LWR yielding strongly diminishing variability [45, 246].

10 Long wavelengths can be neglected, since changes with respect to the lithographic patterns are small.

3.6 Wet Chemical Etching

In order to transfer a lithography pattern into a substrate, etching is required. The most straightforward way is wet chemical etching. In the majority of cases, however, wet chemical etching is isotropic, i. e., it etches with an almost equal rate in vertical and lateral directions. As a result, wet chemical etching leads to a mask undercut, and thus to a reduction of pattern size. On the other hand, advantages of wet chemical etching are that it is simple, relatively cheap and does not require a sophisticated tool set. The major benefit is that wet chemical etching offers vastly different etch rates for different materials, i. e., very high selectivities can be realized. The latter is the reason that wet chemical etching is ideally suited for cleaning processes and—if an appropriate etch stop is available—can be used to thin down nanostructures or remove ultrathin layers from entire substrates in a very controlled way. In the following, a number of typical/useful wet chemical etchants are briefly discussed.

3.6.1 Etching of Oxides

The most prominent example of wet chemical etching is arguably the etching of SiO_2 with respect to silicon using hydrofluoric (HF) acid. Here, a very high selectivity is obtained. There are basically three ways SiO_2 can be etched: with (i) an aqueous solution of HF (such as diluted HF or DHF), (ii) buffered oxide etch (BOE or BHF) and (iii) vapor HF. Concentrated HF usually has an etch rate that is by far too high for typical silicon fabrication and is hence prohibitive for appropriate process control. Therefore, diluted HF is usually used; for instance, 1 % HF etches the native oxide on a silicon (100) surface in approximately 10 s. HF is a relatively weak acid with a pH-value between ~0.9–1.5 (for HF 49 %) and ~1.3–2.3 (for HF 10 %). Nevertheless, it attacks photoresist and leads to underetching and delamination of the resist mask.

Buffered oxide etch is a mixture of NH_4F and HF commercially available in different mixing ratios; a typical mixing ratio is 7:1 (NH_4F(40 wt%):HF(49 wt%) = 87.5 %:12.5 %). The buffering agent NH_4F provides a constant concentration of F-ions during an etch process, and thus yields improved process stability compared to HF. More importantly, BOE(7:1) has a pH-value of 4.5 and since photoresist is not attacked by chemical solutions with pH > 3, BOE is preferred for patterned etching of SiO_2. On the other hand, it was shown that etching with BOE roughens Si(100) surfaces on the atomic scale due to its pH < 7 [2]. This roughening can be avoided with the use of pure, aqueous 40 % NH_4F(aq) leading to atomically smooth Si(111) and Si(100) surfaces (if hydrogen bubble formation can be avoided) [50]. As an example, Figure 3.23 shows atomic force microscopy images of Si (111) surfaces after removing the surface SiO_2 layer with DHF (a) and etching in a mixture of NH_4F + $(NH_4)SO_3$ + NH_3 (10:1:1) (b) [3]. The addition of $(NH_4)SO_3$ reduces residues on the wafer surface and NH_3 increases the pH to ~7.8 yielding an atomically flat Si (111) surface.

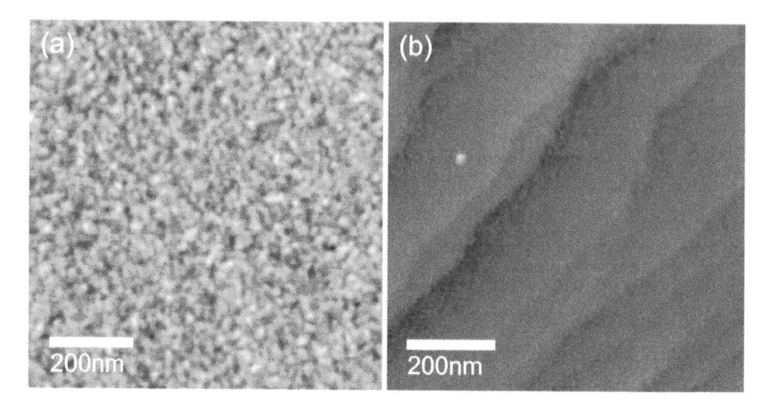

Figure 3.23: Oxide removal from a Si (111) surface with DHF (a) and 15 min etching in $NH_4F + (NH_4)SO_3 + NH_3$ (10:1:1) (b).

A variant of HF etching that is particularly suited for the release of nanostructures is the use of vapor HF. During vapor HF etching, a sample is mounted on a chuck whose temperature can be controlled. Adjusting the temperature in the range of 35 °C to 60 °C enables tuning the condensation of HF vapor on the sample, and thus the etch rate. The benefit of vapor HF is that the sample is immediately dry when removed from the etching and issues with stiction due to the surface tension at the liquid–air interface are avoided. Interestingly, ion implantation into SiO_2 allows strongly modifying the vapor HF etch rate such that a selectivity of up to 150:1 between implanted and nonimplanted SiO_2 can be achieved (in contrast, the selectivity is 3 with aqueous HF etching) as has been demonstrated recently [212].

When etching SiO_2, a hydrogen-terminated, hydrophobic silicon surface is obtained once the hydrophilic SiO_2 is completely removed. The changeover from hydrophilic to hydrophobic is easy to observe, and thus provides a simple yet effective method for process control. Moreover, the hydrogen termination prevents a reoxidation of silicon for a certain amount of time (for a review on surface passivation, see [91]). Usually, when a DI water rinse is carried out after the removal of the SiO_2 reoxidation is prevented for approximately 20–30 min. The reason for this rather short time is that the hydrogen termination is not complete and during the final DI water rinse, part of the dangling bonds are saturated with hydroxyl groups and fluorine that are much less stable thereby enabling the reoxidation. An almost perfect hydrogen termination was reported to be obtained when removing the oxide with a HF/ethanol mixture without DI water rinse [71]. Also, drying with IPA has been reported to yield passivated Si surfaces whose reoxidation is prevented for up to a few days (as already mentioned in Section 3.1.3.3). Such clean silicon surfaces are important for a subsequent epitaxial growth of Si/SiGe but also to enable proper contact formation.

3.6.2 Etching of Silicon Nitride

A major benefit of silicon processing is that there are a number of different materials that can be deposited, patterned and removed selectively with respect to each other. In Figure 3.4(a), a local oxidation of silicon is shown that is the result of using the property of silicon nitride to act as an effective diffusion barrier. After the oxidation, the nitride layer often needs to be removed and this can be done with hot phosphoric acid (to be precise, the top part of the silicon nitride is oxidized during the LOCOS process and this oxynitride layer has to be removed with HF first). Phosphoric acid at a temperature of 165 °C enables etching stoichiometric Si_3N_4 selectively with respect to silicon and SiO_2.[11] However, with pure H_3PO_4 a selectivity between Si_3N_4 and SiO_2 of only ~17 can be realized. Adding sulfuric acid has been shown to be effective to increase the selectivity [226]. Figure 3.24 shows measurements of the etch rate of Si_3N_4 (black dots) and SiO_2 (blue dots) in hot phosphoric acid as a function of the concentration of sulfuric acid proving that a selectivity of ~60 can be obtained with an optimum mixing ratio.

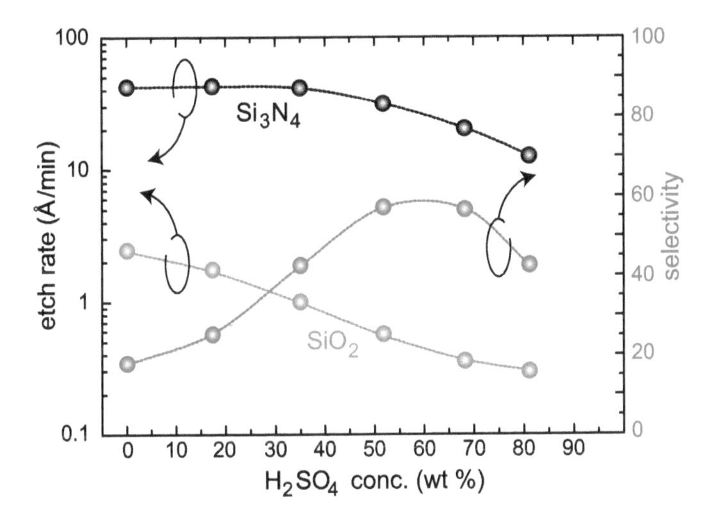

Figure 3.24: Dependence of the Si_3N_4- and SiO_2-etch rates as a function of H_2SO_4 content of a 165 °C H_3PO_4/H_2SO_4 etch solution. The red data points show an optimum selectivity at wt 60 %.

3.6.3 (Digital) Etching of Silicon

Silicon can be etched with a mixture consisting of HNO_3 and HF. This means that the etching occurs via an intermediate oxidation of the silicon and simultaneous HF stripping

11 In the case of a silicon-rich nitride, the selectivity between nitride and SiO_2 may drop to approximately one.

of the generated oxide. As has been discussed in Section 3.2.3, the oxidation of silicon with nitric acid is nearly self-terminating and, therefore, the etch rate of silicon can be tuned by adjusting the HNO_3/HF ratio in the mixture. Since the HF-stripping of the oxide yields a hydrophobic Si surface, citric acid is usually added to the etch mixture in order to improve the wetting of the Si surface thereby improving the uniformity of the etch.

Numerous textbooks are available that discuss wet chemical etching of silicon in detail and the reader is referred to those publications for further details. However, an important point about Si etching deserves more attention: When employed alone, each of the two ingredients of the silicon etch mixture provide a self-terminating process, either oxidation or a virtually infinite etch selectivity. Therefore, splitting the etching into two sequential process steps yields excellent process control over the etch rate. Due to the separation into an oxidizing and a stripping step, the etch process is often called "digital etching" and is particularly suited to thin down silicon (as well as SiGe) nanostructures. For instance, using a room temperature nitric acid an oxide of approximately 1 nm is grown (cf. Figure 3.5(a)) and taking into consideration that the ratio of the consumed Si and the grown SiO_2 is ~0.46, less than 5 Å of Si can be removed during one cycle of the two-step etching providing unprecedented process control. The ultrathin silicon-on-insulator layer shown in Figure 3.5(b) has been thinned down to ~1.7 nm with digital etching. A separation of the etching into two steps is also known in dry etching where it is usually referred to as atomic layer etching (ALE), which will be discussed in Section 3.7.6.

3.6.4 Anisotropic Silicon Etching

While wet chemical etching usually exhibits an isotropic etch behavior, silicon can also be etched anisotropically exploiting the variable etch rates of different crystallographic planes in hydroxide alkaline solutions. Most notably, aqueous KOH solutions are used to etch silicon with etch ratios between the {100} and {111} planes as large as 400:1 [69]. In the case of a (100) silicon wafer, the {111} planes enclose an angle of 54.7° with the surface plane and are aligned along the ⟨110⟩ directions. Hence, when etching (100) silicon wafers with KOH V-shaped grooves appear that are bound by {111} planes as illustrated in Figure 3.25. As a result, the etching basically stops once a V-groove has been fully formed. Due to this self-terminating etch behavior of the {111} planes, the etching undercuts a masking layer until a {111} plane has been prepared. For instance, a circular mask will lead to an inverse pyramid, which can be used to determine the crystallographic orientation of a substrate. Furthermore, in Si(110) wafers, the {111} planes are perpendicular to the wafer surface (if aligned along a [110] direction). Hence, rectangular structures with completely vertical side walls can be etched into silicon which allows fabricating tall fin-structures with high aspect ratio using wet chemical etching (see Figure 3.28, right panels).

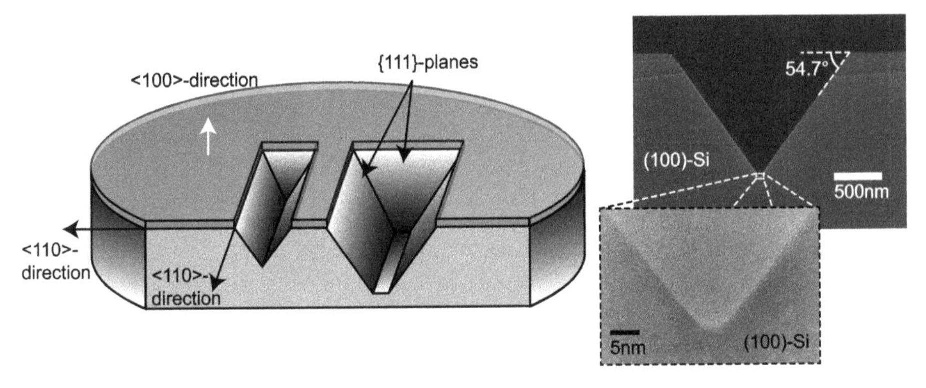

Figure 3.25: Left: Schematic of a Si (100) wafer with V-groove structures etched with anisotropic silicon etching. The right panels show a cross-section electron micrograph and a transmission electron microscope image of the tip region of a V-groove etched into (100) silicon with KOH.

Task 15.
Anisotropic Si etching: Anisotropic silicon etching exploits the very slow etch rate of the {111} planes of silicon in KOH/TMAH. Compute the angle between the {100} and {111} planes and show that it is 54.7°. Figure 2.26 may be helpful in answering the question. Now suppose you want to etch a V-groove in a Si(100) wafer that is 200 nm deep. What is the width of a required mask opening if the etch rate of the {111} planes can be neglected and the mask is not attacked? How do you have to change your mask, when the ratio between the etch rates of the {100} and {111} planes is 50:1?

Typical mask materials for KOH etching are SiO_2 and Si_3N_4 (see [69] for more details). Moreover, silicides can also be used as mask material; in fact, the deposition of a metal on top of a hydrogen passivated silicon surface leads to the formation of an ultrathin silicide layer already at room temperature. This ultrathin layer is sufficient to serve as an etch mask. When designing an appropriate mask the different behavior of convex and concave corners needs to be taken into consideration. In particular, there is no termination of the etching at convex corners and the resulting mask undercut must be accounted for with an appropriate mask design [138].

In general, the etch rate of KOH depends on the concentration and temperature of the KOH solution [69]. Interestingly, the etch rate shows a maximum at a concentration of 20 %. This means that during etching a sufficient amount of KOH solution has to be used and stirring is indispensable in order to avoid concentration gradients. Apart from the etch rate, the concentration also has an impact on the smoothness of the resulting {100} planes: concentrations beyond 40 % are required to obtain mirror-flat {100} planes. Figure 3.25 shows a cross-section electron micrograph of a KOH-etched V-groove showing the 54.7° angle of the {111} planes with respect to the wafer surface. A close-up of the tip region using transmission electron microscopy is displayed in the lower panel of Figure 3.25 showing that a very small radius of curvature of approximately 5 nm can be achieved with KOH etching.

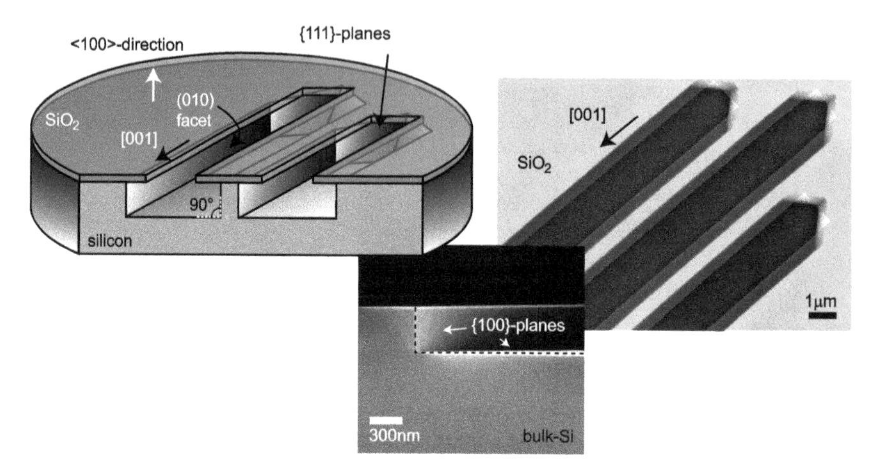

Figure 3.26: Left: Schematic of a (100) wafer with a mask oriented along a [100] direction. Rectangular etch grooves bound by smooth crystallographic {100} planes are obtained when IPA is added to KOH (see center image). The etching does not stop on the {100} planes, and hence mask undercut is obtained (see right electron micrograph).

The etch characteristic of KOH can be modified by adding isopropyl alcohol (IPA) to the KOH solution. The addition of IPA increases the etch ratio between the {100} and {110} planes and reduces the etch rate of high-index planes (such as {221}, {331} and {441}). The latter decreases the mask undercut at convex corners [43, 291]. Moreover, IPA yields smooth {100} planes and if a mask is oriented along a [100] direction, rectangular etch grooves bound by {100} planes are obtained as shown in Figure 3.26 [242]. Since the KOH etch does not stop on {100} planes, mask undercut is observed (left electron micrograph in Figure 3.26).

Potassium diffuses easily through the gate oxide of field-effect transistors leading to a shift of the threshold voltage, and hence to a drift of the electrical characteristics. Therefore, potassium needs to be thoroughly cleaned off the wafer surface after anisotropic silicon etching. This can be done with a 1:1:1 mixture of standard clean 2 (cf. Section 3.1.3.2). Alternatively, tetramethyl-ammonium-hydroxide (TMAH) can be used where the potassium is replaced with the TMA group. TMAH is in widespread use, since it replaces KOH whenever a process should be "CMOS-compatible," i. e., when a possible contamination with potassium should be avoided altogether (for instance, in "metal-ion-free" developers for photoresists). The etch characteristics of TMAH slightly deviate from KOH. For instance, high-index crystallographic planes are etched differently (see [69] for more details). However, although the etch rate of {111} planes with TMAH is higher compared to KOH, the process still yields proper V-groove structures in (100)-oriented silicon. An example is displayed in Figure 3.27 that shows an electron micrograph (center right panel) of a TMAH-etched line pattern. Using AFM, the etched surfaces, both {111} planes as well as {100} planes, are shown to be atomically flat.

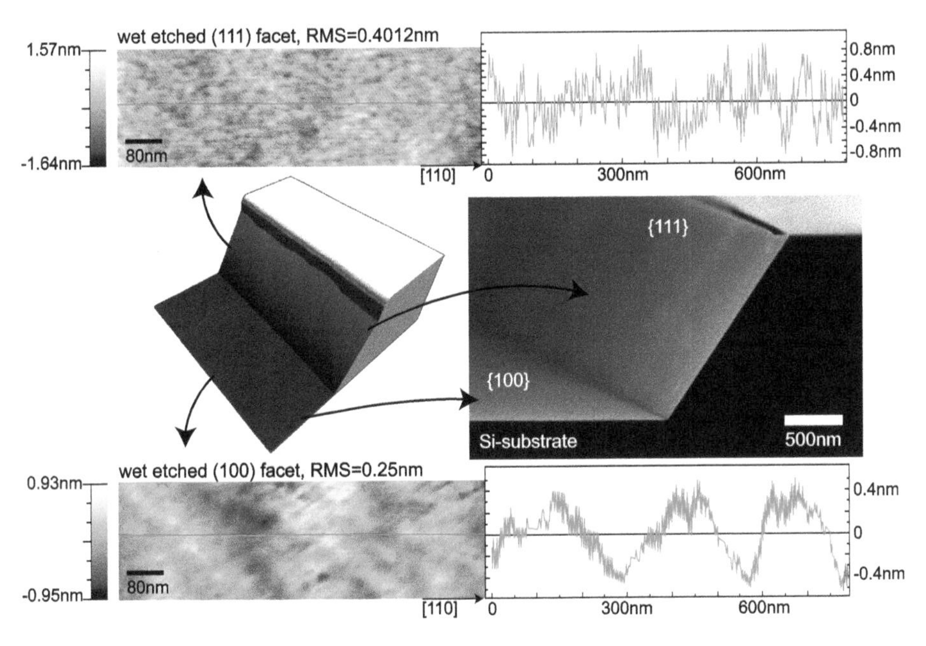

Figure 3.27: Atomic Force Microscopy of TMAH-etched {111} and {100} planes displayed in the central electron micrograph.

A note on TMAH: Working with TMAH requires particular care and attention. Recently, a number of studies have been published that show that TMAH is an extremely hazardous chemical. The hazardousness stems from the absence of alertness due to the unnoticeable skin contact with TMAH. It is even more dangerous than HF because there is no effective medical treatment available and a wrong way of rinsing may even enlarge the exposed area. In fact, several cases of lethal TMAH incidences have been reported [273, 165, 272]. Based on experiments with rats [272], the concentration of TMAH and the exposed body area are the most decisive factors. However, even at low concentrations, i. e., around 1 %, TMAH is a hazardous chemical and needs to be treated accordingly.

The self-terminating property of anisotropic silicon etching can, for instance, be used to deliberately undercut a mask until a {111} plane is prepared. This allows one to reduce line-edge roughness to a minimum as illustrated in the left panels of Figure 3.28; the lower left panel shows a V-groove in Si(100) etched with TMAH where the silicon nitride mask has been underetched. Moreover, the right panels of Figure 3.28 display results of anisotropic silicon etching in Si(110) substrates. The schematic and the left scanning electron micrograph show the realization of a nanowire structure in (110) SOI using KOH etching with ~11 nm width. The side walls of the nanowire are bound by {111} planes such that line-edge roughness existing in the original mask pattern was underetched, and thus strongly reduced. The right SEM image depicts high-aspect ratio fin-structures etched in bulk Si(110) with TMAH. Here, the higher etch rate of TMAH regarding the {111} planes yields a substantial underetching of the SiN mask.

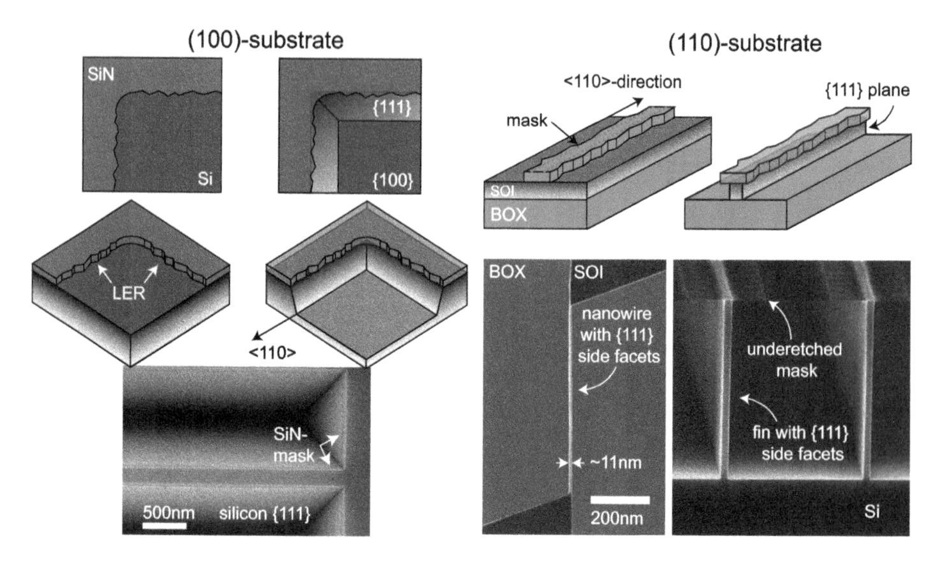

Figure 3.28: Anisotropic Si etching with TMAH/KOH. A V-groove is formed in the case of (100) substrates (left). Underetching of the mask leads to a reduction of LER. Fabrication of a ~11 nm nanowire with KOH etching of (110) SOI (center). The right panel shows a Si-fin etched with TMAH into (110) bulk-Si.

3.7 Dry Etching

When pattern fidelity is required, dry etching is the preferred method to transfer lithographic patterns into a substrate. A wide range of different processes and methods has been developed that rely on the generation of a cold plasma. The various methods enable different etch characteristics and it is this versatility in combination with reproducibility that is the reason for the widespread use of dry etching techniques in semiconductor fabrication. In the present section, the most important techniques are briefly discussed.

3.7.1 Reactive Ion Etching

For reactive ion etching (RIE), a plasma is ignited in a reactor that in its simplest version contains two parallel-plate electrodes. The top electrode is grounded while the bottom electrode is connected via a blocking capacitor to a radio-frequency (RF) generator (cf. Figure 3.29). If a process gas is injected into the reactor, some of the gas molecules or atoms may become ionized. The frequency of the RF generator is chosen such that the light electrons can follow the oscillating electric-field thereby colliding with gas molecules/atoms that may either be excited or ionized, too. At the same time, the frequency of the RF generator is high enough so that the heavy ions cannot follow the oscillating electric field and can be considered to be unaffected. Initially, the plasma in-between the top and bottom electrodes therefore consists of neutral process gas species, fast oscillating electrons and slowly moving ions. Far away from surfaces, the density

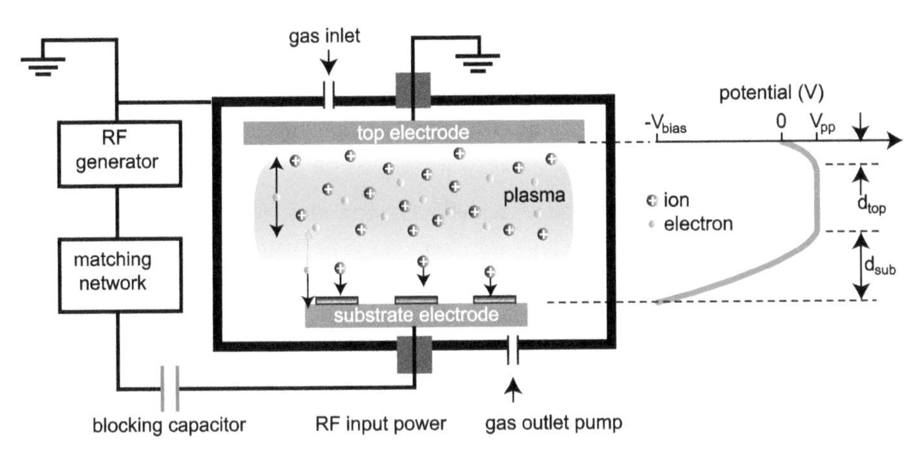

Figure 3.29: Illustration of a simple parallel-plate reactive ion etcher (left). The right panel shows the electrostatic potential with V_{pp} and $-V_{bias}$ being the positive plasma potential and the negative DC bias, respectively.

of ions within the plasma is the same as the density of electrons, such that the plasma is overall neutral. However, electrons in the vicinity of the electrodes or the reactor walls may hit the respective electrode/wall in a one-half period of the RF-oscillation where they get trapped. As a result, the areas around the electrodes get depleted of electrons. Since the bottom electrode is connected to the RF generator via a capacitor, no DC current can flow, and hence the bottom electrode is charged increasingly negative during each half-period of the RF-field where electrons are accelerated toward the bottom electrode; this charge-up will continue until further electrons are completely repelled from the electrode (in other words, the capacitance associated with it is charged). The top electrode on the other hand is connected to ground such that it is not charged up. The resulting electrostatic potential in between the electrodes can be computed with the Poisson equation. Somewhere in the middle between the two electrodes, the plasma is neutral and lies on the so-called plasma potential V_{pp}. Due to the positively charged depletion region at the top electrode, V_{pp} is positive preventing the electrons within the central plasma region from leaving. The negative charge-up of the bottom electrode leads to a substantial bias voltage $-V_{bias}$ building up between the plasma and the bottom electrode (see Figure 3.29). The bias voltage leads to an acceleration of the positively charged ions toward the bottom electrode, and hence the bottom electrode is where the substrates are being placed. The accelerated ions lead to a bombardment of the substrates that is exploited in reactive ion etching in that byproducts are sputtered off the sample surface or chemical reactions are induced.

A proper design of the areas of the top electrode A_{top} and of the substrate electrode A_{sub} helps increasing the bias voltage. Using the dependence $\propto \frac{V^{3/2}}{d^2}$ of the space-charge limited current (an easy-to-follow derivation of this can be found in [253]), and noting that the ion currents toward the top and the bottom (i.e. substrate) electrode should be equal, one obtains $\frac{V_{\text{sub}}^{3/2}}{d_{\text{sub}}^2} = \frac{V_{\text{top}}^{3/2}}{d_{\text{top}}^2}$ where $d_{\text{top,sub}}$ are the widths of the depletion regions at the two electrodes (cf. Figure 3.29) and $V_{\text{top,sub}}$ are the respective voltages between plasma and electrode potential. Noting that the reactor can be considered as two capacitors $C_{\text{sub/top}} = \frac{A_{\text{sub/top}}}{d_{\text{sub/top}}}$ in series (connected through the plasma), one obtains $\frac{A_{\text{sub}}}{d_{\text{sub}}} V_{\text{sub}} = \frac{A_{\text{top}}}{d_{\text{top}}} V_{\text{top}}$. Inserting the relation above finally yields $\frac{V_{\text{sub}}}{V_{\text{top}}} = [\frac{A_{\text{top}}}{A_{\text{sub}}}]^4$. Experimentally, an exponent rather equal to 2 is found. In any case, the bias voltage can be increased if $A_{\text{top}} > A_{\text{sub}}$.

In addition to ionization, further plasma processes include dissociation as well as the generation of radicals, which lead to chemical etching of the substrate if an appropriate process gas is employed. The chemical reaction leads to volatile products that are pumped out of the reactor. Therefore, reactive ion etching always implies a chemical etching component as well as a physical component due to ion bombardment. Choosing appropriate gas mixtures with suitable flux and pressure in the reactor and adjusting the power of the RF generator allows tuning the etch characteristics of RIE over a wide range including the etch rate, etch selectivities as well as the anisotropy of the etch process.

3.7.2 Inductively Coupled Plasma Etching

The density of ions in a plasma is typically very small (approximately a factor 10^3 smaller compared to neutral gas species) [68]. Increased ion densities and increased density of dissociated gas molecules can be realized by increasing the RF power. However, this implies a more negative $-V_{\text{bias}}$, and hence increased ion bombardment, which may lead to a loss of selectivity between substrate and masking material (mask erosion), to a loss of selectivity between different materials within the substrate and to the damage of the etched surfaces because of a disproportionate physical component of the etch process. Since the reason for the formation of the DC bias is the blocking capacitor, shunting it with a variable resistor would enable tuning V_{bias}. This has indeed been investigated (see, e. g., [211]) but leads to a decreased plasma stability. The preferred way to increase the plasma density is therefore adding a second source to the RIE chamber that allows coupling more power into the plasma without increasing the bias voltage. Conceptually, the most straightforward way is an inductively coupled plasma (ICP).

An ICP-RIE reactor is depicted in Figure 3.30 and consists of a RIE chamber where the top electrode has been replaced with a dielectric tube with a coil wrapped around it. The coil creates an oscillating magnetic field that induces alternating RF electric fields. In turn, this yields oscillations of the electrons within the plasma, which lead to an increased rate of ionization and dissociation without necessarily increasing V_{bias}. High etch rates can be obtained and due to the increased ionization/dissociation rate, the pressure in the reactor can be reduced which is a prerequisite to obtain anisotropic etching at relatively low V_{bias} (see the next section).

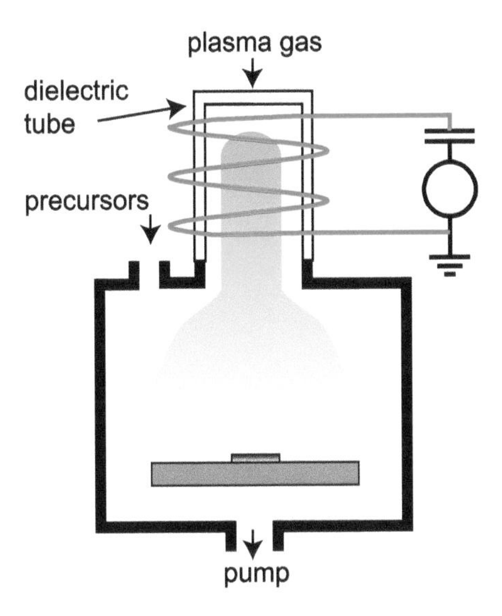

Figure 3.30: Schematic of an ICP-reactive ion etching tool.

3.7.3 Tuning the Anisotropy of Dry Etching

Anisotropic dry etching can, in principle, be achieved by simply exploiting the sputter-etching of the ions accelerated due to the bias voltage V_{bias}. However, as already mentioned above, pure mechanical etching is not selective and may damage the etched surfaces. Therefore, to realize selectivity and anisotropy during dry etching, chemical and physical components need to be combined with appropriate proportion.

Two different mechanisms can be exploited for achieving anisotropic etching (see e. g., [118]). In the first case, the process gas does not lead to spontaneous etching (i. e., the formation of a volatile byproduct) but solely to a modification of the surface. This is, for instance, the case when etching silicon with chlorine: the chlorine merely weakens the Si-Si bonds. However, if ion bombardment is added, the modified surface region will be released and volatile $SiCl_4$ is generated (see Figure 3.31(a)). A similar effect is used when etching silicon with XeF_2. While XeF_2 etches silicon spontaneously, and hence isotropically, ion bombardment renders XeF_2 more reactive and sputters away the byproducts; hence, a directional ion bombardment results in an increased etch rate in this direction and, therefore, to anisotropic etching.

The second method exploits the fact that with appropriate etch chemistry, a passivation layer is formed that is deposited onto the substrate during etching. This passivation prevents the chemical component from attacking the substrate. Consequently, the etching would stop rather quickly. However, with appropriate ion bombardment, the passivation layer is sputtered away on surfaces that are hit by the impinging ions and further etching occurs. If the pressure in the reactor is low, and consequently, the ion flux is perpendicular to the wafer surface, the passivation will remain intact at vertical

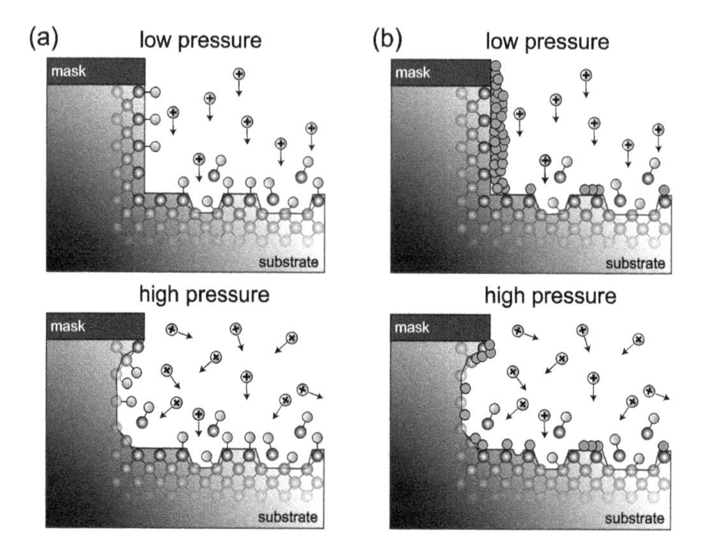

Figure 3.31: Methods to adjust the anisotropy of RIE etching. (a) A chemical reaction yielding a volatile product is induced by ion bombardment. Etching occurs preferentially in the direction of the ion flux. (b) A passivation layer is removed due to ion bombardment enabling further etching in this direction. In both cases, the anisotropy can be modified with the process pressure.

side walls, and hence an anisotropic, vertical etch behavior is obtained as illustrated in Figure 3.31(b). After the etching, an isotropic cleaning plasma (usually an oxygen plasma) is required to remove the passivation from the side walls. A typical process that exploits this behavior is etching silicon with SF_6. While SF_6 etches Si isotropically, adding oxygen to the process gas leads to the formation of a SiO_xF_y passivation layer. Thus, adjusting the oxygen flow allows tuning the anisotropy of the process.

In both cases mentioned so far, a proper vertical etch behavior is obtained if the ion bombardment occurs perpendicular to the surface. This can be ensured by reducing the pressure in the RIE reactor resulting in a sufficiently large mean free path of the ions. Hence, ICP-RIE is ideally suited for this purpose since the pressure can be reduced still providing a high density plasma. In turn, this means that increasing the pressure allows one to randomize the motion of the ions such that they hit the substrate and side walls under a certain angular distribution. As a result, the etching becomes more isotropic, as depicted in the lower panels of Figure 3.31.

3.7.4 Selectivity of Dry Etching

The selectivity of RIE etching can be adjusted with an appropriate process gas mixture. A prominent example is the selective reactive ion etching of SiO_2 with respect to Si. Both materials can be etched with a similar rate using a CF_4 plasma. Adding H_2 to CF_4 reduces the amount of fluorine available for silicon etching (i. e., for the formation of

Table 3.4: Typical process gas (compositions) for reactive ion etching.

Material	Process gas	Material	Process gas
Si/a-Si	1) $CF_4(O_2)$	SiO_2	1) $CF_4(H_2)$
	2) $SF_6(O_2)$		2) CHF_3
	3) HBr	III–V	CH_4/H_2
	4) Cl/BCl_3	SiN	CF_4/O_2
Al	Cl/BCl_3	resist	O_2

volatile SiF_4). As a result, Si etching is suppressed while the etching of SiO_2 is only slightly affected. On the other hand, adding oxygen to CF_4 leads to an increase of the amount of fluorine radicals, and thus to an increase of the Si etch rate. Table 3.4 lists a few possible process gas and gas mixtures for a number of common materials. For detailed recipes, the reader is referred to the vast literature available. Note, however, that each process needs to be adapted to a specific tool.

3.7.5 Plasma Etching

Dry etching can also be carried out in a way that mostly relies on the chemical etching component of the process gas. This is preferable whenever a dry etching process with very high selectivity is desired and pattern fidelity is rather unimportant. A pronounced chemical component can be obtained by grounding the substrate electrode (and by making $A_{top} = A_{sub}$). In this case, only the plasma potential V_{pp} leads to a small acceleration of the ions toward the substrate such that ion bombardment is strongly suppressed.

Completely avoiding ion bombardment is possible with special reactor geometries. The most common one is a barrel reactor as depicted in Figure 3.32. Here, samples are mounted at the center of a quartz barrel. Process gas is injected into the quartz barrel and two electrodes are arranged around the barrel, one connected to an RF generator, the other one grounded. In addition, a perforated Faraday cage is inserted such that only neutral gas species (radicals) enter the central region of the barrel where the sample is mounted. The particular geometry of the barrel reactor leads to an inhomogeneous etching. However, due to the absence of a physical etch component, samples can be overetched for an extended period of time provided the etching is highly selective. The most prominent example is photoresist ashing in an oxygen plasma (Often, a further gas such as nitrogen is added in order to prevent recombination of radicals and thus increase their concentration.). Due to the absence of ion bombardment and the extremely high selectivity of oxygen plasmas, the process can even be carried out as a batch process with multiple wafers at the same time.

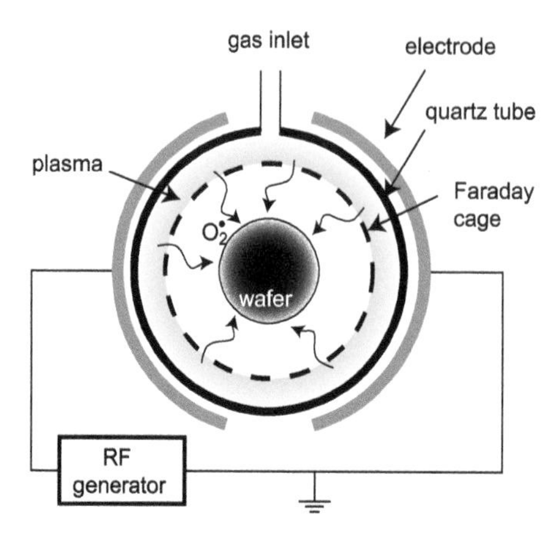

Figure 3.32: Schematic of a barrel reactor with Faraday cage preventing ion bombardment of the sample. The method is particularly suited for resist ashing in an oxygen plasma.

3.7.6 Sequential Etch Process I—Atomic Layer Etching

In Section 3.6.3, digital etching using wet chemistry was discussed. Digital etching provides excellent process control because the etching is split into a self-limiting and a highly selective process step that are cyclic repeated. This idea can also be used in the case of RIE, if the etching of the substrate with a certain gas species works as depicted in Figure 3.31(a). Here, a process gas weakens the bonds of the substrate material and a simultaneous ion bombardment leads to the removal of substrate material. As mentioned above, a typical example of such a process is the etching of silicon with chlorine [18]. A two-step dry etching would then work as depicted in Figure 3.33: Chlorine gas is first injected into the chamber until it has been adsorbed (chemisorbed) at the surface. Since there is no ion bombardment no spontaneous etching occurs. Next, the reactor is evacuated. It is important to remove all process gas within the chamber before the second process step starts. In step 2, argon is injected into the reactor and a plasma is ignited. The choice of argon avoids any chemical etching of the substrate by the process gas (Ar) during this process step. Step 2 leads to an excellently controlled removal of a single layer, if the bias voltage is adjusted appropriately such that the bombardment with Ar ions has sufficient energy to induce a removal of the surface layer in contact with the chlorine while at the same time being low enough to avoid physical etching. Furthermore, an extraordinary anisotropic etch behavior can be obtained if a low pressure Ar plasma is ignited where ions are accelerated only perpendicular to the substrate surface. After the Ar ion bombardment, the chamber is evacuated and a new cycle starts by exposing the sample to Cl gas leading to an unprecedented process control with layer by layer etching.

Results of a kinetic Monte Carlo simulation of such a two-step RIE for different process parameters are displayed in Figure 3.33(b); here, the gray shaded area has been

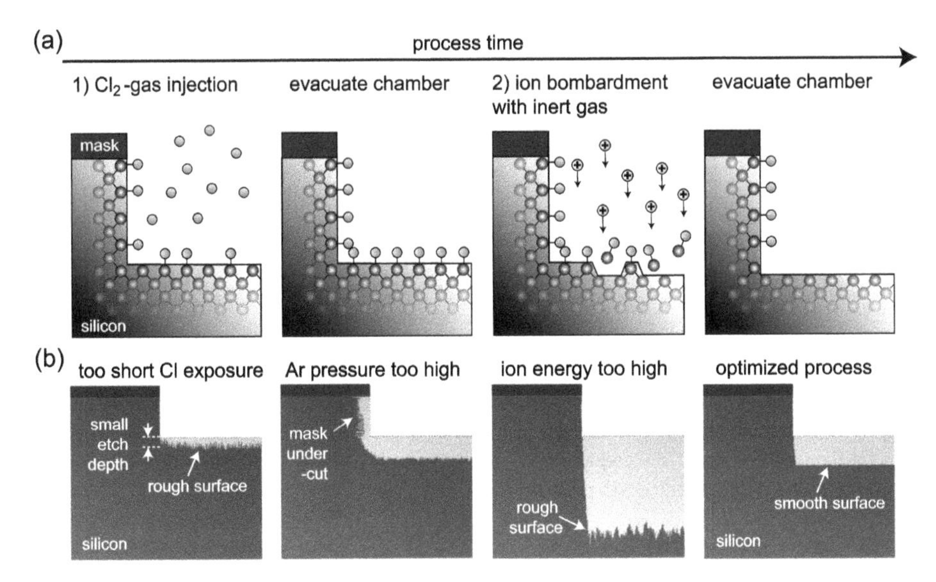

Figure 3.33: (a) Two consecutive process steps (with evacuation steps in-between) enable atomic layer etching (adapted from [17]). (b) Results of a kinetic Monte Carlo simulation with different process conditions. The gray shaded area has been removed, i. e., it shows the cross-section of the substrate prior to ALE.

removed, i. e., it shows the substrate cross-section before and after the etching. A too short Cl-exposure in step 1 yields a small etch depth with very rough surface. A too high Ar pressure leads to a substantial mask undercut. If the Ar ion energy in step 2 is too high, a large etch depth with rough surface is obtained. The result of optimized process parameters is displayed in the right panel of Figure 3.33(b). The kinetic Monte Carlo program to simulate the etch process is accessible through QR code #32.

Due to the similarity with atomic layer deposition (cf. Section 3.8.3.1), this RIE-process is called atomic layer etching (ALE). While special ALE RIE tools are available that allow automated and quick switching between the process steps, ALE can in principle be performed with standard equipment, too [1].

3.7.7 Sequential Etch Process II—Bosch Process

Another method where the splitting of the etching into sequential process steps is used is the so-called Bosch-process. Here, the process is split into a deposition of a passivating layer and two subsequent etch steps. Figure 3.34 (left panels) shows a schematic of the process sequence: in the first step, a thin polymer layer is deposited conformally, which is done with a C_4F_8 plasma at high process pressure. Next, C_4F_8 is evacuated and SF_6 is injected into the chamber. A plasma is ignited with appropriate RF power that results in a $-V_{bias}$ sufficiently high to sputter away the polymer layer: this is done at a low pressure

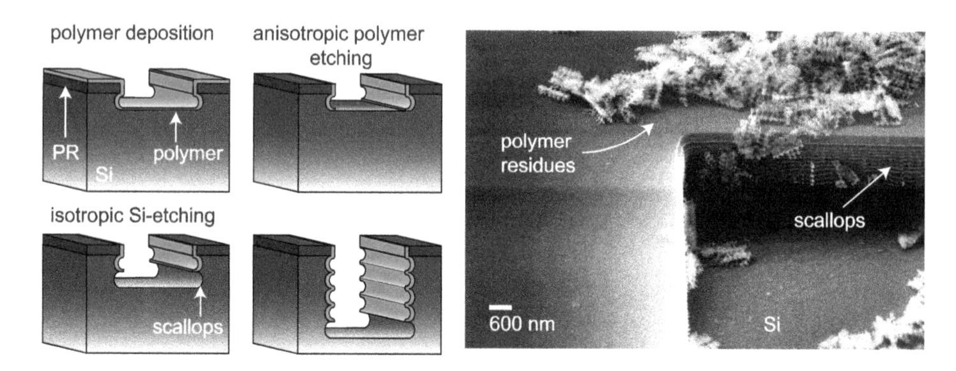

Figure 3.34: Left panel: Schematic of a Bosch-process cycle. (i) Conformal deposition of a polymer with a C_4F_8-plasma, (ii) anisotropic removal of the polymer with ion bombardment (SF_6-plasma) and (iii) isotropic plasma etching with SF_6. The cycle is repeated multiple times. Right panel: Electron micrograph of residues due to overpolymerization after a Bosch process.

to ensure an anisotropic etch behavior leaving vertical areas covered with the polymer. After the removal of the polymer layer, the RF power is switched off in order to avoid ion bombardment, and hence prohibit further physical etching. Thus, only the chemical component of the process acts and leads to an isotropic etch behavior that results in the formation of so-called scallops (cf. Figure 3.34). After evacuation of the chamber, a new process cycle starts again with depositing a polymer layer. Care has to be taken with respect to overpolymerization, which may lead to the formation of silicon-rich residues in the scallops that can hardly be removed with oxygen plasma anymore. The right panel of Figure 3.34 shows a scanning electron microscopy image of such residues that clearly exhibit the shape of the etched side walls (scallops).

The cycles of the Bosch process can be repeated multiple times until the desired etch depth is achieved. Due to the fact that the actual etching of silicon occurs only based on the chemical component, a very high selectivity between silicon and the mask material can be realized. As a result, even rather thin photoresist layers are suitable for large etch depths.

Task 16.
Bosch process: Suppose that the cross-sections of the scallops generated during a Bosch process can be described by a half-circle whose radius is approximately half of the etch depth of the isotropic SF_6 etch step (this is certainly not the case and only a crude approximation). You would like to etch 10 μm deep into a Si substrate and the side-wall roughness due to scallop formation must not be larger than 50 nm. How many Bosch cycles do you need?

An example of a deep reactive ion etching (DRIE) using a Bosch process is shown in the electron micrograph in the left panel of Figure 3.35 where ~26 μm were etched with a 1.6 μm thin photoresist mask. A closer look at the etch flanks reveals the scallop-formation during the Bosch process. Scallops become less pronounced if the duration

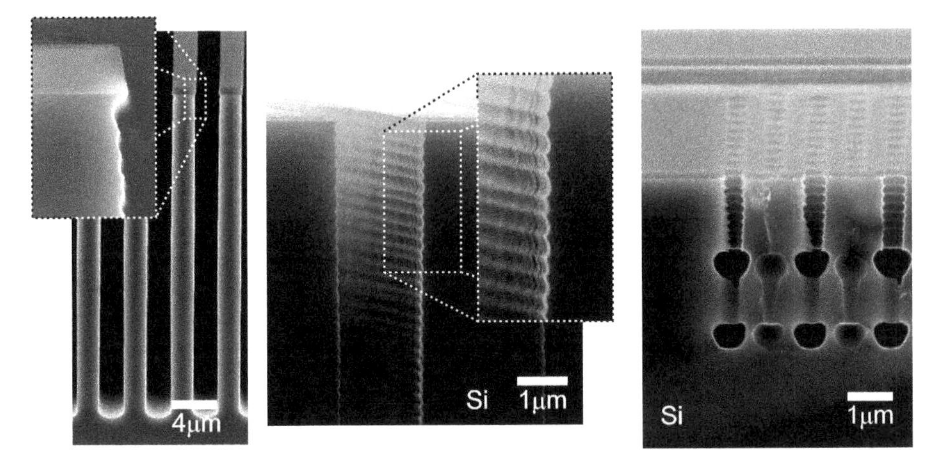

Figure 3.35: Left and center panels: Electron micrographs of deep trenches etched with the Bosch process. Right panel: Bosch process etched with a circular mask. The larger, spherical structures are due to a prolonged isotropic SF_6 etch step.

of the isotropic etch step is reduced at the price of a substantially longer overall process time due to the increased number of evacuation steps. On the other hand, the formation of scallops can also be exploited: The right panel of Figure 3.35 shows circular shaped patterns etched with the Bosch process. Varying the length of the isotropic etch step leads to a pattern as displayed where 9 Bosch cycles with a short isotropic etch step and one long etch cylce lead to the spherical pattern; this is followed by another sequence of multiple short and one long isotropic etch steps. In Section 3.12, we discussed how such a pattern can be used to form voids in a silicon substrate stacked on top of each other that may be used to realize, e. g., pressure sensors.

3.7.8 Issues of Dry Etching—Mask Erosion and Overpolymerization

During dry etching, two mechanisms may lead to distortions of the resulting etch pattern: (i) mask erosion and (ii) overpolymerization. Mask erosion is a severe issue that can strongly affect the outcome of the pattern transfer. If the masking layer has flanks that are not perpendicular to the substrate surface (e. g., as a result of a low resist contrast, see Section 3.5.2.2, or because of resist reflow due to a hard bake, cf. Figure 3.17) and if the etch selectivity is not infinite, mask erosion will lead to a widening of the etch pattern and to nonvertical etch flanks as illustrated in Figure 3.36(a). In fact, depending on the initial mask geometry and the selectivity the angle α of the beveled etch flank can be adjusted as is shown in the two cross-section electron micrographs in Figure 3.36(a). In the two cases displayed, the initial mask geometry was modified with a hard-bake (cf. Figure 3.17) and the selectivity of the etch process was tuned with adding oxygen to a SF_6 plasma. Very shallow etch flanks can be generated with this method.

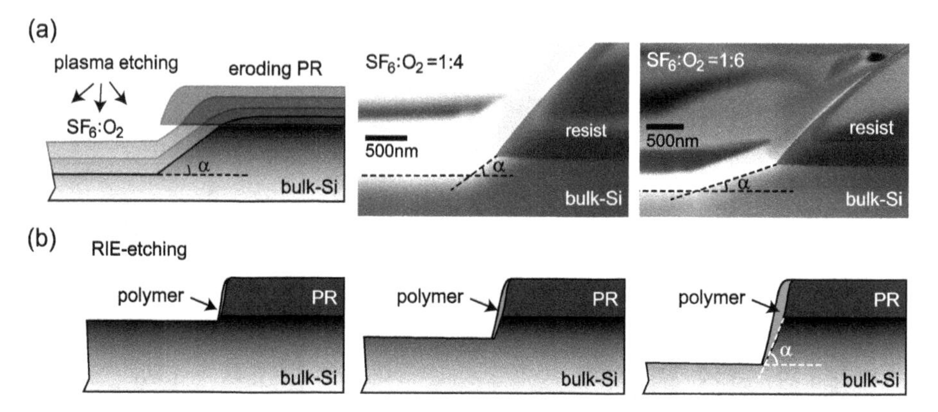

Figure 3.36: (a) Mask erosion leads to a lateral increase of the transferred pattern and a beveled etch flank with angle α. The angle α can be adjusted by annealing the photoresist mask and by decreasing the etch selectivity due to the addition of oxygen to the process gas. (b) Overpolymerization leads to an etch-flank passivation whose thickness increases during etching, resulting in a beveled etch flank.

As discussed above, the deposition of a polymer on the etched side walls facilitates an anisotropic etch characteristic. However, overpolymerization (as already shown when discussing the Bosch process, see Figure 3.34) may increase this side-wall passivation layer during the etching and as a result, a beveled etch flank with a certain angle is obtained as illustrated in Figure 3.36(b). In contrast to mask erosion, overpolymerization leads to a reduced width of the bottom of the etch pattern. Therefore, the resulting etch pattern needs to be compared with the intended mask pattern in order to determine whether mask erosion or overpolymerization led to the loss of anisotropy. Overpolymerization can be mitigated by adding oxygen to the process gas and/or increasing the pressure in the reactor; both are measures that would worsen mask erosion.

3.7.9 Issues of Dry Etching—ARDE, Microloading, Trenching, Grassing

In addition to mask erosion and overpolymerization discussed in the preceding section, further issues may occur that strongly impact the etch result (see Figure 3.37). First, in deep reactive ion etching, the so-called aspect-ratio-dependent etching (ARDE) [89] leads to a lower etch rate of patterns with a smaller lateral extent as shown in the electron micrograph and schematic of Figure 3.37. In the case of the Bosch process, it has been shown that this unfavorable effect can be reduced to less than 2–3 % by adjusting the individual Bosch process steps, i. e., polymer deposition rate, polymer removal rate and the silicon etching rate [161]. Microloading is another encountered issue that refers to the dependence of the etch rate on the density of features. This means that features with equal width are etched at a lower rate if they are placed closer to each other on

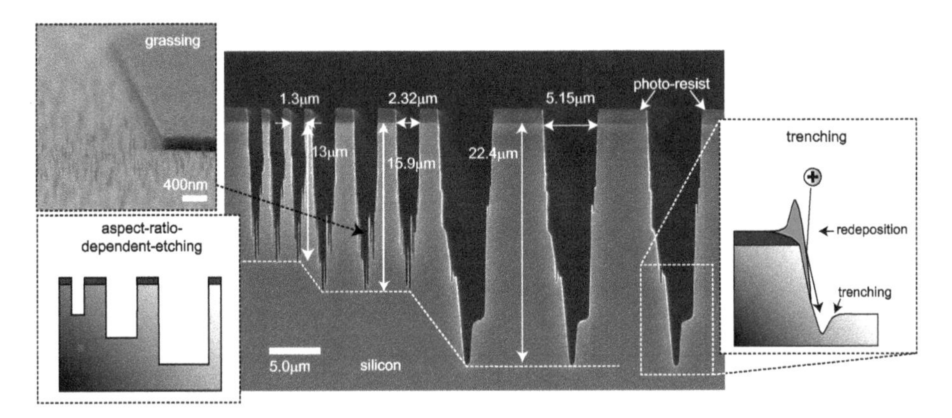

Figure 3.37: Cross-section of a bulk Si sample showing ARDE, grassing and trenching that may occur during reactive ion etching.

the sample. Microloading is due to a local depletion of etch species in areas with higher density of features [102].

During etch processes carried out in inductively coupled plasma tools, the ion density can become rather high and as a result, trenching may occur [201] where the etch rate at the base of the etched side walls is substantially increased leading to an etch profile as shown in Figure 3.37. Trenching stems from a specular reflection of ions hitting the side walls at grazing angles, which leads to a focusing of the etch species [57] close to the side wall of the etched structure. In addition, a redeposition of material at the etched side walls (particularly polymers) may occur at a high ion density in the plasma. In this case, the physical component of etching, i. e., sputtering may become dominant; no volatile byproduct due to a chemical reaction is generated and the sputtered material is redeposited.

Finally, in highly anisotropic etch processes (low pressure, high ion density), redeposited (mask) material, strongly eroded resist or an incomplete removal of the passivation layer result in micromasking that in turn leads to the formation of so-called grass. An example of this grassing is shown in the top left inset in Figure 3.37. Carefully removing resist residues with a DESCUM process, increasing the oxygen content of the process gas (in the case of a polymer used as passivation) or decreasing it (in the case of a SiO_xF_y passivation) and/or slightly increasing the pressure in the RIE reactor help to avoid grass formation.

3.8 Thin Film Deposition

The ability to deposit thin films onto substrates (adding material) is equally important as etching (removing) and together with lithography allows to realize complex, three-dimensional device structures. One distinguishes between chemical vapor deposition (CVD) and physical vapor deposition (PVD). CVD and PVD can be carried out with a num-

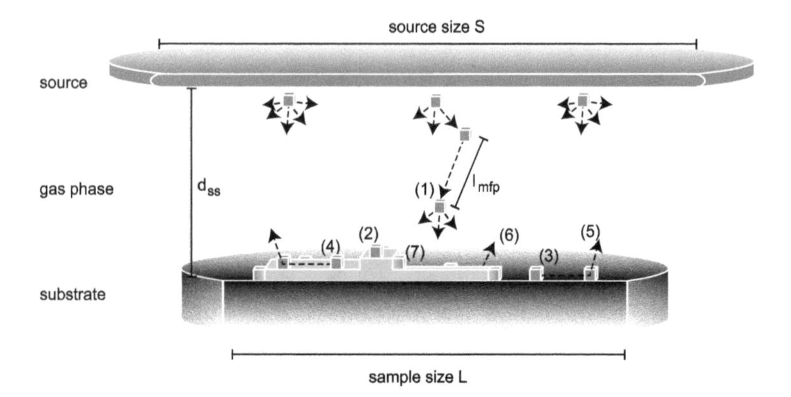

Figure 3.38: Schematic of a general deposition process with transport through the gas phase (1), adsorption (2), diffusion across the substrate (3), (4), desorption (5), evaporation (6) and thin film growth (7).

ber of different techniques that are optimized for particular purposes. Let us first discuss general aspects of thin film deposition that determine the resulting distribution of the film across a substrate.

Figure 3.38 depicts a general deposition process showing that it can be subdivided into the three parts (i) emission from the source, (ii) transport in the gas phase and (iii) deposition on (including diffusion across) the substrate. The source is characterized by its lateral extent S, the angular distribution and velocity with which material is ejected into the gas phase. The gas phase is characterized by the distance d_{ss} between source and substrate as well as the mean free path l_{mfp} in-between two scattering events. When particles (e. g., precursor molecules or evaporant) scatter during their transport through the gas phase, their velocity and directions may be randomized (process denoted (1)). Using kinetic theory, l_{mfp} within the gas phase can be deduced by noting that the interaction volume for a single collision is $l_{mfp} \cdot \pi d^2$ where d is the diameter of the molecule. Hence, the mean distance per collision is $l_{mfp} \propto 1/(\pi d^2 n_V)$ with n_V being the density of molecules in the gas phase. Assuming an ideal gas, $n_V = pN_A/(RT)$ and, therefore, $l_{mfp} \propto T/p$ is a function of temperature T and pressure p [109, 207]. When particles hit the substrate, a number of processes can occur including adsorption (2), diffusion across the substrate (3), (4), desorption (5), evaporation (6) and incorporation (7) (thin film growth) all of which are to a large extent determined by the substrate temperature.

In the following, a number of CVD/PVD deposition methods are discussed. The processes (1)–(7) mentioned above are generic to all of them and the result of the deposition process is a matter of their importance relative to each other. While in some limiting cases a straightforward qualitative forecast of the deposition is feasible, some of the next subsections are fortified with QR codes that allow downloading simulation tools for further exploration of the dependence of deposition results on various process parameters. These simulation tools are based on kinetic Monte Carlo (kMC) calculations (see reference [5], for instance).

3.8.1 Physical Vapor Deposition—Electron-Beam Evaporation

Electron-beam evaporation (EBE) is one of the most widespread metalization techniques because of its simplicity and ability to evaporate a great variety of different materials. During EBE, a current of electrons extracted from a filament and accelerated by a high voltage is bent via an appropriate magnetic field onto a crucible filled with the material that is to be evaporated (see the illustration in Figure 3.41, right). The material is locally heated and will evaporate when a sufficiently high temperature is reached. Due to heating of the target with a focused beam of electrons, EBE allows evaporating materials with high melting temperature and the evaporation occurs from a point-like source. The latter means that a relatively small crucible can be employed making the method potentially cost effective. In addition, crucible and substrate are several tens of centimeters apart from each other (d_{ss}) and the deposition needs to be carried out in a high vacuum chamber. This means that in EBE, $S < L$, $d_{ss} \gg S$ and $l_{mfp} \gg d_{ss}$ and as a result, the evaporated material hits the sample under an angle of 90° with respect to the sample surface. Furthermore, the energy input due to deposited material hitting the resist and sample is in the meV-range and the condensation and radiant heat usually do not significantly heat up the sample. As a result, in most cases the substrate temperature stays below ~100 °C, and thus diffusion across the substrate (see Figure 3.38) can be neglected, i. e., evaporated material will stay at the position where it hits the substrate. This leads to a deposition almost exclusively on lateral areas without coverage of vertical areas (depicted in Figure 3.39(a) and (b)) which is ideally suited for lift-off processes (cf. Figure 3.12).

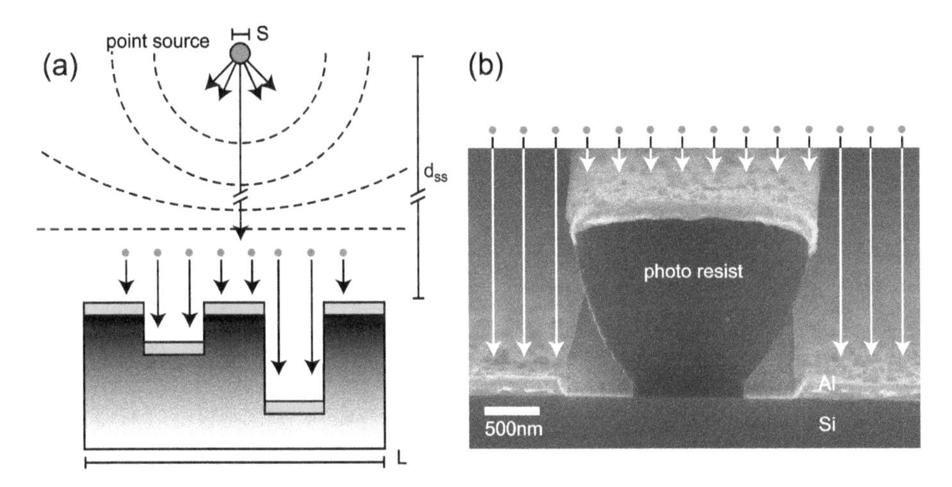

Figure 3.39: (a) Schematic of a EBE deposition process. The source can be considered as a point source yielding a horizontal front of evaporant parallel to the wafer surface. (b) Evaporation of Al onto a Si sample with a resist mask generated with an image reversal process (cf. Section 3.5.2.3).

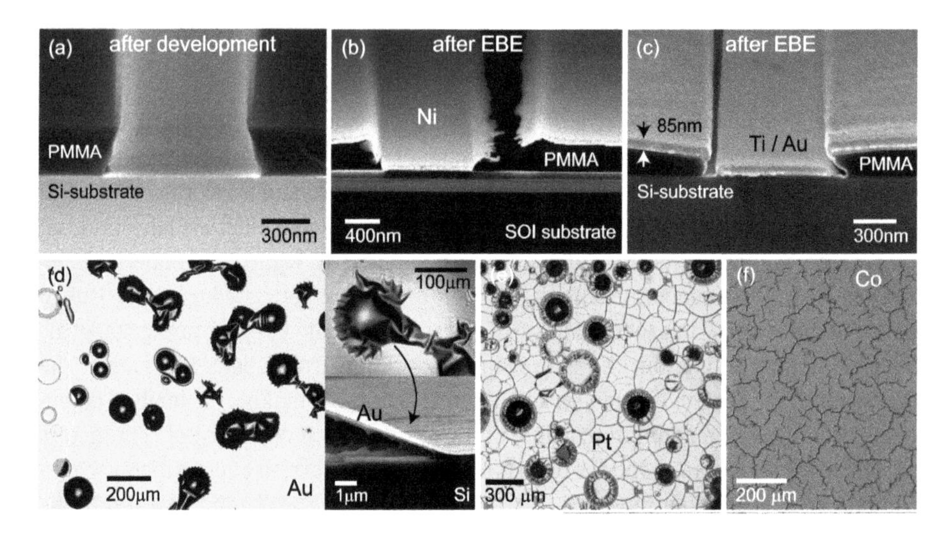

Figure 3.40: Cross-section electron micrograph of a PMMA film after e-beam lithography and development (a) and after EBE of Ni (b) and Ti/Au (c). Resist shrinkage leads to cracks and a degradation of the resist flanks. The lower row shows blistering after EBE of Ti/Au, Pt and Co.

However, frequently encountered issues during EBE are resist shrinkage and blistering of the metallic film particularly when deposited onto resists used for electron beam lithography. Figure 3.40 shows exemplary cross-section electron micrographs of a PMMA resist film after EBL and development (a) and the same film after EBE ((b) and (c)). The shrinkage of the resist and the degradation of the resist flanks are clearly observable. Moreover, (d)–(f) show blistering and strain after Ti/Au, platinum and Co depositions; strain due to resist shrinkage leads to cracks in the metal film. Blistering and resist shrinkage of deposited metallic films may deteriorate the results of a lift-off process or even render it impossible. Both phenomena are not related to thermal stress. The reason for their occurrence is a bombardment of the sample with charged particles, i. e., high-energetic electrons that are back-scattered from the crucible during evaporation as well as ions [210, 243].[12] In the next section, a passive and simple method is briefly discussed that allows avoiding resist shrinkage and blistering.

3.8.1.1 How to Avoid Resist Blistering and Shrinkage

Blistering and resist shrinkage can be avoided by reducing the bombardment of the samples with back-scattered electrons and ions. This can be done with additional electrodes in a parallel-plate capacitor configuration aligned along the path of the charged particle. Applying a sufficiently high voltage deflects positive and negative charges as suggested

12 The generation of X-rays during EBE may be problematic, too [189].

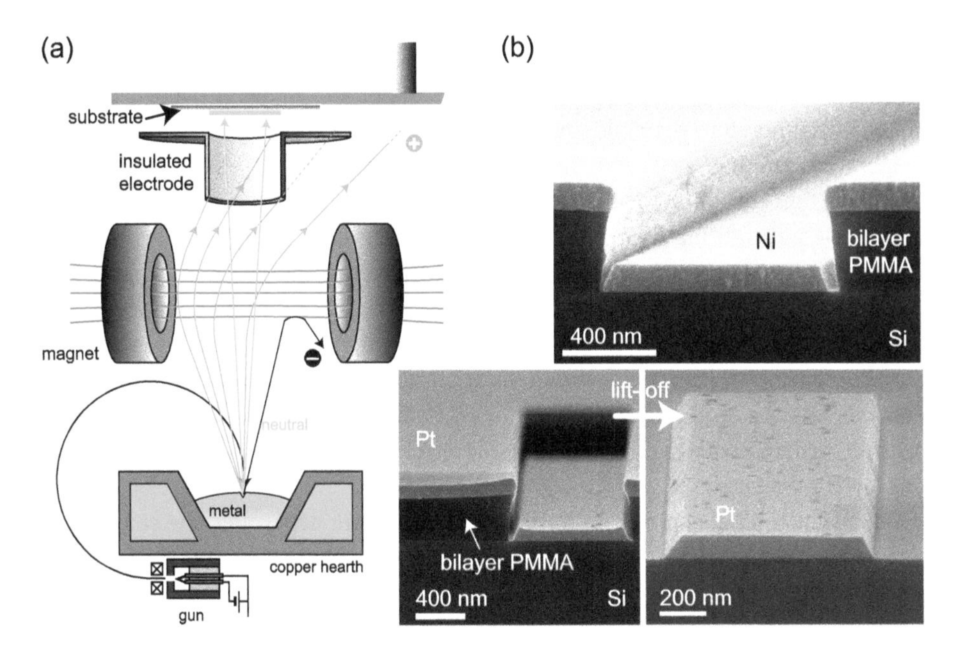

Figure 3.41: (a) Schematic of an EBE system equipped with a strong magnet and an insulated, cylindrical electrode. The magnet bends the path of charged particles so that they cannot reach the sample surface anymore. (b) Blistering and resist shrinkage are completely suppressed and perfect lift-off results can be obtained even in the case of Pt [243].

in [210]. However, relatively large voltages and/or a long capacitor are required including feed-throughs to contact the electrodes.

An attractive alternative to an active electrode is mounting strong permanent magnets arranged as illustrated in Figure 3.41(a). The resulting magnetic field deflects the light electrons strongly away from the sample. The path of the heavier ions, though, is only bent slightly. Therefore, adding a passive, cylindrical electrode with a sufficient length of the cylinder prevents the sample from being bombarded with charged particles [243]. Indeed, the combination of permanent magnets and passive cylinder removes blistering completely during EBE of different materials. As examples, the results of Ni- and Pt depositions are shown in Figure 3.41(b). Because the inverted resist flanks of the PMMA survive the deposition, proper lift-off results are obtained (bottom panel of Figure 3.41(b)).

3.8.2 Physical Vapor Deposition—Sputter Deposition

An alternative PVD method to EBE is sputter deposition. In its simplest form, a plasma is ignited in a vacuum chamber containing two parallel plate electrodes by applying a sufficiently high DC voltage. Ions of an inert gas such as argon will then be accelerated toward the cathode consisting of the target material. The ions knock out material from the target that is deposited onto the substrate lying on the anode. A major benefit of sputtering is therefore the ability to deposit even materials with rather high melting temperature. On the other hand, this may also be a drawback since the material sputtered from the target hits the sample with an energy on the order of a few eV, which may damage the substrate material (for instance, monolayers of 2D materials). Furthermore, damage due to UV irradiation emitted from the plasma may also become an issue.

In many cases, a RF plasma tool is employed since DC sputtering only works with conductive target materials. A RF sputter tool is conceptionally very similar to a parallel-plate RIE reactor. The difference is that the substrate is now mounted on the grounded electrode and the target material is on the electrode that is coupled via the blocking capacitor to the RF power supply such that the developing bias voltage $-V_{bias}$ accelerates ions toward the target.

In a simple parallel-plate reactor design, a significant deposition rate can only be obtained at a relatively high pressure within the reactor. However, as will be discussed below, to obtain a good step coverage a low chamber pressure is required. A higher ion density in the plasma in the case of sputtering can be achieved by placing a permanent magnet at the rear side of the target electrode (cathode). This magnet (cf. Figure 3.42(a)) creates an inhomogeneous magnetic field that forces the electrons to move on spirals toward the cathode thereby increasing the probability to ionize the gas (argon). The resulting inhomogeneous sputtering across the target (see the photography of such a plasma and the resulting inhomogeneously sputtered target in Figure 3.42(b)) is compensated by rotating the sample. This so-called magnetron sputtering yields high deposition rates at low process pressures.

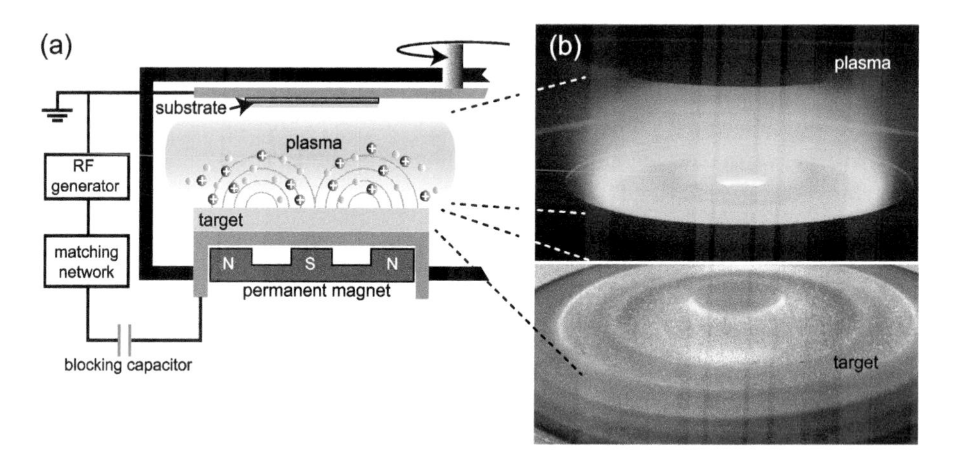

Figure 3.42: (a) Schematic of a RF magnetron sputtering tool. (b) Photography of the plasma (top) during NiCr sputtering. The inhomogeneous plasma due to the magnet is clearly visible leading to an inhomogeneous wear of the target (bottom).

Instead of using an inert gas, sputtering can also be carried out with other process gases such as oxygen or nitrogen. In this case, a reaction will occur between knocked out target material and the process gas, which is called reactive sputtering. Typically, TiN is sputtered reactively with a titanium target and a nitrogen/argon mix as process gas. Reactively sputtered aluminum with oxygen/argon usually leads to substoichiometric alumina; sputtering of an alumina target with a small addition of oxygen (to argon), on the other hand, yields good results.

From the general discussion about thin film deposition above, the deposition characteristics of (magnetron) sputtering can be estimated in the following way: In typical tools, d_{ss} is on the order of centimeters and due to the low pressure $l_{mfp} > d_{ss}$. At the same time, $S > L$, and since deposition is carried out at room temperature, diffusion across the substrate can be neglected to first order. Furthermore, the material ejected from the target assumes a random angle. Thus, the thickness of the deposited layer is determined by the so-called arrival angle θ. The arrival angle refers to the possible angles under which atoms/molecules may arrive at a particular spot on the sample. In two-dimensions, the maximum arrival angle is 180° on extended (horizontal and vertical) planes. Practically, however, $\theta < 180°$ and approximately given by $\theta_1 \approx 2 \cdot \arctan(S/2d_{ss})$. In addition, if a topography needs to be taken into consideration, θ will be even smaller as illustrated in Figure 3.43: in the case of the deep trench with opening d_{op} and depth $d_{depth} = 2 \times d_{op}$, the angle $\theta_3 \lesssim 2 \cdot \arctan(d_{op}/(2d_{depth})) = 2 \cdot \arctan(1/4) \approx 28°$.

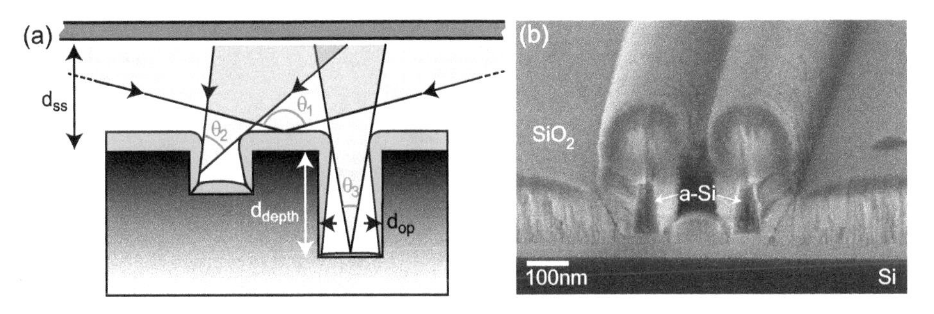

Figure 3.43: (a) Schematic of the sputter deposition process leading to the depicted coating of a topography. The ratio of the arrival angles θ allows estimating the expected film thickness distribution across the sample. (b) Sputter-deposited SiO_2 onto two parallel amorphous Si lines.

The ratio of the arrival angles allows estimating the thickness distribution across the topography. In the example above and assuming $\theta_1 \approx 115°$ (assuming $S = 16$ cm, $d_{ss} = 5$ cm), the thickness on the bottom of the deep trench (right) is only $\theta_3/\theta_1 \approx 24.3\,\%$ of the thickness on horizontal areas of the substrate surface. On the other hand, the part of the side flanks close to the wafer surface and exhibits an arrival angle $\theta_2 > \theta_3$. Thus, a good step coverage of a topography is obtained with sputter deposition if the aspect ratio of the topography $d_{op}/d_{depth} \gtrsim 1$. The right panel of Figure 3.43 shows sputter-deposited

SiO_2. Besides the deposition characteristics expected from the discussion above, a columnar growth due to the low growth temperatures can be seen.

Task 17.
Sputter deposition: Consider two sputter deposition tools. Tool1 has a target with $S = 20$ cm and $d_{ss} = 4$ cm; in tool2 $S = 5$ cm and $d_{ss} = 30$ cm. You would like to deposit on a sample with two trenches. Trench1 has a width of 10 µm and a depth of 5 µm, the width of trench2 is 5 µm and the depth is 30 µm. Determine approximately the ratio of the thickness of the deposited films in the center of the two trenches (bottom) and the middle of the side flank with respect to horizontal areas. Which tool is better suited for a good step coverage and uniform deposition into the trenches?

3.8.2.1 Sputter Deposition for Lift-Off

The good step coverage achievable with magnetron sputtering can also be a disadvantage, particularly in the case of patterning of a metal film with lift-off (cf. Sections 3.5.2.3 and 3.8.1). In this case, the resist side flanks may be covered during the deposition process either preventing a proper lift-off or leading to metallic flakes that stand more or less vertically with respect to the substrate. Figure 3.44(a) and (b) show cross-sectional scanning electron microscopy images of such a behavior. In this example, the aspect ratio of resist opening versus resist thickness (i. e., d_{op}/d_{depth} in the discussion above) is much larger then unity and, therefore, good step coverage is expected. However, if

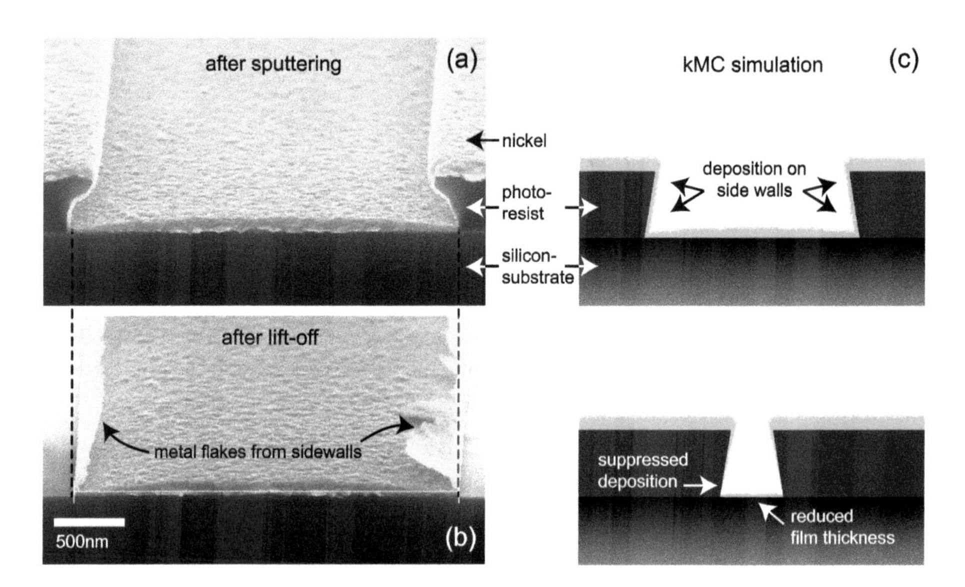

Figure 3.44: (a) Electron micrograph of Ni sputter deposition on a resist mask with $d_{op}/d_{depth} > 1$. The good coverage of the resist side flanks leads to metal flakes remaining after lift-off (b). (c) Result of a kinetic Monte Carlo simulation reproducing qualitatively the experimental results. The lower panel shows that a reduction of d_{op}/d_{depth} enables a lift-off with sputter deposition.

the resist opening is strongly reduced (i. e., $d_{op}/d_{depth} \rightarrow 0$), the arrival angle is substantially decreased. Figure 3.44(c) shows cross-sections of deposition results simulated with a kinetic Monte Carlo approach [5]. The top image qualitatively reproduces the experimental result with coverage of the inverted resist flanks and inhomogenous thickness distribution on the substrate surface. As expected, if the opening of the resist pattern is significantly reduced, there is almost no deposition onto parts of the resist side flanks anymore because of the low arrival angle. This enables a lift-off without metal flakes but at the expense of a much thinner overall film thickness on the sample surface.

3.8.3 Chemical Vapor Deposition—Low Pressure CVD

Chemical vapor deposition is a conceptionally simple process: a flux of precursor gas(es) is injected into a heated furnace (cf. Figure 3.45(a)) and diffuses toward a heated substrate (b). The precursor(s) is(are) adsorbed at the substrate surface where it(they) decompose(s). The reactants then lead to the growth of a layer due to site incorporation into the growing thin film.

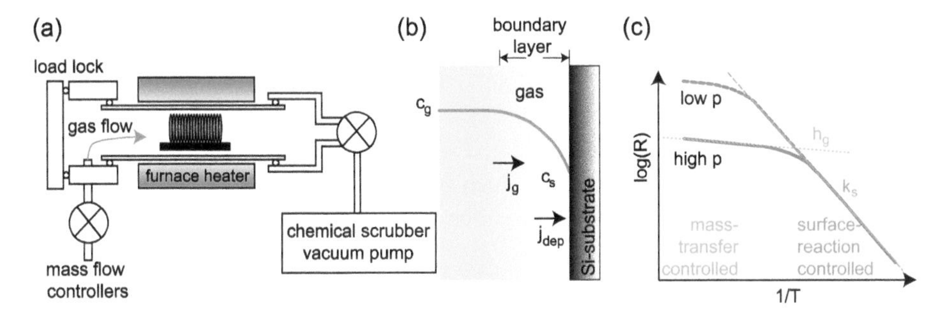

Figure 3.45: (a) Schematic illustration of a LP-CVD furnace. (b) A flux j_g of precursor through the boundary layer leads to a "flux" j_{dep} of thin film growth. (c) Growth rate as a function of $1/T$ in the case of a low and high pressure in the LP-CVD furnace.

In the present case, the source of materials to be deposited can be considered being distributed around the substrate and it is therefore the transport through the gas phase, in particular, the boundary layer close to the substrate surfaces, that matters. Similar to the Deal–Grove model (see Section 3.2.1), the flux of precursor/reactant through the boundary layer can be written as $j_g = h_g(c_g - c_s)$ where h_g is the mass transport coefficient (process (1), Figure 3.38) and $c_{g,s}$ the reactant concentrations at the edge of the boundary layer and the wafer surface. At the surface, a "flux" of materials growth $j_{dep} = c_s k_s$ can be defined where k_s is the surface reaction coefficient. Note that k_s includes all processes (denoted (2)–(7), Figure 3.38) related to adsorption, diffusion, desorption and incorporation of reactant, effectively leading to thin film growth. The continuity

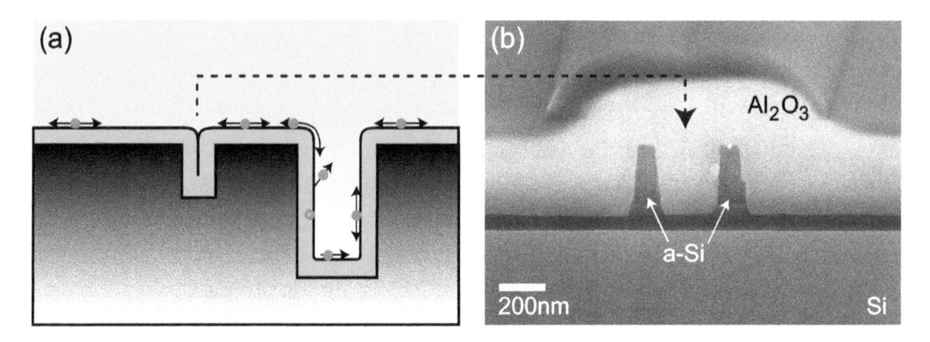

Figure 3.46: (a) Conformal deposition obtained with LP-CVD or ALD. In the case of LP-CVD, the elevated substrate temperature leads to surface migration of the precursor molecules such that an equal thin film growth on horizontal and vertical areas as well as within deep trenches is obtained. (b) Electron micrograph cross-section of Al_2O_3 deposited with ALD on top of two a-Si lines.

of fluxes requires $j_g = j_{\text{dep}} \equiv j$. Solving $j_g = j_{\text{dep}}$ for c_s and inserting this into $j = j_{\text{dep}}$ yields $j = c_s k_s = c_g \frac{k_s h_g}{k_s + h_g}$. With N being the number of atoms incorporated into the growing film per unit volume, the growth rate R is $R = \frac{j}{N} \propto \frac{k_s h_g}{k_s + h_g}$. This means that in the limiting cases $k_s \gg h_g \rightarrow R \propto h_g$ and $k_s \ll h_g \rightarrow R \propto k_s$. In the first case, growth is mass transfer limited, while in the second case it is surface reaction controlled (cf. Figure 3.45(c)). In order to avoid depletion of precursor gas, which would result in an inhomogeneous deposition and in order to facilitate batch processing of many wafers at once, the case $R \propto k_s$ is highly desirable. At the same time, high growth temperatures are preferable to obtain high quality films. In addition, high substrate temperatures facilitate surface migration, which leads to conformal deposition as illustrated in Figure 3.46(a). However, the surface reaction coefficient $k_s \propto \exp(-E_A/k_B T)$ with an activation energy E_A. Consequently, increasing the temperature exponentially increases k_s driving the process into the mass transfer controlled regime. Luckily, $h_g = \frac{D_{\text{diff}}}{d_{\text{bl}}}$, with the diffusion coefficient proportional to $D_{\text{diff}} \propto \frac{T^{3/2}}{p}$, and hence lowering the pressure p in the furnace allows the surface reaction controlled regime to be extended to higher process temperatures (cf. Figure 3.45(c)). Therefore, low pressure(LP)-CVD is the preferred method when high quality films with conformal deposition in a batch process are required.

Typical precursor gases for LP-CVD include silane (SiH_4), oxygen, ammonia, nitrous oxide, dichlorosilane (SiH_2Cl_2) and tetrathyl orthosilicate (TEOS), which enable the conformal growth of high quality polycrystalline Si, SiO_2, Si_3N_4 and oxynitride at different temperatures.

3.8.3.1 Chemical Vapor Deposition—Atomic Layer Deposition

Atomic layer deposition (ALD) is a chemical vapor deposition process where the deposition occurs in the ideal case in discrete steps, i. e., monolayer by monolayer. This is possible because the deposition process due to a reaction of appropriate precursor

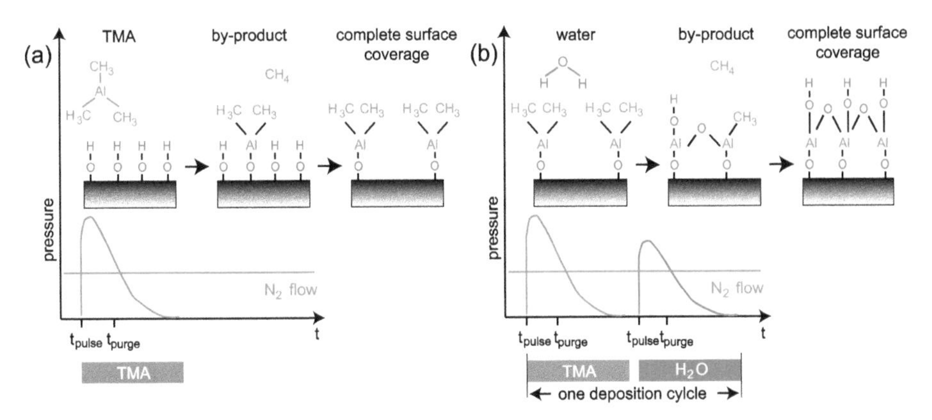

Figure 3.47: Schematic illustration of an atomic layer deposition of Al_2O_3 on SiO_2. Sequential TMA- (a) and H_2O-steps (b) yield a single monolayer of aluminum oxide.

gases is subdivided into sequential process steps. The most prominent ALD process is the deposition of Al_2O_3 with the precursors tetramethylaluminum (TMA) and water. In its simplest form, a reactor for ALD is a container with heated chuck connected to a vacuum pump with two mass-flow-controlled feed lines for the two precursors. Figure 3.47 shows schematically the progression of the ALD deposition of Al_2O_3 on a SiO_2-covered substrate. In the first step, TMA is injected into the chamber. The TMA reacts with the hydroxyl groups present at the SiO_2-surface. As a result, one of the aluminum bonds is connected to oxygen and a volatile CH_4 byproduct is built. When the complete surface is covered, the TMA feed line is closed and all remaining precursor and byproducts are pumped out of the reactor. Next, water is injected into the reactor, which reacts with the remaining two CH_3-groups, again forming CH_4 as byproduct. When all CH_3-groups have reacted, the water feed line is closed and the reactor is evacuated. This process sequence yields a first monolayer Al_2O_3. Since the new surface is again covered with hydroxyl groups, the process can now be cyclic repeated yielding one monolayer after each sequence. ALD is also possible using a remote plasma. Plasma-enhanced ALD (PE-ALD) allows deposition of a broader range of materials with ALD even at low substrate temperatures. ICP sources are usually used for the generation of the plasma.

Apart from the fact that ALD provides unprecedented process control over the thickness of the deposited layer, the separation of the deposition into two sequential steps can lead to a near ideal conformal deposition (cf. Figure 3.46(b)). The reason is simply the fact that the separation into two sequential steps allows the precursor gases to diffuse even into deep trenches and can therefore cover the entire surface (vertical and horizontal) equally while there is no deposition of a full monolayer (due to a lack of the complementary precursor). As a result, ALD is ideally suited for the fabrication of gate stacks in wrap-gate device architectures, spacer formation or vertical capacitors. However, a near ideal conformal deposition can only be achieved if each of the process steps is sufficiently long. This in turn may strongly reduce the throughput during device fabrication. Therefore, carefully adjusting the duration of the ALD steps is necessary in order to obtain optimum results in an acceptable time. QR code #35 provides access to results on ALD obtained with a kMC simulation tool (that can be downloaded) with which the impact of the ALD parameters can be qualitatively investigated.

3.8.4 Chemical Vapor Deposition—Plasma-Enhanced CVD

While LP-CVD or ALD are the preferred methods to obtain high quality, conformal deposition, relatively high temperatures are required in the case of LP-CVD and ALD is only suited for thin films. If deposition of rather thick layers at substantially lower temperatures is necessary, plasma-enhanced CVD (PE-CVD) is the method of choice. PE-CVD exploits a plasma instead of the hot wafer surface to decompose the precursor enabling the thin film growth. To this end, basically the same reactor as in the case of plasma etching can be used with the substrate electrode connected to ground and both electrodes in the parallel-plate reactor having similar size.

While PE-CVD provides high quality films with good uniformity at low temperatures, the deposition is usually not conformal, in particular, if a rather high growth rate is desired. The reason for this is that the lower temperature compared to LP-CVD leads to strongly reduced surface migration. In addition, the relatively high pressure of parallel-plate PE-CVD reactors yields a small l_{mfp}, i. e. a randomization of particle motion in the gas phase, and thus arrival angles up to 270° at convex corners. This leads to a thicker film thickness at such corners and an imperfect deposition in concave corners (cf. Figure 3.48(a)). An experimental example of a PE-CVD deposition is shown in Figure 3.48(b); here, SiO_2 was deposited on trenches in a Si substrate. The deviation from the expected shape of the deposited film cross-section (see inset) is likely to stem from material being sputtered away due to slight ion bombardment.[13] Using a remote plasma source such as an ICP source to densify the plasma, reducing the process pressure and the deposition rate as well as choosing a deposition temperature of 300 °C allows one to improve the step coverage and morphology significantly as is shown in the examples displayed in Figure 3.48(c) and (d) (compare with magnetron sputtered SiO_2, Figure 3.43(b)).

3.9 Damascene Process

Lift-off processes are useful whenever an appropriate etch process for a specific material is not available and, indeed, very narrow contact or gate electrodes can be fabricated (in combination with electron-beam lithography). However, they lack reproducibility,

13 Although the substrate is on ground, the plasma potential leads to an acceleration of ions from the plasma toward the substrate.

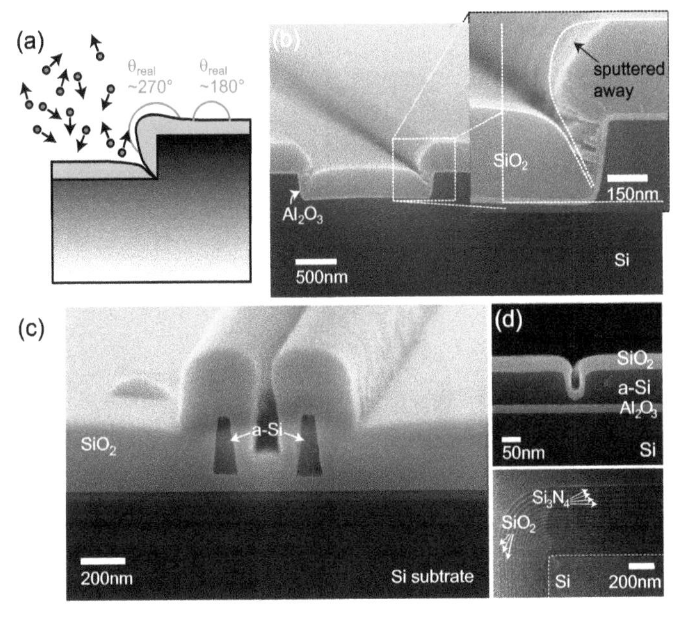

Figure 3.48: (a) Schematic of the deposition characteristic of PE-CVD. (b) PE-CVD grown SiO_2 on top of a trench exhibiting poor step coverage. The dashed white line in the inset shows the expected deposited profile (as in (a)). The deviation is due to material sputtered away due to ion bombardment during the deposition. (c) Remote plasma enhanced (RPE-)CVD deposition of SiO_2 showing a step coverage similar to magnetron sputtering with improved morphology. (d) RPE-CVD with reduced deposition rate of SiO_2 (top) and SiO_2/Si_3N_4 multilayers (bottom) with improved step coverage.

reliability and pattern fidelity. A process that combines all requirements concerning process control but does not require to etch a certain material is the so-called damascene process depicted in Figure 3.49(a). During a damascene process a "mold" is etched into a thin film of a sacrificial material for which appropriate (selective, anisotropic) dry etching exists. After etching and resist removal, the material that was intended to be patterned in the first place is deposited, e. g., with sputter deposition or ALD. Finally, chemical-mechanical polishing (CMP) is utilized to remove the overburden and stop on the sacrificial material. As an example for such a damascene process, Figure 3.49 shows images of 4″ silicon wafers deposited with Al with incomplete polishing (b) and complete polishing (c). In (b), metal residues on the patterns (toward the wafer edge) would short-circuit all devices in this area; adjusting the CMP parameters leads to complete polishing, and hence patterning of the Al is achieved without the necessity of etching aluminum.

A damascene process also facilitates patterning of materials on very small length scales. As an example, Figure 3.50(a) shows an SEM image of a mold etched into silicon with anisotropic Si etching (Section 3.6.4). Subsequently, alternating layers of TiN and Al_2O_3 are deposited with ALD. Finally, CMP is used such that a multielectrode structure is obtained. Figure 3.50(b) shows the structure after CMP. Because of the extremely smooth

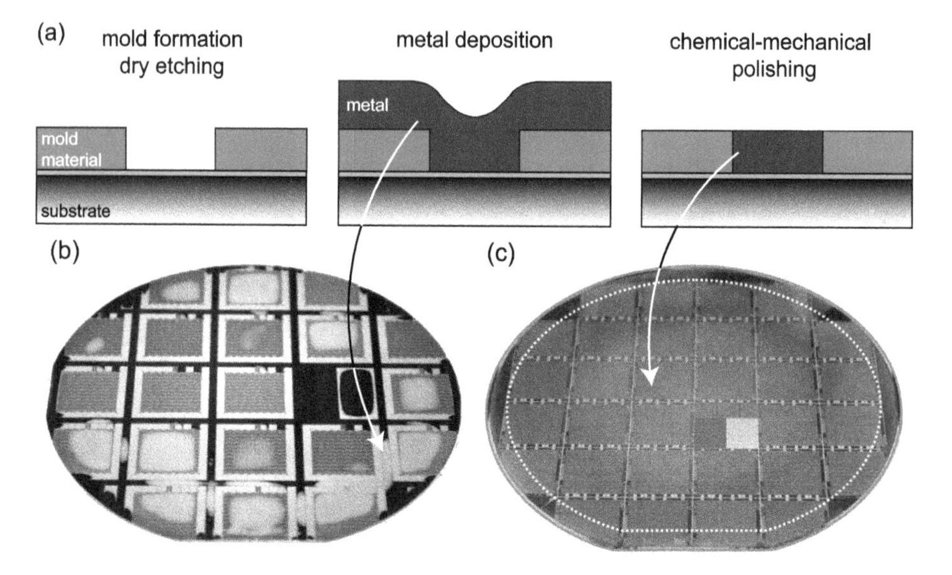

Figure 3.49: Damascene processes to form a gate electrode (a). Images of an incomplete (b) and a complete (c) polishing of a 4″ silicon wafer with an aluminum damascene process [155].

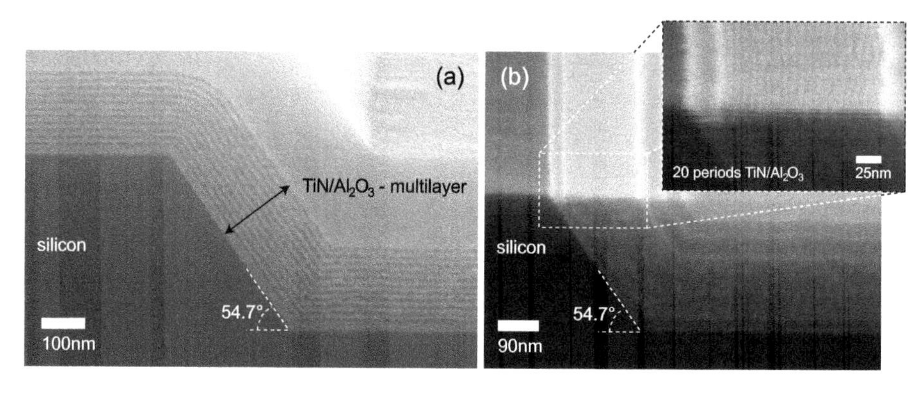

Figure 3.50: Damascene process of a Al_2O_3/TiN multilayer stack deposited with ALD onto a TMAH-etched Si substrate (a). (b) CMP allows the fabrication of sub-5 nm TiN electrodes with sub-5 nm interelectrode distance (see the inset) [225].

etch flanks obtained with anisotropic silicon etching using TMAH (cf. Figure 3.27) and combining this with ALD, a buried multielectrode structure can be realized with sub-5 nm electrode size and sub-5 nm interelectrode distance (see the inset of Figure 3.50(b)) [225]. The damascene process avoids many issues regarding the development and/or availability of appropriate etch processes. In return, it requires reliable CMP processes for different materials.

During chemical-mechanical polishing, a wafer is polished with a slurry that contains abrasive particles (SiO_2 or alumina) and chemical additives. The latter enable a certain degree of selectivity with polishing, e. g., the deposited metal at a higher rate compared to other materials on the sample. In addition, if the overall area of regions where the deposited (metal) film should be removed with polishing is significantly smaller than the area of the wafer where the material is supposed to stay, removing the overburden will occur at a much faster rate than further polishing a flat wafer surface. Therefore, adding supporting structures or removing irrelevant parts of the "mold" on the substrate is a viable way to improve CMP results. Furthermore, a polishing selectivity is achieved by choosing materials with appropriate hardness for the substrate material (i. e., the "mold") if this is possible. At first sight, a large difference in the hardness of mold and polished materials appears beneficial: If the hardness of the substrate material is larger than the hardness of the deposited (electrode) material, polishing will strongly slow down once the electrode material on top of the substrate material has been polished away. This slow-down of the polishing rate can be used for an endpoint detection providing process control. On the other hand, a large difference in hardness promotes scratching, erosion and dishing in the soft material. Dishing is the result of increased polishing in the center of a structure that consists of a softer material than its surroundings (mold) leading to a dish-shaped cross-sectional polishing profile. Dishing is more pronounced in larger structures (see Figure 3.51).

An example of a CMP process with materials strongly differing in their hardness is the generation of aluminum electrodes buried into an oxidized silicon wafer as shown in the lower panel of Figure 3.51(a). Here, ansisotropic Si etching was used to form a

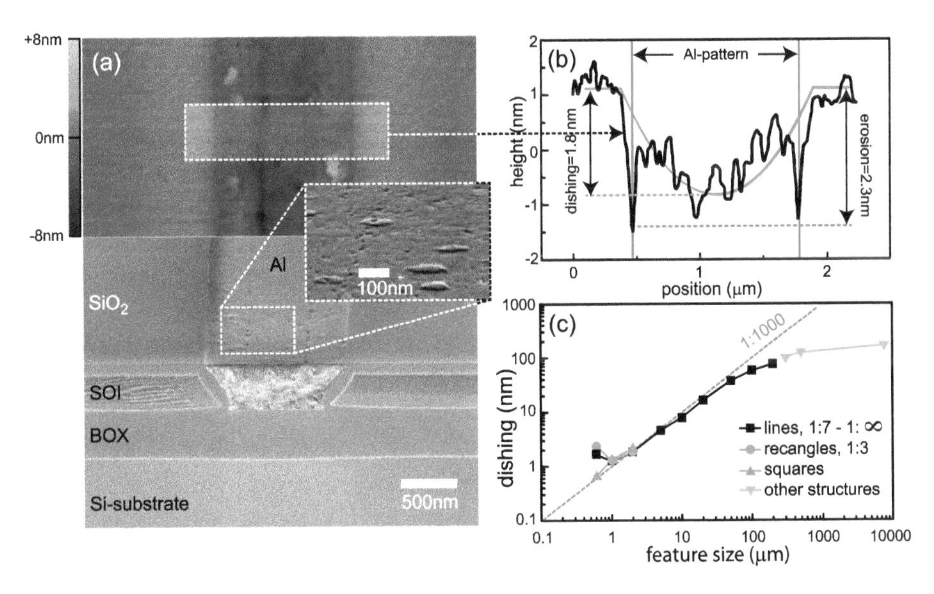

Figure 3.51: (a) AFM image of the polished Al electrode shown in the electron micrograph (lower panel), the inset shows a close-up of the polished Al surface. (b) Averaged height profile extracted from AFM measurements to determine erosion and dishing of the Al electrodes. (c) Dishing for several Al electrode structures of different size [155].

V-groove in a silicon-on-insulator wafer. After thermal oxidation, aluminum is sputtered onto the substrate followed by CMP [197]. The top panel of Figure 3.51(a) shows an AFM image, which was used to determine dishing and erosion of the aluminum electrode shown in (b). Although the hardness of SiO_2 and Al is quite different, a CMP process was found that yields little erosion (~2.3 nm) and dishing (~1.8 nm across approximately 1.6 μm). Indeed, measuring the dishing of several Al electrodes with different size, a maximum dishing of $\sim\frac{1}{1000}$ of the lateral structure size was observed (displayed in (c)) [155].

3.10 Ion Implantation and Activation

Ion implantation is one of the most important and versatile techniques for semiconductor technology. It allows modifying the semiconducting properties locally within nanoscale volumes (see the discussion in Section 4.1). This is accomplished with an appropriate mask on top of a substrate. Choosing a proper mask material, thickness and ion energy, the flux of ions is blocked in the masked regions. As a result, ions are merely injected into the unmasked semiconductor areas. The depth of the dopants can be adjusted with the energy of the ions. Hence, combining multiple successive implantations with different mask materials on the wafer surface sophisticated three-dimensional doping profiles can be realized. In recent years, however, the active volumes in semiconductor devices have become so small that the statistical nature of the implantation process becomes relevant. Variability issues due to a random dopant distribution have since been a matter of intense investigations. Ultimately, random dopant related issues can only be mitigated by either removing all dopants from the device or by avoiding the randomness. The former will be studied in Section 4.4 and Chapter 7. The latter can be accomplished by realizing deterministic implantation. Deterministic implantation is challenging since for a process on industrial scale one needs to create a dense array of ion channels that can be switched on and off and that are enabled to detect the passing of a single ion; so far, a demonstration of such an array has not been accomplished. However, using an AFM tip and drilling an ion channel into it, deterministic doping of individual dopants has been demonstrated successfully [191].

Apart from the statistical nature of ion implantation, there is another important issue: the implanted ions need to be activated, i. e., they need to replace a host atom of the semiconductor in order to donate or accept an electron. To this end, an annealing of the sample after the implantation needs to be carried out. Unfortunately, dopants, in particular boron, will diffuse during the annealing thereby broadening the dopant distribution. Consequently, the activation anneal should be as short as possible for which rapid thermal annealing has been employed that allows annealing with high heating and cooling ramps. However, studying the diffusion of boron in SOI annealed with different thermal budgets reveals that even a spike anneal in a rapid thermal annealing tool leads to substantial diffusion of dopants [164]. Figure 3.52 shows secondary ion mass spectrometry (SIMS) profiles of boron implanted into a 70 nm thick SOI layer. Obviously,

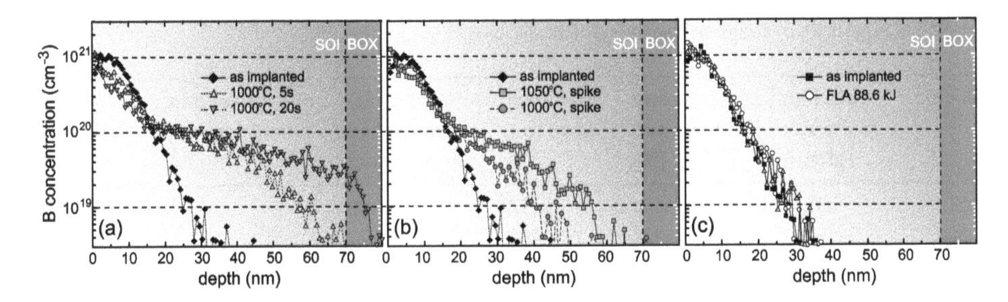

Figure 3.52: Boron distribution (acquired with SIMS) within a silicon-on-insulator layer after ion implantation and annealing with different thermal budgets. (a) Rapid thermal annealing with 20 s and 5 s, (b) spike annealing with 1000 °C and 1050 °C, (c) flash lamp annealing [164].

annealing leads to strong diffusion (a) and even when spike annealing is employed, the profile significantly deviates from the as-implanted distribution (b). Only flash lamp annealing allows one to mitigate diffusion such that the annealed and as-implanted distributions are almost the same (c). The reason for this is that during the μs-short flash lamp annealing only the top part of the substrate is heated to elevated temperatures where dopants have been implanted. The bulk part remains at ~600 °C, and thus acts as efficient heat sink; heating and cooling ramps are hence way faster compared to rapid thermal (spike) annealing.

3.11 Silicidation

In the preceding section, it was discussed that the statistical nature of ion implantation and the diffusion during an activation anneal are problematic in nanoscale devices. Moreover, in Section 4.3.2 it will become clear that dopants may get deactivated in nanostructures giving rise to high parasitic resistances. Hence, it appears desirable to replace dopants with metals that provide high conductivity and atomically abrupt interfaces. In this respect, transition metal silicides are highly attractive. A number of silicides exhibit properties that are compatible and suitable for use in Si(Ge) technology. Table 3.5 shows the most common silicides with typical properties.

A major benefit of silicides is that they can be realized in a self-aligned way. This means that a silicidation process is carried out by opening the areas where the silicide is supposed to be formed. Next, the metal is deposited on top of the entire sample followed by an annealing step. During annealing, the silicide only forms in the areas where the metal was in contact with silicon. Finally, the superficial metal is selectively removed from the sample. To this end, several different acidic etch solutions (for instance piranha) can be used that neither attack the silicide nor Si, SiO_2 or Si_3N_4. As a result, a metal electrode is formed without the need for an etching process or additional lithography. If the process is done as depicted in Figure 7.9 (bottom panels), metal source/drain contacts

Table 3.5: Properties of common transition metal silicides.

Silicide	Thin film resistivity (μΩcm)	Sintering temperature (°C)	Stable on Si up to T (°C)	nm of Si consumed per nm metal	nm of resulting silicide per nm metal	Barrier height to n-Si (eV)
PtSi	28–35	250–400	~750	1.12	1.97	0.84
$TiSi_2$	13–16	700–900	~900	2.27	2.51	0.58
$CoSi_2$	14–20	600–800	~ 950	3.64	3.52	0.65
NiSi	14–20	400–600	~650	1.83	2.34	0.64
$NiSi_2$	40–50	600–800		3.65	3.63	0.66
WSi_2	30–70	1000	1000	2.53	2.58	0.67

self-aligned with respect to the gate are realized (called SALICIDE=self-aligned silicidation).

From the silicides listed in Table 3.5, $TiSi_2$, $CoSi_2$ and NiSi show the lowest resistivity. Moreover, NiSi also consumes the least amount of silicon and can be formed at relatively low temperatures. Therefore, it has received a great deal of attention and we will exclusively concentrate on nickel silicide in the following.

3.11.1 Nickel Silicidation

During nickel silicidation, nickel is the diffusing species and, therefore, lateral diffusion of NiSi needs to be carefully observed. In fact, when working with fully silicided source/drain contacts, the lateral diffusion of NiSi during silicidation can cause severe encroachment of NiSi into the channel. Ultimately, the channel of short channel devices may become short-circuited. As it turns out, the diffusion of nickel silicide strongly depends on temperature and on the geometry of the silicided silicon volume. The left panel of Figure 3.53 shows the silicide diffusion length as extracted from scanning/transmission electron microscopy images, exemplarily shown in the right panels of Figure 3.53. Here, two sets of different nanostructures were silicided: silicon-on-insulator layers (green and yellow data points in Figure 3.53) and silicon nanowires (blue and red data points) [10]. The silicidation length is plotted as a function of the inverse of the SOI-thickness/nanowire diameter. As expected, the diffusion depends on the annealing temperature yielding significantly longer lengths if the SOI is silicided at 450 °C compared to 400 °C. Less expected is the fact that the data clearly shows a stronger dependence of the silicide diffusion on the geometry of the silicided Si nanostructure than on temperature. In fact, if a nanowire is chosen, an annealing temperature of only 280 °C (note that at such a low temperature the resulting silicide is not NiSi but rather Ni_2Si) yields diffusion far above the SOI samples. It is therefore extremely important to control the diffusion of the silicide when working with nickel.

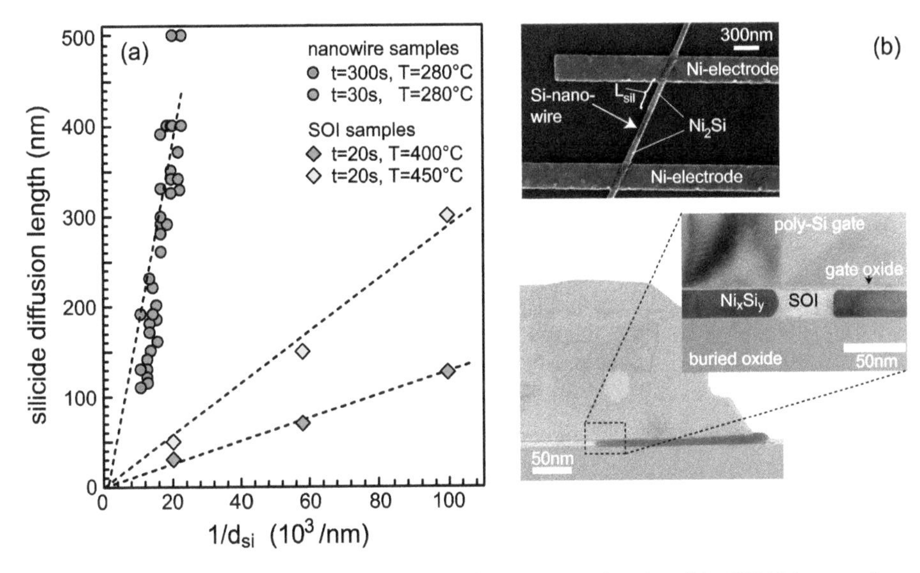

Figure 3.53: (a) Diffusion length during nickel silicide formation as a function of the SOI thickness and the diameter of Si nanowires (both called d_{Si}) for different formation temperatures [10]. (b) Scanning and transmission electron micrographs of silicided Si nanowires and silicon-on-insulator.

Continued silicide diffusion is possible when there is a (quasi-)infinite source of nickel on top of the silicon, which was the case in the experiments shown in Figure 3.53. If, on the other hand, the available nickel is limited, diffusion will stop once the entire nickel has been consumed. Hence, silicide diffusion can be controlled even in nanostructures, when the amount of nickel can be precisely controlled. For silicon-on-insulator, this can be accomplished easily by depositing an appropriate thickness of nickel on top of the silicon (with the factor provided in Table 3.5). If nanowires are being used, deposition of the appropriate amount is not trivial due to the three-dimensional geometry. A suitable way to circumvent excessive silicide diffusion is therefore a two-step silicidation. This means that after nickel deposition, a first low temperature and short silicidation process is carried out. At low temperatures, a Ni-rich phase is formed and thus, removing the superficial nickel, a finite source of Ni is realized, i. e., the superficial nickel within the Ni_2Si. In a second annealing at higher temperatures, NiSi (400–600 °C) or $NiSi_2$ (600–800 °C) then forms without excessive diffusion.

3.11.2 Dopant Segregation During Nickel Silicidation

The last row in Table 3.5 shows that all silicides exhibit a rather large Schottky barrier with respect to the conduction band of silicon (this will be elaborated on in Section 4.6.2). To overcome the barrier and realize a proper Ohmic contact, the semiconductor is usually heavily doped, which yields a thin, highly transmissive Schottky barrier (cf. Figure 4.22). However, in many device applications (such as the emitter of solar cells, fully-depleted SOI MOSFETs, etc.) an inhomogeneous doping profile with a highly doped layer only at the silicide–silicon interface is desirable. In the case of a highly doped interface layer and provided the layer has an appropriate thickness, a strong band bending can be expected, which makes the Schottky barrier thinner thereby promoting the tunneling of carriers through the barrier.

Nonuniform doping profiles can be realized with the technique of dopant segregation during silicidation, which is depicted in Figure 3.54: Consider a bulk silicon sample into which dopants have been implanted. If the sample is coated with nickel and if subsequently the entire implanted volume in the silicon sample is nickel-silicided, the dopants redistribute and are piled up at the silicide–silicon interface. This phenomenon is called dopant segregation during nickel silicidation. The fact that nickel is the diffusing species in the NiSi formation leads to a "consumption" of the implanted silicon volume thereby pushing the dopants in front of the moving NiSi-Si interface resulting in a substantial change of the dopant profile.

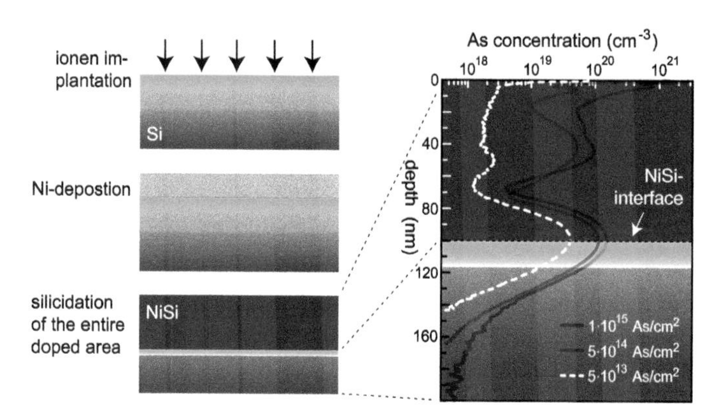

Figure 3.54: Left: Schematics of dopant segregation during nickel silicidation in bulk silicon. The right panel depicts SIMS profiles showing the segregation of arsenic after nickel silicidation.

Whether dopant segregation occurs or not is determined by the diffusivity and solid solubility of the dopants in the silicide and the presence of point defects at the silicide–silicon interface. A significant change of volume is involved when the silicide is formed, which leads to a high strain at the interface. As a result, point defects (self-interstitial or vacancies) can be generated in order to partially relieve the stress. Due to the formation of vacancies, the diffusivity of dopants in silicon is enhanced such that they move toward the interface where they get piled up. With dopant segregation during silicidation, a local dopant concentration higher than the solid solubility can be realized due to the high density of strain-induced point defects.

The occurrence of dopant segregation in the NiSi–silicon system is confirmed by a secondary ion mass spectroscopy (SIMS) investigation of bulk silicon samples, which are implanted with arsenic, and subsequently the whole implanted area is fully silicided. The right panel of Figure 3.54 shows the depth distribution of arsenic as measured by time-of-flight SIMS. The resulting depth profiles indicate that dopant segregation occurs at the NiSi–silicon interface. The As-concentration exceeds 1×10^{20} cm^{-3} for an initial implantation dose larger than 5×10^{14} cm^{-2} (see Figure 3.54, right). The pronounced peak at a depth of ~100 nm corresponds to the silicide–silicon interface indicating that the initially implanted region (with peak at ~10 nm) is totally silicided. Moreover, Figure 3.55(a) shows that the dopants are pushed in front of the silicidation front into the bulk of the substrate similar to a snow-plow maintaining a steep dopant distribution. In addition, there is only a slight reduction of the peak As concentration observable with increasing NiSi thickness. As a result, dopants can be redistributed and piled up at the NiSi interface enabling inhomogeneous and steep doping profiles. The question now is whether these dopants are active keeping in mind that no activation anneal is done after the silicidation (and most of the times also prior to silicidation) and that the silicidation is carried out at rather low temperatures. To clarify this, electrochemical capacitance voltage measurements are required as shown in Figure 3.55(b). Here, the distribution of active dopants in a silicon substrate is shown prior (blue) and after nickel silicidation. One clearly observes the redistribution with a strong increase of the peak concentration. More importantly, these dopants are active proving that dopant segregation can be used to create inhomogeneous and steep doping profiles with high active peak concentrations.

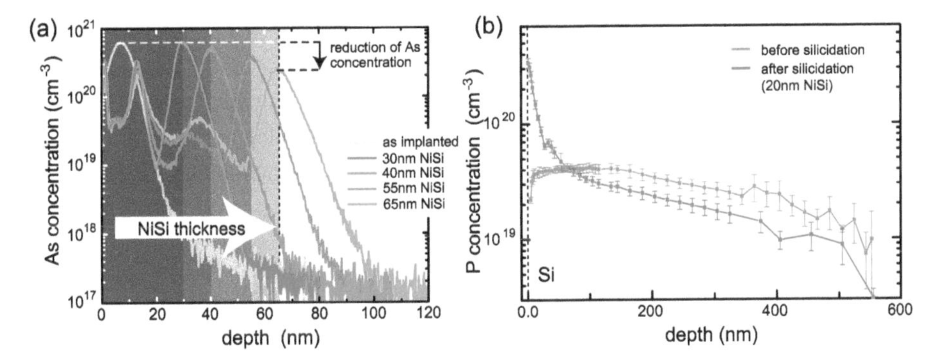

Figure 3.55: (a) SIMS profiles of arsenic segregation for different NiSi thicknesses. (b) Depth profiles of active dopants (phosphorous) in silicon prior (blue) and after nickel silicidation (red).

3.12 Hydrogen Annealing

Annealing structures etched into silicon in pure hydrogen allows rather drastic shape transformations. If properly done, H_2-annealing increases the surface migration of Si atoms and hence leads to a reflow and rearrangement of the Si surface [166]. This is used, for instance, to smoothen etched Si fins and nanowires [67], remove variability

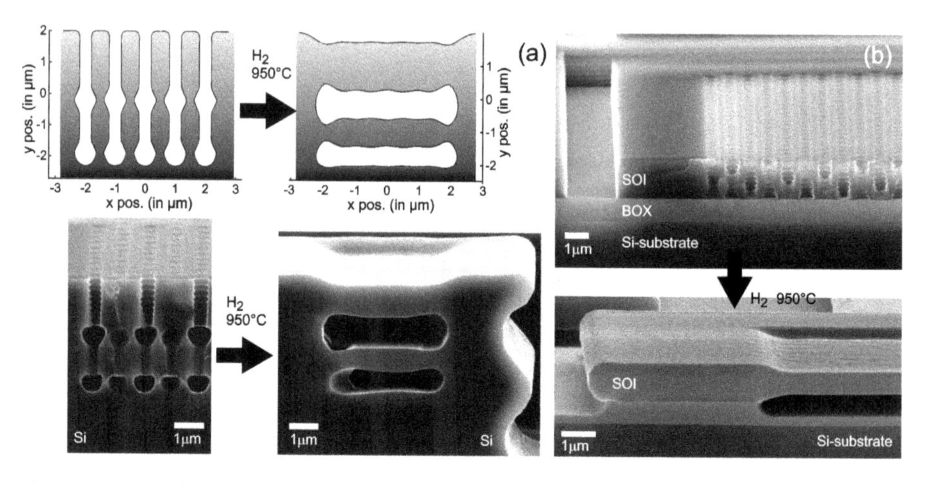

Figure 3.56: (a) Hydrogen annealing of bulk Si. Simulation results utilizing the level-set method (top) and experimental results (bottom) [245]. (b) H_2-annealing of a SOI sample. [245].

and transform a rectangular cross-section of a Si nanowire to a rather circular one (cf. Figure 3.58). At elevated temperatures above ~950 °C, H_2 annealing can lead to a strong reflow enabling spectacular shape transformations. Figure 3.56 shows two such examples: in (a) a series of periodically arranged holes have been etched into bulk silicon with a sequence consisting of multiple short and two long Bosch etch steps that lead to the pattern shown in Figure 3.56(a) (lower panel). Upon hydrogen annealing, a double membrane with two empty spaces in silicon is created. The upper panels show a simulation based on a level-set approach, nicely reproducing the experimental results [245]. In (b), a similar pattern of holes (only short Bosch etch steps) is etched into a SOI wafer. Subsequently, the buried oxide has been removed with HF followed by H_2 annealing. As a result of the reflow (an animation of the effect of hydrogen annealing is provided through QR code #36), a closed, crystalline Si membrane on top of an empty space is obtained.

3.13 Low-Dimensional Semiconductors

As will be discussed in detail in Chapter 5, nanostructures with extremely small dimensions, are ideally suited for the realization of high performance transistor devices in highly integrated circuits. Suitable nanostructures comprise nanowires/nanotubes that can be fabricated with a so-called top-down or bottom-up approach, which will both be discussed briefly in the following sections. In addition, two-dimensional materials such as graphene, transition metal dichalcogenides and black phosphorous have recently attracted a great deal of attention. Therefore, processes specific for the manipulation of 2D materials (such as transfer from a host to a different substrate) will be presented at the end of the chapter.

3.13.1 Top-Down Fabrication of Nanowires

Nanowires can be readily fabricated using a top-down approach with e-beam lithography (cf. Section 3.5.3) and reactive ion etching as illustrated in Figure 3.57. Using silicon-on-insulator, an H-shaped pattern is etched down to the buried oxide followed by the removal of the BOX to release the nanowire from the substrate. However, a lithographic process always leads to line-edge roughness, which becomes increasingly problematic when the size of the nanostructure is scaled down. A viable way to mitigate this is to pattern a larger nanostructure and then trim down the nanowire to the desired size employing thermal oxidation (cf. Figure 3.4, bottom right) and HF-stripping and/or digital etching (cf. Section 3.6.3). In fact, with appropriate thermal oxidation conditions, even a self-terminating process can be exploited that prohibits the nanostructure to become fully oxidized leaving a silicon nanowire core with a few nanometers in diameter [200]. In addition to oxidation, annealing in a pure, low pressure hydrogen atmosphere can be used in order to reshape and smoothen the nanowires before the formation of the gate dielectric (see the preceding section) [166, 67].

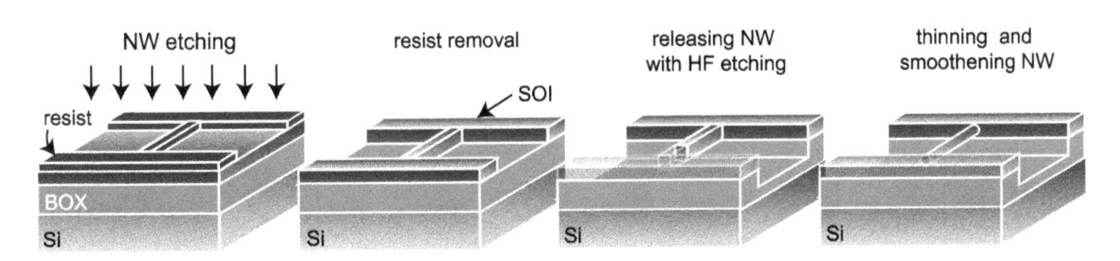

Figure 3.57: Process flow for the top-down fabrication of lateral Si nanowires.

An alternative approach to nanowire fabrication can be implemented with the help of anisotropic silicon etching. This process does not require the use of a lithography method with nanoscale resolution such as EBL [286, 244]. Figure 3.58 shows an example of such a process: First, a thin SiN mask is generated (e. g., with RTN, see Section 3.3) on top of a SOI wafer. After patterning the SiN along a [110]-direction (a), anisotropic silicon etching is employed (Section 3.6.4) (b), followed by a dry oxidation. Since the SiN serves as a diffusion barrier, the Si will only oxidize locally on the exposed {111} crystalline facets (Section 3.2.2) (c). A second mask layer is generated with optical lithography (d) and the SiN is selectively patterned (Section 3.6.2) (e). Subsequently, anisotropic Si etching is carried out a second time (f). Since the first etch flank is covered by the LOCOS oxide, a nanowire with triangular cross-section is obtained. Finally, the LOCOS oxide is removed and the nanowire released from the substrate (g). An oxidation plus HF stripping or hydrogen anneal can then be used to realize circular, smooth nanowires (h).

A very interesting approach that allows the fabrication of a whole array of multiple silicon nanowires is to exploit the scallop formation during the Bosch process [200, 36]: if a (multiple) nanoscale fin-structure(s) is etched into silicon with a Bosch process where

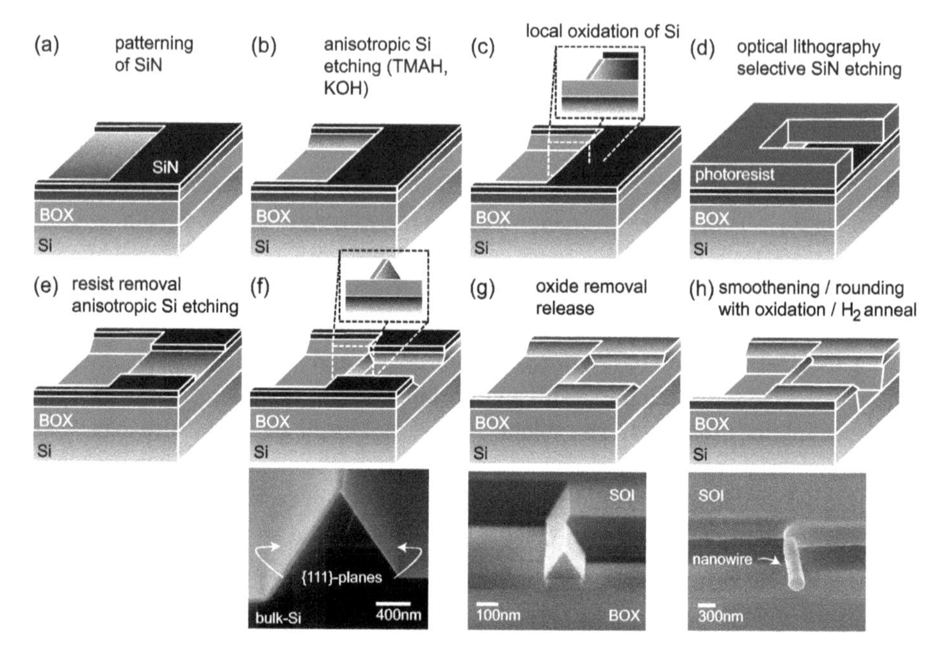

Figure 3.58: Schematic process flow for the top-down fabrication of lateral, suspended Si nanowires realized with a combination of anisotropic silicon etching and local oxidation of silicon [286, 244].

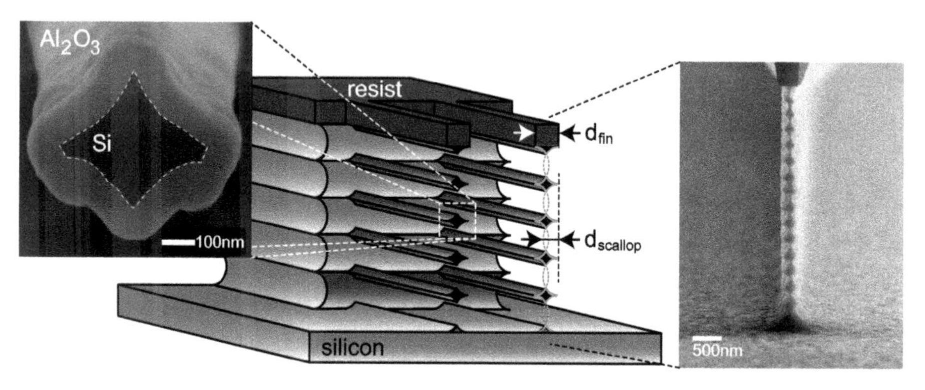

Figure 3.59: Array with vertically stacked nanowires realized exploiting the overlap of Bosch scallops (red dashed lines) when etching fins with $d_{\text{fin}} < 2 \cdot d_{\text{scallop}}$ [200]. The electron micrographs show a cross-section of a single diamond-shaped nanowire (left) and an experimental realization of vertically stacked nanowires (right).

the isotropic etch step leads to a depth of the scallops d_{scallop} on the order of or larger than half of the fin width d_{fin} (cf. Figure 3.59), diamond-shaped patterns appear as illustrated in Figure 3.59. Subsequently, (self-terminating) oxidation and HF-stripping or digital etching in combination with hydrogen annealing can be employed to trim down and round the Si nanostructures [200]. A similar stack of nanowires can also be realized based on Si/SiGe heterostructures (see Section 5.8.3). The addition of self-terminating and self-adjusting processes to top-down fabrication provides unprecedented control over the geometry and position of the nanostructures.

3.13.2 Growth of Nanowires

Nanowires can also be realized with a (bottom-up) growth process. One of the most frequently used approaches is the so-called VLS (vapor-liquid-solid) method [263] which exploits the properties of a eutectic such as the AuSi system; its binary phase diagram is depicted in Figure 3.60. Suppose that a Au nanoparticle is at a temperature of 400–500 °C with a Si content of 30 %. If the Si concentration dissolved in the nanoparticle is continuously increased (e. g., by providing Si through the catalytic decomposing of silane as illustrated in the inset) the liquidus (red line) is crossed horizontally and the AuSi system enters the binary (liquid/solid)phase region. Interestingly, the Au concentration in the solid phase is zero and as a result, a pure Si solid grows out of the Au nanoparticle.

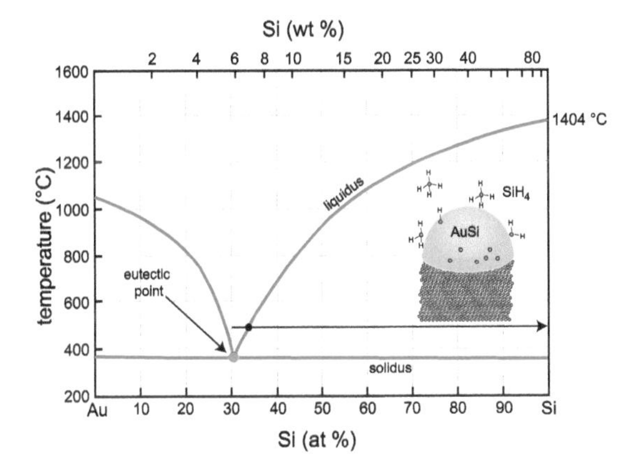

Figure 3.60: The gold-silicon binary phase diagram. The inset shows a schematic of a nanowire growing from a gold nanoparticle due to supersaturation of Si.

If a silicon substrate with Au nanoparticle is mounted into a CVD chamber and, e. g., silane is provided the nanowire will grow at the silicon–nanoparticle interface. Because the solid Si phase within the binary phase region does not contain Au, the gold nanoparticle is not consumed during the growth but "moves" upwards staying at the top of the nanowire. This behavior is shown in Figure 3.61 [230]. Here, a gold-disc is deposited onto a patterned Si/SiO_2 substrate with EBL and lift-off. Annealing the sample yields a gold nanoparticle from which the Si nanowire grows. In this respect, another important property of the eutectic is exploited; the eutectic point of the AuSi system is at a rather low temperature. This means that the VSL growth can be carried out at temperatures significantly below the temperature where the precursor (e. g., silane) spontaneously

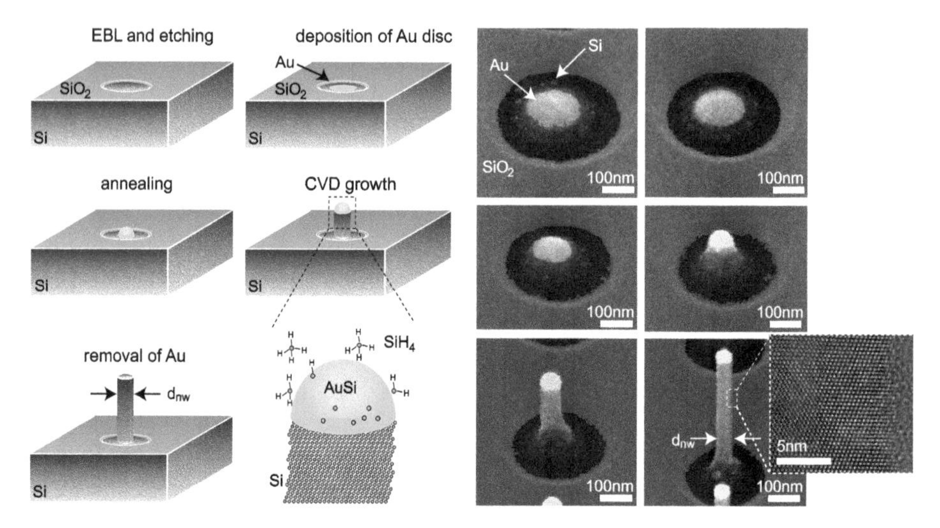

Figure 3.61: Illustration and electron micrographs of a vapor-liquid-solid growth of silicon nanowires using the AuSi eutectic material system [230].

decomposes. As a result, planar Si-growth is avoided, which means that the nanowire diameter will stay (almost) constant during the growth and is determined by the diameter of the Au nanoparticle. After the growth, the gold is etched away, and subsequently the nanowire can be harvested (mechanically scratching it off the sample or by using ultrasound) or further used to realize, e. g., vertical nanowire FETs (see Section 9.2.3).

It is certainly a viable question what the benefit of a nanowire growth is if either harvesting of the nanowires and transfer to another substrate after the growth process or if a lithography process for patterning and positioning of the seed particles is necessary. The answer is that at the nanoscale materials can be grown epitaxially on top of each other that exhibit vastly different lattice constants (cf. [258], for instance). This provides a great deal of opportunities to tailor material properties with appropriate heterostructures along the nanowire axis unparalleled in conventional bulk epitaxial growth. Moreover, increasing the temperature during nanostructure formation, such that planar CVD growth is enabled, opens the possibility to realize axial and radial heterostructures (core-shell nanowires) within the same nanowire. Therefore, nanowire growth is a perfect test bed for the exploration of ways to optimize devices or explore new device functionalities.

3.13.3 2D Materials—Exfoliation and Visibility on Substrates

The first investigations of the electronic transport in a two-dimensional material were carried out by Novoselov and Geim [203] and initiated a new field of research. Soon after the demonstration of monolayers of graphene, a great variety of different, graphene-like two-dimensional materials (2D materials) were found (see Section 2.8). Since then, transition metal dichalcogenides (TMDCs) such as MoS_2 and WSe_2 but also materials such as black phosphorous have attracted a great deal of attention.

The initial realization of graphene relied on exfoliating flakes of mono- and multi-layer graphene with an adhesive tape (often called the "scotch-tape" method) from a block of graphite. After thinning the exfoliated flake, it is transferred to an insulating substrate (usually oxidized silicon) by pressing the tape onto the substrate. Subsequently, the substrate is inspected using optical microscopy to find suitable flakes, preferably exhibiting parts with monolayer thickness. Finding a monolayer of graphene with optical microscopy (OM) on suitable substrates is possible because of multiple reflections and interference of the light at the graphene surface as well as graphene-insulator and insulator–substrate interfaces. In order to obtain a sufficiently high contrast enabling to distinguish a monolayer graphene from a spot on the substrate without any flake and even to distinguish between a mono and a bilayer, an appropriate combination and thicknesses of substrate and insulator need to be chosen. As mentioned above, the most straightforward substrate and insulator are silicon and SiO_2, respectively. However, in many cases other substrate materials and other two-dimensional materials are desirable. Being able to identify the number of layers of a specific 2D flake from OM images is therefore extremely helpful when working with 2D materials.

Figure 3.62 shows optical microscopy images of (a) graphene on a 300 nm SiO_2/Si substrate [154], (b) WSe_2 on a 287 nm SiO_2/Si substrate [196] and (c) WSe_2 on a 7 nm Al_2O_3/Al substrate [198]. Simulations of the visibility are plotted in the figures, too, showing that the color impression can be reproduced very well. This allows the identification of suitable areas with a specific number of layers of a 2D material on a substrate and it

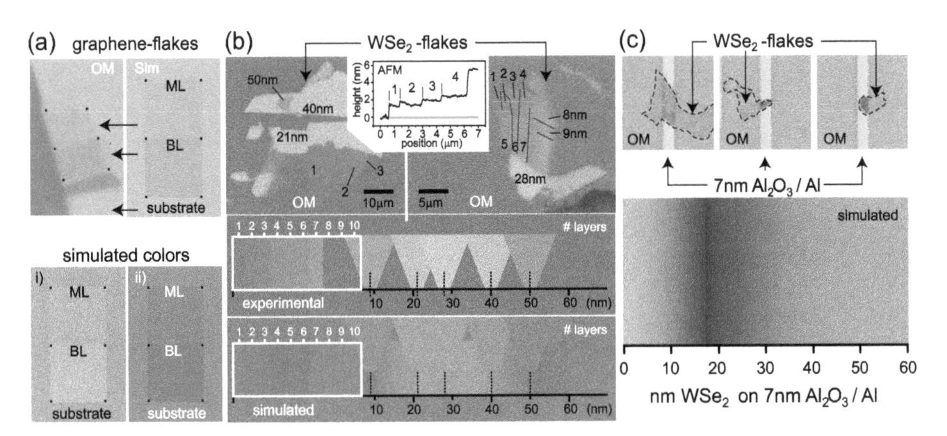

Figure 3.62: (a) Visibility of monolayer (ML) and bilayer (BL) graphene on 300 nm SiO_2/Si (top), simulated color of graphene on (i) 101 nm SiO_2/34 nm $BaTiO_3$/125 nm HfO_2/Si and (ii) 29 nm SiO_2/47 nm Al_2O_3/12 nm $BaTiO_3$/Si [154]. (b) Optical microscopy (OM) of WSe_2 on SiO_2/Si. The color coding was confirmed with AFM (inset). The lower panel compares experimental and simulated colors [196]. (c) OM (top) and simulated visibility of WSe_2 on Al_2O_3/Al [198].

enables the prediction of the dielectric layer sequences of substrates providing optimum visibility such as (i) and (ii) shown in Figure 3.62(a) for graphene [154]. The visibility and color impression for various additional substrates are shown in Figure A.1, Appendix A. A computer program that allows the simulation of the visibility can be downloaded through the QR code #37.

3.13.4 2D Materials Transfer—PVA/PMMA Method

In many cases, the 2D material needs to be transferred from one substrate to another one. For instance, growth of graphene is often done on copper foils, which are obviously conductive and cannot be used as a substrate for a device. Transfer of 2D materials is also required if the material needs to be placed at a specific position on the substrate since simple exfoliation leads to a random distribution of flakes with different thicknesses across a substrate. Finally, if heterostructures consisting of several 2D materials are realized, transfer of several different 2D materials on top of each other is required. There are quite a few different methods of transfer from different substrates. In the present chapter, only the frequently used PVA/PMMA method is described; the reader is referred to [125] that discusses in detail various alternative approaches.

In order to enable the 2D material transfer, an intermediary substrate based on bulk silicon is fabricated as illustrated in Figure 3.63. First, an array of marker structures is etched into the substrate (a). Next, PVA and PMMA are spun on the substrate followed by baking the sample on a hot plate. The marker structures in the Si substrate need to be deep enough so that they are transferred into the dual polymer film. Afterwards, flakes are exfoliated onto the PMMA (b). The color contrast (see preceding section) helps to identify regions with mono-, bi- and multilayer material. As described in more detail below, Raman characterization is then carried out in order to determine the number of

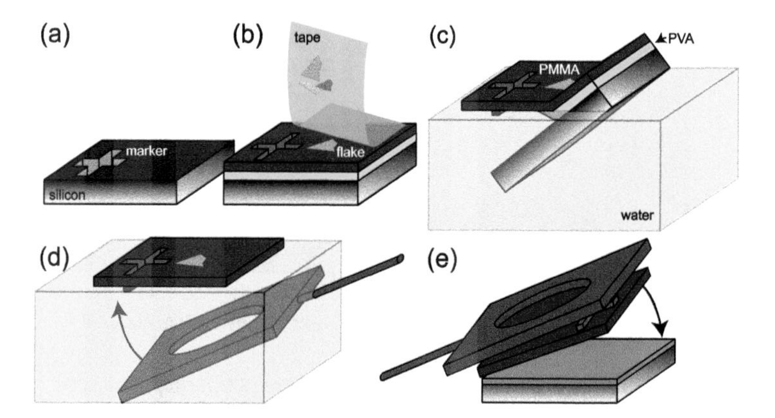

Figure 3.63: Process flow of the PVA/PMMA method for 2D materials transfer. (a) Marker etching, (b) spin-coating of PVA and PMMA, exfoliation of 2D material, (c) dissolution of PVA in water, (d) fishing of the PMMA floating on the water, (e) transfer of the PMMA with the 2D material upside down onto a new wafer.

layers of the respective material unambiguously. After scratching one side of the polymer layer, the sample is carefully put on the surface of water making sure that the sample floats (c). The scratch enables the water to dissolve the PVA starting from the side of the sample with the scratch. When the PVA is completely dissolved, the PMMA sheet floats on the water surface and can be fished with an appropriate device (d). Finally, the PMMA including the 2D material is put upside down onto the new substrate. The hole in the fishing device allows observing the marker structures of the initial silicon substrate that have been transferred into the PMMA. Knowing the position of the 2D material flake with respect to the markers allows aligning the flake with respect to a pattern on the new substrate. Finally, the PMMA is removed, which can in principle be done with acetone or by heating it off in an appropriate furnace.

Figure 3.64 shows optical microscopy images of a PVA/PMMA transfer. The marker structures of the initial substrate are clearly visible in (a) and also in (b), which shows the PMMA upside down on the new substrate (featuring a multielectrode structure). (c) shows the final result where the PMMA has been removed.

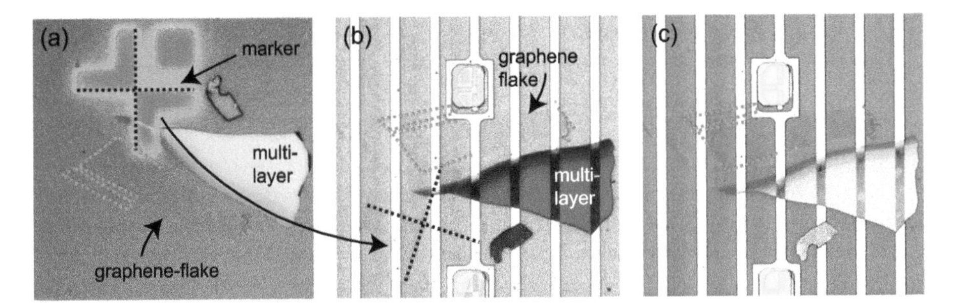

Figure 3.64: Optical microscopy images of a 2D material transfer. (a) Exfoliated graphene on top of a PVA/PMMA coated Si substrate. (b) Transferred graphene and PMMA layer on the new substrate; (c) after the removal of the PMMA.

Although an appropriate substrate allows to distinguish between mono-, bi- and multilayer (cf. Section 3.13.3), an unambiguous determination whether a 2D material flake is a monolayer or not requires a Raman characterization. Exemplarily for the case of graphene on a PVA/PMMA-coated Si substrate, Figure 3.65(a) shows an optical and an atomic force microscopy image (AFM) together with Raman spectra measured at the spots on the graphene flake marked with the colored dots in the OM image. One clearly observes that in the case of the PVA/PMMA substrate it is rather difficult to distinguish between mono-, bi- and multilayer using merely optical contrast. Furthermore, the AFM image does not allow one to distinguish the number of layers either. This is mainly due to the fact that because of the deposition of the flake onto the intermediary PVA/PMMA stack, graphene forms bubbles, which can be as tall as 10 nm, i. e., much larger than the step height of a graphene layer. Therefore, Raman characterization (see main panel of Figure 3.65(a)) is mandatory.

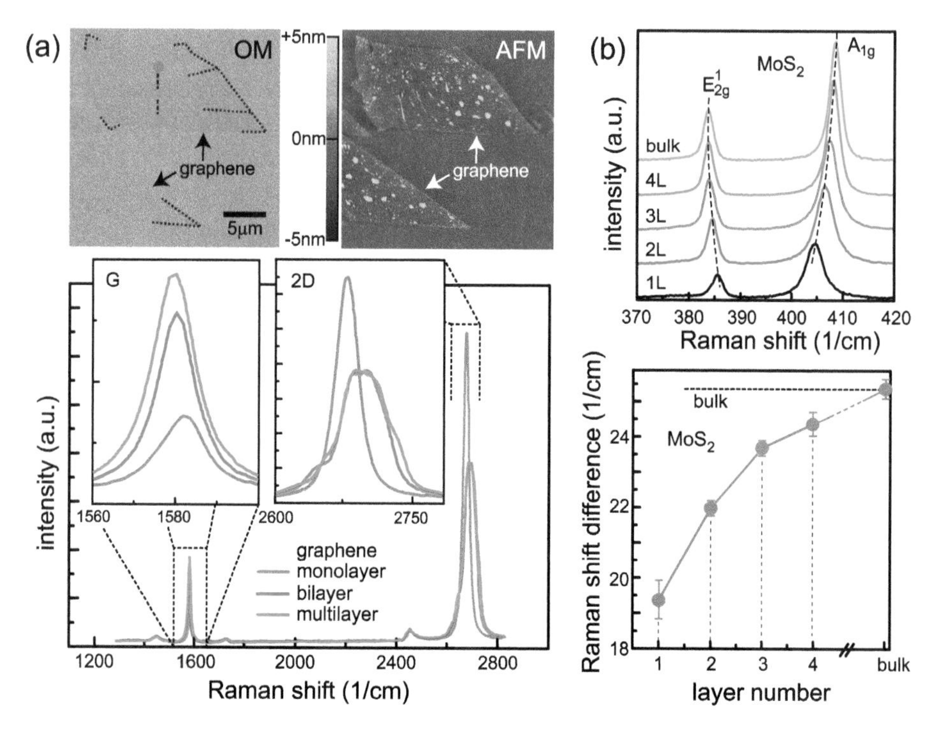

Figure 3.65: (a) Optical microscopy, AFM and Raman characterization of a graphene flake with mono-, bi., and multilayer regions. The colored dots in the OM image show where the respective Raman measurements were carried out. (b) Typical Raman spectra for different layers of MoS_2. The difference between the E^1_{2g}- and the A_{1g}-modes is plotted in the lower panel and allows a clear distinction between mono-, bi- and triple-layers of MoS_2.

Three peaks in the Raman spectrum are characteristic for graphene. The *G*-peak (at ~1581 cm^{-1}) is due to bond stretching in all pairs of atoms of the hexagonal carbon lattice. The *D* and the 2*D*-peaks (at ~1343 cm^{-1} and ~2674 cm^{-1}) can be explained by a double resonance scheme. The 2*D*-peak is key to the differentiation of mono- and bilayer graphene. In the case of a monolayer, a single Lorentzian peak is observed (see Figure 3.65). In contrast, a bilayer shows a 2*D*-peak, which is asymmetric, and exhibits a characteristic shoulder on its lower Raman-shift side [90]. The spectra of graphene exhibiting three layers or more are all very similar and symmetric. As an additional indicator for the identification of a monolayer graphene, the ratio between the heights of the 2*D*- and *G*-peaks can be used, which should be greater than 10 : 1 (cf. Figure 3.65(a)).

Raman spectroscopy allows also the identification and differentiation of the number of layers in TMDCs. Figure 3.65(b) shows the Raman spectra of MoS_2 flakes. Here, the difference between the E^1_{2g} and the A_{1g} modes serves as an indicator of the number of layers present in the investigated flake. The differences of the Raman shifts are plotted as a function of the number of layers in Figure 3.65(b), lower panel. Obviously, one can distinguish between 1–4 layers within the flake.

After the transfer of a flake of 2D material, contacts have to be made to the flake using, e. g., PMMA, electron-beam lithography and lift-off. However, for the transfer process itself, PMMA is used and as a result, the PMMA layer can be utilized for the contact formation. To this end, the EBL is carried out *before* the transfer, i. e., while the flake still resides on the host substrate. The benefit of this is that a stable substrate with marker structures can be used for the lithography while the substrate the flake will be transferred to, can be optimized to facilitate best performance of a targeted device concept. For example, 2D materials show strain-dependent electrical properties and to study this, flexible substrates are required. Processing the sample on a flexible substrate, however, is not a trivial task. Carrying out the e-beam lithography on the intermediary substrate (i. e., exposure through the 2D material into the PMMA), transferring the sample and then develop the PMMA allows using the PMMA as transfer substrate and lithography mask. After final deposition of a metal and lift-off, contact electrodes can readily be fabricated even on flexible (or extremely small) substrates. As an example, Figure 3.66 shows three optical microscopy images of (a) the PVA/PMMA substrate (the dashed lines are the contact areas exposed with 10 keV e-beam lithography), (b) the transferred and developed pattern (note the mirror-inverted electrodes) and (c) the structure after metal deposition (EBE) and lift-off.

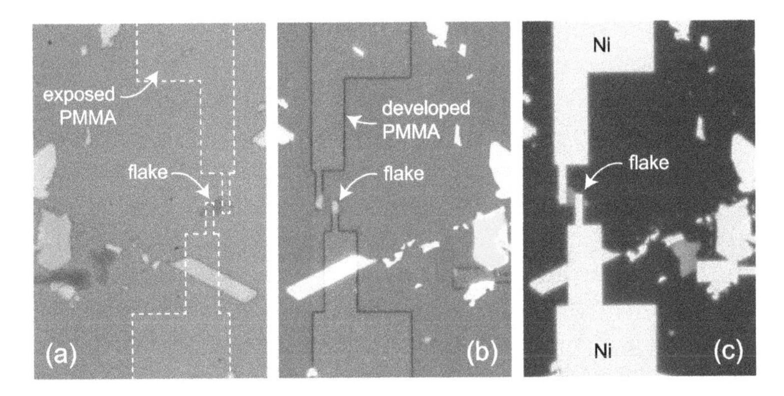

Figure 3.66: 2D material transfer process including electrode patterns written with EBL into the PMMA layer used for the transfer prior to the actual transfer process (a). The development of the PMMA is carried out after the transfer on the new substrates (b). After EBE, a lift-off is carried out (c).

Exercises

Exercises together with solutions are accessible via the QR code.

4 Basic Ingredients for Nanoelectronics Devices

The present chapter introduces the most important ingredients for nanoelectronics devices and in particular field-effect transistors. First, doping of semiconductors is discussed ranging from simple p-n-junctions to studying effects related to doping at the nanoscale and a possible alternative to impurity doping. As a second ingredient, metal-oxide-semiconductor capacitors (MOS capacitor) will be investigated. Finally, metal-semiconductor contacts will be elaborated on in detail followed by a short discussion of heterostructures.

4.1 Doping of Semiconductors

In the case of an intrinsic semiconductor, all electrons in the conduction band have been thermally excited from the valence band. Using the density of states (DOS) in three dimensions (cf. Equation (2.83)) the bulk electron density n can be computed as energy integral of the DOS multiplied with the Fermi distribution function:

$$n = \int_{-\infty}^{\infty} dED_{3D}(E) \cdot f(E_f) \approx \int_{E_c}^{\infty} dE \frac{m^{\star}_{\mathrm{DOS}}}{2\pi^2\hbar^3} \sqrt{2m^{\star}_{\mathrm{DOS}}(E - E_c)} \cdot \exp\left(-\frac{E - E_f}{k_B T}\right) \tag{4.1}$$

where on the right-hand side the Fermi distribution function has been replaced with the Boltzmann approximation. This is justified because in the undoped (intrinsic) case the Fermi level will be located within the band gap with $E_c - E_f \gg k_B T$. Note that degeneracy, e. g., due to the different valleys in the silicon conduction band can be taken into consideration by choosing an appropriate density of states effective mass $m^{\star}_{\mathrm{DOS,c/v}}$ for the conduction or valence band (see Section 2.11.2). The right side of Equation (4.1) is a definite integral and can be rewritten as a product of a prefactor and an exponential term, which yields $n_i = N_c e^{-(E_c - E_f)/k_B T}$. Here, N_c is the so-called effective density of states of the conduction band as has already been stated in Section 2.11.3 (cf. Equation (2.94)). This means that the carrier density appears as being located at the conduction band edge E_c with a certain density given by $N_c = 2(\frac{2\pi m^{\star}_{\mathrm{DOS,c}} k_B T}{h^2})^{3/2}$ multiplied with the Boltzmann factor, i. e., the probability that carriers can be found at E_c. Replacing $E_c - E_f$ with $E_f - E_v$ and N_c with $N_v = 2(\frac{2\pi m^{\star}_{\mathrm{DOS,v}} k_B T}{h^2})^{3/2}$ allows computing the density of holes p in the valence band in the same fashion. With appropriate values for the effective masses, one obtains $N_c \approx 2.81 \times 10^{25}\,\mathrm{m}^{-3}$ and $N_v \approx 1.83 \times 10^{25}\,\mathrm{m}^{-3}$ at room temperature.

Since in the intrinsic case all electrons are thermally excited from the valence band leaving behind a hole, the intrinsic carrier density n_i must be equal to n, which is equal to p and as a result, $n_i = \sqrt{n \cdot p}$. Inserting Equation (4.1) and its equivalent for p yields the well-known relation between n_i and the band gap,

$$n_i = \sqrt{N_c \cdot N_v} \exp\left(-\frac{E_g}{2k_B T}\right), \tag{4.2}$$

https://doi.org/10.1515/9783111054421-004

with $E_g = E_c - E_v$. The exponential dependence of the intrinsic carrier density on the band gap yields rather small n_i at room temperature if the band gap is not small. In fact, in silicon, $n_i \approx 1.5 \times 10^{10}$ cm^{-3}, which means that undoped silicon at room temperature is a rather bad conductor. We can define an intrinsic Fermi level E_f^i using Equation (4.2) by inserting $n_i = N_c \exp(-\frac{E_c - E_f^i}{k_B T})$. Solving for E_f^i yields after some algebra

$$E_f^i = \frac{E_c + E_v}{2} + \frac{k_B T}{2} \ln\left(\frac{N_v}{N_c}\right), \tag{4.3}$$

meaning that the Fermi level in an intrinsic semiconductor is close to midgap; if the effective densities of states in the conduction and valence bands $N_{c,v}$ are equal, E_f^i is exactly at midgap.

Task 18.
Nondegenerate carrier density: Explicitly carry out the computation of the carrier density in a bulk semiconductor in the nondegenerate limit. You may want to use the definite integral $\int_0^\infty \sqrt{x} \exp(-x)\, dx = \sqrt{\pi}/2$.

In order to increase the number of carriers and thereby the conductivity of the semiconductor, doping is necessary. Doping of semiconductors is one of the most important techniques of microelectronics that virtually enabled all of today's electronic devices such as diodes, bipolar and field-effect transistors as well as photodetectors and solar cells. The reason for this importance is that doping allows very large changes in the conductivity of the semiconductor based on electron conduction (n-type) in the conduction band as well as hole conduction (p-type) in the valence band. A very important fact is that these conductivity changes can be accomplished while leaving the band gap mostly intact (apart from a band-gap lowering and band-tailing, see discussion below) because the energy levels of the dopants are close to the conduction band (in the case of donors) or to the valence band (for acceptors).

Let us consider an elemental group IV semiconductor (e. g., silicon or germanium) and assume that the semiconductor is doped n-type (p-type) with a concentration of N_d donors (N_a acceptors). In order to act as a donor (acceptor), a group V (group III) impurity is needed that replaces an atom on a crystal lattice site of the host semiconductor. Since four electrons are needed to form covalent bonds with the adjacent atoms, donors (acceptors) exhibit one superficial (one deficit) electron. A simple model to estimate the ionization energy of the donors (acceptors) is to consider the donor (acceptor) as a hydrogen-like atom (this model is further elaborated on in Section 4.3.1). Essentially, due to the donor (acceptor) being embedded into the dielectric of the host semiconductor and due to a change in effective mass, the ground-state energy of the (hydrogen-like) donor is the donor level E_d and the ionization energy is $E_{\text{ion}} = E_c - E_d$. For better readability, the following discussion will be stated explicitly for donors as dopants; a similar analysis can be carried out for acceptors.

Since typical donors have an E_d close to the conduction band, most dopants are ionized at room temperature providing their superficial electron into the conduction band thereby leaving behind a positively charged donor. The probability of finding ionized donors is meditated by the Fermi distribution function for donors f_{donor} (see Equation (2.72)) or one should rather say $1 - f_{\text{donor}}$, i. e., by the probability that the donor states are empty. If the doping concentration is rather small, such that all dopants can be considered as being isolated from each other (see the next section and Figure 4.2, left panel) the density of states for the dopants is simply given by a delta function (as in Equation (2.78)) multiplied with the dopant density, i. e., $D_{\text{dop}}(E) = N_d\delta(E - E_d)$. As a result, the density of ionized (positive charge) donors is

$$N_d^+ = \int dEN_d\delta(E - E_d)\left(1 - \frac{1}{1 + \frac{1}{2}e^{\frac{E-E_f}{k_BT}}}\right) = \frac{N_d}{1 + 2e^{\frac{E_f-E_d}{k_BT}}} \tag{4.4}$$

where the integration can be carried out right away due to the delta function. Remember that the factor $\frac{1}{2}$ has to be introduced in the Fermi function for donors to account for the two possibilities of occupying the donor state (see the discussion in Section 2.10.1 regarding the derivation of the Fermi function in the case of distinguishable and indistinguishable states). Hence, the particle density of electrons is given by

$$n \approx N_d^+ + p \xrightarrow{p=\frac{n_i^2}{n}} n \approx N_d^+ + \frac{n_i^2}{n} \tag{4.5}$$

where we considered the case without any acceptors in the semiconductor. Expression (4.5) yields a quadratic equation for the determination of n. However, in the case of sufficiently high n-type doping and/or low temperatures $n_i^2/n \ll n$ and, therefore, the term is often neglected resulting in $n \approx N_d^+$. With increasing temperature on the other hand, the carrier density eventually is again dominated by thermal excitation from the valence band such that the contribution of p needs to be incorporated and ultimately yields $n \approx p$ as in the intrinsic case.

Equation (4.5) allows computing the energetic position of E_f as a function of doping concentration and temperature. Explicitly writing down Equation (4.5) in the nondegenerate case, it is clear that $E_f(T)$ needs to be computed numerically by solving

$$N_c e^{-\frac{E_c-E_f}{k_BT}} = \frac{N_d}{1 + 2e^{\frac{E_f-E_d}{k_BT}}} + N_c N_v e^{-\frac{E_g}{k_BT}} \frac{1}{N_c} e^{-\frac{E_f-E_c}{k_BT}} \tag{4.6}$$

where $N_{c,v} \propto (k_BT)^{3/2}$. As discussed already above, for high temperatures the intrinsic case is obtained with E_f lying close to mid-gap. For cryogenic temperatures, on the other hand, $\frac{E_f-E_d}{k_BT} \gg 1$, and hence $n \approx N_d^+ \to N_c e^{-\frac{E_c-E_f}{k_BT}} \approx \frac{N_d}{2} e^{-\frac{E_f-E_d}{k_BT}}$. We can now define a low temperature (LT) carrier density $n_{\text{LT}} = \sqrt{n \cdot N_d^+}$ in the same way as the intrinsic carrier density has been derived above which results in $n_{\mathrm{LT}} = \sqrt{\frac{N_c N_d}{2}} e^{-\frac{E_c - E_d}{2k_B T}}$. This means that (apart from the factor of 2 under the square root) the same equations are obtained as in the intrinsic case. But now, the doping energy level E_d plays the role of the valence band and the difference $E_c - E_d$ the role of the band gap. Hence, at cryogenic temperatures E_f approaches the level (approximately) in the middle between $E_c - E_d$ according to

$$E_f = \frac{E_c - E_d}{2} + \frac{k_B T}{2} \ln\left(\frac{N_d}{2N_c}\right) \tag{4.7}$$

where the second term becomes rather small due to the temperature dependence of N_c and the factor in front of the natural logarithm. $E_f(T)$ is shown schematically in Figure 4.1 for an arbitrary doping concentration N_d. The exact quantitative relation depends of course on the numerical values of N_d; qualitatively, however, the curves will behave as shown. With the same reasoning, E_f will be in between the acceptor level E_a and the valence band E_v in the case of p-type doping (see Figure 4.1, right panel). This means that even small doping concentrations are sufficient to move the Fermi level close to the conduction or valence bands when cooling the sample to sufficiently low temperatures. However, in the case of nondegenerate doping with a finite gap between $E_c(E_v)$ and $E_d(E_a)$ a very small free carrier density will be obtained even in the case that the Fermi level is moved to ~15–25 meV below (above) the conduction (valence) band (since $k_B T = 0.083$ meV at $T = 1$ K). This lack of free carriers in the conduction (valence) band in nondegenerate semiconductors at low temperatures is called “freeze-out” and leads to an insulating behavior at low temperatures.

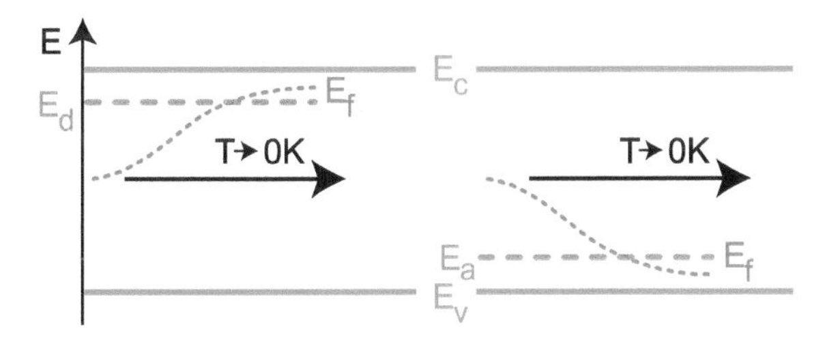

Figure 4.1: Fermi level relative to the conduction and valence bands together with the energy level of the donors E_d and acceptors E_a as a function of temperature. For low T, the Fermi level moves in between the dopant level and the respective band. For high temperatures, the semiconductor becomes intrinsic.

4.1.1 Degenerate Doping Concentration

With ion implantation, it is possible to implant a very high concentration of dopants into a host substrate that can be activated with an appropriate annealing technique (cf. Section 3.10). At sufficiently high doping concentrations, the semiconductor is said to be degenerately doped and behaves similar to a metal. If we assume for the time being that we were able to place the dopants on a regular lattice, then we would simply obtain a superlattice of dopants within the periodic crystal structure of the host solid. As a result, we could use the tight binding approach (see Section 2.4) and expect a band to develop centered around E_d. If we adopt the (simplistic) picture of dopants being hydrogen-like atoms, we expect a cosine band as illustrated, e. g., in Figure 2.16. With further increasing dopant concentration, the overlap between adjacent dopants increases giving rise to a band with larger energetic width (cf. Figure 2.13). At sufficiently high concentration, the band of dopants overlaps with the conduction band of the host crystal. Once this happens, the semiconductor is said to be degenerately doped. In silicon, the concentration of dopants required to reach degeneracy is $\sim 3 \times 10^{19}$ cm^{-3}.

Extending Equation (2.34), the cosine band in three dimensions for a hydrogen-like superlattice of dopants yields $E(\vec{k}) = E_d - 2V_{ss}^{\text{dop}} \cos(k_x a) - 2V_{ss}^{\text{dop}} \cos(k_y a) - 2V_{ss}^{\text{dop}} \cos(k_z a)$ with V_{ss}^{dop} being the overlap between adjacent dopant atoms. This yields a bandwidth of $E_d \pm 6V_{ss}^{\text{dop}}$. When the semiconductor becomes degenerate, $E_c - E_d \approx 6V_{ss}^{\text{dop}}$. For a typical dopant in silicon with an ionization energy of $E_c - E_d \sim 30$ meV, an overlap of $V_{ss}^{\text{dop}} \approx 5$ meV is obtained. In silicon, there are $\sim 5 \times 10^{22}$ atoms per cubic centimeter, and hence at the threshold of degeneracy approximately one out of $\frac{5 \cdot 10^{22}}{3 \cdot 10^{19}} \approx 1667$ silicon atoms is replaced with a dopant meaning that in each direction every $\sqrt[3]{1667} \approx 12$th atom is a dopant. The strong decrease of the overlap V_{ss} of s-orbital-like wavefunctions with increasing distance in between adjacent (dopant) atoms roughly fits the estimated value of V_{ss}^{dop} in the meV range.

Degenerate doping has two major consequences: first, the effective band gap is lowered. In fact, the minimum energy that is required to excite an electron from the valence band into the conduction band is not E_g anymore. Instead, it is sufficient to overcome the gap between the valence band edge and the bottom of the band of dopants. Since the energetic width of the dopant band depends on the dopant concentration, the effective band gap will also depend on the dopant concentration. Second, in the degenerately doped case the band of dopants overlaps with the conduction (valence) band, and as a result, even at lowest temperatures there will be free carriers available and the semiconductor does not "freeze out" anymore.

In real semiconductors, the dopants are of course not on a regular lattice. Ion implantation and annealing or the diffusion of dopants are statistical processes that lead to a random distribution of the dopants. As a result, the dopants do not form a proper band with sharp edges because of the strong disorder of the dopant band (cf. Chapter 2.12.2). Instead, the band will be smeared out giving rise to a density of states of the dopants that is a delta function at low dopant concentration (cf. Figure 4.2) and develops into a Gaussian with increasing N_d (a similar behavior is expected for acceptors). In the case of a heavily doped semiconductor, the band gap is not only narrowed but also smeared out as depicted in Figure 4.2, right panel. This has important consequences for the so-called band-to-band tunnel FETs (cf. Section 9.1.3.5).

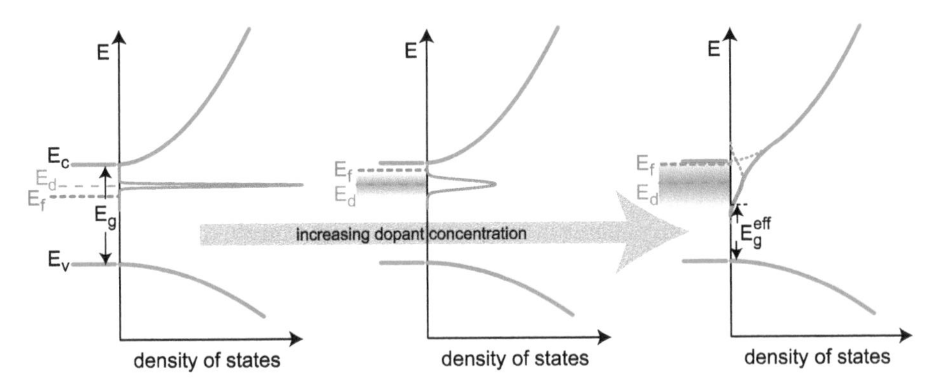

Figure 4.2: Conduction and valence bands together with the dopant band centered around the energy level of individual dopants E_d. In the case of an increasing dopant concentration, the energetic width of the band of dopants increases eventually overlapping with the conduction band. The right parts of the images show the density of states with the $\sqrt{E}$ behavior expected for a three-dimensional semiconductor together with the DOS of the dopant band.

4.2 P-N Junctions

One of the most important ingredients of any semiconductor device is the *p*-*n* junction. Doping neighboring areas with donors and acceptors allows one to create the potential distribution shown in Figure 4.3. In equilibrium, the so-called built-in potential Φ_{bi} is established that is determined by the energetic difference between the conduction bands in the *n*- and *p*-type regions far away from the junction interface. Φ_{bi} is the driving force for carrier separation in solar cells and photo detectors and provides the necessary potential barrier that suppresses carrier flow in transistor devices.

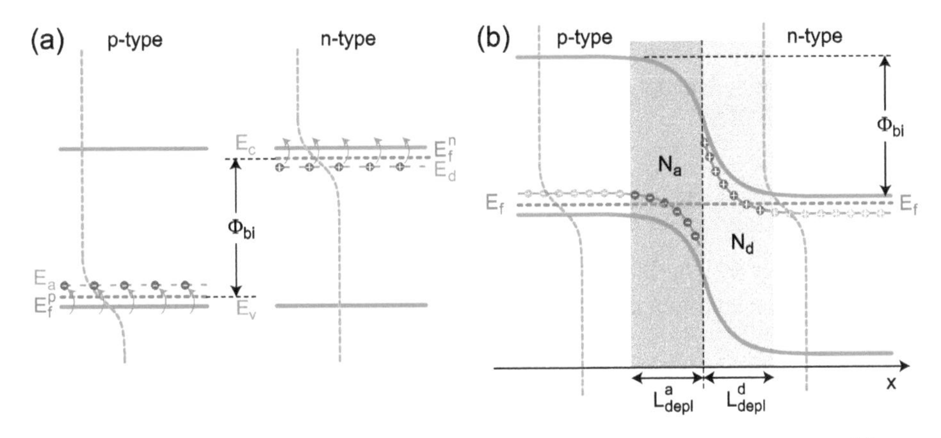

Figure 4.3: (a) *p*-type and *n*-type semiconductors with the energetic position of the Fermi energy $E_f^{p,n}$, and the dopant energy levels E_d for donors and E_a for acceptors. (b) *p*-*n* junction in equilibrium. The depletion zones at the junction with ionized acceptors (negative charge) and donors (positive charge) yield the band bending.

If brought into contact, a diffusion current of electrons will flow from the n-type region to the p-type region and occupy empty states in the conduction *and* valence bands due to the large concentration gradient.[1] In addition, recombination may occur in both regions, since the diffusion of carriers leads to electrons in the conduction band of the p-type region and empty states in the valence band of the n-type part. As a result, there will be a surplus of negative charge in the p-type region and a surplus of positive charge in the n-type region. Solving the Poisson equation shows that the surplus of charge lowers the bands in the n-type region (due to the surplus of positive charge) and lifts the bands in the p-type region. In equilibrium, there needs to be a common Fermi level in both regions and far away from the interface the semiconductor regions need to be neutral and the Fermi level must be on the same level as if they were not in contact with each other. At the interface, therefore, the potential barrier Φ_{bi} builds up that repels the free carriers from the interface region leading to two depletion regions. The charge in both depletion regions is determined by the ionized donors and acceptors. Within the so-called depletion approximation, it is assumed that there is a constant density of ionized donors and acceptors within the respective depletion region of length $L^{a,d}_{\text{depl}}$. Thus, solving the Poisson equation in the p- and n-type regions is straightforward (see Task 19) and simply yields a quadratic dependence on the spatial coordinate x. With the boundary conditions that the electric field needs to be zero at the edge of the depletion zones, the potential needs to be continuous across the p-n junction interface and that we have charge neutrality (i. e., $L^a_{\text{depl}} \cdot N_a = L^d_{\text{depl}} \cdot N_d$) one obtains

$$\begin{aligned} \Phi_n(x) &= \frac{e^2}{2\varepsilon_{\text{si}}\varepsilon_0} N_d \left(L^d_{\text{depl}} - x\right)^2 \quad x \geq 0, \\ \Phi_p(x) &= \Phi_{\text{bi}} - \frac{e^2}{2\varepsilon_{\text{si}}\varepsilon_0} N_a \left(L^a_{\text{depl}} + x\right)^2 \quad x \leq 0, \end{aligned} \tag{4.8}$$

where the depletion lengths in the case of acceptors L^a_{depl} and donors L^d_{depl} are given by

$$L^a_{\text{depl}} = \sqrt{\left(\frac{N_d}{N_a + N_d}\right)\frac{2\varepsilon_{\text{si}}\varepsilon_0\Phi_{\text{bi}}}{N_a e^2}}, \quad L^d_{\text{depl}} = \sqrt{\left(\frac{N_a}{N_a + N_d}\right)\frac{2\varepsilon_{\text{si}}\varepsilon_0\Phi_{\text{bi}}}{N_d e^2}}. \tag{4.9}$$

Since $\Phi_{\text{bi}} = E^n_f - E^p_f$ when the two parts of the semiconductor are not in contact (see Figure 4.3(a)), the built-in potential in the nondegenerate case is simply given by

$$\Phi_{\text{bi}} = E^n_f - E^p_f = k_B T \ln\left(\frac{N_a N_d}{n_i^2}\right). \tag{4.10}$$

---

1 Remember the discussion in Section 2.10. You may equally well think of electrons flowing from the n- to the p-region and holes being injected from the p-type into the n-type region.

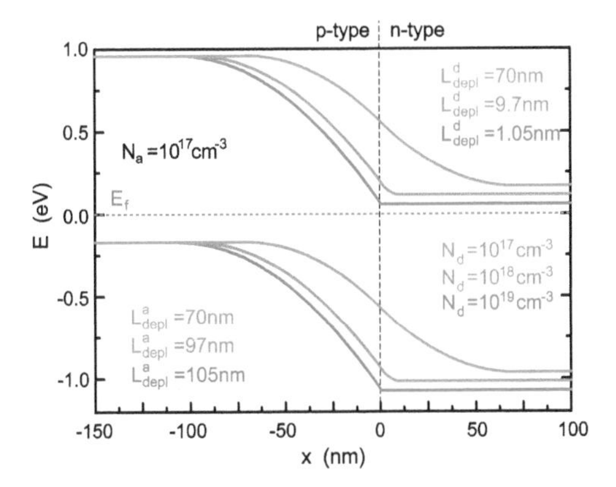

Figure 4.4: Conduction and valence bands of a *p-n* junction with different doping levels (see the figure) in the *p*- and *n* regions giving rise to substantially different depletion lengths $L^{a,d}_{\text{depl}}$.

Figure 4.4 shows the conduction and valence bands of three different *p-n* junctions computed with the depletion approximation (see Task 19 for details) for an increasing doping concentration in the *n*-type region. One clearly sees that in the case of the highest doping a rather small depletion length (see figure) is obtained in the *n*-type region such that it seems that the entire band bending occurs in the *p*-type section. The reason for this is the two orders of magnitude higher density of ionized dopants in the *n*-type region that yields a strong screening over a short depletion length; nevertheless, charge neutrality with $L^a_{\text{depl}}N_a = L^d_{\text{depl}}N_d$ is preserved.

Applying a negative voltage V_{pn} (reverse bias) between *p*- and *n*-type regions increases the built-in potential by separating the Fermi levels of the *p*- and *n*-sides such that $\Phi_{\text{bi}} - eV_{\text{pn}}$ has to be used in the equations above. In this case, a depletion capacitance is associated with the *p-n* junction that can be computed simply with $C_{\text{pn}} = |\frac{\partial Q}{\partial V_{\text{pn}}}|$ where for the charge per area $Q = eN_aL^a_{\text{depl}}(V_{\text{pn}})$ (because of charge neutrality, ionized donors or acceptors can be used interchangeably). Taking the derivative of $L^a_{\text{depl}}(V_{\text{pn}})$ with respect to V_{pn} yields the capacitance per area:

$$C_{\text{pn}} = \left| eN_a \frac{\partial L^a_{\text{depl}}}{\partial V_{\text{pn}}} \right| = \sqrt{\frac{N_aN_d}{N_a + N_d} 2\varepsilon_{\text{si}}\varepsilon_0 e} \frac{1}{2\sqrt{\Phi_{\text{bi}} - eV_{\text{pn}}}}. \tag{4.11}$$

Applying a positive voltage (forward bias), a diffusion capacitance can also be defined, which is due to the free carriers diffusing toward the *p-n* junction interface.

Task 19.
Potential profile of a *p-n* junction: Compute explicitly the expression for the potential profile of a *p-n* junction stated above by solving the Poisson equation using the depletion approximation. Also, verify the expression given for the depletion regions in the *p*-type section L^a_{depl} and the *n*-type region L^d_{depl}.

4.3 Doping at the Nanoscale

In the following chapters, we will deal with nanoscale field-effect transistors. It will become clear that preserving electrostatic control at nanoscale channel lengths requires the use of nanostructures such as nanowires/tubes or two-dimensional materials with an extremely thin channel layer thickness. While ion implantation and in particular the activation of dopants in nanostructures with nanoscale control over the position of the dopants is itself a severe technological challenge, let us assume for the time being that this issue can be handled by employing either in situ doping during the growth of a nanostructure (cf. Section 3.13.2) or an appropriate ion implantation and flash lamp annealing (see Section 3.10). However, at nanoscale dimensions it is not immediately obvious how dopants will behave. In order to assess what difficulties might arise regarding doping of nanoscale structures and what mitigation strategies could be applied, a closer look at doping is necessary, which will be done in the present section.

4.3.1 Hydrogen Model for Activation of Dopants

For first-order estimations of the behavior of dopants, the hydrogen model used already above is employed again. Let us consider phosphorous embedded into a silicon matrix. We seek an approximate expression of the ionization energy $E_{\text{ion}} = E_c - E_d$ (cf. Figure 4.2, left), which can be obtained using Bohr's model of the hydrogen atom and replace $\varepsilon_0 \to \varepsilon_0\varepsilon_{\text{semi}}$ as well as $m_0 \to m^\star$. The latter approximation is well justified in single-band semiconductors such as GaAs but less in silicon [205]. In fact, in GaAs, for instance, the ionization energy of dopants is rather independent of the dopant species, while this is different in silicon. Nevertheless, using $\varepsilon_{\text{si}} = 11.2$ and the light effective conduction band mass of silicon $m^\star = 0.19\, m_0$ in

$$E_{\text{ion}} = \frac{e^4 m^\star}{2(4\pi\varepsilon_0\varepsilon_{\text{si}}\hbar)^2} \tag{4.12}$$

an ionization energy of $E_{\text{ion}} \approx 20.6$ meV is obtained in silicon, which is reasonably close to the actual value, at least for the most common shallow donors. Equation (4.12) suggests that the ionization energy strongly depends on the host material through the dielectric constant. Indeed, Figure 4.5 shows the Coulomb potential $V_{\text{Coul}}(r)$ of a hydrogen atom within a dielectric matrix exhibiting three different dielectric constants ($\varepsilon_{\text{semi}} = 1$, $\varepsilon_{\text{semi}} = 3.9$ and $\varepsilon_{\text{semi}} = 11.2$). One clearly observes a substantially stronger screening of the potential with increasing $\varepsilon_{\text{semi}}$ leading to a significant reduction of the spatial extent (along the radius r) of the dopant atom's Coulomb potential. Together with a lower $m^\star$ within a host solid, this leads to a strong energetic increase of the ground-state level, i. e., it strongly reduces the ionization energy.

Using the simple hydrogen model allows understanding the impact of the dielectric environment and the geometrical size of the silicon nanostructure on the ionization

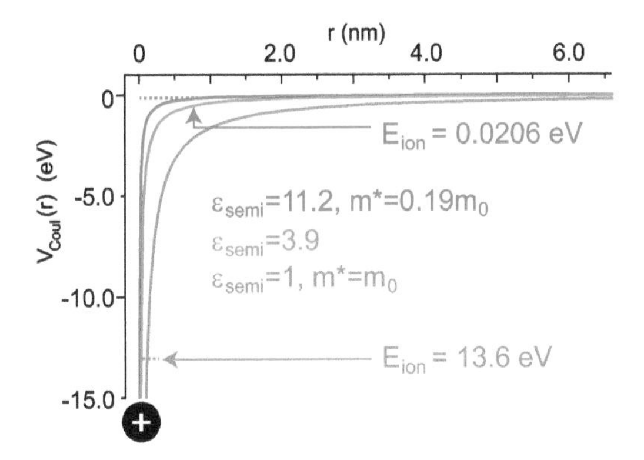

Figure 4.5: Coulomb potential of a single positive charge in three dielectric environments. A higher $\varepsilon_{\text{semi}}$ leads to an increased screening of the potential, which together with a smaller effective mass (red) within a semiconductor yields a significantly reduced ionization energy compared to the vacuum case (blue).

energy of the dopants. It is important to note that in the case of an increasing ionization energy, the dopants become increasingly deactivated since a higher $E_{\text{ion}} = E_c - E_d$ necessarily leads to a substantially reduced probability of activation proportional to $\exp(-\frac{E_{\text{ion}}}{k_B T})$. When scaling down the geometrical size of a nanostructure, the averaged, effective dielectric environment will at some nanoscale dimension change and approach unity with further downscaling of the nanostructure. As a result, an increasing deactivation of dopants due to an increasing E_{ion} is expected. Hence, the doping efficiency in nanoscale devices drops significantly leading to large parasitic resistances in the doped regions such as source and drain in MOSFETs.

In order to sort out the main effects for dopant deactivation and to study their dependence on the geometrical size of the silicon nanostructure, the Poisson equation of a single positive charge (representing the donor hydrogen-like atom) within a material characterized by its dielectric constant has been computed numerically. To this end, the dopant is considered to be at the center of a silicon sphere with radius $r_{\text{si-sphere}}$; a sphere is used here in order to simplify the computation to a one-dimensional calculation along the radius r (cf. Figure 4.6(a)). The silicon sphere is surrounded by a dielectric shell with dielectric constant $\varepsilon_{\text{shell}}$. In addition to the Coulomb potential of the donor atom, a confinement potential of 3 eV at the boundary of the Si sphere has been considered leading to substantial quantization in the case of an appropriately small $r_{\text{si sphere}}$.

After solving the Poisson equation numerically, the resulting potential $V_{\text{Coul}}(r)$ has been used to solve Schrödinger's equation to extract the ground-state energy. Calculations were carried out with three different dielectric environments, namely air with $\varepsilon_{\text{shell}} = 1$, SiO_2 with $\varepsilon_{\text{shell}} = 3.9$ and Si with $\varepsilon_{\text{shell}} = 11.2$, and with or without a 3 eV confinement potential barrier around the silicon sphere; the latter case, i. e., $\varepsilon_{\text{shell}} = 11.2$ and without quantization represents the bulk silicon case.

First, we consider the effect of a locally varying dielectric environment alone, i. e., without additional quantization (the impact of different bulk dielectric environments

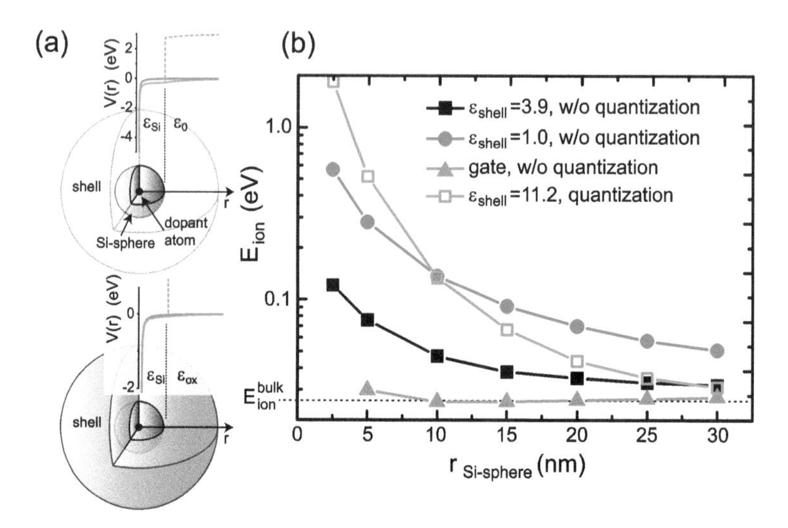

Figure 4.6: (a) Single dopant atom sitting in the center of a Si sphere ($\varepsilon_{si} = 11.2$) with radius $r_{si\ sphere} = 5$ nm. The Coulomb potentials for three different cases are shown in both panels: The red line is the Coulomb potential in bulk silicon, the green line is the potential if the Si sphere is surrounded by a shell of air (top panel) and SiO_2 (bottom panel); note that the energy scales are different in both panels in order to show the potential difference in the case of $\varepsilon_{shell} = \varepsilon_{ox}$. Finally, the green dashed lines show the Coulomb potential for the Si spheres embedded in air and SiO_2 if an additional confining potential barrier of 3 eV is present at the boundary of the Si sphere. (b) Ionization energy E_{ion} of a dopant atom located in the center of a silicon sphere ($\varepsilon_{semi} = \varepsilon_{si} = 11.2$) as a function of the radius of the silicon sphere $r_{si\text{-}sphere}$. Results for different dielectric shells as well as with and without quantization are plotted.

on the potential $V_{Coul}(r)$ is displayed in Figure 4.5, here $\varepsilon_{semi} = 11.2$ is used). As already mentioned above, the absolute value of the energetic position of the ground-state wave function represents the ionization energy E_{ion} since the conduction band is reached at an energy $E = 0$ eV (without additional quantization). Figure 4.6(b) displays E_{ion} as a function of $r_{si\ sphere}$ for a silicon sphere with a shell consisting of air (red line/circles) and SiO_2 (black line/squares). Obviously, the ionization energy strongly increases as the radius of the silicon sphere decreases. In particular, if the dielectric contrast between the silicon sphere and the shell is large, the ionization energy increases rapidly. Next, we discuss the case of quantization alone (green line/hollow squares). To this end, a silicon sphere is considered, which is embedded into a dielectric with the same dielectric constant as silicon but with a confining potential of 3 eV (see, e. g., [49]). In this case, the ionization energy increases stronger at small $r_{si\text{-}sphere}$ compared to the pure dielectric mismatch case leading to rather high values of E_{ion} at the smallest radii of the silicon sphere. The exact quantitative value depends of course on the system that is considered. In the present case, a silicon sphere is considered, which leads to substantially higher quantization energies than one would expect in a nanowire object. In addition, the local changes in dielectric environment will be different in a nanowire compared to the sphere. Nevertheless, qualitatively, the considerations can be transferred to other doped nanostructures and show that deactivation of dopants is expected due to dielectric mismatch already at geometrical sizes of a nanostructure where quantization due to confinement may not play a significant role. In the next section, experiments concerning donor deactivation in silicon nanostructures are discussed and it is indeed shown that this observation is correct.

Dopant deactivation due to dielectric mismatch can be mitigated by embedding the nanostructure into an environment of the same dielectric constant as the host material. However, there is also a different way of accomplishing this: The blue line/triangles in Figure 4.6(b) show E_{ion} as a function of $r_{si\text{-}sphere}$ in the case of a silicon sphere surrounded by a $d_{ox} = 3\,nm$ thin SiO_2 followed by a metallic gate electrode (in the simulation this has been realized by enforcing $V_{Coul}(r = r_{si\text{-}sphere} + d_{ox}) = 0$ with Dirichlet boundary conditions). In the present case, no additional confining potential was assumed so that the curve can be compared with curves for the dielectric mismatch (black and red) cases. Interestingly, the ionization energy slightly drops and approaches the bulk value when $r_{si\ sphere}$ is reduced and then increases for even smaller $r_{si\ sphere}$. This increase of E_{ion} can be disregarded here since it stems from an unwanted confinement because the computational domain is truncated at the metal surface. However, for a radius, where quantization can be neglected E_{ion} decreases. The reason for this peculiar behavior is that the presence of the gate electrode with high carrier density leads to additional screening of the Coulomb potential. The smaller $r_{si\text{-}sphere}$ the closer is the screening electrode to the dopant atom, and thus E_{ion} decreases. In fact, without the unwanted confinement the ionization energy can even become smaller than the bulk Si value because the additional screening due to the close proximity of the gate electrode is comparable to an effective dielectric with higher ε_{semi} than 11.2. As a result, deactivation due to a dielectric mismatch can be avoided by wrapping nanostructures into appropriate gate electrodes.

4.3.2 Deactivation of Dopants in Nanoscale Structures

In the preceding section, the deactivation of dopants due a dielectric mismatch of the semiconducting nanostructure and its surrounding was studied. The results obtained suggest that deactivation occurs at geometrical sizes of the nanostructure well above where quantum confinement has an impact. This is important because already at such geometrical sizes a deactivation may occur that will lead to parasitic resistances particularly of the source/drain extensions of nanoscale transistors.

In order to study the impact of dielectric mismatch in more detail and the geometrical dimensions of nanostructures at which it occurs, a nanostructure is needed where the amount of dopants can be controlled while the size of the nanostructure is reduced. An appropriate method appears to be in situ doping of silicon nanowires during the growth process using the vapor-liquid-solid method in a chemical vapor deposition tool (see Section 3.13.2) [263]. Therefore, silicon nanowires with different diameters are grown using gold particles as seed with diameters in the range between 10 nm and 100 nm [28, 230]. For the growth of the n-type doped Si-nanowires, silane and phosphine are used as precursor gases. A growth temperature of 440 °C is used to realize nanowires that are taper-free and do not exhibit a polysilicon coating (see Section 3.13.2 and [28]). Doping the nanowires during the growth ensures that they contain the same volume doping concentration independent of the nanowire diameter. After the growth, the nanowires are transferred on an oxidized silicon substrate and are contacted with titanium/gold electrodes using electron beam lithography and a lift-off process to facilitate four-point probe measurements.

Figure 4.7 shows an electron micrograph of the nanowire on top of an oxidized silicon substrate together with a schematic of the four-point probe measurement configuration. Since the surfaces of the nanowires are not passivated, they exhibit a rather large density of interface states, which results in a donut-shaped depletion region at the edge of the nanowire. Estimating this interface state density and calculating the resulting depletion region allows one to compute the area through which current flows through the nanowire. This area is given by πr_{elec}^2 where the "electronic radius" r_{elec} is introduced. Using r_{elec} the resistivity of the nanowires (which is a function of the physical radius r_{nw} determined by the gold seed nanoparticles) can be rescaled.

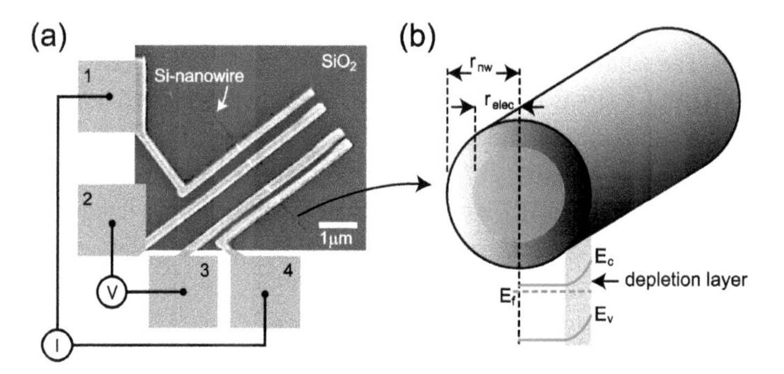

Figure 4.7: (a) Electron micrograph of a VLS-grown, in situ doped Si nanowire with four contacts to facilitate four-point probe measurements [28]. (b) Schematic of the nanowire with depletion region at the edge of the wire that results from interface states present at the nanowire surface. Subtracting the depletion length from the physical radius r_{nw} yields the electronic radius r_{elec}.

Figure 4.8(a) shows the resistivity ρ_s obtained from the four-probe measurements as a function of r_{elec} for three different doping concentrations. In all three cases, the resistivity strongly increases when the diameter of the nanowire is scaled down. Since the depletion at the edge of the nanowire has already been eliminated by rescaling the resistivity using r_{elec} and because quantization due to confinement plays no role at the diameters considered here, the reason for the increase of the resistivity is the large dielectric mismatch between the silicon nanowire and its surrounding ($\varepsilon_{Si} = 11.2$, $\varepsilon_{air} = 1$). As discussed above, the dielectric mismatch leads to an increasing ionization energy E_{ion}, and hence to increasingly deactivated dopants with decreasing nanostructure size

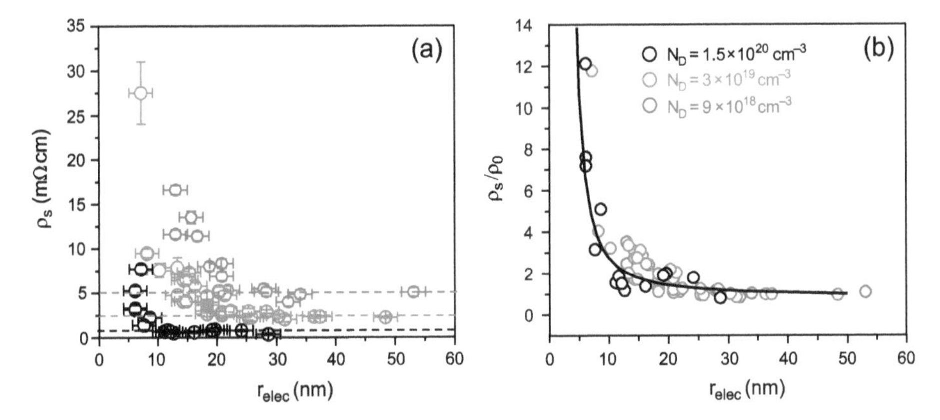

Figure 4.8: (a) Resistivity versus electronic radius r_{elec}. The experimental data was rescaled by subtracting the length of the depletion layer (that develops due to the existence of a rather large interface-state density) from the physical radius of the nanowire. This shows that the strong increase for all three doping concentrations is not a result of depletion but is due to the dielectric mismatch between nanowire and air. The dashed horizontal lines show the expected resistivity without deactivation. (b) Normalized resistivity as a function of electronic radius. The solid line belongs to a theoretical calculation [66, 28].

(cf. Figure 4.6(b)). While the effect of the dielectric mismatch on the ionization energy decreases toward the center of the nanowire, its impact extends over several nanometers, and thus dopant deactivation is observed in nanowires with relatively large diameters [28]. Similar results have been published in [214] where it was found that the dielectric surrounding increases significantly the ionization energy of a single dopant in a nanoscale silicon FET. Moreover, dopant deactivation has recently been shown to appear in doped InN nanowires and is hence not restricted to silicon nanowires [30].

In Figure 4.8(b), the normalized resistivity is plotted as a function of r_{elec} where ρ_s is normalized to the resistivity belonging to the largest diameter ρ_0 (which is basically the resistivity without deactivation). All data points of the three doping concentrations lie on the same curve showing that the mechanism of the deactivation is universal and does not depend on the particular doping concentration; the straight line in Figure 4.8(b) was calculated using the theory regarding E_{ion} in Si nanostructures given in [66], which is in excellent agreement with the experimental data.

It has already been mentioned above that the deactivation of dopants in nanostructures due to a dielectric mismatch can in principle be avoided by putting a dielectric on top of the nanostructure with approximately the same dielectric constant. Indeed, depositing Al_2O_3 on top of the Si nanowire, depicted in Figure 4.7(a), it was shown that a decrease of the resistance of the nanowires could be obtained [28]. However, in a nanowire architecture as illustrated in Figure 5.18, substantially increased parasitic capacitances $C_{\text{s,d}}^{\text{par}}$ are obtained when the source/drain extensions are covered with a high-k material that in the best case exhibits the dielectric constant of silicon.

As a result, the dielectric mismatch requires finding the optimum trade-off between parasitic resistance versus parasitic capacitances of the source/drain extensions.

Further downscaling of the channel length L of field-effect transistors requires a further decrease of the diameter d_{nw} of the nanowire (or channel layer thickness of nanostructures in general, see Section 5.6) in order to maintain electrostatic integrity of the devices. Therefore, quantum confinement will inevitably occur and—as discussed above—may result in a severe deactivation of dopants due to strong quantum confinement. Interestingly, in silicon nanostructures with very small feature size (below ~3 nm), impurity atoms are not necessary any more in order to realize doping. The growth of a thin SiO_2- or SiN-layer is sufficient to shift the conduction and valence bands such that an effective n-type and p-type doping is obtained [159]. This phenomenon will be discussed in detail in the next section.

4.4 NESSIAS as an Alternative to Impurity Doping

At Si/SiO_2 and Si/Si_3N_4 interfaces, an interesting quantum chemical effect occurs that is due to an interface charge transfer (ICT) driven by the difference in the ionicity of bond of silicon and the two anions O and N (see the thorough treatment and explanation of the effect in [156]). In the case of Si/SiO_2, charge is moved toward oxygen resulting in shifting the electronic structure to lower energies [153, 160, 157]; in the following, this effect is called nanoscopic electronic structure shift induced by anions at surfaces (NESSIAS) [156]. NESSIAS modifies the bands on a very small spatial length scale of $\lambda_N \approx 1.5 \pm 0.2$ nm [160, 156], and thus has been mostly overlooked so far. But if nanoscale volumes of silicon are considered, it becomes relevant and significantly changes the electronic structure when compared to the expectation.

If we consider NESSIAS as a potential drop at a Si/SiO_2 interface that is exponentially screened on the length scale λ_N, then we can estimate the size of a silicon nanovolume when NESSIAS becomes relevant: if a nanowell or nanosheet with two interfaces are considered, the thickness of the nanosheet needs to be less than ~$2\lambda_N$ to have a measurable impact on the electronic structure, which matches previous theoretical findings [158]. In the case of a nanowire with four interfaces, it is expected that NESSIAS starts to become relevant if the nanowire diameter is less than ~$4\lambda_N$. In the following, the case of a Si nanosheet will be discussed in detail since this is supported by experimental data.

In a nanosheet or quantum well of thickness d_{qw}, the potential across the nanosheet can be approximated as $\Phi_{IoB}(\exp(-x/\lambda_N) + \exp(-(d_{qw} - x)/\lambda_N))$, which is schematically shown (white line) in the right panel of Figure 4.9. If the thickness d_{qw} of the quantum well is reduced (left panel), the overlapping regions where NESSIAS is present (~λ_N) move the entire conduction band downwards (the red dashed line in the main panel is the maximum potential at $d_{qw}/2$). However, vertical quantization in the nanosheet (cf. Section 2.2) leads to a competing increase of the conduction band (illustrated by the green dashed line). The overall resulting potential is shown by the blue line in Figure 4.9.

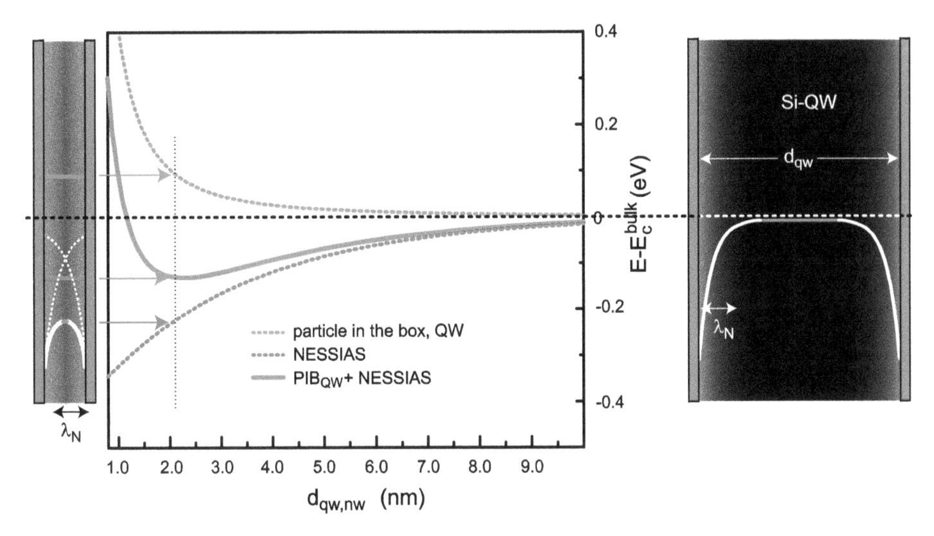

Figure 4.9: Schematic explanation (left and right panel) and resulting dependence of the effective conduction band edge (blue line) as a function of d_{qw} due to the competition of NESSIAS (red dashed line) and quantization (green dashed line) in a silicon quantum well.

Interestingly, this simplistic treatment shows a minimum of the potential at $d_{qw} \approx 2\,\text{nm}$ where the conduction band is moved even below the bulk value (dashed black line).

In order to study the NESSIAS effect experimentally, ultrathin single-crystalline silicon nanosheets were realized with digital etching (see Section 3.6.3) of silicon-on-insulator substrates (for details, see [159]). After thinning, NAOS is used to grow a SiO_2 in a self-terminating way (Section 3.2.3). As a result, the ultrathin Si nanosheet is embedded in SiO_2 (NAOS and BOX); a transmission electron microscopy image of such a nanosheet is shown in Figure 3.5(b). To embed the nanosheet in SiN, rapid thermal nitridation (~3 nm) is used after the thinning and HF stripping, followed by the deposition of PE-CVD grown SiN and SiO_2. The sample is then wafer-bonded (see Section 3.4) onto an oxidized wafer with hydrophilic bonding. Finally, the handle wafer and the BOX of the original SOI wafer are removed and a second SiN is grown with RTN [78].

Synchrotron X-ray absorption spectroscopy at total flourescence yield (XAS-TFY) as well as ultraviolet photoemission spectroscopy (UPS) measurements were carried out at the BACH and BaDelPh beamlines in Trieste in order to measure the conduction and valence band edges as a function of d_{qw}. The results for the nanosheet embedded in SiO_2 are plotted in Figure 4.10(a) (blue data points) and show that at $d_{qw} \approx 2\,\text{nm}$, the conduction band is moved ~200 meV below the bulk conduction band, which is equivalent to a very high n-type doping. In this respect, it is important to note that at $d_{qw} \approx 2\,\text{nm}$ the mobility of carriers is expected to be substantially degraded [252]. However, because of the four interfaces in a nanowire structure, it is plausible that NESSIAS yields a similarly high shift of the conduction band (and hence equivalence to doping) in a nanowire with

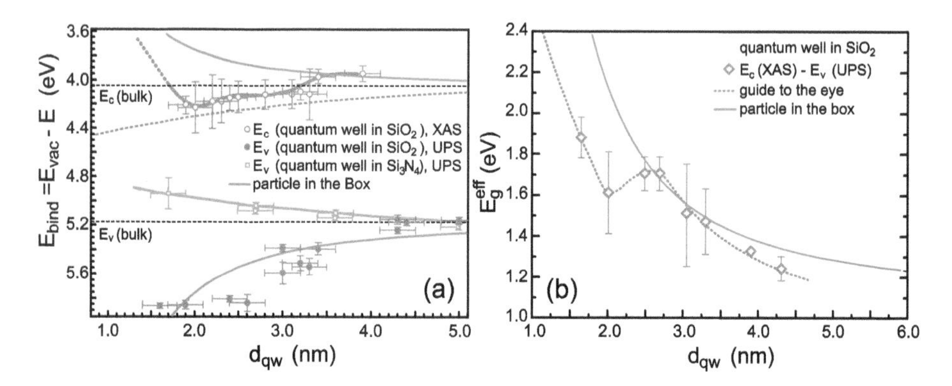

Figure 4.10: (a) Conduction and valence band edges of ultrathin single-crystalline Si embedded into SiO_2 as measured with synchroton XAS-TFY and UPS (blue data points and line) [159, 157, 156]. The bulk values are shown as the black dashed lines. The green and the red dashed lines show the evolution of the band edges with d_{qw} due to quantization and NESSIAS, respectively. The orange data points show the valence band edge of nanosheets embedded in SiN. (b) E_g of a Si nanosheet embedded in SiO_2 as a function of d_{qw} extracted from the measurements shown in (a).

4–5 nm diameter,[2] which corresponds to a nanosheet with $d_{qw} \approx 2$ nm. At the same time, quantization at 4–5 nm is not expected to counteract the benefit of NESSIAS.

While encapsulation into SiO_2 yields an alternative to heavy *n*-type doping, density functional theory calculations suggested that a coating of silicon with SiN leads to an upshift of the band structure, which is equivalent to *p*-type doping. For an experimental verification of the effect of SiN coating, the ultrathin Si nanosheets embedded in SiN fabricated with wafer bonding (see above) were measured with synchrotron UPS. Figure 4.10(a) shows the valence band edge extracted from the measurement as a function of d_{qw} and indeed an upshift of the valence band edge above the bulk value is observed, which is equivalent to a high *p*-type doping.

It was discussed earlier that even in heavily doped silicon, the density of impurity atoms is still rather small leading to variability at the nanoscale. The coating of Si nanovolumes with SiO_2 and SiN provides a large (and increasing with increasing surface to volume ratio) number of bonds. Therefore, variability comparable to random dopant effects is not expected in NESSIAS. Moreover, NESSIAS is to a large extent temperature-independent. Therefore, freeze-out at low temperatures and deactivation are avoided. As a result, highly conducting, ultrathin *n*-type and *p*-type nanowires without impurity doping may become feasible.

Apart from the fact that a simple encapsulation of ultrathin silicon into SiO_2 and SiN yields a band structure shift equivalent to heavy doping, another very interesting

2 When estimating the impact of NESSIAS in the nanosheet, the potential in the nanosheet center at $d_{qw}/2$ has been assumed such that this potential is reduced $\propto e^{-d_{qw}/2\lambda_N}$. In a nanowire, the same NESSIAS is expected for a diameter of $2d_{qw}$.

point can be inferred from the measurements displayed in Figure 4.10(a). Since the XAS-TFY and UPS measurements of the SiO_2-encapsulated samples yield the conduction and valence band edges, E_g as a function of d_{qw} can be extracted, which is displayed in Figure 4.10(b). In contrast to the analysis in Section 2.2.2 based on a particle-in-the-box model, a nonmonotonic behavior is observed, which is due to the competition between carrier confinement and NESSIAS.

4.5 Metal-Oxide-Semiconductor Capacitor

The metal-oxide-semiconductor (MOS) capacitor is another central ingredient of a (nano)electronic device and, therefore, deserves an in-depth consideration. In the present section, the discussion will be carried out based on silicon as semiconducting material due to its widespread use. However, the general concepts and conclusions are valid with appropriate modifications also for other semiconducting materials.

The discussion of the MOS capacitor is exemplarily carried out considering a *p*-type substrate. The MOS capacitor has the simple structure illustrated in Figure 4.11(a): On top of the *p*-type silicon substrate there is a gate dielectric of thickness d_{ox} with dielectric constant ε_{ox} and a metallic gate electrode. Applying a voltage V_{gs} between gate electrode and substrate yields different band situations that will be discussed in the following. For simplicity, it is assumed that the metal exhibits a work function such that at zero gate-substrate voltage V_{gs}, the bands in the silicon substrate at the gate dielectric–silicon interface are flat as illustrated in Figure 4.12(a). For the following discussion, any gate leakage current is neglected.

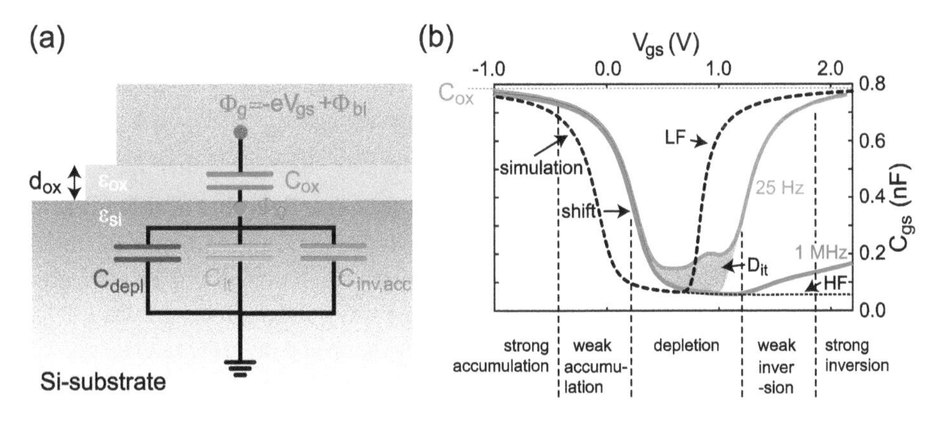

Figure 4.11: (a) Schematic illustration of a simple MOS capacitor with parallel and series combination of the depletion capacitance C_{depl}, the interface capacitance C_{it}, the inversion(accumulation)-layer capacitance $C_{inv,acc}$ and the geometrical oxide capacitance C_{ox} in a MOS capacitor. (b) Experimental high- (HF, blue curve) and low-frequency (LF, red curve) capacitance voltage characteristics of a Si/SiO_2 MOS capacitor. The black dashed line is a simulation.

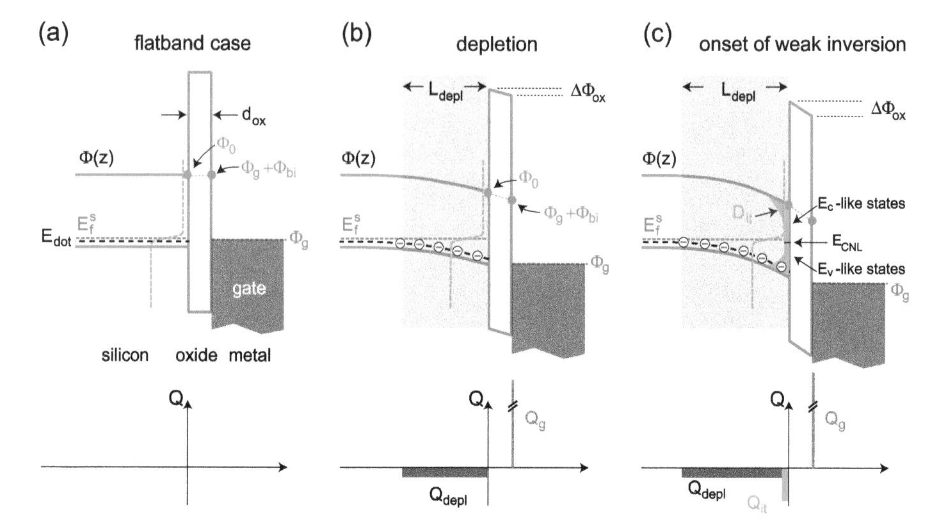

Figure 4.12: (a) Metal-oxide-semiconductor capacitor in the case of flat band. (b) For positive gate voltages a depletion layer consisting of negatively charged ionized acceptors builds up. In real (nonideal) MOS capacitors, there is a nonzero density of interface states D_{it} illustrated with the light blue area at the silicon-gate dielectric interface (c), which is U-shaped and yields conduction band-like states close to the conduction band and valence band-like states for energies close to the valence band edge; E_{CNL} refers to the charge neutrality level.

The overall gate-substrate capacitance C_{gs} is determined by the change of the charge on the capacitor—which is equal to the negative of the charge within the semiconductor $-Q_{semi}$—with respect to a change of the applied voltage at the gate electrode:

$$C_{gs} = -\frac{\partial Q_{semi}}{\partial V_{gs}}. \tag{4.13}$$

The most widespread method to determine C_{gs} experimentally is to measure the impedance of a MOS capacitor by applying a gate voltage $V_{gs} = V_{DC} + V_{AC}\sin(\omega t)$ with $|V_{AC}| \ll |V_{DC}|$ and $\omega = 2\pi f$. The current of the gate electrode is then $I_g = \frac{dQ_g}{dt} = -\frac{dQ_{semi}}{dt} = -\frac{dQ_{semi}}{dV_{gs}}\frac{dV_{gs}}{dt}$. The second term $\frac{dV_{gs}}{dt} = \omega V_{AC}\cos(\omega t)$ and the first term is the gate capacitance. Figure 4.11(b) shows simulated (black dashed line) and experimental (blue and red lines) $C_{gs} - V_{gs}$ curves; the blue curve shows the results at low and the red curve at a high-frequency f of the small signal, oscillating gate voltage superimposed to V_{DC} (see above). At low-frequency (LF), states within the band gap (interface states) can be charged and discharged, and hence contribute to C_{gs}. In the high-frequency case (HF), the occupation of these states does not change with the oscillating gate voltage, and hence C_{gs} drops. The difference (gray shaded area) in the depletion region is a measure for the interface density of states. Compared to simulations, the experimental LF characteristics show a shift to more positive gate voltages, which may be due to fixed charge within the oxide and the work function of the gate electrode.

To calculate C_{gs}, we need to compute the charge density Q_{semi}. Looking at Figure 4.12(b) and (c), it is obvious that V_{gs} drops partly across the gate dielectric and the remainder across the semiconductor. Since the energetic position of the bands in the semiconductor will be crucial for the following discussion, voltages will be transferred into potential energies, which basically implies multiplying the voltages with $-e$. The potential energy drop across the oxide is then simply $\Phi_0 - (\Phi_g + \Phi_{bi})$ where Φ_g is the gate potential and Φ_0 the surface potential of the conduction band at the silicon-gate dielectric interface. Note that here, the built-in potential Φ_{bi} takes the work function of the metal, the position of the Fermi level in the doped silicon and the electron affinity of the semiconductor into account ensuring the assumed flat band conditions at $V_{gs} = 0\,\text{V}$. Using the chain rule, Equation (4.13) can then be written as

$$C_{gs} = -\frac{\partial Q_{semi}}{\partial \Phi_0}\frac{\partial \Phi_0}{\partial \Phi_g}\frac{\partial \Phi_g}{\partial V_{gs}} = C_{semi}\frac{\partial \Phi_0}{\partial \Phi_g}. \tag{4.14}$$

In this equation, the last factor $\frac{\partial \Phi_g}{\partial V_{gs}}$ simply yields $-e$ and together with the first factor the capacitance $C_{semi} = (-e)\left(-\frac{\partial Q_{semi}}{\partial \Phi_0}\right)$ is obtained. To determine C_{semi}, we need to compute all charges within the semiconductor. There are basically three main contributions to the total charge Q_{semi}, which are (i) the depletion charge Q_{depl} due to ionized dopants (acceptors in the present case), (ii) the interface charge Q_{it} due to the occupation of traps, unsaturated bonds etc. and (iii) the inversion charge Q_{inv} due to mobile carriers in the conduction band for large V_{gs} or accumulation charge Q_{acc} for negative V_{gs} in the valence band. Each of these contributions will be discussed in further detail below. However, general statements regarding the MOS capacitance can be made prior to the detailed analysis.

Since for positive V_{gs}, the total charge in the semiconductor is $Q_{semi} = Q_{depl} + Q_{inv} + Q_{it}$ the capacitance C_{semi} is given as a sum of three components according to

$$\begin{aligned} C_{semi} &= (-e)\left(-\frac{\partial Q_{semi}}{\partial \Phi_0}\right) = e\frac{\partial Q_{depl}}{\partial \Phi_0} + e\frac{\partial Q_{inv}}{\partial \Phi_0} + e\frac{\partial Q_{it}}{\partial \Phi_0} \\ &= C_{depl} + C_{inv} + C_{it}. \end{aligned} \tag{4.15}$$

Because the overall gate-substrate capacitance C_{gs} is a series combination of the geometrical oxide capacitance C_{ox} and the capacitance related to the charge in the semiconductor C_{semi}, one obtains $C_{gs} = \frac{C_{semi}C_{ox}}{C_{semi}+C_{ox}}$ (see Figure 4.11(a)). Comparing this with Equation (4.14), the factor $\frac{\partial \Phi_0}{\partial \Phi_g}$ can be identified to be $C_{ox}/(C_{semi} + C_{ox})$ and using Equation (4.15) this yields

$$\frac{\partial \Phi_0}{\partial \Phi_g} = \frac{C_{ox}}{C_{depl} + C_{inv} + C_{it} + C_{ox}}. \tag{4.16}$$

Rewriting the above expression (by multiplying both sides with $\delta\Phi_g$), it states that a change of gate potential $\delta\Phi_g$ is translated into a change of surface potential $\delta\Phi_0$ depending on the ratio of the geometrical oxide capacitance and the sum of all capacitances. This relation will be very important when considering the switching behavior of metal-oxide-semiconductor field-effect transistors in Chapter 5 and it will also be central in metal-semiconductor contacts (cf. Section 4.6).

4.5.1 Depletion Capacitance

When a positive voltage is applied at the gate, the Fermi level of the gate will be moved to lower energies, which results in a bending of the bands within the silicon. This band bending is a consequence of the negatively charged acceptors (with density N_a) whose states are moved further below the Fermi level of the semiconductor. In turn, this yields an increase of the occupation probability of the acceptor states while pushing away free mobile holes with positive charge that counterbalance the negative space charge. Thus, a negative space or depletion charge is present within the depletion length L_{depl} (counted from the gate dielectric-silicon interface).

Employing the depletion approximation (see Section 4.2), the charge density due to ionized acceptors is considered being constant between $z = 0$ and $z = L_{\mathrm{depl}}$ and zero otherwise. This enables an analytic calculation of the band bending by simply integrating the Poisson equation twice yielding a parabolic behavior. As a result, the depletion charge (per unit area) is $Q_{\mathrm{depl}} = -eN_a \cdot L_{\mathrm{depl}}$. Here, it was assumed that all acceptors are ionized within the depletion region L_{depl}, which itself is a function of the surface potential (and hence the gate voltage) Φ_0. From Poisson's equation, one obtains $\Phi(z) = -e^2 \frac{N_a}{2\varepsilon_0\varepsilon_{\mathrm{si}}}(z - L_{\mathrm{depl}})^2$ with $\Phi(z = 0) = \Phi_0$, and hence the well-known expression for $L_{\mathrm{depl}} = \sqrt{\frac{2\Phi_0\varepsilon_0\varepsilon_{\mathrm{si}}}{|e|^2 N_a}}$.[3] As a result, the MOS capacitor is similar to a p-n junction (cf. Section 4.2) where the charge on the n-side (the gate, i. e., Q_g) is very high (it is a metal) such that it can be considered as being δ-shaped (see Q_g in panels (b) and (c) in Figure 4.12) and, moreover, the positive and negative charges are spatially separated by the gate oxide. Eventually, the depletion capacitance is given by

$$C_{\mathrm{depl}} = (-e)\frac{\partial(-eN_a L_{\mathrm{depl}})}{\partial\Phi_0} = |e|^2 N_a \frac{2\varepsilon_0\varepsilon_{\mathrm{si}}}{|e|^2 N_a}\frac{1}{2}\sqrt{\frac{|e|^2 N_a}{2\varepsilon_0\varepsilon_{\mathrm{si}}\Phi_0}} = \varepsilon_0\varepsilon_{\mathrm{si}}\frac{1}{L_{\mathrm{depl}}} \tag{4.17}$$

where the charge, and consequently, the capacitance are computed per area. Equation (4.17) shows that the depletion capacitance can be considered as a simple parallel plate capacitor with L_{depl} being the "oxide" thickness and $\varepsilon_{\mathrm{si}}$ the dielectric constant

3 Note that in order to distinguish the potential energy from the voltage, it has been denoted with capital Greek letters in contrast to the nomenclature used in Chapter 2.

of the "oxide". Due to the $1/\sqrt{N_a}$-dependence of the depletion length on the acceptor concentration, $C_{\text{depl}} \propto \sqrt{N_a}$.

Note the difference between the depletion capacitance C_{pn} of the *p-n* junction given in Equation (4.11) and C_{depl} in a MOS capacitor (Equation (4.17)), whereas C_{depl} is the capacitance due to a change of the depletion charge with respect to a change of the surface potential at the gate dielectric interface, C_{pn} has been computed as the change of depletion charge with respect to a change of the voltage applied to the *p-n* junction.[4]

4.5.2 Interface-States Capacitance

At the surface of a bulk semiconductor, there is a high density of surface states due to dangling bonds. For instance, depending on the crystallographic plane the surface atom density of silicon varies between $6.78 \cdot 10^{14}$ cm^{-2} and $9.59 \cdot 10^{14}$ cm^{-2} (cf. Figure 2.26) resulting in a high surface density of states. A major benefit of silicon is that a SiO_2 can be grown on the surface that facilitates a reduction of the density of states at the Si/SiO_2 interface D_{it} by five orders of magnitude (see Section 3.2).

If the D_{it} remaining after the growth of SiO_2 is considered to be continuously distributed across the band gap, a charge neutrality level E_{CNL} (see Figure 4.12(c)) can be defined, which refers to the energetic position of the Fermi level at the semiconductor surface E_f^s that yields an equal occupation of conduction-band- and valence-band-like interface states, and hence a neutral interface is obtained if $E_{\text{CNL}} = E_f^s$. E_{CNL} represents the top (bottom) of the valence (conduction) band-like interface states and is thus the energy relevant for computing the interface charge. Hence, $Q_{\text{it}} = -e\int_{E_{\text{CNL}}}^{\Phi_0} D_{\text{it}}(E) f(E_f^s) + e\int_{\Phi_0-E_g}^{E_{\text{CNL}}} D_{\text{it}}(E)(1-f(E_f^s))$ where Φ_0 is the conduction band at the MOS-interface (cf. Figure 4.12). In this expression, the first (second) term is the negative (positive) interface charge due to an occupation of the conduction(valence) band-like states with electrons(holes).

When applying different gate voltages at a MOS capacitor, the Fermi level E_f^s within the semiconductor remains unaltered (if leakage between semiconductor and gate electrode is negligibly small) and the conduction/valence bands are moved relative to the Fermi level. However, since the interface states are located within a tiny region at the Si/SiO_2 interface, moving the bands with respect to E_f^s yields (almost) the same change of surface charge (up to a sign change) as if E_f^s was moved relative to constant conduction/valence bands. As a result, $\frac{\partial}{\partial \Phi_0} = -\frac{\partial}{\partial E_f^s}$ and, therefore, the capacitance (per area since D_{it} is the density-of-interface states per energy and area) C_{it} related to the interface charge Q_{it} can be written as

4 A similar result as in Equation (4.17) would be obtained for the *p-n* junction if we computed $C_{\text{pn}} = \frac{\partial e^2 N_a L_{\text{depl}}^a}{\partial \Phi(x=0)}$, where $\Phi(x=0)$ is the potential profile of the *p-n* junction at the junction interface.

$$C_{\text{it}} = e\frac{\partial Q_{\text{it}}}{\partial \Phi_0} = |e|^2 \int_{E_{\text{CNL}}}^{\Phi_0} dE\, D_{\text{it}}(E)\left(-\frac{\partial}{\partial E_f^s} f(E_f^s)\right)$$
$$+ (-|e|^2) \int_{\Phi_0 - E_g}^{E_{\text{CNL}}} dE\, D_{\text{it}}(E)\left(-\frac{\partial}{\partial E_f^s}[-f(E_f^s)]\right)$$
$$= |e|^2 \int_{\Phi_0 - E_g}^{\Phi_0} dE\, D_{\text{it}}(E)\left(-\frac{\partial}{\partial E_f^s} f(E_f^s)\right). \tag{4.18}$$

In Section 2.10, it was shown that the negative derivative of $f(E_f^s)$ with respect to the Fermi energy E_f^s can be approximated with a delta function (cf. lower panel of Figure 2.44). Using this in Equation (4.18) yields

$$C_{\text{it}} = |e|^2 \int_{\Phi_0 - E_g}^{\Phi_0} dE\, D_{\text{it}}(E) \frac{1}{k_B T} \frac{e^{\frac{E-E_f^s}{k_B T}}}{(1 + e^{\frac{E-E_f^s}{k_B T}})^2} \approx |e|^2 \int_{\Phi_0 - E_g}^{\Phi_0} dE\, D_{\text{it}}(E)\delta(E - E_f^s) = |e|^2 D_{\text{it}}(E_f^s). \tag{4.19}$$

If there is negligible inversion or accumulation charge on the MOS capacitor and if $C_{\text{depl}} \ll C_{\text{ox}}$, Equation (4.16) can be written as

$$\frac{\partial \Phi_0}{\partial \Phi_g} = \frac{1}{1 + \frac{C_{\text{it}}}{C_{\text{ox}}}} \approx \frac{1}{1 + \frac{|e|^2 d_{\text{ox}} D_{\text{it}}(E_f^s)}{\varepsilon_0 \varepsilon_{\text{ox}}}}. \tag{4.20}$$

This relation will play a prominent role when discussing metal-semiconductor contacts and the appearance of Fermi level pinning further below.

4.5.3 Density-of-States or Quantum Capacitance

For sufficiently large gate voltages, the conduction band is moved below the intrinsic Fermi level, meaning that the silicon surface layer changes from p-type to n-type behavior, which is called inversion. For even larger gate voltages, there will be a substantial amount of free charge in the conduction band at the silicon-gate dielectric interface. In the following, the charge in this inversion layer is calculated enabling the computation of the inversion-layer capacitance (often called density of states or quantum capacitance) C_{inv}. As an example of a nontrivial dispersion relation, the case of the silicon conduction band is discussed. Therefore, we need to know how the quantization within the inversion layer modifies the six ellipsoids (cf. Figure 2.28(b)). Due to its importance, the formation of the Si inversion layer is also explained in the video accessible through QR code #41.

For large gate voltages, the conduction band at the Si/SiO_2 interface appears like a triangularly-shaped potential barrier consisting of the band gap of silicon and the oxide barrier (see Figure 4.13, left panel). In order to compute the density of mobile charge in the inversion layer, the Schrödinger equation needs to be solved. The MOS capacitor is considered to have a very large width W and length L, so that the potential distribution $\Phi(x, y, z) = \Phi(z)$ within the semiconductor can be considered as being independent of x and y. Hence, a separation ansatz for the wavefunction is appropriate with $\psi_{\vec{k}}(x, y, z) = \phi_{k_x}(x)\varphi_{k_y}(y)\eta_{k_z}(z)$. Since the potential $\Phi(z)$ does not depend on x and y, the wavefunctions $\phi_{k_x}(x)$ and $\psi_{k_y}(y)$ are simply plane waves and we only have to compute the wavefunction along the z-direction.

The quantization within the triangular potential well (see white dahed line in the inset of the left panel in Figure 4.13) yields discrete energy levels along the z-direction. Solving the Schrödinger equation in a triangular potential well results in so-called Airy functions for $\eta_{k_z}(z)$ and a quantization energy of

$$E_n = \left(\frac{e^2 \mathcal{E}_{\text{si}}^2 \hbar^2}{2m^\star}\right)^{1/3} a_n \quad \text{with } a_n \approx \left(\frac{3\pi}{2}\left(n - \frac{1}{4}\right)\right)^{2/3} \tag{4.21}$$

where $\mathcal{E}_{\text{si}} = \frac{1}{|e|}\frac{\partial\Phi(z)}{\partial z}|_{z=0}$ is the electric field of the triangular potential and $n = 1, 2, \ldots$ an integer number. The eigenenergies increase with decreasing effective mass and vice versa, similar to the quantization energies in a particle-in-the-box (cf. Section 2.1.2). As discussed at length in Section 2.2.1, the quantization in the z-direction yields 2D subbands with a free motion of carriers along x and y starting from $\Phi_0 + E_n$, with E_n given in Equation (4.21). From Figure 4.13, we can infer that there are two ellipsoids or "valleys" (red), which have the long axis along the z-direction and four valleys (green) lie within the x–y plane (recall that in a cubic lattice such as silicon k_x, k_y, k_z are aligned with x, y, z directions in real space). In order to obtain the 2D subbands in the conduction band inversion layer of silicon, the two types of ellipsoids can be considered separately. The red ellipsoids have the heavy effective mass of $m_h^\star = 0.92m_0$ in quantization direction (i. e., z-direction) whereas the green valleys(ellipsoids) have the light effective mass $m_l^\star = 0.19m_0$ in quantization direction. Thus, we expect two sets of quantization energies, and hence two sets of 2D subbands stemming from the red and green valleys.

The right panels of Figure 4.13 show the six ellipsoids of the silicon conduction band (top left). Since the k_x- and k_y-directions are equivalent, the $k_{x,y,z}$ coordinate system can be tilted so that the k_y direction is aligned perpendicular to the plane of the page. Due to confinement there are only discrete k_z values allowed and if the energy is increased, it becomes clear that a 2D subband can only appear if the ellipsoids cross the dashed black lines, which indicate an allowed k_z^n-state. Since the red valleys expand much more quickly along the k_z-direction (because the heavy mass is aligned in this direction), the two red valleys provide the first 2D subband at an energy of $E = E_1$, which is degenerate twice. If the energy increases to $E = E_2$, the green valleys will also result in a 2D subband, which is four times degenerate since four green valleys cross the black dashed line of

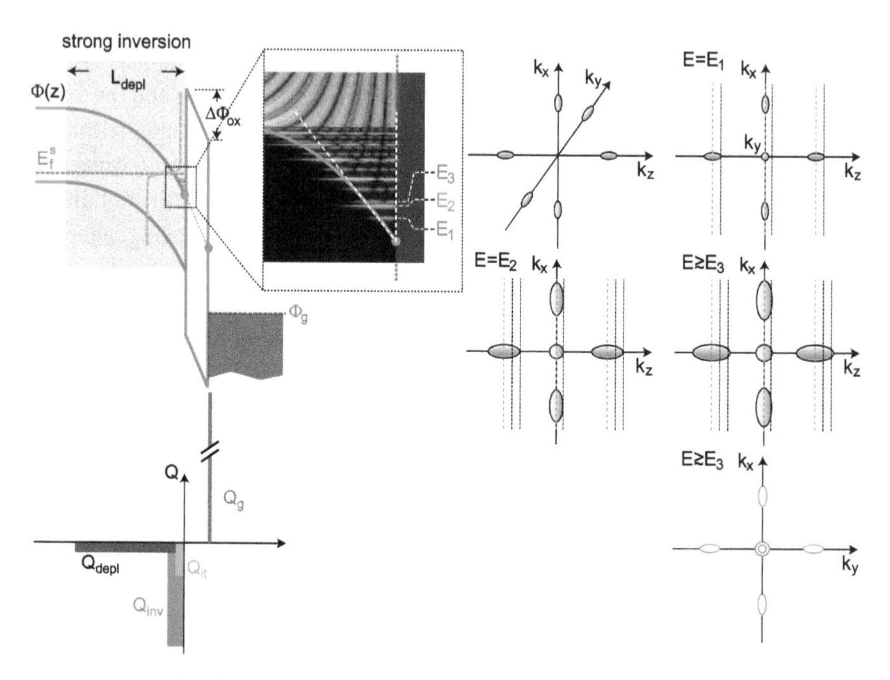

Figure 4.13: Metal-oxide-semiconductor capacitor in the case of strong inversion (left panel). The six-fold degeneracy of the silicon conduction band splits into two red and four green valleys due to quantum confinement in the approximately triangular (illustrated with the white dashed line) potential well. The inset shows a plot of the local DOS of the red and green valleys.

allowed k_z-values. Eventually, at a slightly higher energy $E = E_3$ the red valleys lead to a second 2D subband, which is again degenerate twice.

Projecting the constant energy surfaces onto the k_x–k_y plane at the energy $E \gtrsim E_3$ yields the image shown in the bottom right panel of Figure 4.13. Here, the four ellipsoids from the green valleys and two circles from two 2D subbands stemming from the red valleys are shown. Since the red valleys have the heavy effective mass in the quantization direction, the 2D dispersion relation will be a rotational paraboloid explicitly given by

$$E_{\mathrm{red}}(k_x, k_y, n) = \frac{\hbar^2(k_x^2 + k_y^2)}{2m_l^\star} + \left(\frac{e^2\mathcal{E}_{\mathrm{si}}^2\hbar^2}{2m_h^\star}\right)^{1/3}\left(\frac{3\pi}{2}\left(n - \frac{1}{4}\right)\right)^{2/3}. \tag{4.22}$$

The green valleys, on the other hand, have the light effective mass in the direction of quantization, and thus exhibit the light and heavy effective masses along k_x/k_y giving rise to an ellipsoid (cf. bottom right panel in Figure 4.13). Because the ellipsoids lie on the k_x- and k_y-axes, silicon shows isotropic conduction properties (see Section 2.11.2). Replacing the effective masses $m_l^\star$ and $m_h^\star$ with a 2D density-of-states effective mass $m_{\mathrm{DOS}}^\star = \sqrt{m_l^\star m_h^\star}$ allows for writing the dispersion relation of the green valleys E_{green} in the same form as Equation (4.22) with $m_{\mathrm{DOS}}^\star$ instead of $m_l^\star$.

Finally, the electric field $\mathcal{E}_{\text{si}}$ at the silicon-gate dielectric interface can be related to the gate voltage by noting that the dielectric displacement needs to be continuous across the interface, i. e., $\varepsilon_{\text{si}}\mathcal{E}_{\text{si}} = \varepsilon_{\text{ox}}\mathcal{E}_{\text{ox}}$ with the electric field in the oxide being equal to $\mathcal{E}_{\text{ox}} = \frac{1}{|e|}\frac{\Phi_0-(\Phi_g+\Phi_{\text{bi}})}{d_{\text{ox}}}$.

After these considerations, we are now in a position to compute the carrier density $n(x, y, z)$, which in equilibrium is given by the energy integral of the product of the local density of states (see inset of Figure 4.13, left panel) and the Fermi distribution function. Since the wavefunctions in x- and y-directions are plane waves, the local density of states depends only on z (cf. Equation (2.79) and Task 12 for details on $m^\star_{\text{DOS}}$) and becomes

$$D(E,z) = \sum_{k_x,k_y,n} \frac{1}{W \cdot L}\delta(E - E(k_x,k_y,n))|\eta_n(z)|^2 = \sum_n \frac{m^\star_{\text{DOS}}}{\pi\hbar^2}|\eta_n(z)|^2\Theta(E - E_n), \tag{4.23}$$

i. e., a sum over the two-dimensional density of states due to the different red and green subbands. Here, $\Theta(E - E_n)$ is the Heaviside function resulting in a 2D DOS contribution for each subband E_n. As a result, one obtains the following expression for the charge density in the inversion layer of the conduction band of a silicon (100) MOS capacitor:

$$\begin{aligned} Q_{\text{inv}}(z) = (-e)n_{\text{inv}}(z) = 2(-e) \times &\sum_{n,\text{red valleys}} \int_{E_n}^{\infty} dE \frac{m_l^\star}{\pi\hbar^2}|\eta_n(z)|^2 f(E_f^s) \\ + 4(-e) \times &\sum_{j,\text{green valleys}} \int_{E_j}^{\infty} dE \frac{\sqrt{m_l^\star m_h^\star}}{\pi\hbar^2}|\eta_j(z)|^2 f(E_f^s). \end{aligned} \tag{4.24}$$

Note that the prefactors 2 and 4 account for the degeneracy of the red and green valleys. Next, since at the moment we are not interested in the spatial dependence of the carrier density on z, we can integrate over the z-coordinate and obtain the carrier density per area. The integration can be readily carried out since the wavefunction $\eta(z)$ is considered to be normalized, and thus $\int dz|\eta(z)|^2 = 1$. Moreover, the integration over energy can also be carried out due to the constant two-dimensional density of states (cf. Section 2.11.3). However, we can also cast Equation (4.24) into the usual form $n = \int_{-\infty}^{\infty} dE D_{\text{eff}}(E - \Phi_0) f(E_f^s)$ with an effective density of states given by

$$D_{\text{eff}}(E - \Phi_0) = 2\sum_n \Theta(E - E_n)\frac{m_l^\star}{\pi\hbar^2} + 4\sum_j \Theta(E - E_j)\frac{\sqrt{m_l^\star m_h^\star}}{\pi\hbar^2} \tag{4.25}$$

where again the Heaviside step function $\Theta(E - E_{n,j})$ is used that is zero for $E < E_{n,j}$ and unity otherwise. This allows for writing the inversion layer or density of states capacitance in the form

$$C_{\text{inv}} = -|e|^2 \frac{\partial}{\partial \Phi_0} \int_{\Phi_0}^{\infty} dE\, D_{\text{eff}}(E - \Phi_0) f(E_f^S) = -|e|^2 \int_0^{\infty} dE\, D(E) \frac{\partial}{\partial \Phi_0} f(E + \Phi_0 - E_f^S). \quad (4.26)$$

In the final step in Equation (4.26), the transformation of variables $E \rightarrow E - \Phi_0$ was carried out in order to remove the dependence on Φ_0 from the lower bound of the integration as well as from D_{eff}. As a result, the derivative $\frac{\partial}{\partial \Phi_0}$ only applies to the Fermi distribution function, which can be computed analytically. As has been done in the preceding section, the derivative of the Fermi distribution function is approximated with the delta function (cf. Section 2.10), i. e., $\frac{\partial f(E+\Phi_0-E_f^S)}{\partial \Phi_0} \approx -\delta(E - (E_f^S - \Phi_0))$. The integration can then be carried out trivially and one obtains for the inversion-layer capacitance

$$C_{\text{inv}} \approx |e|^2 D_{\text{eff}}(E_f^S - \Phi_0). \quad (4.27)$$

This means that the inversion-layer capacitance is approximately proportional to the density of states (hence the frequently used name density of states capacitance).

Looking at the MOS capacitor within the conduction band of the (100) silicon surface an important implication can be extracted: for increasing gate voltage, an increasing number of two-dimensional subbands are moved toward the Fermi energy, meaning that the effective density of states in the inversion layer steadily increases with increasing gate voltage. As a result, at some arbitrary V_{gs} the density of states capacitance will become larger than all other capacitances in the MOS capacitor, i. e., $C_{\text{inv}} \gg C_{\text{ox}} + C_{\text{it}} + C_{\text{depl}}$. In this case, the gate capacitance is

$$C_{gs} = \frac{C_{\text{ox}}(C_{\text{inv}} + C_{\text{it}} + C_{\text{depl}})}{C_{\text{ox}} + C_{\text{inv}} + +C_{\text{it}} + C_{\text{depl}}} \approx \frac{C_{\text{ox}} C_{\text{inv}}}{C_{\text{inv}}} = C_{\text{ox}}. \quad (4.28)$$

This can also be seen in the experimental curve in Figure 4.11(b) that approaches C_{ox} at large V_{gs}. So, the reason why in classical textbooks (e. g., to compute the MOSFET characteristics within the gradual channel approximation, see Section 5.1) the gate capacitance is simply replaced with the geometrical oxide capacitance is the high density of states in a silicon inversion layer. However, there are cases where C_{inv} may become rather small and C_{ox} will be the dominant capacitance. In this case, (assuming C_{depl} and C_{it} can be neglected), $C_{\text{ox}} \gg C_{\text{inv}}$ leading to the peculiar case where $C_{gs} \approx C_{\text{inv}}$. In literature, this limit is often called the quantum-capacitance limit and will be further discussed in Chapter 5.9.1. Interestingly, in the quantum-capacitance limit one can measure the density of states simply by carrying out a capacitance-voltage measurement of the MOS capacitor.

4.5.4 Accumulation Capacitance

If the gate voltage of the MOS capacitor considered here (with the flat-band case at $V_{gs} = 0\,\text{V}$) is negative, the valence band is moved toward and eventually above the

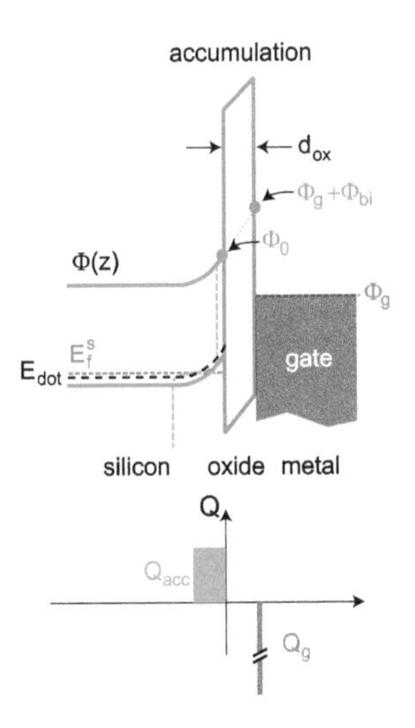

Figure 4.14: Metal-oxide-semiconductor capacitor in accumulation. In the case of a *p*-type substrate, negative gate voltages move the valence band above E_f^s giving rise to a large accumulation density of holes.

Fermi level E_f^s. Hence, the acceptors will remain empty and, therefore, neutral since their occupation probability strongly decreases. However, at the Si/SiO_2 interface a large accumulation of holes occurs as illustrated in Figure 4.14. Knowing the dispersion relation in the valence band the charge density can be computed in the same fashion as has been done in Section 4.5.3, namely summing over the different 2D subbands. The silicon valence band is twice degenerate with one band exhibiting a heavy hole mass $m_{hh}^\star = 0.49m_0$ and one band with a light effective hole mass $m_{lh}^\star = 0.16m_0$. In addition, a split-off band can be found 0.044 eV below the valence band edge with an effective hole mass of $m_{sh}^\star = 0.29m_0$. For sufficiently small energies (i. e., below the valence band), all three bands can be considered as being isotropic. As a result, the capacitance associated with the accumulation layer C_{acc} is similar to the one obtained in the conduction band. The only difference is the effective density of states where the different effective masses have to be used and each band has a degeneracy factor of unity. With the same reasoning as above, an increasing number of 2D subbands will contribute to the density of states in the valence band, and thus $C_{\text{acc}} \gg C_{\text{ox}}$ will eventually be reached. Therefore, $C_{\text{gs}} \approx C_{\text{ox}}$ for large negative gate voltages as has been found in the case of strong inversion (see also Figure 4.11(b) for negative V_{gs}).

4.5.5 Gate Dielectrics with High Dielectric Constant

It has been mentioned a few times above that one of the major advantages of silicon as the material of choice for the realization of highly integrated circuits is the existence of a native oxide, SiO_2, which fulfills all requirements for a proper gate dielectric. SiO_2 is an excellent insulator exhibiting a large band gap of ~9 eV with appropriate conduction and valence band offsets ΔE_c and ΔE_v (see Figure 4.16). Equally important, it facilitates a very strong reduction of the interface density of states such that $C_{ox} \gg C_{it}$ can be achieved, which is of utmost importance for proper switching of MOSFETs (cf. Section 5.2.2). In addition, silicon dioxide can be grown in a straightforward way either with wet or dry thermal oxidation (cf. Section 3.2). However, as will be discussed in detail in the next chapter, scaling down the dimensions of metal-oxide-semiconductor field-effect transistors requires a strong increase of C_{ox} in order to maintain appropriate electrostatic gate control of the device. In this respect, the major drawback of SiO_2 is the low value of its dielectric constant with $\varepsilon_{ox} = 3.9$. As a result, extremely thin SiO_2 thicknesses are required that eventually lead to a strong increase of leakage current due to direct tunneling of carriers through the gate insulator; in Section 5.9.3, this will be further elaborated on with simulations. Because of the exponential dependence of the direct tunneling current on the thickness of the gate dielectric, gate leakage becomes increasingly deleterious as d_{ox} is scaled down and becomes intolerable for $d_{ox} \lesssim 1.5$ nm.

Therefore, gate dielectric materials with substantially higher dielectric constant ε_k are in use nowadays. The idea behind this is rather simple: If a gate dielectric with a higher dielectric constant is used, the same value of geometrical oxide capacitance C_{ox} can be obtained with a thicker insulator d_k that yields an exponentially suppressed gate leakage current. In order to facilitate a simple comparison between SiO_2 and the use of an alternative gate dielectric material, the so-called "effective oxide thickness" (EOT) was introduced. The EOT is the thickness of the gate dielectric if it was SiO_2 yielding the same capacitor values as in the case of ε_k:

$$C_{ox} = \frac{\varepsilon_0 \varepsilon_{ox}}{d_{ox}} = \frac{\varepsilon_0 \varepsilon_k}{d_k} \rightarrow \text{EOT} = \frac{\varepsilon_{ox}}{\varepsilon_k} d_k \tag{4.29}$$

where again C_{ox} is the geometrical oxide capacitance per area. As a result, the physical thickness d_k can be a factor $\varepsilon_k/\varepsilon_{ox}$ larger still leading to the same capacitance value but the leakage current due to direct tunneling will be exponentially suppressed.

The integration of materials with high ε_k value is a delicate issue due to a number of reasons:

- When depositing an alternative gate dielectric an interfacial SiO_x or metal silicate is likely to form such that an interface layer of thickness d_i builds up. In addition, even if highly doped polysilicon is used as the gate electrode, a (gate-voltage-dependent) depletion layer of thickness L_{depl}^{gate} exists at the interface between the gate electrode and dielectric (illustrated in Figure 4.15(a)). In this case, the gate stack consists of three geometrical capacitors in series, namely $C_k = \frac{\varepsilon_0 \varepsilon_k}{d_k}$, $C_i = \frac{\varepsilon_0 \varepsilon_i}{d_i}$ and $C_{depl}^{gate} = \frac{\varepsilon_0 \varepsilon_{si}}{L_{depl}^{gate}}$. As a result, the effective oxide thickness is given by

$$\text{EOT} = \frac{\varepsilon_{ox}}{\varepsilon_k} d_k + \frac{\varepsilon_{ox}}{\varepsilon_i} d_i + \frac{\varepsilon_{ox}}{\varepsilon_{si}} L_{depl}^{gate}. \tag{4.30}$$

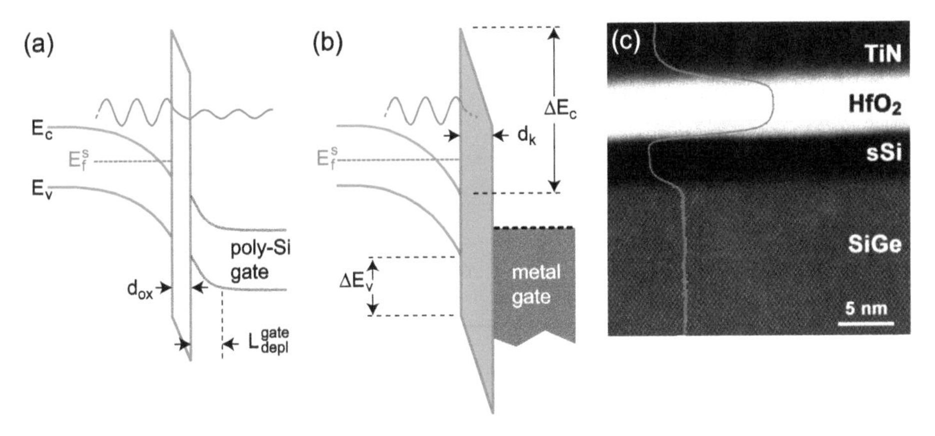

Figure 4.15: (a) MOS capacitor consisting of a poly-Si gate electrode with depletion length L_{depl}^{gate} and an ultrathin SiO_2 that leads to direct tunneling. (b) MOS capacitor with a high-ε_k material of physical thickness d_k and a metal gate. (c) Transmission electron microscopy image of a TiN/HfO_2 gate stack on strained Si on SiGe (M. Luysberg, FZ-Jülich).

It is clear that the last two terms limit the achievable values of EOT, and hence d_i and L_{depl}^{gate} need to be made as thin as possible.

- The contribution L_{depl}^{gate} can be made negligible by replacing poly-Si as gate electrode with a metal (cf. Figure 4.15(b) and (c)). To circumvent issues with the thermal budget (e. g., during the activation anneal of implanted dopants), the replacement-gate technique using a damascene process (see Section 3.9) was invented. Avoiding the formation of an interfacial layer, d_i is substantially more involved. On the one hand, if one was able to completely remove d_i, the smallest EOT would be obtained. However, this usually leads to a higher density of interface states, and thus potentially a strong increase in C_{it}. On the other hand, the formation of an interfacial oxide layer reduces C_{it} but yields a minimum possible EOT $= d_i$ (where $\varepsilon_i = \varepsilon_{ox}$ was assumed for simplicity). Therefore, obtaining extremely thin interface layers requires a very careful surface treatment prior to the deposition of the high ε_k material.
- The value ε_k of the gate dielectric needs to be as large as possible but at the same time appropriate insulation is needed. In this respect, the trend of a decreasing band gap and decreasing conduction band offset ΔE_c with increasing ε_k (displayed in Figure 4.16) limits the amount of useful materials. In addition, possible gate dielectric materials need to be compatible with the fabrication processes and must be thermally stable on silicon.

The family of hafnium-oxide-based materials, specifically HfO_2, $HfSi_xO_y$, HfO_xN_y and $HfSiO_xN_y$, have been selected and implemented in the most advanced CMOS chips [94]. The ε_k-value of pure HfO_2 amounts to 22, while those of other Hf-based compounds have lower values.

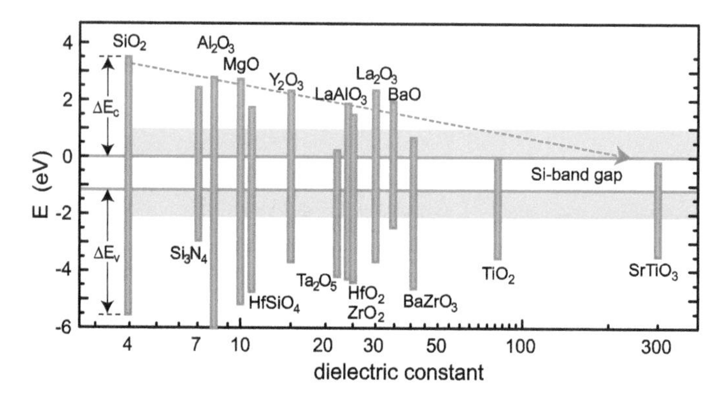

Figure 4.16: Band gap and conduction/valence band offsets $\Delta E_c/\Delta E_v$ with respect to the silicon band gap (green straight lines) of various gate dielectric materials as a function of their dielectric constant. The light blue region illustrates the minimum energy range required for a proper gate dielectric ensuring appropriate insulation. The general trend of a reduced band gap and conduction band offset with increasing ε_k can be seen.

4.6 Metal-Semiconductor Contacts

Metal-semiconductor contacts are one of the most important ingredients for electronic devices. Any device is only useful if it can be connected to the "outside world" to inject or extract charges from the semiconductor. As a result, a metal-semiconductor (MS) contact will always be present and, therefore, a tremendous amount of work has been devoted to MS contacts. Due to the advent of "novel" 2D materials such as transition-metal dichalcogenides (see Chapter 10) but also in rather classical semiconductors such as silicon (for instance, to realize passivated contacts in solar cells [227]), germanium or III–V compound semiconductors the realization of proper contacts is today still a very important topic.

When a metal and a semiconducting material are brought into contact, there will be an exchange of charge between the two materials. This ensures that in equilibrium there will be a constant Fermi level throughout the MS-contact. In addition to this requirement, there is a second condition that needs to be fulfilled. Each material has its own work function Φ_m, i. e., the energy that is required in order to remove an electron from it. The work function is usually measured with respect to the Fermi level. In a semiconductor, however, the Fermi level may lie within the band gap where no states can be found, and hence, in such materials the electron affinity χ_{semi} is used. χ_{semi} is the energy needed to remove an electron from the conduction band of the semiconductor. Now, the second condition to be fulfilled at a metal-semiconductor contact is that the work function of the metal and the electron affinity of the semiconductor are at the same energy at the interface. As a result, in ideal metal-semiconductor contacts the energy difference between the Fermi level in the metal and the conduction band in the semiconductor is $\Phi_{\text{SB}} = \Phi_0 - E_f = \Phi_m - \chi_{\text{semi}}$; this potential barrier is called the Schottky barrier. If the work

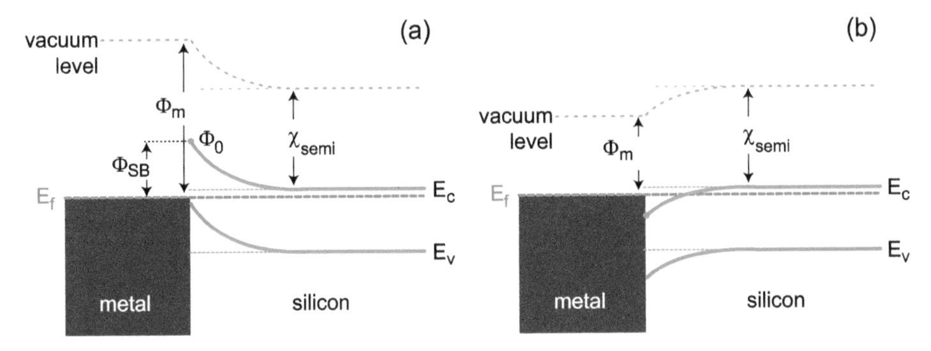

Figure 4.17: Ideal metal-semiconductor contact in the case of a large work function (a) and a small work function (b) of the metal. The requirements of a constant Fermi level and that the work function and the electron affinity need to be at the same energy at the interface allows in principle Ohmic contacts to the valence band (a) and the conduction band (b).

function of the metal is so large that $\Phi_{SB} \gtrsim E_g$, the band bending in the semiconductor leads to an Ohmic contact formation to the valence band as depicted in Figure 4.17(a). On the other hand, if Φ_m is small resulting in $\Phi_{SB} = \Phi_m - \chi_{semi} \lesssim 0$ then a band bending in the other direction occurs enabling an Ohmic contact to the conduction band (cf. Figure 4.17(b)).

Figure 4.18 shows the work function of various materials (values are extracted from [247]). Obviously, suitable metals that straddle a rather large range of different work functions can be identified (for instance, the red marked metals in Figure 4.18). This means that if metal-semiconductor contacts were ideal and behaved according to what is called the Schottky–Mott limit one would just need to choose an appropriate metal and the issue of Ohmic contact formation to a semiconductor would be resolved. For instance, a low work function metal such as scandium or magnesium and even aluminum

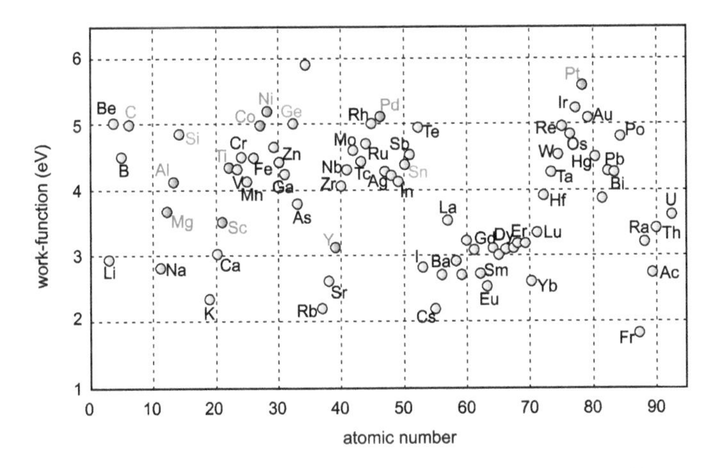

Figure 4.18: Work function of various metals; the values were extracted from [247].

would be useful to realize contacts to n-doped silicon, whereas platinum or palladium would be appropriate for contacting p-doped silicon. However, metal-semiconductor contacts are unfortunately far from being ideal.

4.6.1 Fermi Level Pinning

When depositing different metals with different work functions on the same semiconductor, one finds in almost all cases a phenomenon called "Fermi level pinning." This means that the MS contacts show similar electrical properties irrelevant of the metal that was deposited onto the semiconductor; the Fermi level appears to be pinned at a specific energy (or within a small energy range), which for most technologically useful semiconductors lies within the band gap. As a result, a Schottky barrier builds up at MS contacts and carriers need to tunnel through this barrier and/or need to be thermally exited over this barrier in order to be injected into the semiconductor. This strongly deteriorates the electrical behavior of devices due to a high contact resistance (cf. Chapter 7).

There has been a decade-long debate about the origin of Fermi level pinning. An obvious cause may be dangling bonds at the surface of a bulk semiconductor. However, dangling bonds as the main course does, for instance, not fit to the experimental observation of Fermi level pinning at the interface between metals and 2D materials (cf. Chapter 10). A model that is able to explain the experimental observations is the model of "metal-induced gap state" (MIGS) [199]. MIGS stem from evanescent waves penetrating the (classically forbidden) band gap of the semiconductor from the metal side. As a result, a high density of interface states is obtained continuously distributed across the band gap. At the metal-semiconductor interface, there will certainly also be defects and dangling bonds that contribute to the overall density of interface states. And if the density of these defects/bonds becomes very large, they will have an impact on Fermi level pinning and the formation of Schottky barriers at the metal-semiconductor interfaces. However, the density of the MIGS is usually higher and they penetrate deeper into the semiconductor, and hence are the dominant factor for Fermi level pinning. As an example, Figure 4.19 shows the local density of states at a GaAs-Au MS contact (measured with scanning tunneling spectroscopy) that was fabricated in ultrahigh vacuum to avoid any impact of defects, impurities, etc. The MIGS can clearly be observed leading to a substantial density of states at the interface across the entire band gap [115]. Figure 4.19(b) shows the computed local density of states of a MS contact within the band gap of the semiconductor (for details on the computation, see Chapter 6). The MIGS can be identified and can be subdivided into conduction band-like and valence band-like states (as has already been done above, see Section 2.4.7), separated by the branching energy E_{br} (cf. Figure 2.20). If E_g is large enough (top panel), the charge neutrality level E_{CNL} is approximately at E_{br} since the density of MIGS within the relevant energy range ($\sim 4k_BT$) above and below E_{br} is similar. As a result, Fermi level pinning will occur close to E_{br}. If

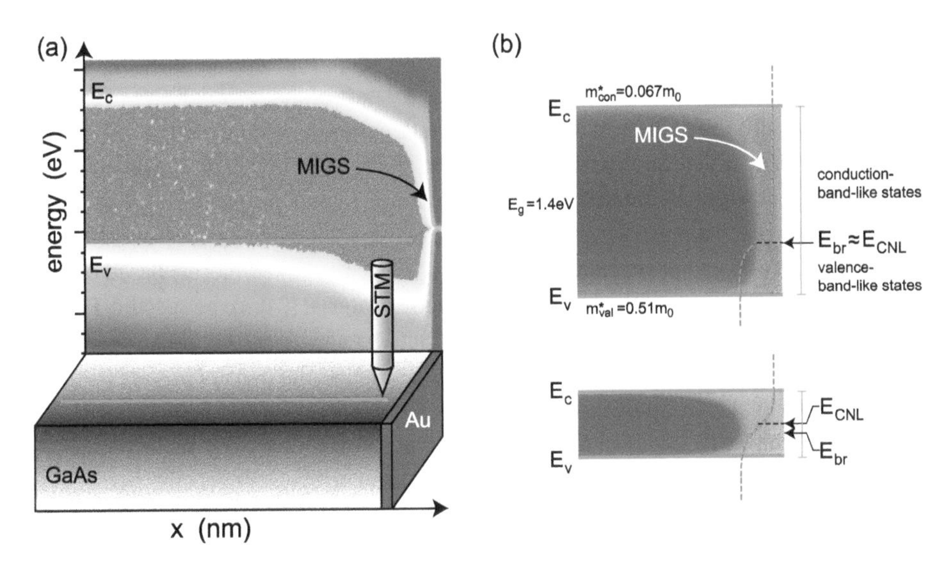

Figure 4.19: (a) Local density of states of a GaAs-Au contact measured with scanning tunneling microscopy showing MIGS at the MS interface (M. Wenderoth, University of Göttingen) [115]. (b) Computed local DOS of MS contacts in the case of GaAs (top) and a semiconductor with small E_g but same effective masses (bottom). For better visibility, the conduction and valence band-like MIGS are colored, separated by E_{br}.

the band gap is rather small or if further bands need to be considered, E_{CNL} may deviate from E_{br} (lower panel).

The Fermi level at the MS interface must be close to the charge neutrality level since otherwise a large interface charge will be present shifting conduction and valence bands so as to reduce the interface charge. The Schottky-barrier $\Phi_{\mathrm{SB}} = \Phi_c - E_f$ is thus determined by

$$\Phi_{\mathrm{SB}} = S_{\mathrm{MIGS}}(\Phi_m - \chi_{\mathrm{semi}}) + (1 - S_{\mathrm{MIGS}})E_{\mathrm{CNL}} \quad \text{with } S_{\mathrm{MIGS}} = \frac{\partial \Phi_{\mathrm{SB}}}{\partial \Phi_m} \tag{4.31}$$

where the so-called slope parameter S_{MIGS} is between 0 and 1. Equation (4.31) means that if $S_{\mathrm{MIGS}} = 1$, the Schottky barrier changes one to one with Φ_m which is the ideal MS contact depicted in Figure 4.17. On the other hand, if $S_{\mathrm{MIGS}} \to 0$, Φ_{SB} is determined by the charge neutrality level and becomes independent of Φ_m, i. e., the Fermi level is pinned at E_{CNL} resulting in $\Phi_{\mathrm{SB}} = E_{\mathrm{CNL}}$. The factors S_{MIGS} and E_{CNL} that determine the MS contact are further elaborated on in the next sections.

4.6.1.1 Charge Neutrality Level

In order to compute the charge neutrality level, we could in principle use the Green's function approach detailed in Section 2.12.3 utilizing the tight-binding recipe given in Section 2.7. The Green's function approach allows even to incorporate dangling bonds, surface reconstruction, etc. by first computing iteratively a semiinfinite contact and using the surface Green's function of this to couple it via appropriate self-energies to the layer containing the dangling bonds. The imaginary part of the retarded Green's function then provides the density of states and from this the charge neutrality level can in principle be determined. Such a calculation is certainly rather involved. However, it has been shown that although the complex band structure is a bulk property, E_{CNL} can be related to it as has been done in Figure 4.19(b).[5] As a result, a detailed model of the interface is not required and an appropriate value for E_{CNL} can be obtained merely based on the (complex) band structure of the bulk semiconductor material [64]. To be specific, E_{CNL} is found by varying the Fermi level taking the conduction and valence band-like character of the complex band structure (separated by E_{br}) as well as the density of states within conduction and valence bands into account and requiring zero net charge.

As an example, let us consider the 1D two-band model discussed in Section 2.4.7. With a nonzero coupling $V_{sp} \neq 0$, the complex band structure connects the two bands (in the following, these bands are referred to as conduction and valence bands). The density of states within each band can be obtained with the computed dispersion relation utilizing, e. g., the smearing method (Section 2.12.1). The complex band structure is (if an analytical computation is not possible) calculated by finding the complex κ that belongs to a real energy within the band gap. In this case, the smearing method does not provide appropriate results for the DOS since it yields a density of states that approaches zero at E_{br}. While this is indeed expected since $D(E) \propto dk/dE$ (see, e. g., [75], p. 4), here we would like to use the complex band structure to mimic the situation at a metal-semiconductor interface. Therefore, the fact that the density of states in 1D is proportional to $1/k \rightarrow 1/\kappa$ is used. The latter is strictly valid only in the case of a quadratic dispersion (i. e., in the vicinity of the band edges) but is used within the entire band gap. Hence, because wavefunctions decay exponentially into the band gap according to $e^{-\kappa x}$, the decay can be associated with a length scale $\delta = \kappa^{-1}$. The larger κ, the smaller will the density of MIGS be. This has important implications for Fermi-level pinning as discussed below.

Figure 4.20 shows conduction and valence bands together with the complex band structure of the 1D two-band model. In (a), the coupling strength between s- and p_x-orbitals is varied. A weak coupling (red) leads to larger κ-values and also to a E_{br} closer to the valence band compared to the case with (three times) stronger coupling (green). As a result, a stronger suppression of the DOS within the band gap is obtained in the red case. In Figure 4.20(b), the same 1D two-band model is computed. However, in this case, two two-band models are added that exhibit different band gaps. This mimics the case where the valence band has a light (LH) and a heavy hole (HH) band but there is only a single conduction band (with much lighter $m^{\star}$) as found in low band gap III–V semiconductors. In this case, the complex band structure of the heavy hole mass band is connected to an

5 In fact, in Figure 4.19(b), Flietner's dispersion relation was used as a simple way to take the bulk complex band structure into account.

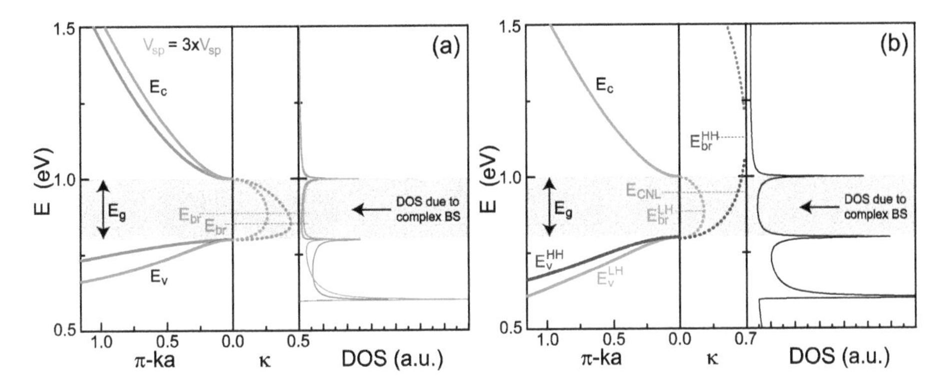

Figure 4.20: (a) 1D two-band model with complex band structure and DOS for two different coupling strengths V_{sp} between s-(valence) and p_x-(conduction) band. (b) Two two-band models with complex band structure and DOS to mimic a semiconductor exhibiting a light (LH) and heavy-hole (HH) valence band.

energetically higher lying band. However, since the branching energy of the latter band is at a higher energy, the complex band structure of HH contributes substantially to the hole density at the interface, which shifts E_{CNL} closer to the conduction band.

4.6.1.2 Slope Parameter for Fermi-Level Pinning

The impact of MIGS on Fermi-level pinning can be described by the slope parameter S_{MIGS} (cf. Equation (4.31)) that is equal to the change of the Schottky-barrier height with changing work function of the metal. With the knowledge about the MOS capacitor, we can now derive an expression for S_{MIGS}.

A metal-semiconductor can be considered as a MOS capacitor with extremely thin insulator. Changing the work function of the contact metal will therefore result in a change of the band bending in the semiconductor that depends on the magnitude of the density of MIGS, similar to the impact of a variation of the applied gate voltage in the MOS capacitor. Hence, we can replace $\frac{\partial \Phi_0}{\partial \Phi_g}$ in Equation (4.20) with $\frac{\partial \Phi_0}{\partial \Phi_m}$ and since the surface potential Φ_0 (see Figure 4.17(a)) equals the Schottky barrier (tacitly setting $E_f = 0$) we obtain an expression for S_{MIGS} very similar to Equation (4.20) [56]. If the density of interface states D_{it} of the MOS capacitor is replaced with the density of metal-induced gap states D_{MIGS} and if the geometrical oxide capacitance is interpreted as the capacitance associated with the decay length $\delta = \kappa^{-1}$ of the MIGS around E_{CNL} (i. e., $C_{ox} = \frac{\varepsilon_{ox}}{d_{ox}} \to \frac{\varepsilon_{si}}{\delta}$), we obtain

$$S_{MIGS} = \frac{\partial \Phi_{SB}}{\partial \Phi_m} = \frac{1}{1 + \frac{e^2 \delta \times D_{MIGS}}{\varepsilon_0 \varepsilon_{si}}} \tag{4.32}$$

in the case of a metal-semiconductor(silicon) contact. As already mentioned above, the decay length $\delta = \kappa^{-1}$ depends on the complex band structure of the semiconductor. As a result, S_{MIGS} appears to be completely determined by properties of the semiconductor.

In fact, in the two-band example above, the DOS within the band gap (see Figure 4.20) was computed with $D_{MIGS} \propto \kappa^{-1}$. Since for not too small band gaps, E_{CNL} is roughly at the branching energy where κ is maximal and because κ_{max} is approximately proportional to E_g the product $\delta \times D_{MIGS}$ is expected to be proportional to E_g^{-2} [97]. If we rewrite Equation (4.32) as $1/S_{MIGS}-1$, we obtain the factor $\frac{e^2 \delta \times D_{MISG}}{\varepsilon_0 \varepsilon_{si}}$ and if the value for $1/S_{MIGS}-1$ is extracted for various materials including insulators and different semiconductors and plotted as a function of the band gap of the material, one obtains the graph shown in Figure 4.21. Here, the green data points belong to the linear axis and the blue to the double logarithmic; the red data points represent MoS_2 and $MoTe_2$.

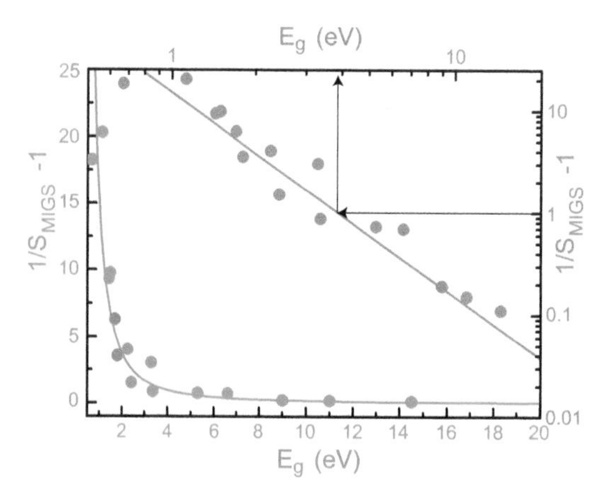

Figure 4.21: $1/S_{MIGS} - 1$ extracted from literature as a function of E_g of different insulators and semiconductors. The green data belong to the linear plot, the blue to the logarithmic plot. Both show excellent agreement with a fit (straight lines) $\propto 1/(E_g)^2$. The red marked data points belong to MoS_2 and $MoTe_2$.

Both plots show excellent agreement with a fit $1/S_{MIGS} - 1 \approx 15/(E_g)^2$ (straight lines), which means that indeed $\delta \times D_{MISG} \propto E_g^{-2}$. This strong dependence on the band gap is also the reason why Fermi-level pinning is not observed in proper insulators with sufficiently large E_g. On the other hand, for any semiconductor with a reasonable band gap Fermi-level pinning is expected to occur. The Fermi level is pinned around the charge neutrality point E_{CNL} (see Figure 4.12) that can be extracted from the complex band structure. The impact of the work function of the metal is taken into consideration via Equation (4.31) using Equation (4.32) giving rise to a substantial Schottky barrier in most MS contacts.

4.6.2 MS Contacts to Highly Doped Semiconductors

Fermi-level pinning leads to a substantial Schottky barrier at metal-semiconductor interfaces, which is prohibitive for any reasonable device functionality. There are basically two approaches to resolving the issue: first Fermi-level depinning, and second strong "thinning" of the Schottky barrier in order to increase the tunneling probability such that carriers can be injected with a transmission close to unity. Fermi-level depinning will be discussed in the succeeding section, so let us concentrate on the second approach.

Thinning the potential barrier is easily accomplished by heavily doping the semiconductor in contact with the metal. With the depletion approximation[6] used in Section 4.2 and assuming strong Fermi-level pinning yielding a constant Schottky barrier Φ_{SB}, the potential distribution at the metal-semiconductor interface is parabolic within the depletion region (cf. Figure 4.22) where the depletion length is related to the doping concentration N_d and the Schottky-barrier height according to $L_{depl} = \sqrt{\frac{2\varepsilon_0\varepsilon_{si}\Phi_{SB}}{e^2 N_d}}$. Using the WKB approximation, the transmission probability for carriers to tunnel through the SB can be computed in the following way: Within the classically forbidden region, a plane wave solution of the Schrödinger equation yields $\phi_k(x) = Ae^{ikx}$ with a purely imaginary wavenumber $k = i\sqrt{\frac{2m^*(\Phi(x)-E)}{\hbar^2}}$. The wavefunction $\phi(x)$ at $x + dx$ is then given by $\phi(x+dx) = Ae^{-k(x+dx)} = \phi(x)e^{-kdx}$. If the classically forbidden region extends from $x = 0$ to $x = d_{WKB}$, the wavefunction will be $\phi(d_{WKB}) = \phi(0)\exp(-\int_0^{d_{WKB}} \sqrt{\frac{2m^*(\Phi(x)-E)}{\hbar^2}}dx)$. The transmission probability $T(E)$ is then simply given by the ratio of the absolute squares of the transmitted wavefunction (i. e., ϕ at $x = d_{WKB}$) and the incident wavefunction (at $x = 0$):

$$T(E) \approx \frac{|\phi(d_{WKB})|^2}{|\phi(0)|^2} = \exp\left(-2\int_0^{d_{WKB}} dx\sqrt{\frac{2m^*(\Phi(x)-E)}{\hbar^2}}\right). \tag{4.33}$$

An analytic solution for $T(E)$ can in principle be obtained by inserting the parabolic potential distribution computed within the depletion approximation into Equation (4.33). However, a more handy expression can be calculated by approximating the parabolic potential distribution with a triangular potential barrier (as illustrated in Figure 4.22) whose height is Φ_{SB} and its width approximately $\frac{3}{4}L_{depl}$. As illustrated in the right panel of Figure 4.22 (black dashed lines), the full L_{depl} would underestimate the tunneling probability whereas using the slope of $\Phi(x)$ at the Schottky barrier leads to a triangular barrier with width $\frac{1}{2}L_{depl}$, which overestimates the transmission. As a result, the transmission probability is approximately given by

$$T(E) \approx \exp\left(-\sqrt{\frac{2m^*}{\hbar^2}}\sqrt{\frac{2\varepsilon_0\varepsilon_{si}}{e^2 N_d}}\left(\frac{\Phi_{SB}-E}{\Phi_{SB}-\Phi_0}\right)^{3/2}\right). \tag{4.34}$$

6 In the case of heavy, i. e., degenerate doping, the depletion approximation is not necessarily fully justified but allows for obtaining a fairly accurate picture in terms of trends and dependencies of the MS contact on the device parameters.

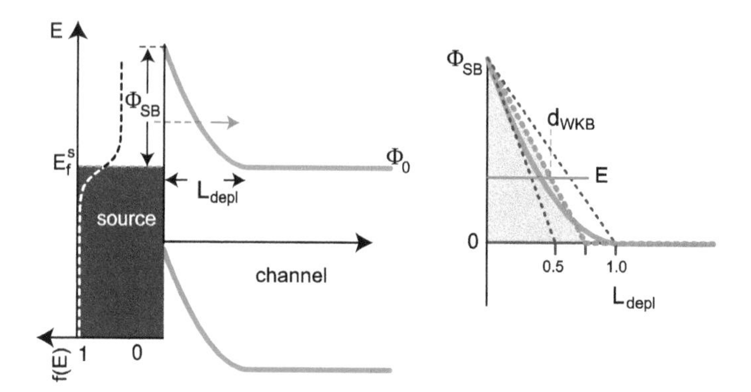

Figure 4.22: Metal-semiconductor contact. A highly doped semiconductor yields a very thin Schottky barrier. Carriers can tunnel through the thin SB with a probability close to unity facilitating Ohmic contact formation to the semiconductor.

At $E = 0$ (i. e., at the Fermi level in the MS contact), with a typical Schottky barrier of ~0.6 eV, an (electron) tunneling effective mass for silicon of ~$0.19 m_0$ one obtains $T \approx 0.009$ at a doping concentration of $N_d = 1 \cdot 10^{20}$ cm^{-3} and $T \approx 0.066$ at $3 \cdot 10^{20}$ cm^{-3}. Apart from the uncertainties of the exact numerical value of T due to the approximate calculation, this clearly shows that very high dopant concentrations are necessary in order to guarantee Ohmic contact behavior. Since a doping concentration of $N_d = 3 \cdot 10^{20}$ cm^{-3} approaches the solubility limit for most dopants, a further reduction of the contact resistance can only be obtained with a reduction of the Schottky-barrier height Φ_{SB} itself. This, however, requires a depinning of the Fermi level.

4.6.3 Fermi-Level Depinning with Ultrathin Insulators

It has been discussed above that MIGS are responsible for Fermi-level pinning and this prevents realizing Ohmic contacts to the conduction and valence bands with an appropriate choice of the work function of the metal contact. Therefore, a removal of the MIGS allows depinning the Fermi level. This can be done by deliberately inserting an insulator of thickness d_{iso} with band gap E_g^{iso} in between the metal and the semiconductor in order to suppress the MIGS. Such an insulator must fulfill a number of requirements: (i) its band gap needs to be large enough to avoid a repinning at a different energy, (ii) it must have an appropriate thickness in order to suppress MIGS and simultaneously allow low contact resistances, (iii) the interface between the semiconductor and the insulator should exhibit a low D_{it} (e. g., due to dangling bonds) and (iv) one should be able to control the fabrication.

The first point (i) is only necessary if a single insulator is desired that enables *n*- and *p*-type devices to be realized with an appropriate metal. A first guess for suitable insulators can be obtained, if a range of available work functions of $\Delta\Phi_m \approx 2$ eV is assumed (see Figure 4.18). In order to contact conduction and valence bands of silicon, $\Delta\Phi_m$ must be sufficient to move the surface potential Φ_0 through the entire band gap of the semiconductor, i. e., $\Delta\Phi_0 = 1.12\,\text{eV}$ in the case of silicon. As a result, $S_{\text{MIGS}} \approx \frac{\Delta\Phi_0}{\Delta\Phi_m} = 1.12/2 \approx 0.5$. Figure 4.21 can then be used to extract suitable band gaps of the insulator. With $1/S_{\text{MIGS}} - 1 \approx 2 - 1 = 1$, it is clear that $E_g^{\text{ins}} \geq 4\,\text{eV}$ (see black arrows in Figure 4.21).

A more rigorous analysis can be carried out based on Equation (4.32) [97]: if the surface including the insulator is considered as a series combination of capacitors, the ratio $\delta/(\varepsilon_0\varepsilon_{\text{si}})$ in Equation (4.32) needs to be replaced with $\delta_{\text{si}}/(\varepsilon_0\varepsilon_{\text{si}}) + d_{\text{iso}}/(\varepsilon_0\varepsilon_{\text{iso}})$. Moreover, the density of MIGS in silicon is replaced with $D_{\text{MIGS}}^{\text{si}} = D_{\text{MIGS}}^{\text{iso}} \exp(-d_{\text{iso}}/\delta_{\text{iso}})$ where $\delta_{\text{si,iso}}$ are the decay lengths in silicon and the insulator, respectively, and $D_{\text{MIGS}}^{\text{iso}}$ is the density of MIGS at the metal-insulator interface. Since, both, $\delta_{\text{iso}} \propto 1/E_g^{\text{iso}}$ and $D_{\text{MIGS}}^{\text{iso}} \propto 1/E_g^{\text{iso}}$, the slope parameter $S_{\text{MIGS}}^{\text{MIS}}$ of the combined metal-insulator-semiconductor contact becomes a function of E_g^{iso} and the thickness d_{iso}:

$$S_{\text{MIGS}}^{\text{MIS}} = \left(1 + \frac{e^2 D_{\text{MIGS}}^{\text{iso}} e^{-d_{\text{iso}}/\delta_{\text{iso}}} (\varepsilon_{\text{iso}}\delta_{\text{si}} + \varepsilon_{\text{si}} d_{\text{iso}})}{\varepsilon_0 \varepsilon_{\text{iso}} \varepsilon_{\text{si}}}\right)^{-1}. \tag{4.35}$$

This equation suggests that either a large d_{iso} or a large E_g^{iso} (or both) is preferable since $S_{\text{MIGS}}^{\text{MIS}} \to 1$ in this case. However, it is clear that both measures eventually prohibit a proper, Ohmic contact formation. If we use the WKB approximation (cf. Equation (4.33)) and assume, for simplicity, a constant potential barrier of approximately $E_g^{\text{iso}}/2 - E_g^{\text{si}}/2$, then $T_{\text{WKB}} \propto \exp(-\sqrt{E_g^{\text{iso}} - E_g^{\text{si}}}\, d_{\text{iso}})$. This shows, that in terms of tunneling probability, it is preferable to increase E_g^{iso} instead of d_{iso} since the exponent in T_{WKB} depends linearly on d_{iso} but only as the square root of E_g^{iso}. The bottom line is that for contact formation a (likely ultrathin) d_{iso} has to be chosen such that Fermi-level pinning is sufficiently suppressed while still allowing a decent tunneling probability into the conduction or valence bands.

Thermally grown silicon nitride is an excellent option to serve as an insulator to depin the Fermi level in metal-silicon contacts. Reasons for this include: first, that the band gap of silicon nitride is large enough to avoid (re)pinning at the SiN-metal interface. Second, SiN can be grown in a rapid thermal annealing process using NH_3 with an almost self-limiting growth kinetics (cf. Section 3.3) that enables excellent process control and reproducibility of ultrathin SiN layers. Furthermore, since silicon nitride is an excellent diffusion barrier, even an ultrathin SiN prevents oxidation of the silicon and, therefore, allows reliable processing.[7] Finally, the energetic alignment of the band gap with respect to the silicon band gap is favorable in that it is aligned approximately mid-gap with respect to silicon (see Figure 4.16). As a result, choosing an appropriate SiN thickness,

7 An approximately 3 nm thin thermally grown SiN even prevents rapid thermal oxidation as shown in Figure 3.4.

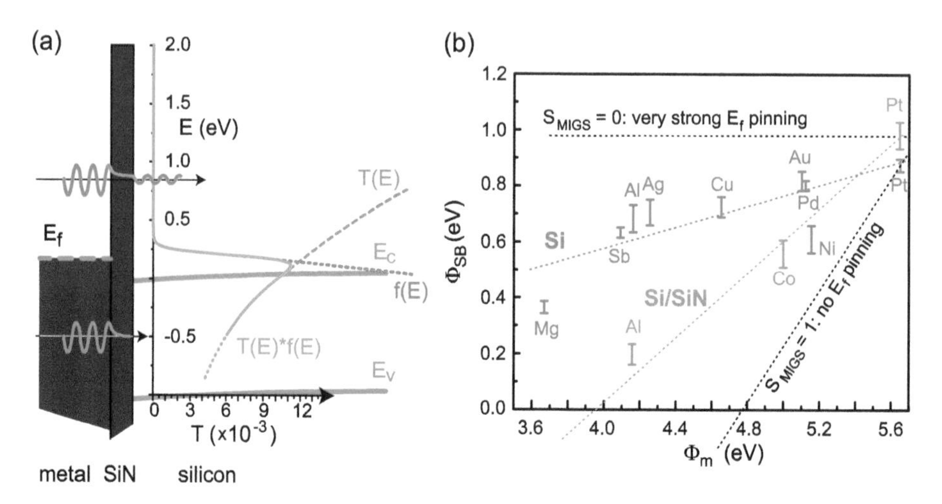

Figure 4.23: (a) Schematic of the metal-SiN-silicon contact. The blue dashed line represents a WKB approximation of the transmission probability $T(E)$, the red dotted line is the Fermi- distribution function and the orange line is the product of both. In the present calculation, a SiN thickness of 8 Å, $m^* = 0.2m_0$ and $E_g^{\mathrm{SiN}} = 5$ eV was assumed. (b) Measured Schottky barriers as a function of Φ_m for various metal-Si (red) [247] and metal-SiN-Si contacts (green) [223].

metal-induced gap states are strongly reduced such that the Fermi level is depinned at the silicon surface. Replacing a potential barrier (i. e., the Schottky barrier) with another potential barrier (the SiN) does at first sight not appear reasonable. However, to explain the rationale behind this in a bit more detail consider Figure 4.23(a). Here, the tunneling probability $T(E)$ through the SiN layer is plotted as a function of energy. $T(E)$ was computed with the WKB approximation and it clearly shows that it increases exponentially for larger energies. This means that the MIGS that pin the Fermi level approximately around mid-gap (at E_{CNL}) are substantially stronger suppressed than electronic states that would inject, for instance, into the conduction band. As a result, with appropriate SiN thickness, MIGS can be suppressed while the injection into the conduction band is much less affected. Thicknesses of the SiN in the range of 4–8 Å are best suited for depinning the Fermi level.

With SiN depinning layer, contact formation to the conduction band as well as to the valence band is enabled using metals with appropriate work functions. An example of n- and p-type devices realized with two different metallic source/drain contacts with low and high Φ_m will be discussed in Section 7.3. Figure 4.23(b) shows experimentally measured Schottky-barrier heights of metal-Si contacts as a function of Φ_m (the respective metals are stated in the figure). The red data points (taken from [247]) show the expected strong Fermi level pinning with $S_{\mathrm{MIGS}} \approx 0.19$. On the other hand, with a 8 Å SiN layer (green data), Fermi level pinning is substantially suppressed resulting in $S_{\mathrm{MIGS}} = 0.58$ [223].

4.6.4 Transfer Length of Contacts

A metal-semiconductor contact is often made in a way that the metal covers a part of the semiconductor and the actual contact is distributed across a certain length l_{con}. The contact can then be modeled by considering it as a distributed resistor network as illustrated in Figure 4.24 where it is assumed that there is only a variation of the contact properties along the x-direction. If the entire contact is subdivided into n small sections of length δx with $l_{\mathrm{con}} = n \cdot \delta x$, then in each section three resistances can be defined belonging to the metallic lead (r_{met}), to the semiconductor (r_{semi}) and to the coupling between the metal and the semiconductor (r_t) mediated, for instance, by a tunneling process either through a Schottky barrier in the usual MS-contact case or through an ultrathin SiN layer if depinning of the Fermi level is used. All resistances are in units Ω·m and need to be divided by the width of the respective contact part in order to obtain the resistance. For instance, the resistance of the metal contact lead is given by $r_{\mathrm{met}} = \rho_m / d_{\mathrm{met}} \cdot \delta x$ with ρ_m being the specific resistivity of the metal and d_{met} its thickness. A similar expression can be found for r_{semi} and r_t, depending in the particular case considered.

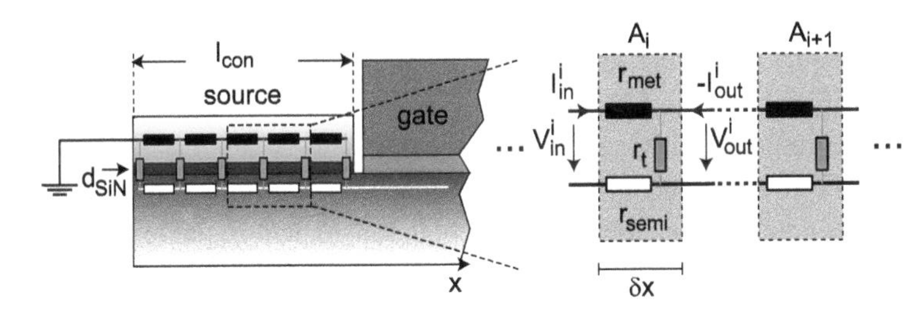

Figure 4.24: Schematic of the typical contact geometry where a metal covers a part of the semiconductor over a length l_{con}. The contact can be regarded as a distributed resistor network with the sheet resistances of the contact metal r_{met} and of the semiconductor r_{semi}. The tunneling resistance r_t can be computed with the WKB approximation. The right panel shows the part of the distributed resistor network that is repeated along the length of the contact.

Using four-pole theory, it is straightforward to determine the impedance matrix Z_i for one particular section of the resistor network to be

$$\mathsf{Z}_i = \begin{pmatrix} r_{\mathrm{met}} + r_{\mathrm{semi}} + r_t & r_{\mathrm{semi}} + r_t \\ r_t & r_t \end{pmatrix} \tag{4.36}$$

from which the transmission matrix A_i for one section (gray area in Figure 4.24, right panel) can be computed. The overall matrix for the entire network is then simply the product of the matrices for each of the n sections $\prod_i \mathsf{A}_i$. Since all sections are alike, this yields A^n. An analytical calculation can be carried out by first diagonalizing the matrix A yielding $\mathsf{D} = \mathsf{S}^{-1}\mathsf{A}\mathsf{S}$ where S is the orthogonal transformation consisting of the eigenvectors of A. As a result,

$$\mathsf{A}^n = \mathsf{S}\mathsf{D}^n\mathsf{S}^{-1} = \mathsf{S}\begin{pmatrix} \lambda_1^n & 0 \\ 0 & \lambda_2^n \end{pmatrix}\mathsf{S}^{-1} \tag{4.37}$$

where $\lambda_{1,2}$ are the eigenvalues of the matrix that can be obtained by solving the characteristic polynomial. The total contact resistance r_{con} can then be read off from the matrix A^n: the inverse of the (2,1)-element of A^n divided by the width of the contact is the total contact resistance. If r_{met}, r_{semi} and r_t are constant throughout the contact of length l_{con}, the contact resistance can be computed analytically yielding $r_{\mathrm{con}} \propto l_T \coth^{-1}(l_{\mathrm{con}}/l_T)$ where l_T is the so-called transfer length. As a result, if $l_{\mathrm{con}} > l_T$ the contact resistance saturates and, therefore, a further increase of the contact length does not yield any benefit. On the other hand, this means that the minimum contact length should be $l_{\mathrm{con}}^{\mathrm{min}} \approx l_T$, since otherwise the contact resistance strongly increases.

4.7 Heterostructures

Heterostructures are a unique possibility to tailor the potential profile within a device to enable or improve a certain functionality. Prominent examples include diode lasers and high-electron mobility transistors. When two semiconductors are brought into contact, the band line-up is obtained with the same requirements already stated above: In equilibrium, a single Fermi level is obtained and at the interface the work functions (electron affinities) of the two materials need to match. Doing so, the three different scenarios schematically shown in Figure 4.25 are possible: in a so-called type I heterostructure, the band gap of the material with the small E_g is completely straddled by the band gap of the second material. A conduction or valence band offset $\Delta E_c/\Delta E_v$ can be used to create carrier confinement in, e. g., the triangular potential shown in the left panel. In so-called type II heterostructures, either a staggered or a broken band line-up (center and right panels) is possible. Type II heterostructures will be discussed in the framework of band-to-band tunnel FETs in Chapter 9.

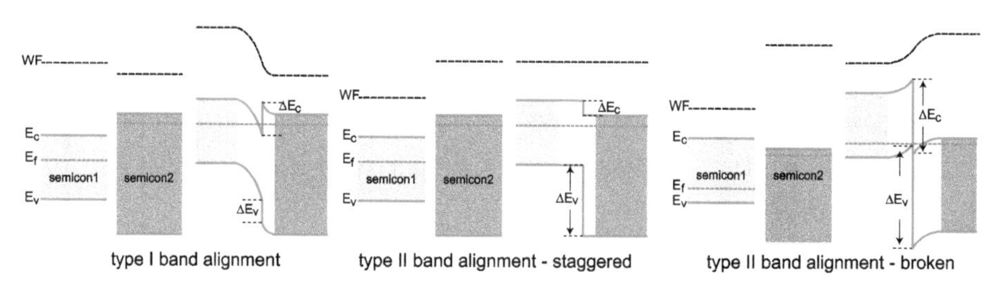

Figure 4.25: Possible heterostructure band line-ups between two different semiconductors.

Exercises

Exercises together with solutions are accessible via the QR code.

5 Metal-Oxide-Semiconductor Field-Effect Transistors

In the present chapter, the working principles of metal-oxide-semiconductor field-effect transistors will be discussed. The focus of the present chapter is on nanoscale MOSFETs based on low-dimensional nanostructures such as ultrathin-body silicon-on-insulator and nanowires. However, before going into more detail it is instructive to start the discussion with recapitulating the classical textbook version of MOSFET functionality based on the gradual channel approximation.

5.1 Operation Principles—Gradual Channel Approximation

Figure 5.1(a) shows a cross-section of a conventional, planar, single-gate silicon MOSFET. Note that the following discussion will be restricted to n-type MOSFETs; the extension to p-type devices is, however, straightforward with appropriate modifications (such as changing polarity of voltages, etc.). Two highly n-doped source and drain regions are separated by a channel of length L within the p-doped silicon substrate. On top of the substrate, a gate electrode consisting of highly doped polysilicon or a metal is placed and insulated from the substrate by a gate dielectric with dielectric constant ε_{ox} and physical thickness d_{ox}. The gate electrode has a width W and the gate length is considered to be equal to the channel length L. A coordinate system is chosen with the x-direction along the channel, y along the width and the z-direction is oriented perpendicular into the substrate. W is assumed to be very large such that the potential distribution in the device is independent of the y-coordinate.

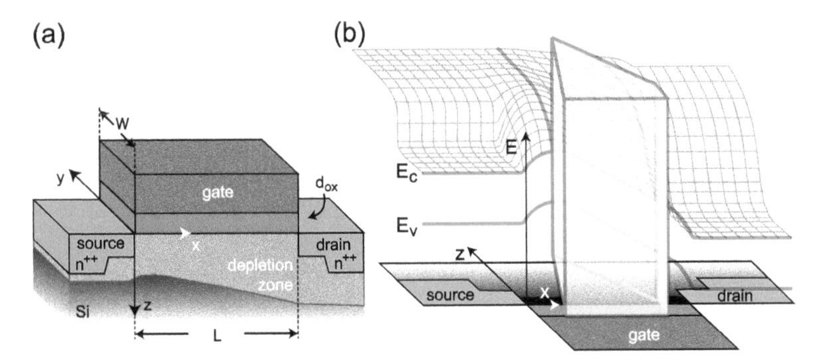

Figure 5.1: (a) Schematic of a conventional n-type, planar, single-gate MOSFET with channel length L and width W. (b) Potential distribution in the conduction band of the MOSFET with applied voltages.

The current density j in such a transistor is $j = (-e)n(x,z)v(x)$ with the carrier density $n(x,z)$ and the carrier velocity $v(x)$ both being independent of y due to the assumption of a very wide transistor. Furthermore, the current flowing through the transistor

https://doi.org/10.1515/9783111054421-005

will consist of inversion charge (cf. Section 4.5) induced by a sufficiently high gate voltage. The carriers of the inversion charge occupy 2D subbands that are a result of confinement in the triangular potential well between substrate and gate insulator such that $n(x, z)$ depends on x and z. Due to the confinement, there is certainly no current flow in the z-direction, and hence $v(x)$ solely depends on x. The current density j flows through an area of size W times the channel thickness (which is a not very well-defined quantity in the z-direction) and the total current is obtained by integrating over the y- and z-coordinates. Carrier confinement ensures that the wavefunctions exponentially drop in the z-direction (cf. Figure 4.13, left panel) and, therefore,

$$I = \int_0^W dy \int_0^\infty dz\, en(x,z) \cdot v(x) = WQ^{\square}(x) \cdot v(x) \tag{5.1}$$

where $Q^{\square}$ is the charge per unit area. The carrier velocity $v(x)$ is proportional to the electric field $\mathcal{E}$ according to $v(x) = \mu\mathcal{E}$ with μ being the carrier mobility (cf. Section 2.13.1). In Section 4.5.3, it was found that in the strong inversion case of silicon the gate capacitance can be replaced with the geometrical oxide capacitance. Therefore, the charge per unit area $Q^{\square}$ can be written as $C_{\mathrm{ox}}^{\square}(V_{\mathrm{gs}} - V_{\mathrm{th}})$, where $V_{\mathrm{gs}} - V_{\mathrm{th}}$ is the so-called gate overdrive with V_{gs} being the applied gate-source voltage and V_{th} the threshold voltage.[1] The threshold voltage is the gate voltage when strong inversion is obtained. Here, the threshold voltage is taken to be the gate voltage necessary to move the energy level of the first 2D subband in the inversion layer to the same energy as the Fermi level E_f^s in source (cf. Figure 4.5.3). Thus, it is clear that the gate overdrive $V_{\mathrm{gs}} - V_{\mathrm{th}}$ drives the MOSFET well into strong inversion, and hence into the on-state.

The gate overdrive $V_{\mathrm{gs}} - V_{\mathrm{th}}$ applies only at the source end of the channel. At the drain end, on the other hand, the drain-source bias V_{ds} leads to an effective reduction of the overdrive $V_{\mathrm{gs}} - V_{\mathrm{th}} - V_{\mathrm{ds}}$, and in general, along the channel, one obtains $V_{\mathrm{gs}} - V_{\mathrm{th}} - V(x)$. Then noting that $\mathcal{E} = \frac{dV}{dx}$ one finally obtains

$$\begin{aligned} I = WC_{\mathrm{ox}}^{\square}\mu(V_{\mathrm{gs}} - V_{\mathrm{th}} - V(x))\frac{dV}{dx} &\rightarrow I\,dx = WC_{\mathrm{ox}}^{\square}\mu(V_{\mathrm{gs}} - V_{\mathrm{th}} - V(x))dV \\ \rightarrow \int_0^L I\,dx = \int_0^{V_{\mathrm{ds}}} WC_{\mathrm{ox}}^{\square}\mu(V_{\mathrm{gs}} - V_{\mathrm{th}} - V)dV. \end{aligned} \tag{5.2}$$

Integrating both sides of the equation and noting that due to the requirement of current continuity in an ideal MOSFET I is independent of x, we finally obtain

1 There are only so many Latin and Greek letters, which leaves us with the dilemma that the potential in the Hamiltonian is denoted with the Latin V in order to distinguish it from the Greek Φ, which is mostly used for the wavefunction. Here, however, V is used for voltages.

$$I_d = \mu \frac{W}{L} C_{\text{ox}}^{\square} \left(V_{\text{ds}}(V_{\text{gs}} - V_{\text{th}}) - \frac{V_{\text{ds}}^2}{2} \right), \tag{5.3}$$

where the current is now denoted as the drain current I_d. In the case of a small V_{ds}, a Taylor expansion of Equation (5.3) yields $I_d = \mu \frac{W}{L} C_{\text{ox}}^{\square}(V_{\text{gs}} - V_{\text{th}})V_{\text{ds}}$. If the drain–source bias is increased and reaches $V_{\text{ds}} = V_{\text{gs}} - V_{\text{th}}$, the gate voltage overdrive at the drain end of the channel vanishes and the channel is said to be pinched off. The simple equation for the carrier density above tells that the charge at pinch-off goes to zero and for further increased V_{ds} the pinch-off point moves toward the source contact. Of course, the carrier density cannot become zero, since if it was zero the current would vanish, which violates current continuity. Hence, the carrier density becomes only very small and this reduction of carrier density is compensated with a higher carrier velocity in order to fulfill current continuity. It will be discussed in Section 5.2.3 in the framework of the Landauer formalism (cf. Section 2.13.2) that this is indeed the case. Anyway, for a bias voltage $V_{\text{ds}} \geq V_{\text{gs}} - V_{\text{th}}$ the current remains constant. In this saturation regime, I_d depends quadratically on V_{gs} according to

$$I_d^{\text{sat}} = \mu \frac{W}{L} C_{\text{ox}}^{\square} \frac{(V_{\text{gs}} - V_{\text{th}})^2}{2}. \tag{5.4}$$

The left panel of Figure 5.2 shows the I_d–V_{ds} curves for several V_{gs} obtained with the gradual channel approximation. The insets illustrate the pinch-off of the channel at the drain end when $V_{\text{ds}} = V_{\text{gs}} - V_{\text{th}}$ and the appearance of current saturation due to shifting the pinch-off point in the channel toward source when the bias increases. The right panel shows exemplarily one transfer characteristic for a drain-source bias where the MOSFET is in current saturation. Below V_{th}, the current drops to very small values close to zero and increases for larger V_{gs} according to Equation (5.4).

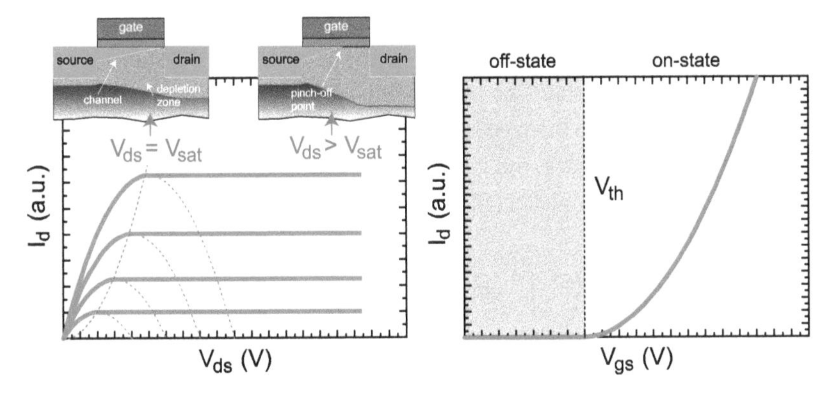

Figure 5.2: Output (left) and transfer (right) characteristics of an ideal, conventional long-channel MOSFET according to the gradual-channel approximation. The insets illustrate the pinch-off of the channel at the drain end for sufficiently large V_{ds} leading to current saturation.

5.2 Nanoscale Transistors with Ballistic Transport—From 1D to 2D to Bulk MOSFETs

The gradual channel approximation allows deriving a closed expression for the current through a MOSFET as a function of the applied voltages as well as of the geometrical and material parameters. As such, it is capable of describing the general functionality of large-scale, bulk-Si MOSFETs and provides guidelines on how to obtain performance improvements. However, it is certainly inadequate to describe effects such as quantization, tunneling of carriers, etc. as expected in the truly nanoscale MOSFETs of current and future technology generations. Therefore, instead of going top-down and refining the GCA, let us start bottom up and describe field-effect transistors based on a microscopic, i. e., quantum mechanical picture. To this end, a transistor is considered that is supposed to be short enough such that scattering can be neglected and electronic transport can be considered as being ballistic. Moreover, let us start with a device based on a nanowire/tube small enough such that current transport is one-dimensional.

The exact potential distribution in the channel needs to be computed by solving the Poisson equation within the device, which will be done in Section 5.5. For the time being, it is sufficient to note that because of the *n*-*p* junction at the source channel and the *p*-*n* junction at the channel-drain interfaces there must be a potential maximum in the channel that will be denoted with Φ_f^0 (as in the discussion of the MOS capacitors) and it is this potential maximum and its dependence on the terminal voltages that determines the electrical behavior of the device.

Figure 5.3(a) schematically shows the conduction band profile along the direction of current transport for a nanowire transistor with cross-section $W \times W$ small enough, so that confinement leads to the formation of discrete 1D subbands.[2] The Landauer expression for a 1D subband, derived in Section 2.13.2, can now be used to compute the current flow through such a device. The net current is given by the difference between carriers injected from the source over the potential maximum Φ_f^0 and those injected from the drain flowing toward the source (cf. Figure 2.53). These two carrier populations are depicted in Figure 5.3(a) in red and blue for carriers injected from source occupying the right branch of the dispersion and injected from drain occupying the left part of the dispersion relation, respectively. Hence, the current for the first 1D subband depicted in Figure 5.3(a) is given by (cf. Equation (2.117))

$$I_d^{1D} = \frac{2e}{h} \int_{\Phi_f^0 + E_1}^{\infty} dE(f_s(E_f^s) - f_d(E_f^d)). \tag{5.5}$$

2 Quantization is for clarity only shown along the *y*-direction but is supposed to also include the *z*-direction.

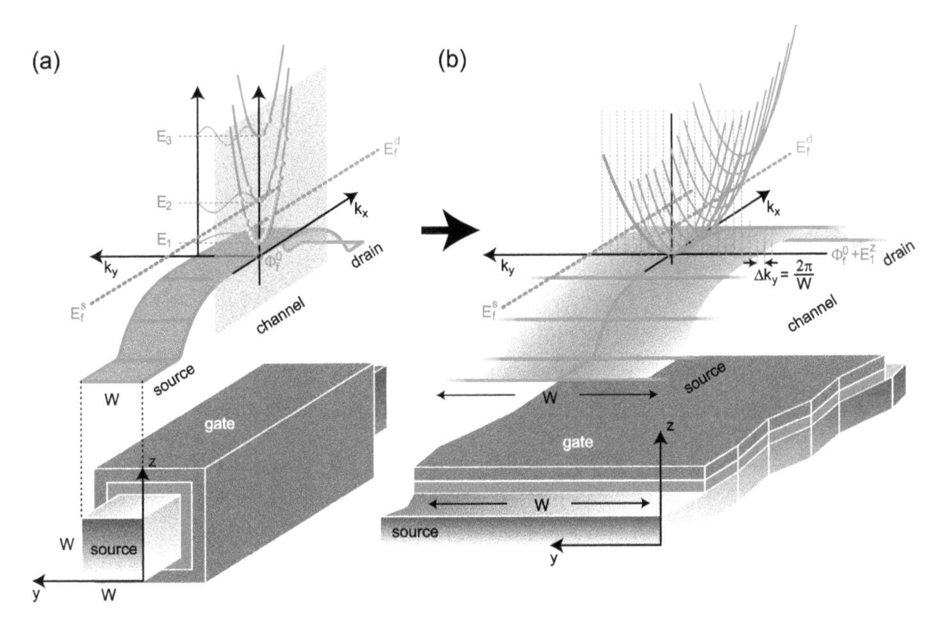

Figure 5.3: (a) Confinement in a nanowire FET with cross-section $W \times W$ leads to discrete 1D subbands with quantization energies $E_{1,2,...}$. (b) 2D MOSFET with very large width W along the y-direction. Only the first 2D subband in z-direction with quantization energy E_1^z contributes. The 2D current is obtained by summing over independent 1D-subbands.

The one-dimensional case allows obtaining a closed expression of the drain current I_d since the integral in Equation (5.5) can be solved analytically (cf. Task 13 and Section 2.11.3). Before explaining where gate and drain-source voltages are hidden in the equation, an extension to higher-dimensional devices will be derived.

Pure 1D transport in FETs is difficult to obtain at room temperature even in carbon nanotube FETs (see Chapter 8). Consequently, in most nanowire/-tube devices at least a few 1D subbands contribute to the current and need to be taken into account. The total current in the ballistic case (which implies no interband scattering such that the subbands can be treated as independent) is simply a sum over the discrete 1D subbands (max. = 3 in the case depicted in Figure 5.3(a)) leading to

$$I_d^{\text{multi-mode}} = \sum_{n=1}^{\text{max.}} d_{\text{deg}}^n \frac{2e}{h} \int_{\Phi_f^0+E_n}^{\infty} dE(f_s(E_f^s) - f_d(E_f^d)) \tag{5.6}$$

where each of the subbands starts at the energy $\Phi_f^0 + E_n$. The energy E_n is the energy of the respective 1D subband that is a consequence of confinement in y- and z-directions. If the cross-section of the nanowire is $W \times W$ and a particle-in-the-box quantization is used the subbands are at energies $E_n = E^{m,l} = \frac{\hbar^2\pi^2}{2m^* W^2} m^2 + \frac{\hbar^2\pi^2}{2m^* W^2} l^2$ as has been discussed in detail in Section 2.2.2. The factor d_{deg}^n in Equation (5.6) is the degeneracy factor of the nth

1D subband. For instance, the second subband is twice degenerate since $E_2 = E^{21} = E^{12}$, and hence $d^2_{\text{deg}} = 2$ (cf. Section 2.2.2).

If the width W in one spatial direction (say y) is made very large whereas the confinement is such that in z-direction only the first subband is occupied, a two-dimensional MOSFET is obtained as illustrated in Figure 5.3(b). In such an idealistic, infinitely wide device with ballistic electronic transport, there is no electric field component in the y-direction. As a result, the momentum $p_y = \hbar k_y$, and thus k_y, is conserved. Consequently, the 2D transport can be considered to consist of a multitude of 1D subbands that are independent of each other, and the current is therefore obtained by summing over these independent 1D subbands depicted in Figure 5.3(b). Each 1D subband starts at the energy $\Phi_f^0 + E_1^z + \frac{\hbar^2 k_y^2}{2m^\star}$ where E_1^z is the first quantization energy due to carrier confinement in the z-direction and $\frac{\hbar^2 k_y^2}{2m^\star}$ is the dispersion relation in the y-direction. If periodic boundary conditions in y-direction are used, the sum over k_y can be transferred into an integral as has been described in the info-box in Section 2.11.1. To this end, a factor of $1 = \frac{\Delta k_y}{2\pi/W}$ is inserted in the integral. Then $W \to \infty$, which implies $\Delta k_y \to dk_y$, and thus

$$I_d^{2\text{D}} = \frac{2e}{h} \int \underbrace{dk_y}_{=\frac{d\epsilon}{\hbar}\sqrt{\frac{m^\star}{2\epsilon}}} \frac{W}{2\pi} \int\limits_{\Phi_f^0 + E_1^z + \frac{\hbar^2 k_y^2}{2m^\star}}^{\infty} dE(f_s(E_f^s) - f_d(E_f^d)). \tag{5.7}$$

Using the dispersion relation $\epsilon = \frac{\hbar^2 k_y^2}{2m^\star}$, variables are changed from k_y to ϵ and, since the integral over dE can be carried out analytically (see Section 2.11.3), one eventually arrives at

$$I_d^{2\text{D}} = W \frac{2e}{h^2} \sqrt{\frac{m^\star}{2}} k_B T \int\limits_0^{\infty} d\epsilon \frac{1}{\sqrt{\epsilon}} \ln\left(\frac{1 + \exp(\frac{E_f^s - (\Phi_f^0 + E_1^z + \epsilon)}{k_B T})}{1 + \exp(\frac{E_f^d - (\Phi_f^0 + E_1^z + \epsilon)}{k_B T})} \right), \tag{5.8}$$

where the final integration over ϵ needs to be computed numerically. As expected, the expression for $I_d^{2\text{D}}$ is proportional to the width W of the transistor. Furthermore, in contrast to the 1D expression, the current in 2D depends explicitly on the material properties of the semiconductor through $\sqrt{m^\star}$. Interestingly, it appears as if $I_d^{2\text{D}}$ drops with decreasing effective mass. This, however, is only partly true since Φ_f^0 (see discussion below) depends on the charge in the channel and, therefore, also on the density of states, which is $\propto m^\star$ in 2D. As a result, the dependence of $I_d^{2\text{D}}$ on $m^\star$ is more involved than immediately apparent from the $\sqrt{m^\star}$-dependence. On the other hand, if perfect gate control over Φ_f^0 is obtained (see Section 5.9.1) Φ_f^0 becomes independent of the charge, and in this case the drain current of a 2D FET will indeed be deteriorated with decreasing DOS (i. e., decreasing $m^\star$) in the channel [265].

Finally, having calculated the 2D current $I_d^{2\text{D}}(E_1^z)$ as a function of the subband energy E_1^z, it is in principle straightforward to extend the expression covering multiple 2D subbands that carry the current as appropriate for a bulk MOSFET. For instance, quantization in the z-direction due to the (approximately) triangular potential of an inversion layer (see Figure 4.13) leads to the formation of 2D subbands with energies denoted E_n^z (Equation (4.21)), and thus one arrives at

$$I_d^{\text{bulk}} = \sum_{n=1}^{N} I_d^{2\text{D}}(E_n^z). \tag{5.9}$$

5.2.1 Dependence on Terminal Voltages—Top-of-the-Barrier Model

Looking at the expression in Equation (5.5) it is not immediately obvious how the current depends on the applied drain-source and gate voltages. Considering the fact that $E_f^d = E_f^s - |e|V_{\text{ds}}$, it is clear that V_{ds} appears in the drain Fermi distribution. But what about the dependence of Φ_f^0 on V_{gs} and how does V_{ds} impact Φ_f^0? To answer this question, we would need to solve the Poisson equation (self-consistently) in the device, which has to be done numerically. However, using the so-called "top-of-the-barrier" model [218], we can further elaborate on the working principles of MOSFET without carrying out such a calculation.

In the "top-of-the-barrier" model, the transistor is considered to consist of capacitors depicted in Figure 5.4(a). Here, the conduction band of an n-type MOSFET along the direction of current transport is shown; similar considerations can certainly also be carried out for p-type MOSFETs. The capacitors related to the source-channel p-n junction C_s, the channel-drain junction C_d as well as the geometrical oxide capacitance C_{ox} are all connected to the potential maximum Φ_f^0 (i. e., the top of the barrier). Φ_f^0 in

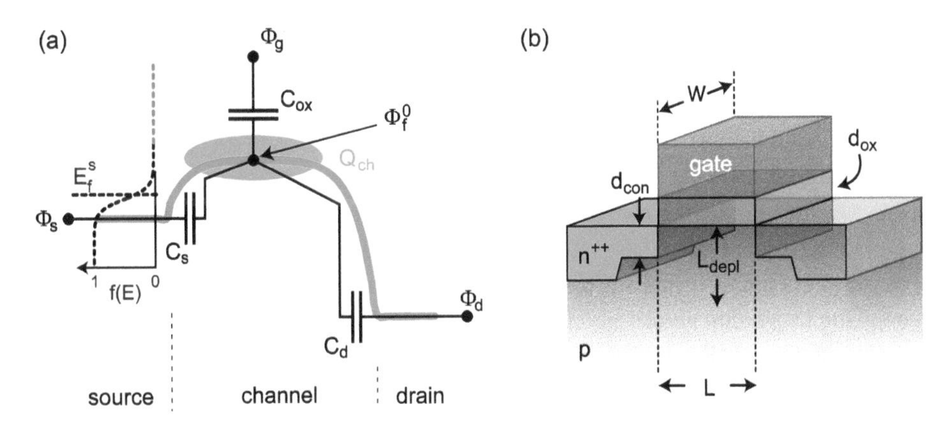

Figure 5.4: (a) Conduction band along the direction of current transport in an n-type MOSFET (green line). Various capacitors can be defined and the charge in the channel is shared among the capacitors C_s, C_d, and C_{ox}. (b) Schematic of a conventional MOSFET showing the simple approximation of C_d as parallel-plate capacitor in-between the two green areas of source and drain.

turn determines the overall charge within the channel $Q_{\rm ch}$ that consists of the contributions from the depletion charge and the inversion charge as well as a possible contribution due to interface charge of the MOS capacitor (cf. Section 4.5), i. e., $Q_{\rm ch} = Q_{\rm depl} + Q_{\rm inv} + Q_{\rm it}$. On the other hand, $Q_{\rm ch}$ is shared among the three capacitors C_s, C_d and $C_{\rm ox}$, and thus $Q_{\rm ch} = Q_s + Q_d + Q_g$. The charge on the various capacitors can be stated explicitly; for instance, the charge on the gate capacitor Q_g is given by the difference between surface- and gate-potentials multiplied with the geometrical oxide capacitance, i. e., $-\frac{1}{e}(\Phi_f^0 - \Phi_g - \Phi_{\rm bi})C_{\rm ox} = Q_g$, where Φ_g and $\Phi_{\rm bi}$ are the gate and built-in potentials (note that in Section 4.5 $Q_g = Q_{\rm ch}$ because there were no source/drain electrodes). Similar relationships are obtained in the case of the remaining charge contributions: $-\frac{1}{e}(\Phi_f^0 - \Phi_s)C_s = Q_s$, $-\frac{1}{e}(\Phi_f^0 - \Phi_d)C_d = Q_d$ for the charges on the source and drain capacitors. Inserting this into $Q_{\rm ch} = Q_s + Q_d + Q_g$ yields

$$eQ_{\rm ch} = (\Phi_g + \Phi_{\rm bi} - \Phi_f^0)C_{\rm ox} + (\Phi_s - \Phi_f^0)C_s + (\Phi_d - \Phi_f^0)C_d. \tag{5.10}$$

Solving this equation for Φ_f^0 results in

$$\Phi_f^0(C_{\rm ox} + C_s + C_d) = (\Phi_g + \Phi_{\rm bi})C_{\rm ox} + \Phi_s C_s + \Phi_d C_d - eQ_{\rm ch}. \tag{5.11}$$

With $C_\Sigma = C_{\rm ox} + C_s + C_d$, an expression for Φ_f^0 is obtained:

$$\Phi_f^0 = \frac{C_{\rm ox}}{C_\Sigma}(\Phi_g + \Phi_{\rm bi}) + \frac{C_s}{C_\Sigma}\Phi_s + \frac{C_d}{C_\Sigma}\Phi_d - \frac{eQ_{\rm ch}}{C_\Sigma}. \tag{5.12}$$

If we knew how $Q_{\rm ch}$ depends on Φ_f^0, Equation (5.12) could be solved for Φ_f^0 and this in turn could be inserted into the expressions for the drain current as derived for transistors in various dimensions above. While in equilibrium the calculation of $Q_{\rm ch}$ is rather straightforward (cf. Section 4.5.3), electronic transport in a transistor requires a nonequilibrium expression of the charge (see Chapter 6 for a proper treatment based on the nonequilibrium Green function formalism). However, as a first-order approximation, we can separate the carriers into two populations: injected from the source with positive group velocity occupying the right branch of the dispersion relation, and carriers injected from the drain with negative group velocity occupying only the left branch of the dispersion. Due to ballistic transport, the two carrier populations will not mix and occupy the dispersion according to the Fermi level of the terminal they were injected from (as depicted in Figure 5.3(a)). As a result, the density of states within the channel can simply be divided by two and the carrier density is then computed with the Fermi distributions of source and drain.

As an example of such a calculation, a 2D transistor is considered consisting of an isotropic material with effective mass m^* yielding a DOS equal to $\frac{m^*}{\pi\hbar^2}$ (cf. Section 2.11.1). To simplify the argumentation, the transistor is considered to be very long, so that C_s

and C_d can be neglected. Furthermore, depletion and interface-state charges are also considered to be negligible. As a result, Equation (5.12) boils down to

$$\Phi_f^0 = \Phi_g + \Phi_{\text{bi}} - \frac{eQ_{\text{ch}}}{C_{\text{ox}}} \tag{5.13}$$

and Q_{ch} consists only of mobile inversion charge. The two contributions of $Q_{\text{ch}} = Q_{\text{ch}}^s + Q_{\text{ch}}^d$ due to injection from source and drain are given by

$$Q_{\text{ch}} \approx \underbrace{(-|e|) \int_{\Phi_f^0+E_1^z}^{\infty} dE \frac{1}{2} \frac{m^\star}{\pi\hbar^2} f_s(E_f^s)}_{Q_{\text{ch}}^s} + \underbrace{(-|e|) \int_{\Phi_f^0+E_1^z}^{\infty} dE \frac{1}{2} \frac{m^\star}{\pi\hbar^2} f_d(E_f^d)}_{Q_{\text{ch}}^d} \tag{5.14}$$

where $E_f^d = E_f^s - |e|V_{\text{ds}}$ and E_1^z is again the quantization energy of the first 2D subband. Equation (5.14) can be solved analytically (cf. Equation (2.95)) and provides the required relation between Φ_f^0 and the terminal voltages V_{ds} and V_{gs}:

$$\Phi_f^0 = \Phi_g + \Phi_{\text{bi}} - \frac{e^2}{C_{\text{ox}}} \frac{m^\star k_B T}{2\pi\hbar^2} [\ln(1 + e^{\frac{E_f^s - (\Phi_f^0 + E_1^z)}{k_B T}}) + \ln(1 + e^{\frac{E_f^d - (\Phi_f^0 + E_1^z)}{k_B T}})] \tag{5.15}$$

When solved for Φ_f^0 and inserted into the expressions for the drain current, a self-consistent description of the current as a function of the terminal voltages that depends also on the geometrical and material properties is obtained. Moreover, the impact of channel length scaling, depletion as well as interface charges, can be taken into consideration when employing Equation (5.12) and using the appropriate density of states for the calculation of the inversion charge. This yields a rather comprehensive description of a nanoscale MOSFET with ballistic transport.

Often, one is only interested in the change of the surface potential with changing gate and/or drain potentials. In this case the variation $\delta\Phi_f^0$ has to be computed resulting in

$$C_\Sigma \delta\Phi_f^0 = C_{\text{ox}} \delta\Phi_g + C_d \delta\Phi_d - \frac{e\partial Q_{\text{ch}}}{\partial \Phi_f^0} \delta\Phi_f^0, \tag{5.16}$$

where $\delta\Phi_s = 0$, since the potential in source Φ_s is usually set to ground and stays constant. The term $\frac{e\partial Q_{\text{ch}}}{\partial\Phi_f^0}$ can be written as $\frac{e\partial Q_{\text{inv}}}{\partial\Phi_f^0} + \frac{e\partial Q_{\text{it}}}{\partial\Phi_f^0} + \frac{e\partial Q_{\text{depl}}}{\partial\Phi_f^0}$ and the individual summands are already known to be the inversion-layer capacitance C_{inv}, the interface-state capacitance C_{it} and the depletion capacitance C_{depl}. Moving the term $\frac{e\partial Q_{\text{ch}}}{\partial\Phi_f^0}\delta\Phi_f^0 = (C_{\text{inv}} + C_{\text{it}} + C_{\text{depl}})\delta\Phi_f^0$ to the left side of Equation (5.16) and dividing both sides by $C_\Sigma + C_{\text{inv}} + C_{\text{it}} + C_{\text{depl}}$ finally yields

$$\delta\Phi_f^0 = \frac{C_{\text{ox}}}{C_{\text{ox}} + C_{\text{it}} + C_{\text{depl}} + C_s + C_d + C_{\text{inv}}} \delta\Phi_g$$

$$+\frac{C_d}{C_{\text{ox}}+C_{\text{it}}+C_{\text{depl}}+C_s+C_d+C_{\text{inv}}}\delta\Phi_d. \tag{5.17}$$

This expression states that changing Φ_g (i. e., changing the gate-source voltage) and/or changing Φ_d (i. e., modifying the drain-source bias) leads to a change of the potential maximum Φ_f^0 and, therefore, of the current I_d.

Using expression (5.17), we can already obtain a number of insights and even scaling rules regarding proper MOSFET operation. As an example, the single-gate, planar, bulk-like MOSFET depicted in Figure 5.4(b) is considered. In the case of such a MOSFET, the various capacitors can be stated explicitly (at least approximately) with simple geometrical expressions. As a result, a scaling rule is obtained, which appears to be oversimplified but which is consistent with the more elaborate analysis carried out in Section 5.5.

A well-functioning MOSFET is required to be exclusively switched by the gate voltage. Hence, Φ_f^0 should not be impacted by Φ_d, which is the case when the capacitance ratio in front of $\delta\Phi_d$ in Equation (5.17) vanishes, i. e., when

$$C_d \ll C_{\text{ox}}+C_{\text{depl}}+C_s+C_d+C_{\text{inv}}. \tag{5.18}$$

Since in the off-state of the transistor there is very little inversion charge in the channel, the capacitance C_{inv} can be neglected. Furthermore, C_s is smaller than the remaining capacitors and if we assume an excellent gate dielectric leading to a very low interface density of states Equation (5.18) becomes $C_d \ll C_{\text{ox}}+C_{\text{depl}}+C_d$.

In the bulk MOSFET depicted in Figure 5.4(b), the following geometrical relations can be used to provide approximate expressions for the capacitors C_{ox}, C_{depl} and C_d: The geometrical oxide capacitance is simply the parallel-plate capacitor $C_{\text{ox}}=\varepsilon_0\varepsilon_{\text{ox}}\frac{W\cdot L}{d_{\text{ox}}}$, the depletion capacitance can be approximated with $C_{\text{depl}}=\varepsilon_0\varepsilon_{\text{si}}\frac{W\cdot L}{L_{\text{depl}}}$ where L_{depl} is the depletion length of the depletion region in the p-doped bulk silicon (cf. Equation (4.17)). Finally, the drain capacitance C_d is approximated by considering the parallel-plate capacitor in between the two green-shaded areas of the source/drain doping profiles resulting in $C_d=\varepsilon_0\varepsilon_{\text{si}}\frac{W\cdot d_{\text{con}}}{L}$ where d_{con} is the depth of the contacts. Inserting the expressions for the capacitors eventually yields

$$\begin{aligned}\varepsilon_{\text{si}}\frac{W\cdot d_{\text{con}}}{L} &\ll \varepsilon_{\text{ox}}\frac{W\cdot L}{d_{\text{ox}}}+\varepsilon_{\text{si}}\frac{W\cdot L}{L_{\text{depl}}}\\ \rightarrow L &\gg \underbrace{\sqrt{\frac{\varepsilon_{\text{si}}d_{\text{con}}d_{\text{ox}}}{\varepsilon_{\text{ox}}+\varepsilon_{\text{si}}d_{\text{ox}}/(L_{\text{depl}})}}}_{\lambda}.\end{aligned} \tag{5.19}$$

Here, λ represents a "natural" length scale of the transistor under consideration. In order to avoid an impact of the drain bias on the potential barrier Φ_f^0 within the channel, and thus to fulfill the inequality (5.18), λ needs to be made sufficiently small. From Equation (5.19), it is obvious that this can be accomplished by reducing the physical oxide thickness d_{ox}, reducing the contact depth d_{con}, increasing the dielectric constant of the gate oxide ε_{ox} but also by decreasing the depletion length L_{depl}. The latter can in principle be achieved by increasing the doping concentration within the channel. However, as will become clear when discussing the switching of MOSFETs, increasing the channel doping may become detrimental to the off-state performance of the devices.

5.2.2 Off-State

Let us now concentrate on the off-state performance of a transistor, i. e., when the gate voltage is below the threshold voltage V_{th} (cf. Figure 5.2). To simplify the discussion, a 1D nanowire FET is considered here. However, the findings are also valid for the 2D and bulk device (see Task 20). Since in the off-state, Φ_f^0 lies substantially above the Fermi energy (of source and drain), we can replace the Fermi distribution functions with the Boltzmann approximation. As a result, the current depends exponentially on Φ_f^0, which in turn is a function of the gate potential Φ_g (cf. Equation (5.12)). One obtains

$$I_d \approx \frac{2e}{h} \int_{\Phi_f^0}^{\infty} dE\, f_s(E) \overset{(\Phi_f^0 - E_f^s) \gg k_B T}{\approx} \frac{2e}{h} k_B T \exp\left(-\frac{\Phi_f^0(\Phi_g) - E_f^s}{k_B T}\right). \tag{5.20}$$

Here, V_{ds} has been assumed large enough, such that carrier injection from the drain is negligible because it is exponentially smaller than the contribution from source.

The switching behavior of a MOSFET is characterized by the so-called inverse subthreshold slope S. The inverse subthreshold slope has the unit mV/dec, i. e., it states the gate voltage change needed in order to change the current through the transistor by one order of magnitude, which is illustrated in Figure 5.5, right panel. At a constant V_{ds}, S is given by

$$S = \left(\frac{\partial \log(I_d)}{\partial V_{gs}}\right)^{-1} = \ln(10)\left(\frac{\partial I_d}{\partial V_{gs}} \frac{1}{I_d}\right)^{-1}. \tag{5.21}$$

Inserting Equation (5.20) into Equation (5.21) allows one to compute a closed expression for S in the following way. First, note that using the chain rule

$$\frac{\partial I_d}{\partial V_{gs}} = \underbrace{\frac{\partial I_d}{\partial \Phi_f^0}}_{=-\frac{I_d}{k_B T}} \cdot \underbrace{\frac{\partial \Phi_f^0}{\partial \Phi_g}}_{=\frac{C_{ox}}{C_\Sigma}} \cdot \underbrace{\frac{\partial \Phi_g}{\partial V_{gs}}}_{=-e}, \tag{5.22}$$

where the first term is computed using Equation (5.20) and the third term is simply the relation between the gate voltage and gate potential. The second term stems from Equation (5.17) at constant V_{ds} (i. e., $\delta\Phi_d = 0$). Since the off-state is considered (i. e., $C_{inv} \approx 0$)

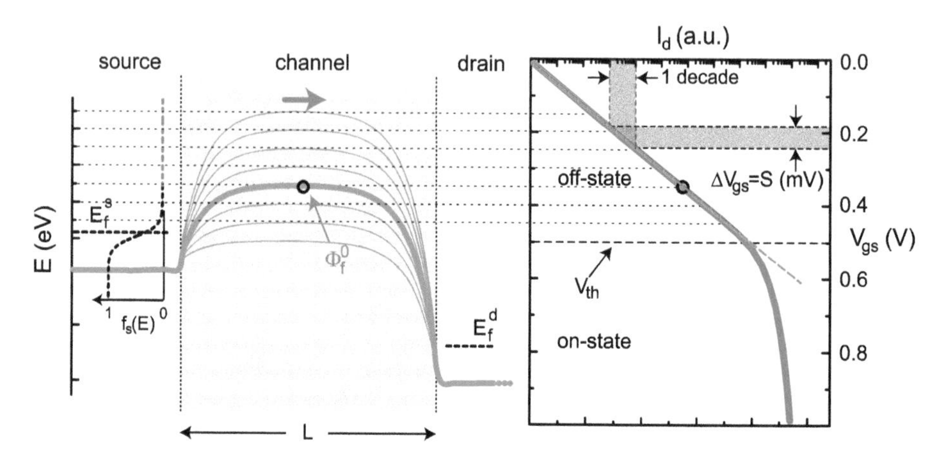

Figure 5.5: Conduction band profiles along current transport of a *n*-type MOSFET for constant V_{ds} and varying gate voltage V_{gs}. The right panel displays a semilogarithmic I_d–V_{gs} curve with exponential increase of the current in the device's off-state. The gray shaded projections onto the I_d and V_{gs} axes show how the inverse subthreshold slope S is determined.

and neglecting again C_s, one finds $\frac{\partial \Phi_f^0}{\partial \Phi_g} = \frac{C_{\text{ox}}}{C_{\text{ox}}+C_{\text{depl}}+C_{\text{it}}+C_d}$. Inserting this expression into Equation (5.21) finally yields

$$S = \frac{k_B T}{|e|} \ln(10) \cdot \frac{\partial \Phi_g}{\partial \Phi_f^0} = \frac{k_B T}{|e|} \ln(10) \left(\frac{C_{\text{ox}} + C_{\text{depl}} + C_{\text{it}} + C_d}{C_{\text{ox}}} \right). \tag{5.23}$$

The inverse subthreshold slope S should be as small as possible in order to ensure that the transistor can be switched with a voltage interval as small as possible. The reason for this is that the smaller S the smaller can the supply voltage of an integrated circuit be, and hence the lower the power it will consume (see Section 5.2.2 for more discussion on this topic). It is thus clear that $\frac{C_{\text{depl}}+C_{\text{it}}+C_d}{C_{\text{ox}}}$ should be as small as possible, which can only be achieved with $C_{\text{ox}} \gg C_{\text{depl}} + C_{\text{it}} + C_d$. The discussion related to the natural length scale λ in the preceding section can now be completed; while a high doping concentration yields steep *p-n* junctions, and thus eventually prevents an impact of Φ_d on Φ_f^0, it results in a large C_{depl} that may lead to an unfavorable ratio between C_{ox} and C_{depl} ultimately leading to a higher inverse subthreshold slope.[3] The same is true for the ratio C_d/C_{ox} and $C_{\text{it}}/C_{\text{ox}}$. In all cases, C_{ox} needs to be the largest capacitance and if this is the case, $\delta\Phi_0 = \delta\Phi_{g,}$ meaning that perfect gate control with a one-to-one change of the potential maximum in the channel with gate potential is obtained (see the video

3 This is the reason for the complicated ion implantation profiles with super steep retrograde channel implants, etc. in bulk MOSFETs facilitating a suppression of leakage currents through the bulk while giving rise to a small C_{depl}.

provided through the QR code #43 for additional information). Hence, Equation (5.23) reduces to the famous expression yielding 60 mV/dec at room temperature:

$$S = \frac{k_B T}{|e|} \ln(10) \approx 60\,\text{mV/dec}. \tag{5.24}$$

Equation (5.24) is a remarkable result since it states that the minimum achievable inverse subthreshold slope only depends on temperature but neither on geometrical nor on material properties of a device. The reason for this fact is the injection of carriers from a thermally broadened Fermi distribution function. In turn, this means that any FET relying on the field-effect-controlled modulating of carrier injection from a thermally broadened Fermi distribution function will exhibit at best 60 mV/dec at room temperature and there is nothing one can do about this. However, in Chapter 9 device concepts will be discussed that potentially allow one to overcome the 60 mV/dec limit by circumventing the injection of carriers from a thermally broadened Fermi distribution.

Task 20.
Off-state of a 2D MOSFET: Compute the inverse subthreshold slope in a two-dimensional MOSFET based on the Landauer formalism (Equation (5.8)) and prove that a 2D MOSFET indeed shows the same 60 mV/dec limit as a 1D transistor. To simplify the calculation, you may assume that a drain-source bias has been applied, which is large enough so that the contribution of drain to the net current can be neglected.

Figure 5.5 illustrates the switching behavior of MOSFETs discussed so far. The left panel shows conduction band profiles for various gate voltages. Carriers are injected from source (again, V_{ds} has been assumed large enough such that the drain contribution is negligible) and only carriers with energies above Φ_f^0 can contribute to the current. When Φ_f^0 is moved down with increasing V_{gs}, an increasing fraction of the source Boltzmann tail (red part) contributes to I_d giving rise to an exponential increase of the current. In the semilogarithmic I_d–V_{gs} plot shown in the right panel of Figure 5.5, this yields a linear current increase with inverse subthreshold slope S.

5.2.3 On-State—Output Characteristics

In the present section, the on-state and in particular the I_d–V_{ds} characteristics will be discussed. For simplicity, a 1D transistor is considered as in the preceding section. Moreover, it is again assumed that C_{ox} is substantially larger than C_{depl} and C_d (as well as C_{it} and C_s). As a result, the potential maximum is given by Equation (5.13). Let us now assume that a certain potential maximum Φ_f^0 has been computed self-consistently (for instance, by solving a 1D equivalent of Equation (5.15)). Note that in the on-state, when there is a substantial charge Q_{ch} in the channel, the potential maximum Φ_f^0 in the channel will become dependent on Φ_d although $C_{ox} \gg C_d$ has been assumed. The reason for this is the drain contribution to the charge Q_{ch} (cf. Equation (5.14)). For small V_{ds}, the channel charge is approximately given by its equilibrium value. However, even for large V_{ds} the charge is reduced at most by a factor of two (when only the branch with positive group velocity of the dispersion relation is occupied with carriers). Therefore, the dependence of Φ_f^0 on Φ_d plays only a role in the transition region between low bias and saturation of the device and will therefore not be considered for the time being.

The drain current I_d^{1D} as a function of the bias V_{ds} behaves as depicted in Figure 5.6. That is, for small bias, the current increases linearly and saturates for larger V_{ds}, which can be understood as follows: Applying a (positive) V_{ds} between source and drain moves the Fermi level of drain E_f^d energetically downwards leading to $E_f^d = E_f^s - |e|V_{ds}$. For small V_{ds}, the drain Fermi distribution can be Taylor-expanded resulting in $f(E_f^d) \approx f(E_f^s) - \frac{1}{k_B T}\frac{\partial f(E_f^s)}{\partial E}(-|e|V_{ds})$. Hence, the integral in Equation (5.5) becomes

$$I_d \approx \frac{2e}{hk_BT}\int_{\Phi_f^0}^{\infty} dE \frac{\partial f(E_f^s)}{\partial E}(-|e|V_{ds}) = \frac{2e^2}{h}\frac{1}{1+\exp(\frac{\Phi_f^0 - E_f^s}{k_BT})}V_{ds} \tag{5.25}$$

and indeed the drain current depends linearly on the bias for small V_{ds}. If, on the other hand, the bias is made so large that E_f^d is moved substantially below Φ_f^0, the carrier injection from drain into the channel drops according to the Boltzmann tail of the drain Fermi distribution function, i. e., $\propto \exp(-\frac{\Phi_f^0 - (E_f^s - |e|V_{ds})}{k_BT})$. As a result, for sufficiently high V_{ds}, there will be a negligible injection of carriers from the drain into the channel, and

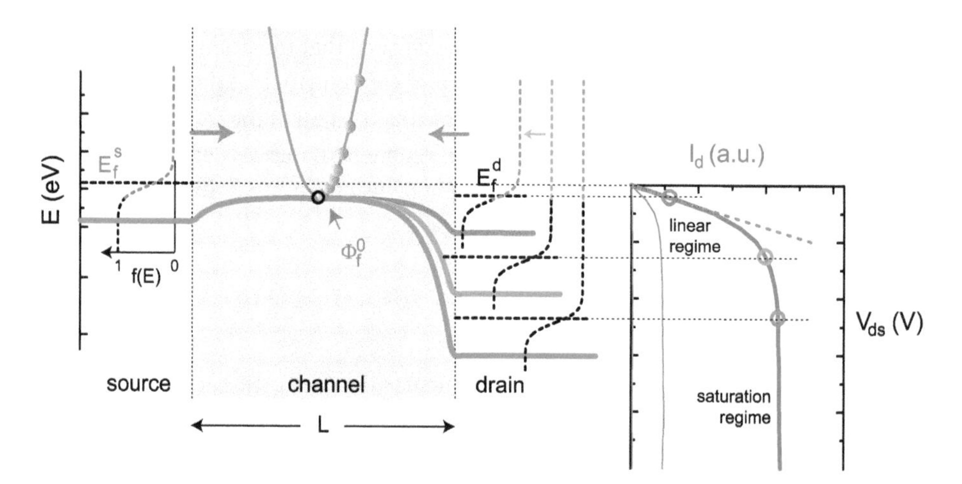

Figure 5.6: Conduction band of a MOSFET along the direction of current transport for three different V_{ds} (left panel). A small bias leads to a linear increase of the current while at larger bias current saturation occurs because of source exhaustion: only the right branch of the dispersion relation with positive carrier velocity is occupied in this case as illustrated in the left panel. The right panel shows I_d–V_{ds} curves for different V_{gs}.

hence current saturation occurs due to source exhaustion giving rise to the saturation current:

$$I_d^{\text{sat}} = \frac{2e}{h}\int_{\Phi_f^0}^{\infty} dEf(E_f^S) = \frac{2e}{h}k_BT\ln\left[1+\exp\left(\frac{E_f^S-\Phi_f^0}{k_BT}\right)\right]. \tag{5.26}$$

In Section 5.1, current saturation was explained with the pinch-off of the channel at the drain end for sufficiently high V_{ds}. Here, we see that current saturation is due to source exhaustion and is thus rather a matter of source than a property of drain. However, the pinch-off can also be understood within the framework of the Landauer equation: When deriving the Landauer current formula (cf. Section 2.13.2), it was found that $I_d = 2e/h\int dEv(E)D(E)/2\,(f_s - f_d)$ and it turned out that the product of the energy-dependent carrier velocity $v(E)$ and (half of) the 1D-density of states $D(E)/2$ is equal to $1/h$. For our consideration here, this means that the pinch-off at the drain is due to the fact that the potential bends downwards into the drain contact leading to a high electric field in the drain end. This high electric field accelerates the carriers and they acquire kinetic energy (the overall energy is of course constant due to the ballistic transport we consider here). As a result, the carriers become faster but the current stays constant because when carriers move up in energy in a quadratic dispersion relation, the density of states is reduced by the same amount as carriers are accelerated. Therefore, the carrier density (not the current) at the drain end indeed drops substantially, which can be interpreted as a "pinch-off" of the channel. Figure 5.7 illustrates this scenario schematically.

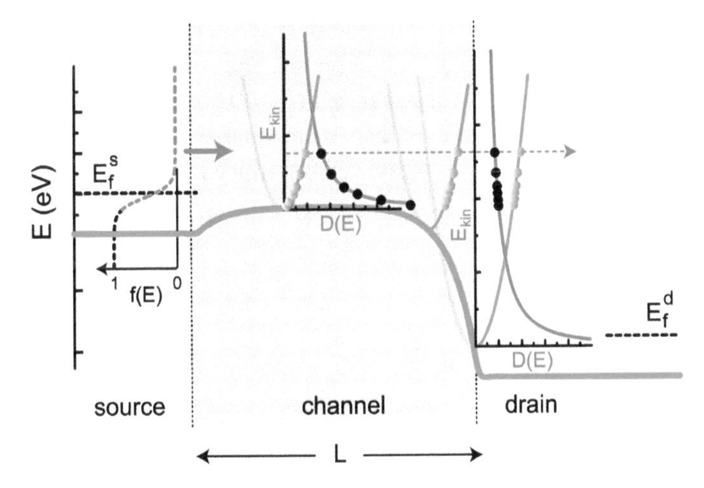

Figure 5.7: Ballistic transport at high drain-source bias. Due to the lack of inelastic scattering, the energy of the carriers remains unchanged leading to a strong increase of their kinetic energy at the drain end. Since the density of states (red curve) drops the same amount as carrier velocity increases the current remains constant. The reduced density of states (position of the black dots) leads to a reduced carrier density at the drain end, which can be interpreted as a pinch-off of the channel.

Here, the green parabolas are the dispersion relation and the red curves the density of states. The red dots illustrate the carriers and the black dots their value of the DOS reflecting the strongly reduced carrier density at the drain end.

5.2.4 On-State—Transfer Characteristics

From our discussion so far, the transfer characteristics of a MOSFET in its on-state are obvious—provided the dependence of Φ_f^0 on the gate voltage has been calculated—and have been mentioned explicitly for 1D and 2D devices (see above). However, there is one point concerning multimode transport that should be mentioned.

Suppose we know how Φ_f^0 depends on the gate voltage, then the integration of the 1D Landauer expression can be carried out, leading to

$$I_d = \frac{2e}{h} \int_{\Phi_f^0}^{\infty} dE(f_s(E_f^s) - f_d(E_f^d)) = \frac{2e}{h} k_B T \ln\left(\frac{1 + \exp(\frac{E_f^s - \Phi_f^0}{k_B T})}{1 + \exp(\frac{E_f^d - \Phi_f^0}{k_B T})} \right). \tag{5.27}$$

As a result, a transfer characteristic is obtained as illustrated in Figure 2.54: if Φ_f^0 is moved on the same energetic level as E_f^s, the MOSFET is in its on-state and current increases approximately linearly. In the case of small V_{ds}, the current will saturate for larger gate voltages (see Figure 2.53). However, if the gate voltage is large enough that a second, third, etc. subband can contribute to the current, a stepwise increase of the current is expected if the energetic subband spacing ΔE^{sb} is substantially larger than $\sim 4 \times k_B T$ and larger than $|eV_{\text{ds}}|$ as has already been discussed in Section 2.13.3. However, in real MOSFETs multimode transport is usually not observed. The most obvious reason is a subband spacing, which is too small. For instance, in the 2D MOSFET discussed in Section 5.2, the subband spacing $\Delta E^{\text{sb}} \rightarrow 0$, and hence the number of contributing subbands continuously increases with increasing gate voltage, which leads to a continuously increasing drain current.

5.3 Impact of Scattering

Up to now, transport has been considered as being ballistic. However, the impact of scattering can be incorporated into the Landauer formalism by introducing a probability for transmission through the channel $T_{\text{ch}}(E)$. Without mentioning it explicitly, we assumed so far that $T_{\text{ch}}(E) = 0$ for energies below Φ_f^0 and $T_{\text{ch}}(E) = 1$ in the case $E \geq \Phi_f^0$. Hence, generalizing the Landauer expression for current flow, I_d can be written as

$$I_d = \frac{2e}{h} \int_{-\infty}^{\infty} dE T_{\text{ch}}(E)(f_s(E_f^s) - f_d(E_f^d)). \tag{5.28}$$

For the time being, tunneling through the channel will be neglected (see Section 5.9.4 for further discussion), but for energies above the potential maximum Φ_f^0 the impact of scattering can be included with the transmission probability [182, 184]

$$T_{\mathrm{ch}}(E) = \frac{l_{\mathrm{mfp}}(E)}{L + l_{\mathrm{mfp}}(E)}, \tag{5.29}$$

where l_{mfp} is the mean free path for scattering and L is the channel length. Equation (5.29) allows for a smooth transition from the ballistic (where $l_{\mathrm{mfp}} \to \infty$ and hence $T_{\mathrm{ch}}(E) \to 1$) to the scattering regime of electronic transport (with $l_{\mathrm{mfp}} \ll L \to T_{\mathrm{ch}}(E) \to l_{\mathrm{mfp}}/L$); for details on the derivation of Equation (5.29), the reader is referred to [60, 183]. Equation (5.29) can be understood when considering the 1D case at $T = 0$ K, which led to Equation (2.118). Inserting $T_{\mathrm{ch}}(E)$ then yields $I_d = \frac{2e^2}{h} T_{\mathrm{ch}} MV$, with M being the number of 1D modes. Rearranging this equation to compute the resistance results in

$$R_{\mathrm{tot}} = \frac{V}{I_d} = \frac{h}{2e^2} M^{-1} (T_{\mathrm{ch}})^{-1} = \underbrace{\frac{h}{2e^2 M} \frac{L}{l_{\mathrm{mfp}}}}_{R_{\mathrm{cond}}} + \underbrace{\frac{h}{2e^2 M} \frac{l_{\mathrm{mfp}}}{l_{\mathrm{mfp}}}}_{R_{\mathrm{cont}}} \tag{5.30}$$

where $R_{\mathrm{cond}} \propto L/l_{\mathrm{mfp}}$ and R_{cont} are the expected resistance of the conductor (channel) and the contact resistance, respectively.

5.4 Optimizing the Performance of MOSFETs

With the knowledge about the working principles of MOSFETs, the question can now be answered how the performance of MOSFETs can be improved. To this end, it is not sufficient to look at an individual device, since improving the on-current of a MOSFET can be accomplished easily by, e. g., increasing C_{ox} or even simpler, by choosing a larger width. Rather, the devices have to be regarded in a circuit environment. In this case, the current of one device is used to charge the gate of a succeeding transistor and as such, the mere on-current is not a good measure. An appropriate figure of merit for the on-state performance is the gate delay, also called “CV-over-I” metric, $\tau = \frac{C_g V_{\mathrm{dd}}}{I_d^{\mathrm{on}}(V_{\mathrm{gs}}=V_{\mathrm{ds}}=V_{\mathrm{dd}})}$ where V_{dd} is the supply voltage of the circuit. Thus, τ represents the ratio of the charge required on the gate to realize a certain on-current. Optimized MOSFETs should make the gate delay as small as possible. Inserting the saturation current from the GCA and setting $C_g \approx C_{\mathrm{ox}}^{\square} WL$, one obtains $\tau \propto \frac{L^2 V_{\mathrm{dd}}}{\mu(V_{\mathrm{dd}}-V_{\mathrm{th}})^2}$. Note that $C_{\mathrm{ox}}^{\square}$ and W drop out of this expression as they should (to first order); increasing both yields a higher I_d^{on} but also a gate capacitance C_g of the succeeding device in an integrated circuit (IC) that is increased by the same amount.

An improved performance can in principle be obtained by increasing either V_{dd} and/or the gate overdrive $V_{\mathrm{dd}} - V_{\mathrm{th}}$. However, the performance of the transistors should be improved while at the same time the power consumption of the circuit (consisting of those devices) should be reduced as much as possible. The right part of Equations (5.31) is an approximate expression for the power consumption P of highly integrated circuits. We have for τ and P,

$$\tau = \frac{C_g V_{dd}}{I_d^{on}} \propto \frac{L^2 V_{dd}}{\mu(V_{dd} - V_{th})^2}, \qquad P = \underbrace{A \cdot C_{tot} \cdot f \cdot V_{dd}^2}_{\text{dynamic}} + \underbrace{I_{leak} \cdot V_{dd}}_{\text{static}}. \tag{5.31}$$

As already mentioned, V_{dd} is the supply voltage of the IC. C_{tot} is the total load capacitance (including all device-related capacitances as well as all capacitances due to metal leads, interconnects, etc.) and f is the clock frequency of the integrated circuit.

Optimized devices/circuits must reduce τ and P at the same time. Hence, increasing V_{dd} is obviously not an option: while τ would be reduced, a larger V_{dd} strongly increases the dynamic part of P. On the other hand, decreasing V_{dd} to save power leads to a performance loss due to a reduction of I_d^{on} (by $\approx$50 % in the case shown in Figure 5.8). Furthermore, increasing the gate overdrive $V_{dd} - V_{th}$ by reducing V_{th} is also not an option. The reason for this is the minimum inverse subthreshold slope of MOSFETs of 60 mV/dec at room temperature (cf. Section 5.2.2); a reduced V_{th} would shift the transfer characteristics "rigidly" along the gate voltage axis toward lower V_{gs}, which leads to an exponential increase of the off-state leakage current I_{leak} as depicted in Figure 5.8. Therefore, the best choices to improve the MOSFET performance are downscaling of the channel length L and increasing the carrier mobility μ. In particular, downscaling has been very attractive since it allows putting a larger number of better performing devices onto the same chip area. The next sections are therefore devoted to the effects occurring when scaling down MOSFETs. However, as scaling becomes increasingly difficult and will hit the physical limit in the very near future, increasing the mobility by incorporating novel

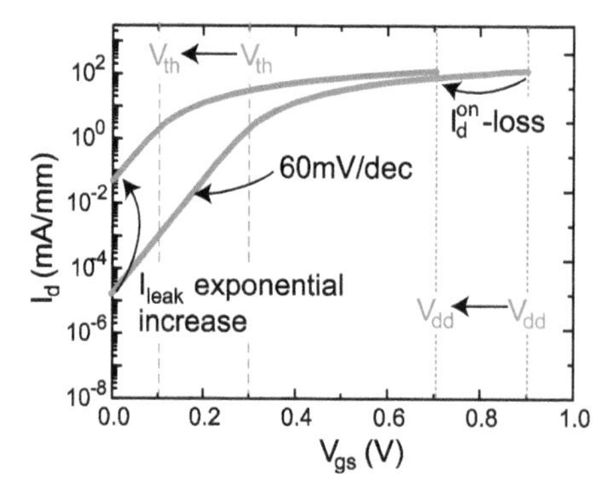

Figure 5.8: Transfer characteristics of a conventional MOSFET with optimum inverse subthreshold slope at room temperature (green curve). Reducing V_{dd} either leads to a loss of on-state performance or to an exponential increase of leakage (blue curve) due to S_{min} = 60 mV/dec.

materials into existing technology has been a second path used by industry to improve transistor performance.

5.5 A Simple Model for the Electrostatics

While the top-of-the-barrier model used in preceding sections provided a number of insights into the functionality and performance of MOSFETs, it appears to be rather simplistic neglecting the exact potential distribution. The electrostatics within a MOSFET has to be computed by solving Poisson's equation and since a MOSFET is a three-dimensional object, in principle a 3D Poisson equation needs to be solved. However, considering a very wide, planar MOSFET, solving a 2D Poisson equation is sufficient. We have

$$\frac{\partial^2 \Phi(x,z)}{\partial x^2} + \frac{\partial^2 \Phi(x,z)}{\partial z^2} = -\frac{e^2(n(x,z) \pm N_{\mathrm{dop}})}{\varepsilon_0 \varepsilon_{\mathrm{si}}} \tag{5.32}$$

where $n(x,z)$ is the density of inversion charge and N_{dop} is the density of ionized dopants (donors or acceptors) and Φ the potential energy. Although reduced to a 2D equation, solving Equation (5.32) requires numerical methods, which will be presented in Chapter 6. Fortunately, the electrostatics of thin-body devices, such as silicon-on-insulator FETs and nanowire transistors, can be reduced to a one-dimensional modified Poisson equation that provides a rather accurate description of the electrostatics. The solution of this 1D modified Poisson equation can then be used to obtain a more accurate, semi-analytical description of the device behavior. In Chapter 6, this Poisson equation will also be used for numerical simulations of nanoscale FETs.

To reduce the electrostatics to a one-dimensional equation, the approach by Young [284] and Yan [278] (see also Auth and Plummer [19]) is used. In the following, the derivation for a planar single-gate MOSFET is explicitly stated. The main approximation used in this respect is a quadratic expansion of the potential distribution in the z-direction (see Figure 5.9), which yields

$$\Phi(x,z) \approx c_0(x) + c_1(x)z + c_2(x)z^2. \tag{5.33}$$

This approximation is reasonable since, similar to the discussion in Section 5.1, the dependence of n on z can be neglected. And in this case, the potential is essentially determined by the charge in the depletion region in the MOS capacitor (gray shaded area in Figure 5.9). Using the depletion approximation, a solution of the Poisson equation is obtained by simply integrating twice over the constant charge density due to the ionized dopants, and as a result a quadratic potential is expected as illustrated with the light blue parabola in Figure 5.9.

Equation (5.33) requires one to determine the three "constants" (with respect to z) c_0, c_1 and c_2. In the following, we are interested in the surface potential $\Phi_f(x)$ at $z = 0$

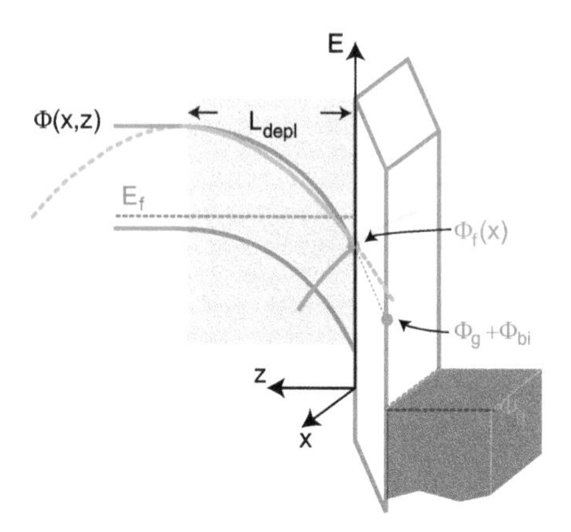

Figure 5.9: Conduction and valence bands within a MOS capacitor (green curves). The light blue line is a quadratic approximation to the potential distribution along the *z*-direction.

i. e., $\Phi(x,z)|_{z=0} = \Phi_f(x)$. At $z = 0$, Equation (5.33) yields $\Phi_f(x) = c_0(x)$, so the first "constant" has already been determined. Next, at the interface between the silicon substrate and the gate dielectric, the dielectric displacement needs to be continuous leading to the boundary condition: $\varepsilon_{\text{si}}\mathcal{E}_{\text{si}} = \varepsilon_{\text{ox}}\mathcal{E}_{\text{ox}}$. The electric field $\mathcal{E}_{\text{si}}$ within the silicon at $z = 0$ is given by $(-e)\mathcal{E}_{\text{si}} = -\frac{\partial\Phi(x,z)}{\partial z}|_{z=0}$. From Equation (5.33), we obtain $\frac{\partial\Phi(x,z)}{\partial z}|_{z=0} = c_1(x) + 2c_2(x)z|_{z=0} = c_1(x)$. Furthermore, the electric field within the gate dielectric is $(-e)\mathcal{E}_{\text{ox}} = -\frac{\Phi_f(x)-(\Phi_g+\Phi_{\text{bi}})}{d_{\text{ox}}}$ (cf. Figure 5.9). Putting this together allows for the determination of the second constant $c_1(x) = \frac{\varepsilon_{\text{ox}}}{\varepsilon_{\text{si}}}\frac{\Phi_f(x)-(\Phi_g+\Phi_{\text{bi}})}{d_{\text{ox}}}$. Finally, at $z = L_{\text{depl}}$ the depletion zone ends and the potential becomes horizontal with zero electric field. Hence, $\frac{\partial\Phi(x,z)}{\partial z}|_{z=L_{\text{depl}}} = c_1(x)+2c_2(x)L_{\text{depl}} \approx 0$, which leads to $c_2(x) = -\frac{1}{2L_{\text{depl}}}c_1(x) = -\frac{\varepsilon_{\text{ox}}}{2L_{\text{depl}}d_{\text{ox}}\varepsilon_{\text{si}}}(\Phi_f(x)-(\Phi_g+\Phi_{\text{bi}}))$. Inserting c_0, c_1 and c_2 into Equation (5.33) results in an approximation for the two-dimensional potential. However, while the *z*-dependence is now determined, the dependence on *x* is not. This means that Equation (5.33) with the constants c_0, c_1 and c_2 needs to be inserted into the Poisson equation (5.32) and evaluated at $z = 0$ to obtain the *x*-dependence of the surface potential $\Phi_f(x)$.

The second derivative of $\Phi(x,z)$ with respect to x evaluated at $z = 0$ results in $\frac{\partial^2\Phi(x,z)}{\partial x^2}|_{z=0} = \frac{\partial^2\Phi_f(x)}{\partial x^2}$ and the second derivative with respect to z yields $\frac{\partial^2\Phi(x,z)}{\partial z^2}|_{z=0} = 2c_2(x) = -\frac{\varepsilon_{\text{ox}}}{L_{\text{depl}}d_{\text{ox}}\varepsilon_{\text{si}}}(\Phi_f(x)-(\Phi_g+\Phi_{\text{bi}}))$. Inserting this into Equation (5.32), one obtains the following 1D modified Poisson equation:

$$\frac{\partial^2\Phi_f(x)}{\partial x^2} - \frac{\Phi_f(x)-(\Phi_g+\Phi_{\text{bi}})}{\lambda^2} = -\frac{e^2(n(x)\pm N_{\text{dop}})}{\varepsilon_0\varepsilon_{\text{si}}} \tag{5.34}$$

where $\lambda = \sqrt{\frac{\varepsilon_{\text{si}}}{\varepsilon_{\text{ox}}}L_{\text{depl}}d_{\text{ox}}}$. If the carrier density is assumed to be constant, Equation (5.34) can be solved analytically with the ansatz $\Phi_f(x) = Ae^{x/\lambda} + Be^{-x/\lambda} + C$ which can easily be verified by inserting the ansatz into Equation (5.34). This means that potential variations are screened on the length scale λ. In particular, the *p-n* junctions at the source-channel and channel-drain interfaces are determined by λ even if the channel is intrinsic. The reason for this screening is the proximity of the gate electrode: The gate electrode is considered as being metallic and as such provides a sufficiently large amount of free carriers to ensure a flat potential. Reducing, e. g., the physical oxide thickness d_{ox} brings this equipotential electrode closer to the silicon surface enforcing a flat potential within the silicon in the vicinity of the silicon-gate dielectric interface. Therefore, screening is obtained not only due to charge within the semiconductor (ionized dopants in the depletion zones of *p-n* junctions, mobile charge, etc.) but also with charge spatially separated (by a gate dielectric) from the channel of the device under consideration (see Section 5.7 for more discussion on screening lengths and also Section 4.3.1).

While the modified Poisson equation has been derived based on a bulk Si MOSFET, we only used this fact to obtain a suitable boundary condition for the electric field within the substrate to determine the parameter $c_2(x)$. The derivation is thus not bound to bulk MOSFETs. For instance, if a device based on silicon-on-insulator with an SOI layer thickness of d_{SOI} is considered and the thickness of the buried oxide d_{BOX} is rather thick, then the electric field at the SOI-buried oxide interface is $\propto 1/d_{BOX}$, and hence approximately zero. In this case, the same modified Poisson equation is obtained; the only difference is the screening length, which is now given by $\lambda = \sqrt{\frac{\varepsilon_{si}}{\varepsilon_{ox}} d_{ox} d_{SOI}}$. Furthermore, in a double-gate MOSFET (see Task 21), the electric field will be zero exactly in the middle of the channel layer due to symmetry reasons and, therefore, an equivalent derivation of the modified Poisson equation can be carried out. As a result, in the case of different device architectures, the modified Poisson equation is always given by Equation (5.34), and the specific device architecture is reflected only in the expression for λ. Figure 5.12 shows different device layouts and the associated screening lengths.

Task 21.
Modified 1D Poisson equation: Compute the modified one-dimensional Poisson equation (i) in the case of a double-gate silicon MOSFET with channel thickness d_{ch} and gate oxide thickness d_{ox} and (ii) in the case of a wrap-gate transistor architecture with a silicon nanowire with a diameter of d_{nw} and a gate oxide thickness of d_{ox} using cylindrical coordinates.

5.6 Scaling and the Appearance of Short-Channel Effects

Figure 5.10 shows schematic cross-sections of a long- and a short-channel *n*-type bulk MOSFET together with conduction profiles in the devices' off-state. In the case of the long-channel device, the channel length L_{long} is substantially larger than the spatial extent of the channel-electrode *p-n* junctions, i. e., the screening length λ. A potential profile is then obtained (white line in Figure 5.10) where the potential maximum Φ_f^0 is deter-

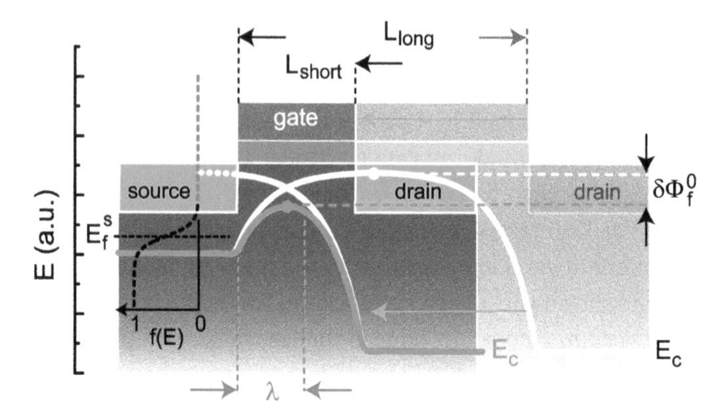

Figure 5.10: Conduction band E_c of a long-channel MOSFET (white line) and in the case of a short-channel device (green). The overlapping source-channel and channel-drain *p-n* junctions in the short-channel transistor lead to a reduction of the potential maximum in the channel when L_{short} and λ are of the same order. The reduced barrier leads to strongly increased leakage.

mined merely by the applied gate voltage. Scaling this device to a channel length L_{short} on the order of λ eventually leads to overlapping *p-n* junctions at the source-channel and channel-drain interfaces. This in turn results in a reduction $\delta\Phi_f^0$ of the potential maximum as illustrated with the green potential profile in Figure 5.10. Since Φ_f^0 determines the injection of carriers from the source Fermi distribution function (as long as tunneling can be neglected), a reduced barrier leads to an exponentially increased leakage current in the off-state of the MOSFET.

The overlap of the *p-n*-junctions in an (inadequately scaled) short-channel MOSFET manifests itself in a number of measurable performance degradations called short-channel effects (SCEs), which are illustrated in Figure 5.11. First, a representative I_d–V_{gs} curve of the short-channel device is displayed in the right panel of Figure 5.11(a) (red curve) and shows a significantly larger inverse subthreshold slope S, which leads to an exponential increase of the leakage current. Furthermore, the threshold voltage of the short-channel device is reduced compared to the long-channel counterpart (blue curve), which is called threshold-voltage roll-off. The right panel of Figure 5.11(b) displays output characteristics of the short-channel device showing a loss of current saturation for larger V_{ds}.

SCEs can all be traced back to the overlap of the source-channel/channel-drain *p-n* junctions as illustrated in the left panels of Figure 5.11. In case (a), the overlap results in a reduced barrier and hence to a shift of V_{th}. In addition, the gate control is reduced: in the present case $\delta\Phi_f^0 < \delta\Phi_g$ and as a result, $S = \frac{k_B T}{|e|} \ln(10) \frac{\partial \Phi_g}{\partial \Phi_f^0} > 60$ mV/dec. Moreover, the potential maximum is not only modified by Φ_g but also by the drain potential Φ_d (cf. Equation (5.17)), which is responsible for the loss of current saturation. This so-called drain-induced-barrier lowering (DIBL) is experimentally measurable as $\delta V_{\text{th}}/\delta V_{\text{ds}}$ and

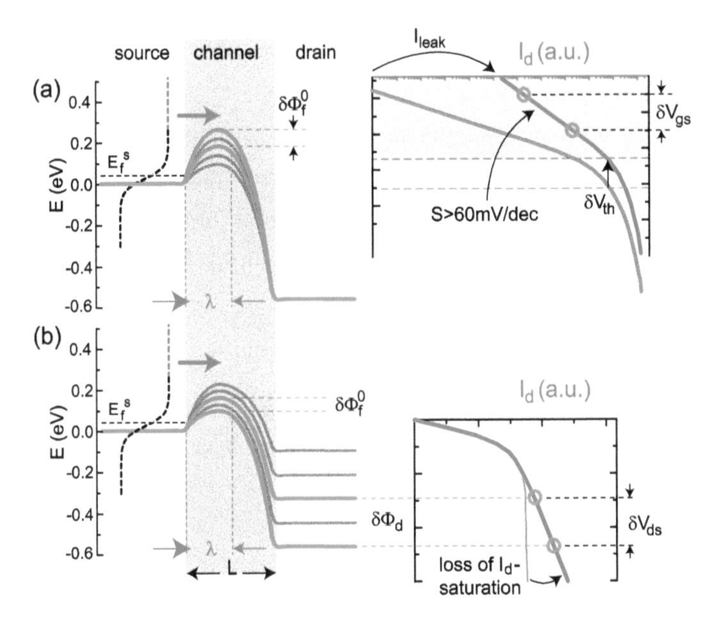

Figure 5.11: (a) Potential profile and transfer characteristics of a conventional MOSFET showing strong SCE (red curve) that lead to a deteriorated S, a strongly increased off-state leakage and a reduction of V_{th} compared to a device without SCEs (blue curve). (b) Potential profile and output characteristic of a conventional MOSFET with strong SCE (red curve). Drain-induced-barrier lowering results in a loss of current saturation.

given in mV/V. Looking at the conduction band profiles, it is obvious that DIBL stems from the reduction $\delta\Phi_f^0$ with changing $\delta\Phi_d$.

From the discussion of the top-of-the-barrier model, the appearance of SCEs was actually expected. The loss of gate control, i. e., $\delta\Phi_f^0/\delta\Phi_g$ for constant Φ_d, as well as DIBL = $\delta\Phi_f^0/\delta\Phi_d$ for constant Φ_g can be computed from Equation (5.17) to be $\frac{\delta\Phi_f^0}{\delta\Phi_g} = \frac{C_{\mathrm{ox}}}{C_{\mathrm{ox}}+C_d}$ and $\frac{\delta\Phi_f^0}{\delta\Phi_d} = \frac{C_d}{C_{\mathrm{ox}}+C_d}$ (neglecting all other capacitances). Moreover, it has also already been mentioned above that ensuring $C_{\mathrm{ox}} \gg C_d$ yields $\delta\Phi_f^0/\delta\Phi_g \to 1$ and $\delta\Phi_f^0/\delta\Phi_d \to 0$. This means SCEs are suppressed if $C_{\mathrm{ox}} \gg C_d$.

Based on the modified Poisson equation, we can now put the appearance of SCE and their suppression on a firmer basis by solving Equation (5.34) in the off-state of the device. To simplify the notation, an intrinsic channel is considered. Note, however, that a constant dopant density can simply be incorporated into the built-in potential Φ_{bi}. Inserting the ansatz $\Phi_f(x) = Ae^{x/\lambda} + Be^{-x/\lambda} + C$ into Equation (5.34) yields $C = \Phi_g + \Phi_{\mathrm{bi}}$. Moreover, since the source is set to ground, the boundary condition $\Phi_f(x = 0) = 0$ results in $B = -A-(\Phi_g+\Phi_{\mathrm{bi}})$ and with $\Phi_f(x = L) = \Phi_d = -eV_{\mathrm{ds}}$ one obtains $A = \frac{\Phi_d+(\Phi_g+\Phi_{\mathrm{bi}})(e^{-L/\lambda}-1)}{2\sinh(L/\lambda)}$. Putting things together, the potential profile along the direction of current transport becomes [278]

$$\Phi_f(x) = \frac{\sinh(x/\lambda)}{\sinh(L/\lambda)}(\Phi_d + (\Phi_g + \Phi_{bi})(e^{-L/\lambda} - 1) + \Phi_g(1 - e^{-x/\lambda})). \tag{5.35}$$

The potential maximum Φ_f^0 can now be computed with $\frac{d\Phi_f(x)}{dx} = 0$, which yields the position x_0 resulting in $\Phi_f^0 = \Phi_f(x = x_0)$. It is straightforward to show that $x_0 = \frac{\lambda}{2}\ln[-(1+\frac{\Phi_g+\Phi_{bi}}{A})]$. Although in a device with SCE, L and λ may be of similar order, we can still assume $L > \lambda$ such that $\sinh(L/\lambda) \approx e^{L/\lambda}$. Using this approximation, one finally obtains

$$x_0 \approx \frac{L}{2} + \frac{\lambda}{2}\ln\left(\frac{\Phi_g + \Phi_{bi}}{\Phi_g + \Phi_{bi} - \Phi_d}\right), \tag{5.36}$$

which states that at $\Phi_d = 0$, the potential maximum is in the middle of the channel and moves towards source when the bias is increased. This effect is stronger in devices with a large λ. Next, inserting x_0 yields

$$\Phi_f^0 \approx \Phi_g(1 - 2e^{-\frac{L}{2\lambda}}) + \Phi_d e^{-\frac{L}{2\lambda}}. \tag{5.37}$$

As a result, the inverse subthreshold slope S and DIBL in devices with SCE are

$$S \approx \frac{k_B T}{|e|}\ln(10)\left(\frac{1}{1 - 2e^{-\frac{L}{2\lambda}}}\right), \quad \text{DIBL} = \frac{\partial\Phi_f^0}{\partial\Phi_d} \approx e^{-\frac{L}{2\lambda}}. \tag{5.38}$$

5.7 Screening Lengths in Nanoscale FETs

It was briefly discussed in Section 5.5 that the screening length λ is the relevant length scale for potential variations. However, this is only true in the case that there is no additional mobile charge (within the channel, i. e., if the device is in the off-state). Additional screening can be taken into consideration in the following way: Consider Equation (5.34) and assume that a small amount of charge δn is added leading to a change $\delta\Phi_f$ of the potential. As a result, Equation (5.34) can be written as[4]

$$\frac{\partial^2(\Phi_f + \delta\Phi_f)}{\partial x^2} - \frac{(\Phi_f + \delta\Phi_f) - (\Phi_g + \Phi_{bi})}{\lambda_{ch}^2} = \frac{e^2(n + \delta n)}{\varepsilon_0\varepsilon_{ch}} \tag{5.39}$$

Since Φ_f and n are supposed to fulfill Poisson's equation (5.34), and since $\delta n = \frac{\partial n}{\partial\Phi_f}\delta\Phi_f$, Equation (5.39) becomes

$$\frac{\partial^2(\delta\Phi_f)}{\partial x^2} - \frac{\delta\Phi_f}{\lambda_{ch}^2} = \frac{e^2}{\varepsilon_0\varepsilon_{ch}}\frac{\partial n}{\partial\Phi_f}\delta\Phi_f. \tag{5.40}$$

4 In order to distinguish the different screening mechanisms, the subscript "ch" for λ due to the device architecture is used in the following, i. e., $\lambda_{ch} = \sqrt{\frac{\varepsilon_{ch}}{\varepsilon_{ox}}d_{ch}d_{ox}}$ in the case of a single-gate SOI transistor.

As an example, let us consider the case of a moderate carrier concentration with $n(x) \propto \exp(\Phi_f/k_BT)$ such that $\frac{\partial n}{\partial \Phi_f} = n/k_BT$. Hence, the right-hand side of Equation (5.40) is $\frac{e^2 n}{\varepsilon_0 \varepsilon_{ch} k_B T}\delta\Phi_f$. Moving the right-hand side on the left side of the equation and noting that the Debey length is $\lambda_{debey} = \sqrt{\frac{\varepsilon_0 \varepsilon_{ch} k_B T}{e^2 n}}$, an overall screening length λ_{tot} can be defined by

$$\lambda_{tot} = \frac{1}{\sqrt{\frac{1}{\lambda_{ch}^2} + \frac{1}{\lambda_{debey}^2}}}. \tag{5.41}$$

Equation (5.41) is generally valid, i. e., the way how screening due to different mechanisms is combined (see, e. g., [42]). For instance, it also yields the appropriate screening in the case of a double-gate FET when both gates are regarded to provide the screening of a single-gate SOI transistor architecture. Furthermore, Equation (5.19) can also be converted into the same form. Moreover, it can also be used in the case of degenerate doping where λ_{debey} needs to be replaced with the Thomas–Fermi screening length. An example of this will be briefly discussed when dealing with device simulations in Section 6.1.1.

5.8 Ultrathin-Body Field-Effect Transistors

From the considerations of the preceding sections, it is now clear that short-channel effects are suppressed if $\lambda_{ch} \ll L$. An appropriate measure can be found by requiring maximum tolerable values for S and DIBL and then use Equation (5.38) to find a relation between λ_{ch} and L. For instance, if $S < 75$ mV/dec is stipulated, $L \geq 5 \times \lambda_{ch}$, which is reasonable keeping in mind that λ_{ch} is the length scale on which potential variations are screened. As already mentioned above, λ_{ch} depends on the actual device layout considered (cf. Task 22). Figure 5.12 displays different device layouts together with the associated screening length λ_{ch}. Obviously, in all cases, a thin d_{ox}, a gate dielectric with high ε_{ox} and an ultrathin channel layer are preferable in order to keep λ_{ch} small and enable transistors with short channel lengths L but without SCEs. In fact, the best suited device architecture is the gate-all-around nanowire transistor based on a nanowire with ultrasmall diameter.[5] In the following sections, experimental examples of such ultrathin-body transistors are discussed. The fabrication techniques used to realize the devices will only be summarized briefly; details can be found in Chapter 3.

5 This is one of the main reasons why carbon nanotubes are perhaps the best suited nanostructures for ultimately scaled field-effect transistors.

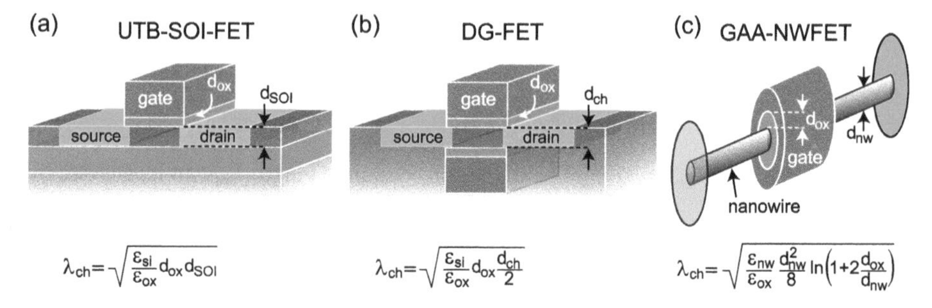

Figure 5.12: Three different ultrathin-body device architectures and their natural length scale λ_{ch}. (a) Ultrathin-body silicon-on-insulator MOSFET with a single gate (UTB-SOI-FET), (b) double-gate MOSFET (DG-FET) and (c) a gate-all-around nanowire-FET (GAA NW-FET).

5.8.1 Single-Gate Ultrathin-Body SOI-MOSFETs

Silicon-on-insulator (SOI) is a very attractive substrate platform to realize ultrathin-body field-effect transistors that can be scaled to very short-channel lengths. Thinning down the top silicon layer to a desired thickness can be done very precisely (for instance, using "digital etching" as explained in Section 3.6.3). Furthermore, using ultrathin SOI has an additional benefit. If the SOI thickness is made substantially smaller than the depletion length L_{depl} at $V_{gs} = V_{th}$, the SOI is said to be fully depleted (FD). As a result, the depletion capacitance $C_{depl} = (-e)\frac{\partial Q_{depl}}{\partial \Phi_f^0}$ (cf. Equation (4.17)) vanishes in FD-SOI. Therefore, an ideal off-state behavior with $C_{depl} = 0$, and consequently, $S = 60$ mV/dec is obtained.

Figure 5.13 shows a cross-section transmission electron microscopy image of an ultrathin-body SOI transistor [13, 14]. In order to realize very short-channel lengths L without highly sophisticated lithography a combination of epitaxial silicon growth and anisotropic wet chemical etching with KOH (cf. Section 3.6.4) is used. Molecular beam epitaxy (MBE) is employed to grow a highly n-doped (antimony with a concentration of

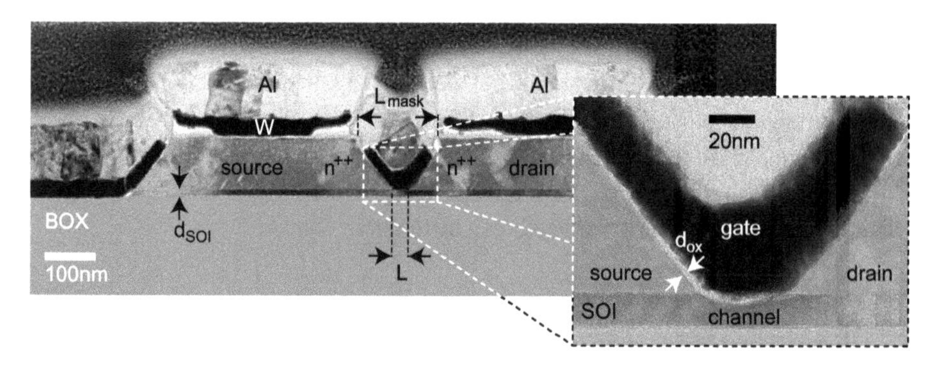

Figure 5.13: Transmission electron micrograph of a single-gate UTB SOI MOSFET fabricated with anisotropic silicon etching. The inset shows a close-up with a minimum channel length of 36 nm [13, 14].

~10^{20} cm^{-3}) silicon layer on a nominally undoped (100) SOI substrate with d_{SOI} = 15 nm. The high doping concentration in the epitaxial layer is necessary since this layer will play the role of the source and drain contacts in the final device. MBE growth and the use of antimony ensure that an abrupt interface between the contacts and the SOI layer is achieved and that dopant diffusion due to thermal treatments during the fabrication process is suppressed as much as possible. A SiO_2 hardmask is then grown and electron-beam lithography is used to pattern appropriate parallel line patterns aligned along the ⟨110⟩-crystalline direction. Subsequently, a V-groove is etched using KOH. The initial width of the line patterns is adjusted with respect to the thickness of the epitaxial layer in a way that ensures that the tip of the V-groove cuts through the highly doped epitaxial layer. This way, source and drain regions are formed separated by a channel within the nominally undoped SOI.

Figure 5.13 reveals that the epitaxial layer shows substantial crystalline defects that lead to an imperfect V-groove formation. Nevertheless, single-gate SOI transistors could be realized with three different channel lengths of $L \approx$ 36 nm, 46 nm and 56 nm. Figure 5.13 also shows that the tip area is actually not a tip but rather flat. The reason for this is the ex situ cleaning procedure of the substrate prior to the epitaxial growth: stripping the native oxide on top of the SOI with HF leads to an incomplete hydrogen passivation of the silicon surface and the hydrophobic nature makes the surface prone to adsorption of carbon and oxygen contaminants (see the discussion in Section 3.1). After V-groove formation the samples are thoroughly cleaned with SC2 (cf. Section 3.1.3.2) followed by the wet thermal growth of a 2.6 nm thin SiO_2 gate oxide [6]. Finally, a tungsten gate is patterned with electron-beam lithography and wet etching in H_2O_2 followed by the patterning of aluminum contact leads.

Figure 5.14 shows output (left panel) and transfer (right panel) characteristics of the three SOI-MOSFETs. It is clearly visible that the transistor with the longest-channel length of L = 56 nm shows the least short-channel effects. This device exhibits current saturation (orange curves in Figure 5.14) in the output characteristics and the steepest inverse subthreshold slope. From the device characteristics, an inverse subthreshold

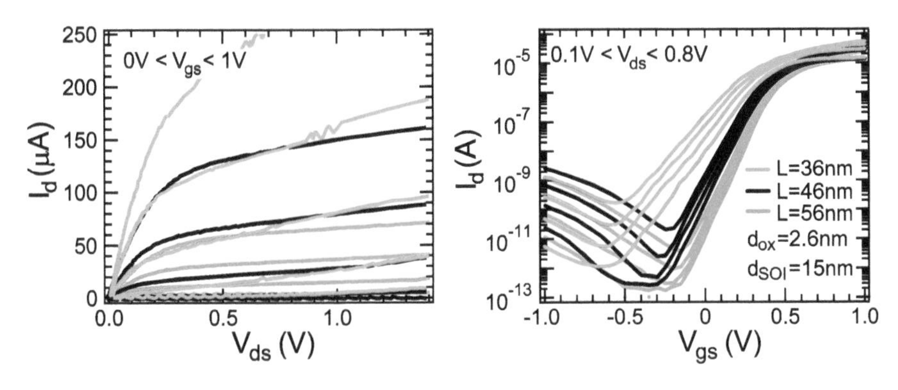

Figure 5.14: Output (left) and transfer (right) characteristics of V-groove SOI-MOSFETs [13].

slope of $S = 80\,\mathrm{mV/dec}$ and a value for drain-induced barrier lowering of ~90 mV/V can be extracted. Reducing the channel length (keeping d_{ox} and d_{SOI} constant) strongly increases short-channel effects. For the device with $L = 46\,\mathrm{nm}$, an inverse subthreshold slope $S = 90\,\mathrm{mV/dec}$ and DIBL ≈ 125 mV/V are obtained and finally in the shortest device $S = 160\,\mathrm{mV/dec}$ and DIBL = 214 mV/V.

The experimental figures of merit can now be compared with our expectation. In the present case, the screening length $\lambda_{ch} = \sqrt{\frac{\varepsilon_{si}}{\varepsilon_{ox}} d_{SOI} d_{ox}}$ turns out to ~10.6 nm. From Equation (5.38), we obtain theoretical inverse subthreshold slopes of $S_{theo}^{56\,nm} = 70\,\mathrm{mV/dec}$, $S_{theo}^{46\,nm} = 77\,\mathrm{mV/dec}$ and $S_{theo}^{36\,nm} = 95\,\mathrm{mV/dec}$ that underestimate the experimentally observed values. However, there is a significant uncertainty in the geometrical parameters of the device. For instance, d_{ox} was determined on a reference bulk Si sample; from Figure 5.13, the oxide thickness could also be larger, d_{SOI} thinner, and although the temperature budget was kept low, dopants could have diffused into the channel, thereby reducing the effective channel length. The experimentally found values of DIBL are also consistently underestimated with $\mathrm{DIBL}_{theo}^{56\,nm} = 71\,\mathrm{mV/V}$, $\mathrm{DIBL}_{theo}^{46\,nm} = 114\,\mathrm{mV/V}$ and $\mathrm{DIBL}_{theo}^{36\,nm} = 183\,\mathrm{mV/V}$. Qualitatively, however, the measured data is consistent with our analysis above of increasing SCE when downscaling the devices.

5.8.2 Double-Gate FinFETs

In the preceding section, strong SCEs are observed in the UTB-SOI MOSFET with a channel length of 36 nm. In order to mitigate the SCEs, the gate control needs to be improved. Since d_{ox} and d_{SOI} are already rather thin, this is accomplished best by using so-called multigate devices. In the case of a double-gate transistor as depicted in Figure 5.12(b), the screening length is reduced by a factor of $\sqrt{2}$ compared to the single-gate SOI-MOSFET. However, the fabrication of such a planar double-gate FET is extremely difficult due to the required alignment accuracy of the two gates with respect to each other. Therefore, so-called FinFETs (see Figure 5.15(b), (c)) were invented where the channel is rotated by 90° such that a double-gate layout can be realized by placing the two gates to the left and right of the fin. This can be done with a single lithography and etch step, and thus the issue of properly aligning the two gates is avoided. The challenges of the FinFET are the fabrication of appropriately thin fins as well as the fact that the MOS surface is now at the vertical, etched flank of the fin instead of the highly perfect substrate surface. However, for both issues solutions were found. First, spacer lithography (also called self-aligned double patterning, SADP) is used to realize the required, extremely thin hard masks for fin etching (see Section 3.5.4). SADP comes with the additional benefit that the conformal deposition employed for spacer formation can be used to reduce the line edge roughness while at the same time the line width roughness can be made negligibly small [45]. Second, hydrogen annealing (Section 3.12) allows smoothening the vertical etch flanks such that mobility degradation due to reactive ion etching is avoided.

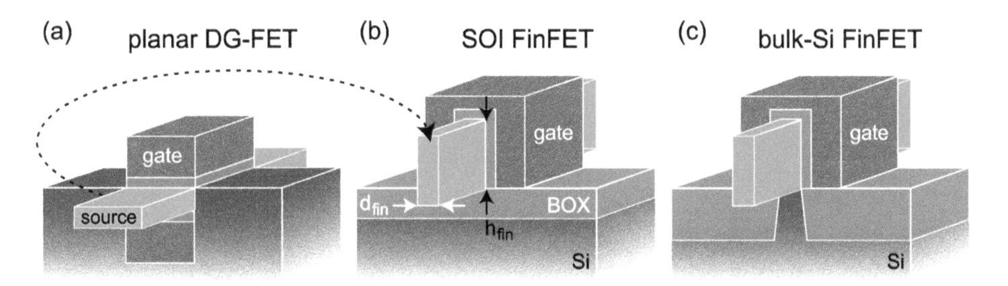

Figure 5.15: (a) Tilting the planar double-gate MOSFET (a) by 90° yields a FinFET that can be realized either on SOI (b) or on bulk Si substrates (c).

5.8.3 Multigate Nanosheet and Nanowire FETs

With continued scaling, the ratio λ_{ch}/L becomes unfavorable again at some channel length L since a reduction of the fin thickness becomes increasingly difficult, and thus SCE reappear. To mitigate this, the height of the fin could be reduced such that the device evolves into a triple-gate FET exhibiting a λ_{ch} approximately $\sqrt{3}$ smaller compared to a single-gate device. However, reducing the fin height comes with a decreased absolute value of the transistor current. A clever way to circumvent this is the realization of nanosheet or nanoribbon FETs as illustrated in Figure 5.16(a) and (b). To this end, alternating layers of Si and SiGe are grown epitaxially on top of each other where the SiGe serves as sacrificial interlayer. Epitaxial growth allows to grow single crystalline layers with excellent control over the thickness of the layers. In a next step, a fin is etched out of the multilayer stack and then the SiGe layers are removed selectively (the remaining silicon nanosheets are certainly connected to a larger contact area not shown in the figure). As a result, multiple crystalline nanosheets or nanoribbons are formed on top of each other whose thickness is controlled by the epitaxial growth. In contrast to a FinFET, this means that the silicon channel is again titled by 90°. The width of the fin etched into the epitaxial material stack in combination with the number of nanosheets determines the effective width of the overall transistor. Finally, a gate dielectric and gate-all-around electrode are realized with conformal deposition (e. g., ALD, Section 3.8.3.1).

Ultimate scalability in terms of the suppression of SCEs is obtained in gate-all-around (GAA) nanowire FETs giving rise to the smallest possible λ_{ch}. Realizing such small screening lengths requires very low nanoscale diameters of the nanowires. This can be realized with etching very narrow fins into the Si/SiGe multilayer stack and again removing the sacrificial SiGe as depicted in Figure 5.16(c). However, in such small nanostructures, deactivation of dopants (see Section 4.3) in the source/drain extensions is likely to deteriorate the device performance. The NESSIAS effect discussed in Section 4.4 may be a solution to this issue.

Multiple nanowire MOSFETs are usually needed in parallel in order to provide the necessary drive currents required in a circuit environment. This can be accomplished by an appropriate number of pairs of Si/SiGe layers of the epitaxial stack as is shown in

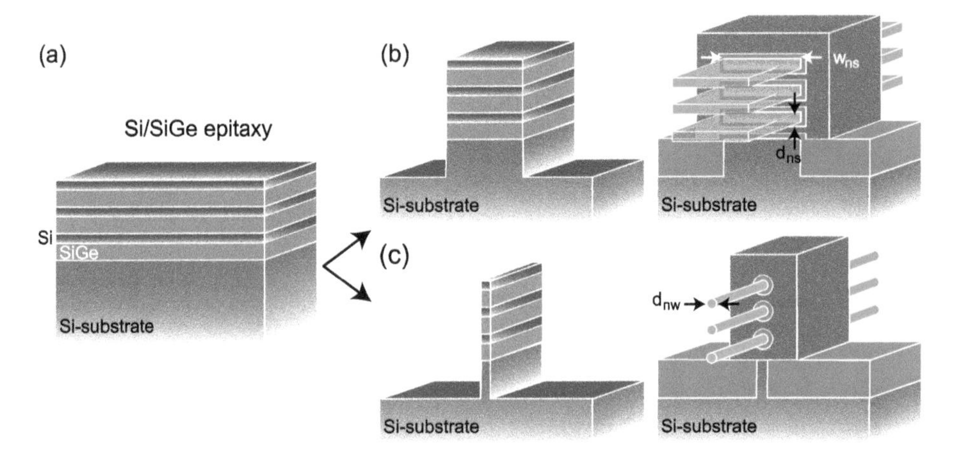

Figure 5.16: (a) Epitaxially grown Si/SiGe layer stack where the SiGe serves as sacrificial layer to realize either multiple, vertically stacked, horizontally oriented nanosheet (b) or nanowire (c) FETs.

Figure 5.16(c) resulting in several vertically stacked nanowires. With sacrificial oxidation and/or hydrogen annealing, nanowires with cylindrical cross-section are obtained. However, manufacturing such vertically stacked nanowires is challenging.[6] Therefore, multiple nanowire FETs are often fabricated laterally as is shown in the top panel of Figure 5.17 [290]. The lower panel shows an individual wrap-gate MOSFET consisting of a Si nanowire with ~8 nm diameter, a HfO_2 gate dielectric and a TiN wrap-gate electrode as can be seen in the transmission electron microscopy close-up shown in the bottom right panel [150]. Although the area consumed by a lateral arrangement is too large for real applications, it allows a straightforward fabrication (cf. Section 3.13.1, Figure 3.57) and enables investigating benefits and requirements of GAA nanowire FETs.

As already mentioned, the gate-all-around (or wrap-gate) device layout yields the smallest screening lengths λ_{ch} (cf. Figure 5.12(c)), and thus allows for the shortest-channel lengths of MOSFETs without suffering from SCEs. In the present case of the GAA nanowire FET with ~8 nm diameter and HfO_2 a very small λ_{ch} of approximately 2–3 nm is obtained, which would enable scaling the device down to a channel length of $L \approx 10$–15 nm without significant SCEs.

Instead of a lateral or vertical arrangement of horizontally oriented nanowires, they can also be grown (e. g., with a vapor–liquid–solid approach, see Section 3.13.2) or etched out of a volume material vertically. Benefits of such a vertical arrangement are an easier fabrication of the gate-all-around and, most importantly, they can be arranged in a bundle that allow reducing parasitic capacitances as will be discussed below.

6 As an alternative to the Si/SiGe heteroepitaxial substrate, reactive ion etching of a fin with the Bosch process in combination with digital etching may be used (see Figure 3.59).

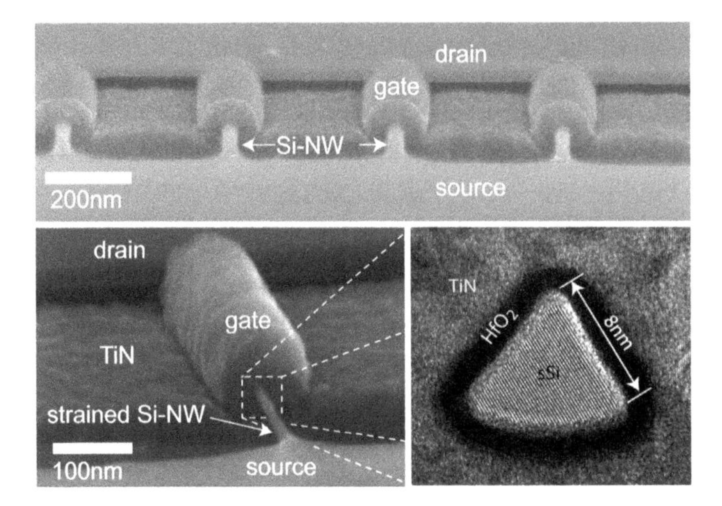

Figure 5.17: Top panel: electron micrograph of several parallel, lateral gate-all-around nanowire FETs (see Figure 3.57 for the fabrication process). The lower left panel shows a close-up of a GAA nanowire FET fabricated based on strained silicon-on-insulator with a diameter of ~8 nm. The bottom right panel shows a transmission electron micrograph of the same device with HfO_2 as gate dielectric and a TiN gate electrode (Q. T. Zhao, FZ-Jülich) [150, 290].

5.9 Ultimate Scaling of MOSFETs

5.9.1 Classical and Quantum Capacitance Limits

In an electrostatically well-behaved device where C_d and C_{depl} can be neglected, the total gate capacitance in the transistor's on-state is given by $C_g = \frac{C_{\mathrm{ox}}C_{\mathrm{inv}}}{C_{\mathrm{ox}}+C_{\mathrm{inv}}}$ where $C_{\mathrm{inv}} = |e|\partial Q_{\mathrm{ch}}/\partial\Phi_f^0$ is the inversion-layer capacitance (or density of states or quantum capacitance) introduced in Section 4.5.3. C_{inv} is (approximately) proportional to the density of states and it has been discussed earlier, that the value of C_{inv} is roughly in-between the full DOS and half of it, depending on the source/drain bias. For convenience, the expression for C_{inv} is repeated here[7]

$$C_{\mathrm{inv}} = (-e)\frac{\partial Q_{\mathrm{ch}}}{\partial\Phi_f^0} \propto -e^2\int dE\, d_{\mathrm{neq}}D(E)\,\frac{\partial f_s(E_f^s)}{\partial E} \approx e^2D(E_f^s - \Phi_f^0) \tag{5.42}$$

where the factor d_{neq} is between 0.5 (large V_{ds}) and unity if $V_{\mathrm{ds}} \to 0$.

In a classical, bulk(Si-) MOSFET, the density of states will increase with increasing gate voltage due to the fact that an increasing number of 2D subbands will be populated with carriers (cf. Section 4.5.3). As a result, $C_{\mathrm{inv}} \gg C_{\mathrm{ox}}$ at some gate voltage, and hence $C_g \approx C_{\mathrm{ox}}$ as was found already earlier. In this case, the change of surface potential Φ_f^0

7 Note that all capacitances are meant as per area in two dimensions and per length in 1D.

with changing gate voltage will approach zero and the MOS capacitor of the transistor behaves like a standard (parallel plate in 2D) capacitance. However, this situation can be very different when a semiconductor is used that exhibits a very small density of states. For instance in graphene (cf. Section 2.12.4), the DOS linearly depends on energy and approaches zero at the Dirac point. As a result, in a graphene FET, the relation between C_{ox} and C_{inv} will be reversed compared to a conventional bulk transistor (at least within a certain gate voltage range). Moreover, in one-dimensional systems, the DOS is proportional to $1/\sqrt{E_f - \Phi_f^0}$, i. e., it drops with increasing V_{gs}. In both cases (graphene around the Dirac point, or 1D nanowires at large gate voltages), the so-called quantum-capacitance limit (QCL) can be reached with

$$C_g = \frac{C_{\text{ox}} C_q}{C_{\text{ox}} + C_q} \overset{C_q \ll C_{\text{ox}}}{\approx} \frac{C_{\text{ox}} C_q}{C_{\text{ox}}} = C_q \tag{5.43}$$

where the nomenclature has been changed from C_{inv} to C_q in order to stress that the QCL is considered here.

The QCL has a number of interesting consequences. First, since $C_g \approx C_q \approx e^2 D(E_f^s - \Phi_f^0)$ a capacitance voltage characterization allows measuring the density of states of the channel material. Indeed, this has been carried out by Chen and Appenzeller [39] who showed a linear dependence of the gate capacitance in a graphene MOS capacitor for small V_g reflecting the linear dependence of the dispersion relation and hence the DOS in graphene (cf. Section 2.12.4 and Figure 2.50(b)). Next, in the case of a constant V_{ds} (i. e., $\delta\Phi_d = 0$) Equation (5.17) results in

$$\delta\Phi_f^0 = \frac{C_{\text{ox}}}{C_{\text{ox}} + C_q} \delta\Phi_g \overset{C_{\text{ox}} \gg C_q}{\approx} \delta\Phi_g \rightarrow \Phi_f^0 = \Phi_g + \text{const.} \tag{5.44}$$

This means that we obtain the same one-to-one dependence of the surface potential on the gate potential as in the off-state of a well-scaled transistor. Hence, the surface potential and the gate potential are related to each other simply by a constant that contains work function differences, etc. The implication of this is that in a 1D nanowire FET in the QCL, the current as a function of gate-source and drain-source bias can be stated explicitly as

$$\begin{aligned} I_d^{\text{QCL}} &= \frac{2e}{h} \int_{\Phi_g + \text{const.}}^{\infty} dE\left(f_s(E_f^s) - f_d(E_f^d)\right) \\ &= \frac{2e}{h} k_B T \ln\left(\frac{1 + \exp\left(\frac{E_f^s - (-eV_{\text{gs}} + \text{const.})}{k_B T}\right)}{1 + \exp\left(\frac{(E_f^s - eV_{\text{ds}}) - (-eV_{\text{gs}} + \text{const.})}{k_B T}\right)} \right). \end{aligned} \tag{5.45}$$

This is a remarkable result, since there is no dependence on the material properties anymore. The ingredients to get there are: (i) 1D electronic transport, (ii) ballistic transport and (iii) the quantum-capacitance limit. In principle, one-dimensionality and ballistic transport can be realized in carbon nanotubes. Furthermore, since CNTs exhibit an extremely small diameter, a gate-all-around device architecture would allow the realization of very large geometrical oxide capacitances C_{ox}, and hence the QCL could be reached. An interesting additional aspect is that at large V_{gs}, the exponential terms in Equation (5.45) dominate, and thus a linear dependence of I_d^{QCL} on V_{gs} is obtained. Such a linear behavior is, for instance, required for an amplification of RF signals without distortion. With conventional MOSFETs, this has to be realized with some circuit overhead leading to significant parasitic power consumption of the amplifier. A linear dependence of I_d on V_{gs} is therefore highly desirable particularly for mobile communication devices.

5.9.2 Reducing Parasitic Capacitances with Nanowire Bundles

When scaling down the transistor devices in an integrated circuit to the nanoscale, the RF performance of the IC may ultimately be determined by parasitic capacitances and not by the (device-) intrinsic gate capacitance C_g. The reason is a continuously reduced C_g (per length) with MOSFET width such that eventually, the parasitics overwhelm the gate capacitance. However, when the width is reduced so far that the device is scaled to the limit where only a single 1D subband contributes, a constant gate capacitance (per length) is reached in such a nanowire device. While being constant, C_g is nevertheless very small and in a typical lateral nanowire device architecture as shown in the top panel of Figure 5.17, the parasitic capacitances between the gate electrode and the extended source/drain contacts $C_{s,d}^{par}$ can become rather large. In particular, if issues with dopant activation play a role (cf. Section 4.3.2), the length of the source/drain extensions needs to be as small as possible aggravating the issue with parasitic capacitances. In this context, it is necessary to distinguish between local source/drain parasitics, $C_{s,d}^{par}$, common to each nanowire device, and global parasitics such as capacitances associated with the wiring of the chip. As will become clear below, using a nanowire bundle configuration as depicted in Figure 5.18 (left panel) allows for reducing the impact of local and global parasitic capacitances.

Dominance of the local capacitances (C_g as well as $C_{s,d}^{par}$) over the global parasitics is obtained by simply making the bundle large enough. Reducing the impact of $C_{s,d}^{par}$ requires some deeper discussion. If the nanowires are closely packed as illustrated in Figure 5.18, left panel, each individual nanowire device is surrounded by six neighboring transistors. As a result, each individual nanowire MOSFET can be approximated as a cylindrical device (see Figure 5.18). The symmetry of the arrangement suppresses contributions of fringing fields to the parasitic capacitances, because of the zero electric field condition at the edge of the nanowires device (i. e., the edge of the device depicted in the close-up in the center of Figure 5.18) as shown in the right panel of Figure 5.18.

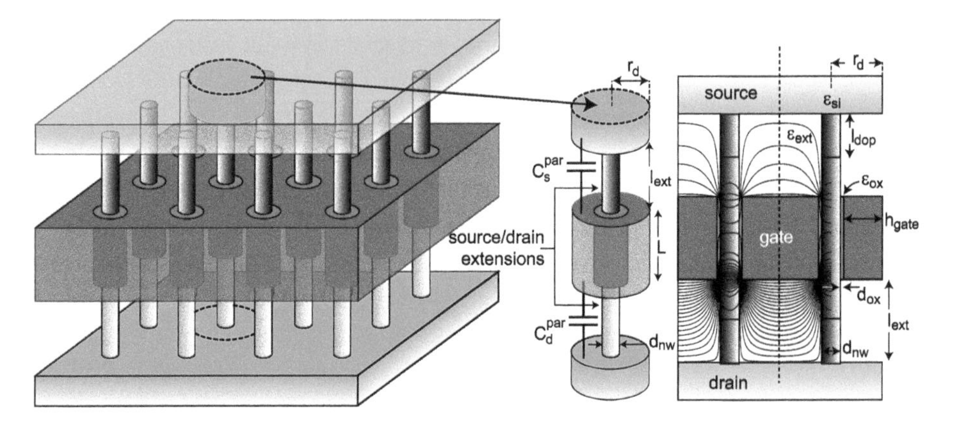

Figure 5.18: Schematic illustration of a GAA nanowire MOSFET bundle (left). The close proximity of the nanowire devices with a specific device being surrounded by six adjacent ones leads to zero electric field in between adjacent transistors due to symmetry reasons. This truncation of the fringing fields in the source/drain underlap regions reduces parasitic capacitances significantly.

Here, equipotential lines within one nanowire MOSFET are displayed showing the zero-field condition at the edges.

Before carrying out a deeper analysis based on simulations, let us start with a simplified discussion and disregard the capacitance contribution between the source/drain extensions of the nanowire and the gate electrode. In this case, $C_{s,d}^{\text{par}}$ are approximately given by the parallel-plate capacitor made up of the two opposite discs of the gate and source/drain electrodes, respectively. The radius r_d of the disc is $r_d = d_{\text{nw}}/2 + d_{\text{ox}} + h_{\text{gate}}$ (cf. Figure 5.18), and hence $C_{s,d}^{\text{par}} \approx \varepsilon_0 \frac{\pi r_d^2}{l_{\text{ext}}}$. If h_{gate}, d_{ox} and d_{nw} are reduced this leads to a strong reduction of $C_{s,d}^{\text{par}}$ but since the nanowire FET is scaled into the quantum limit with a single 1D subband contributing, the (intrinsic) gate capacitance of the device approaches a constant. Moreover, due to the quadratic dependence of $C_{s,d}^{\text{par}}$ on r_d, it is expected that reducing r_d enables shorter l_{ext}, and hence more compact devices while still obtaining reduced local parasitic capacitances. This ultimately allows reducing the relative importance of $C_{s,d}^{\text{par}}$ compared to C_g. At the same time, a larger number of nanowire FETs can be integrated onto the same area such that eventually the global parasitics become less relevant. The nanowire bundle geometry thus represents the ultimate scaling potential for nanowire transistors. In this respect, a vertical layout (see preceding section) is preferable since employing suitable top-down fabrication methods allows the realization of dense arrays of nanowires.

The simple argument regarding the reduction of $C_{s,d}^{\text{par}}$ due to considering it as disc-shaped parallel-plate capacitor is certainly oversimplifying and a detailed calculation is necessary in order to study the benefits of the nanowire bundle configuration. To do so, self-consistent Poisson–Schrödinger computations using the nonequilibrium Green function formalism are employed (see Chapter 6 for details). While a 1D approximation of the electronic transport is chosen, the electrostatics is computed exploiting the radial symmetry of the cylindrical transistor layout displayed in Figure 5.18. At the edge of the device, Neumann boundary conditions with zero electric field are assumed as required by the symmetry discussed above; details on the finite difference solution of the Poisson equation with dielectric boundaries can be found in Chapter 6.

In order to show the reduction of the local parasitic capacitances $C_{s,d}^{par}$, three different wrap-gate nanowire devices with (i) $L = 15$ nm, $l_{ext} = 15$ nm, $h_{gate} = 7.5$ nm, $d_{nw} = 9$ nm, $d_{ox} = 3$ nm, (ii) $L = 10$ nm, $l_{ext} = 10$ nm, $h_{gate} = 5$ nm, $d_{nw} = 6$ nm, $d_{ox} = 2$ nm and (iii) $L = 5$ nm, $l_{ext} = 5$ nm, $h_{gate} = 2.5$ nm, $d_{nw} = 3$ nm, $d_{ox} = 1$ nm are simulated. The total capacitance C_{tot} is determined from $C_{tot}V_{dd} = Q(V_{ds} = 0, V_{gs} = V_{dd}) - Q(V_{ds} = V_{dd}, V_{gs} = 0)$ where the supply voltage is assumed to be $V_{dd} = 0.5$ V. The doping of the source/drain extensions is considered to be 10^{20} cm^{-3}. A small section of 2.5 nm at the interface between source/drain extension and gate is considered to be undoped.

Figure 5.19 shows the total capacitances C_{tot} extracted from the simulations of the three devices as a function of the length l_{ext} of the source/drain extension. First of all, in the case of $L = 15$ nm, C_{tot} increases when $l_{ext} < 20$ nm due to increasing $C_{s,d}^{par}$ that eventually surpass C_g. However, reducing the distance between the nanowire MOSFETs (i. e., reducing d_{ox}, h_{gate} and d_{nw}) strongly decreases $C_{s,d}^{par}$ such that in the case of the device with $L = 10$ nm, C_{tot} is dominated by C_g until l_{ext} is less than 10 nm. For the shortest device with $L = 5$ nm, the parasitic capacitances have become irrelevant. This reconfirms our discussion above and shows that the nanowire bundle configuration indeed represents the ultimate scaling potential for nanowire transistors. It is obvious that reaching this limit requires ultrathin gate dielectrics that are prone to breakdown and leakage (see next section). Furthermore, extremely thin nanowire diameters are necessary. In this respect, our discussion about the NESSIAS (cf. 4.4) effect as an alternative for impurity doping is highly relevant and may allow the realization of appropriately thin GAA nanowire FETs.

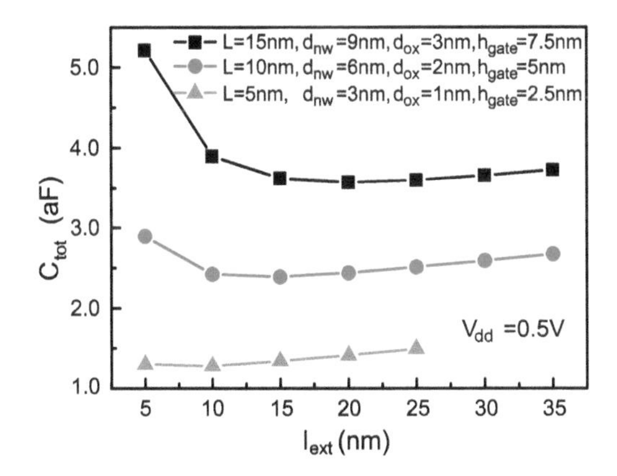

Figure 5.19: Total capacitance $C_{tot} = C_g + C_s^{par} + C_d^{par}$ of a gate-all-around nanowire MOSFET within a nanowire bundle configuration for the different geometries given in the figure. Refer to Figure 5.18 for the various device parameters.

5.9.3 Tunneling Through the Gate Dielectric

Ultimately scaled MOSFETs need very small effective oxide thicknesses to prevent off-state leakage due to SCEs. Eventually, however, tunneling through the gate dielectric becomes an issue leading again to off-state leakage if the physical thickness of the insulator is made too thin. While an estimate of the gate leakage can be obtained using the WKB approximation, in a MOSFET the particular band profile along the current transport direction needs to be taken into consideration (cf. Figure 5.1(b)). Due to the potential drop toward drain, this part of the channel contributes most to the gate leakage, which in turn depends on the applied bias and gate voltage. As a result, simulations are required in order to study the contribution of gate leakage currents (for details see Chapter 6).

Figure 5.20 shows transfer characteristics of an ultrathin-body SOI MOSFET obtained from self-consistent nonequilibrium Green's function formalism simulations at two different bias voltages. An extremely thin SiO_2 of $d_{ox} = 0.8$ nm was chosen here; thus, substantial gate leakage is expected. The gate leakage current (green dashed line for $V_{ds} = 0.2$ V and orange dashed line for $V_{ds} = 0.6$ V) changes its direction and, therefore, the diagram is split into two sections in order to show it on a logarithmic scale. The images on the left depict the device at three gate voltages. In the case of small V_{gs} (top panel), the drain current is dominated by the gate-drain leakage I_{gd}, i. e.,

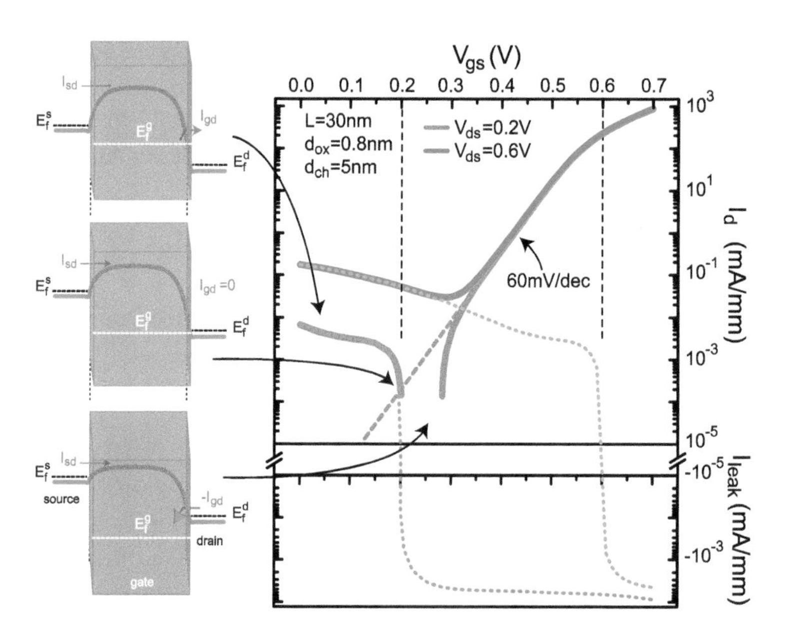

Figure 5.20: Simulated transfer characteristics of an ultrathin-body SOI MOSFET with strong gate leakage for $V_{ds} = 0.2$ V (blue curve) and $V_{ds} = 0.6$ V (red curve). The dashed lines show the gate leakage current, which becomes negative when $V_{gs} > V_{ds}$. In the case of $V_{ds} = 0.2$ V, this leads to a part in the transfer characteristic with seemingly $S < 60$ mV/dec.

$I_d = I_{sd} + I_{gd} \approx I_{gd}$, where I_{sd} refers to the source-to-drain current. When $V_{gs} = V_{ds}$ (center panel), then $I_d = I_{sd}$ since $I_{gd} = 0$. For larger V_{gs} (bottom panel), the gate-drain current becomes negative $I_{gd} \to -I_{gd}$ and as long as $|I_{ds}| < |I_{gd}|$ the drain current will be negative with $I_d = I_{sd} - I_{gd} < 0$. Eventually, when the transistor is switched into its on-state, the source-to-drain current dominates, and thus $I_d \approx I_{sd}$. If the regime $V_{gs} > V_{ds}$ falls into the off-state of the device (as is the case for $V_{ds} = 0.2\,\text{V}$ in Figure 5.20, blue curve), a peculiar situation occurs: around $V_{gs} \approx 0.3\,\text{V}$ the inverse subthreshold seemingly becomes steeper than 60 mV/dec. From the consideration of the gate leakage, it is clear that this happens simply because of tunneling through the gate dielectric. Moreover, it is clear from Section 5.2.2 that $S < 60\,\text{mV/dec}$ is impossible in a conventional MOSFET at room temperature. In Chapter 9, however, when steep slope transistors are discussed, gate leakage can be deceiving since if such devices are optimized a steep slope is expected. In this case, particular care has to be taken with respect to gate leakage and its impact in order to avoid misinterpretations of experimental data.

5.9.4 Direct Source-To-Drain Tunneling

In ultimately scaled MOSFETs, the channel length will become so small that direct source-to-drain tunneling can become an issue that eventually limits the scalability, and hence deserves further consideration. As a criterion for a properly working MOSFET, one could require that the leakage through the device because of direct source/drain tunneling at $V_{gs} = 0\,\text{V}$ is smaller than the current due to thermal emission of carriers over the potential barrier in the channel. Based on this criterion, a minimum possible channel length L_{min} for a particular transistor can be estimated in the following way. Since for ultimate scalability λ_{ch} needs to be very small, the exact potential distribution is neglected and instead we assume a step-function potential distribution as illustrated in Figure 5.21. Let us consider a one-dimensional nanowire FET such that the current

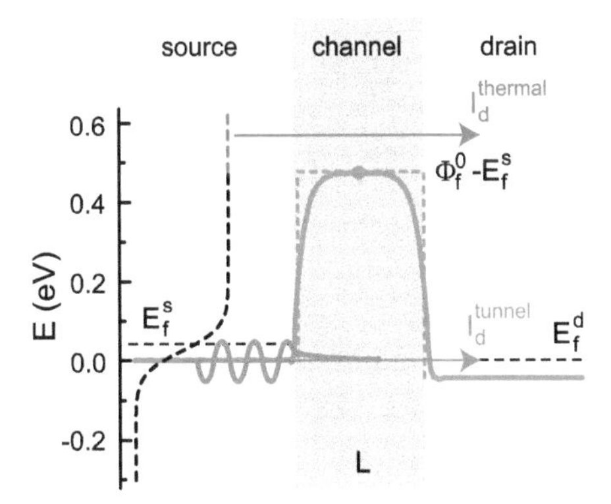

Figure 5.21: Direct source/drain tunneling may become important in ultrasmall channel MOSFETs. Proper functionality of the MOSFET is obtained as long as $I_d^{thermal} > I_d^{tunnel}$ at zero gate voltage.

can be computed with the Landauer expression I_d^{1D}. In the following, a small drain-source bias will be assumed in order to obtain analytic expressions for the two current components. The thermal emission current can be computed right away and although a small bias is considered the drain contribution to the current is negligible; this is justified because the source/drain Fermi distribution functions can be replaced with their Boltzmann approximations. Therefore,

$$I_d^{\text{thermal}} \approx \frac{2e}{h} \int_{\Phi_f^0}^{\infty} dE \exp\left(-\frac{E - E_f^s}{k_B T}\right) = \frac{2e}{h} k_B T \exp\left(-\frac{\Phi_f^0 - E_f^s}{k_B T}\right). \tag{5.46}$$

The tunneling current on the other hand is obtained using Equation (5.28) with a transmission probability $T(E)$ for the tunneling through the potential barrier within the energy range between the conduction band in the source (set to zero) and the potential maximum Φ_f^0. As a result, I_d^{tunnel} is approximately

$$I_d^{\text{tunnel}} = \frac{2e}{h} \int_0^{\Phi_f^0} dE T(E)(f_s(E_f^s) - f_d(E_f^d)) \approx \frac{2e^2}{h} \int_0^{\Phi_f^0} dE T(E)\left(-\frac{\partial f_s}{\partial E}\right) V_{\text{ds}} \tag{5.47}$$

since the difference $f_s(E_f^s) - f_d(E_f^d)$ can be Taylor expanded yielding $-\frac{\partial f_s}{\partial E}(-e)V_{\text{ds}}$. The lower bound of the integration was chosen because the band gap in the source prevents current flow below the conduction band (i. e., below $E = 0$ eV). The derivative of the Fermi distribution function can be approximated with a delta function and as a result, the integration over energy yields $T(E_f^s)$. The transmission probability, in turn, can be computed with the WKB approximation leading to the following approximate expression for the source/drain tunneling:

$$I_d^{\text{tunnel}} \approx \frac{2e}{h} V_{\text{ds}} \exp\left(-\frac{2L}{\hbar} \sqrt{m^\star (\Phi_0 - E_f^s)}\right). \tag{5.48}$$

As already mentioned above, for proper device functionality one may require that $I_d^{\text{thermal}} \geq I_d^{\text{tunnel}}$. Assuming $V_{\text{ds}} \approx k_B T$ allows a direct comparison of the terms in the exponential factors of Equations (5.46) and (5.48) and it follows that $\frac{\Phi_f^0 - E_f^s}{k_B T} \leq \frac{2L_{\text{min}}}{\hbar}\sqrt{m^\star(\Phi_f^0 - E_f^s)}$. Rewriting finally leads to an expression for a minimum possible channel length $L_{\text{min}} \geq \frac{\hbar}{2k_B T}\sqrt{\frac{\Phi_f^0 - E_f^s}{2m^\star}}$. For instance, if $\Phi_f^0 - E_f^s = 0.55$ eV as with intrinsic silicon, an effective mass of $0.023 m_0$ as in InAs leads to a minimum channel length of ~27 nm. Replacing silicon with high mobility III–V materials is therefore questionable since it may limit further downscaling. In this respect, the relatively heavy effective masses of 2D materials such as transition metal dichalcogenides are much more attractive and reasonable than they might appear.

Task 22.
Direct source/drain tunneling: Compute the transmission probability $T(E)$ in the case of a step-function like potential distribution with a constant potential barrier of Φ_0 and a thickness of L (as illustrated in Figure 5.21). To this end, use the WKB approximation.

Exercises

Exercises together with solutions are accessible via the QR code.

6 Device Simulation

In the present chapter, a simulation framework will be described that is used frequently in the book to study the electronic behavior of field-effect transistors. Since we are dealing with nanoscale devices, quantum effects play an important role in understanding the device properties. Therefore, self-consistent Poisson–Schrödinger simulations using the nonequilibrium Green's function formalism (NEGF) are introduced in order to compute the electronic transport properties of the devices. To simplify matters, the bulk part of the chapter will deal with one-dimensional devices such as carbon nanotubes and nanowires devices. The electrostatics of such transistors can be described using the one-dimensional modified Poisson equation (Equation (5.34)) derived in the preceding chapter. This facilitates conceptually simple and easy to implement simulations that are computationally efficient and yield reasonable results in a manageable time even on laptop computers. In addition, several add-ons to the simulation framework such as scattering via Buettiker probes and multiple independent subbands will be discussed. Eventually, an extension to simulating two- and three-dimensional devices will be briefly introduced that may serve as starting point for more elaborate simulations.

6.1 Poisson's Equation in 1D Devices

In the present section, the discretization of the one-dimensional Poisson equation derived in Section 5.5 will be discussed. Using conformal mapping, the applicability of the modified Poisson equation can be extended to incorporate fringing fields in the source/drain extensions.

6.1.1 Finite Difference Discretization

In Section 5.5, a one-dimensional modified Poisson equation (cf. Equation (5.34)) was derived that allows describing the electrostatics of ultrathin-body devices such as SOI-, nanotube- or nanowire transistors. In this one-dimensional modified Poisson equation, the third term that gives rise to a screening of potential variation on the length scale λ_{ch}, takes the specific device geometry into consideration. For single-gate, double-gate and gate-all-around field-effect transistors, the appropriate screening lengths are explicitly stated in Figure 5.12. Therefore, the specific device architecture considered is irrelevant for the discretization of the Poisson equation.

Discretization can be carried out in a straightforward way by considering a finite difference lattice with lattice constant a as shown in the top panel of Figure 6.1. In a first step, we need to approximate the second derivative of $\Phi_f(x)$ with respect to x with appropriate difference quotients. To this end, we can either choose the forward or the backward finite difference, i. e., $\frac{d\Phi_f(x)}{dx} \approx \frac{\Phi_f(x_{j+1})-\Phi_f(x_j)}{a} = \frac{\Phi_f^{j+1}-\Phi_f^{j}}{a}$ or $= \frac{\Phi_f^{j}-\Phi_f^{j-1}}{a}$, which can

https://doi.org/10.1515/9783111054421-006

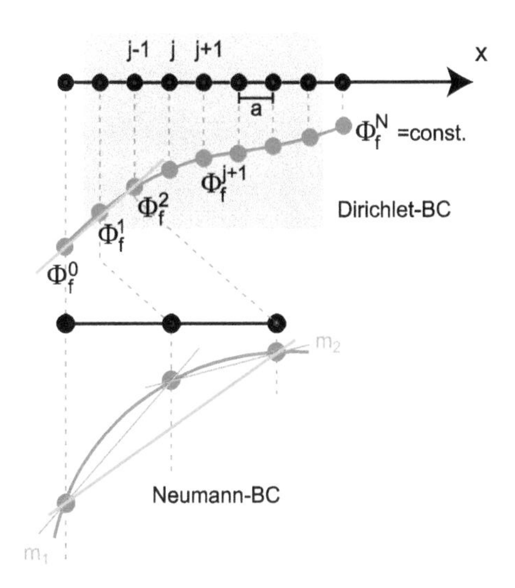

Figure 6.1: Discretization of $\Phi_f(x)$ on a regular finite difference grid with lattice constant a. The lower panel illustrates how an appropriate expression for the electric field is obtained.

give vastly different values if the curvature of $\Phi_f(x)$ is large; an example is shown in the lower panel of Figure 6.1 where the two slopes m_1 and m_2 are quite different. However, in the case of the second derivative we obtain a symmetric expression by taking the derivative of the forward difference quotient and then choosing the backward difference for the first term and the forward difference for the second:

$$\frac{d}{dx}\left(\frac{\Phi_f^{j+1}-\Phi_f^j}{a}\right)=\frac{\frac{\Phi_f^{j+1}-\Phi_f^j}{a}-\frac{\Phi_f^j-\Phi_f^{j-1}}{a}}{a}=\frac{\Phi_f^{j+1}-2\Phi_f^j+\Phi_f^{j-1}}{a^2}. \tag{6.1}$$

Discretizing the terms proportional to $1/\lambda_{\mathrm{ch}}^2$ and the charge in Equation (5.34) is straightforward yielding $\Phi_f(x)\to\Phi_f^j$ and $n(x)\to n_j$. As a result, the following discrete form of the modified Poisson equation is obtained:

$$\frac{\Phi_f^{j+1}-2\Phi_f^j+\Phi_f^{j-1}}{a^2}-\frac{\Phi_f^j-\Phi_{\mathrm{bi}}-\Phi_g}{\lambda_{\mathrm{ch}}^2}=-\frac{e^2 n_j}{\varepsilon_0\varepsilon_{\mathrm{ch}}} \tag{6.2}$$

where Φ_{bi} and Φ_g are again the built-in and the gate potential, respectively.

Equation (6.2) can be written in matrix form, which results in a tridiagonal matrix M applied to the vector that contains the values of the potential energy Φ_f^j at the different sites j of the finite difference lattice. Such a tridiagonal matrix can be solved efficiently with, e. g., the Thomas algorithm. To set up the full modified Poisson equation, there are two ingredients missing: first, since we would like to simulate field-effect transistors, the source and drain contacts need to be taken into consideration. Second, appropriate boundary conditions are required for a unique solution.

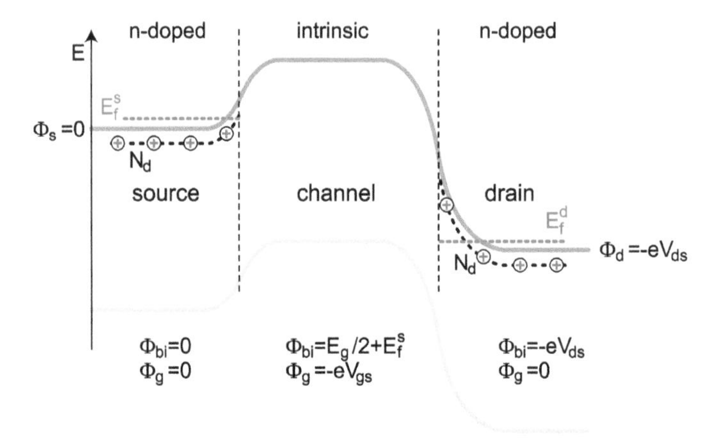

Figure 6.2: Conduction and valence band along the direction of current transport in an n-type MOSFET. The channel is considered to be intrinsic; source and drain are doped with a donor concentration of N_d.

Figure 6.2 displays the conduction band (and valence band in light green) profile along the direction of current transport in an n-type FET. For simplicity, we assume that in source and drain all donors are fully ionized (a more elaborate consideration needs to take the occupation of the donor levels into consideration as was done in Section 2.10) constituting a constant and fixed positive charge. The source contact is considered to be connected to ground such that the potential Φ_s in source sufficiently far away from the source-channel interface is put to $\Phi_s = 0$ eV. Since a source-drain bias V_{ds} is applied, the potential in drain Φ_d is considered to be at an energy of $\Phi_d = -eV_{\text{ds}}$. Note that in a real experiment it is *not* the (conduction) band that is moved by applying a voltage difference V_{ds} between the source and drain. Instead, the Fermi levels in the source and the drain are separated by the terminal voltages, and as a result of this, charges will rearrange in order to ensure charge neutrality deep within the source and drain contacts; hence, the bands move to appropriate energies. This will be discussed a bit further in connection with appropriate boundary conditions when dealing with self-consistency.

In Section 6.2.5, the Newton–Raphson scheme will be introduced in order to find a self-consistent solution. The Newton–Raphson scheme, however, is not globally convergent. In order to help obtaining a self-consistent solution, the modified Poisson equation can be set-up in a way, which yields a potential distribution that is already close to the real solution without knowledge of the charge distribution within the device. To this end, Equation (6.2) will be used throughout the entire device and we therefore have to distinguish between source, channel and drain in the following way: To ensure $\Phi_s = 0$ at the boundary, the built-in and gate potentials in source need to be zero. In the case of Φ_g^{source}, this is obvious since there is no gate electrode in source (apart from fringing fields that are neglected here but will be discussed in Section 6.1.2); $\Phi_{\text{bi}}^{\text{source}} = 0$ follows from the fact that deep inside the source contact the curvature term (second derivative) is zero and since the density of negative electronic charge equals the density of fixed ionized donors N_d, charge neutrality applies yielding $\Phi_s = 0$. The potential in drain, on the other hand, will be at $\Phi_d = -eV_{\text{ds}}$, because deep inside the drain contact charge neutrality applies as well, we should obtain $\Phi_d = -eV_{\text{ds}}$. This is indeed ensured when we implement a built-in potential $\Phi_{\text{bi}}^{\text{drain}} = -eV_{\text{ds}}$ in the drain contact. As in source, $\Phi_g^{\text{drain}} = 0$ since there is no gate electrode in drain. It is important to keep in mind that we actually botched here: assuming a built-in potential within the contacts is equivalent to placing a gate electrode there that keeps the potential at the potential given by Φ_{bi}. In the source and the drain, there is usually no additional gate electrode, though. However, if highly doped source/drain contacts are assumed (which is usually the case) the screening due to charges within the contacts will in most cases be significantly more effective than the screening provided by the (nonexisting) gate electrode (see discussion in Section 5.7 on combining screening lengths). Implementing the simulations as described, allows reaching self-consistency quicker and in a more stable fashion. Finally, it is assumed that the channel is intrinsic meaning that the source-channel *n*-*i* junction gives rise to a built-in potential. In this case, Φ_{bi} will be approximately at mid-gap (cf. Equation (4.3)) $\Phi_{\text{bi}} \approx \frac{E_g}{2} + E_f^s$, where E_g and E_f^s are the band gap and the Fermi energy in source, respectively.[1] Of course, this choice requires a gate metal with appropriate work function, which is implicitly included in Φ_{bi}.

Appropriate boundary conditions (BC) are required in order to solve the Poisson equation. In principle, we could fix the first point of the finite difference lattice to the source potential Φ_s and the last point to Φ_d. Such fixed boundary conditions are called Dirichlet BC. The implementation of the Dirichlet BC is rather simple (illustrated in Figure 6.1 at the last lattice point at $j = N$) and we would set $\Phi_f^{j=0} = 0$ and $\Phi_f^{j=N} = \Phi_d = -eV_{\text{ds}}$ leaving us with $N - 2$ equations to be solved. However, as has already been mentioned above, physically the applied terminal voltages do not determine the position of the bands. The bands move to their position to ensure charge neutrality at the source and drain boundaries (to this end, a self-consistent solution is necessary). Because the effective charge (negative free carrier density $-e \cdot n$ minus positive ionized donors $e \cdot N_d$) at the source/drain boundaries must be zero, the electric field at the boundaries is zero, too. Therefore, we need to consider how an electric field $\mathcal{E}$ can be used as a boundary condition.

BC that impose an electric field are called Neumann BC and can be implemented in the following way: The electric field is simply the first derivative of the potential with respect to the spatial coordinate(s). However, we already discussed above that we could choose between forward and backward difference, which can be substantially different.

1 Note the difference between Fermi level and Fermi energy: While the Fermi level is the energetic position where the Fermi level lies with respect to a chosen energy scale, the Fermi energy in heavily *n*-doped (*p*-doped) contacts is the position of the Fermi level above (below) the conduction(valence) band. For instance, while the Fermi level in the drain is at $E_f^d = E_f^s - eV_{\text{ds}}$, the Fermi energy in the source and drain are the same due to the (symmetric) doping.

In order to obtain a symmetric second-order expression, consider again Figure 6.1. To incorporate the Neumann BC into our calculation, a virtual point is added to the finite difference lattice at $j = 0$. At the left boundary, the forward and backward finite difference are shown that lead to slopes m_1 and m_2 centered around the point $j = 1$. The best approximation to the real electric field is thus the average of the two slopes, i. e., $\frac{m_1+m_2}{2}$. Inserting the finite differences yields $\frac{1}{2}(\frac{\Phi_f^0-\Phi_f^1}{a} + \frac{\Phi_f^1-\Phi_f^2}{a})$, and hence the electric field at position $j = 1$ is $\mathcal{E}_1 = -\frac{1}{e}\frac{\Phi_f^0-\Phi_f^2}{2a}$. The expression for $\mathcal{E}_1$ can then be solved for the unknown $\Phi_f^0 = (-e)2a\mathcal{E}_1 + \Phi_f^2$ such that the discrete Poisson equation centered at $j = 1$ becomes

$$\frac{(-e)2a\mathcal{E}_1 - 2\Phi_f^1 + 2\Phi_f^2}{a^2} + \frac{\Phi_f^1}{\lambda_{\text{ch}}^2} = -\frac{e^2(n_1 - N_d)}{\varepsilon_0\varepsilon_{\text{ch}}}. \tag{6.3}$$

The first term on the left can now be moved onto the right side of the equation and incorporated into the charge term; a similar expression can be set-up for drain.

It was mentioned above that the set of N equations can be cast in the form of a matrix equation according to $\mathrm{M}\vec{\Phi} = -\frac{e^2}{\varepsilon_0\varepsilon_{\text{ch}}}\vec{n}$ that can be solved. In order to write down this matrix equation explicitly, a finite difference lattice consisting of two points in source, two points in the channel and two points in drain is considered. If we further assume that Neumann BC apply in source and drain with $\mathcal{E}_{s,d} = 0$ due to charge neutrality, the resulting discretized Poisson equation is explicitly given by

$$\begin{pmatrix} -\frac{2}{a^2}+\frac{1}{\lambda_{\text{ch}}^2} & \frac{2}{a^2} & 0 & 0 & 0 & 0 \\ \frac{1}{a^2} & -\frac{2}{a^2}+\frac{1}{\lambda_{\text{ch}}^2} & \frac{1}{a^2} & 0 & 0 & 0 \\ 0 & \frac{1}{a^2} & -\frac{2}{a^2}+\frac{1}{\lambda_{\text{ch}}^2} & \frac{1}{a^2} & 0 & 0 \\ 0 & 0 & \frac{1}{a^2} & -\frac{2}{a^2}+\frac{1}{\lambda_{\text{ch}}^2} & \frac{1}{a^2} & 0 \\ 0 & 0 & 0 & \frac{1}{a^2} & -\frac{2}{a^2}+\frac{1}{\lambda_{\text{ch}}^2} & \frac{1}{a^2} \\ 0 & 0 & 0 & 0 & \frac{2}{a^2} & -\frac{2}{a^2}+\frac{1}{\lambda_{\text{ch}}^2} \end{pmatrix} \cdot \begin{pmatrix} \Phi_1 \\ \Phi_2 \\ \Phi_3 \\ \Phi_4 \\ \Phi_5 \\ \Phi_6 \end{pmatrix} = \begin{pmatrix} -\frac{e^2(n_1-N_d)}{\varepsilon_0\varepsilon_{\text{ch}}} \\ -\frac{e^2(n_2-N_d)}{\varepsilon_0\varepsilon_{\text{ch}}} \\ -\frac{e^2(n_3)}{\varepsilon_0\varepsilon_{\text{ch}}} + \frac{\Phi_g+\Phi_{\text{bi}}}{\lambda_{\text{ch}}^2} \\ -\frac{e^2(n_4)}{\varepsilon_0\varepsilon_{\text{ch}}} + \frac{\Phi_g+\Phi_{\text{bi}}}{\lambda_{\text{ch}}^2} \\ -\frac{e^2(n_5-N_d)}{\varepsilon_0\varepsilon_{\text{ch}}} + \frac{-eV_{\text{ds}}}{\lambda_{\text{ch}}^2} \\ -\frac{e^2(n_6-N_d)}{\varepsilon_0\varepsilon_{\text{ch}}} + \frac{-eV_{\text{ds}}}{\lambda_{\text{ch}}^2} \end{pmatrix} \tag{6.4}$$

where the gate potential Φ_g needs to be considered only at the points $j = 3, 4$ belonging to the channel; $\Phi_d = -eV_{ds}$ only appears on the drain side (points $j = 5, 6$). The Neumann BC with zero electric field are accounted for by simply changing the (1, 2)- and $(N, N-1)$-elements of the matrix M in Equation (6.4) from $1/a^2$ to $2/a^2$.

The matrix equation (6.4) can be solved for the potential profile $\Phi_f(x_j)$ provided that the carrier density $n(x_j)$ is known. Since $n(x_j)$ needs to be computed using the Schrödinger equation that itself depends on the potential profile $\Phi_f(x_j)$, a self-consistent solution must be found. However, before the quantum mechanical calculation of the charge density and self-consistency will be discussed, the one-dimensional Poisson equation will be further elaborated on.

6.1.2 Conformal Mapping

In the source and drain regions, there will be fringing fields from the gate electrode that may have an impact on the potential distribution within the contacts (illustrated in Figure 6.3, top left). These fringing fields in the source/drain regions can be taken into consideration employing the technique of conformal mapping. As a result, a spatially dependent $d_{ox}(x)$ is obtained such that the gate impact on source and drain can be incorporated properly using Equation (5.34). This significantly speeds up simulations since one does not have to solve a 2D/3D Poisson equation.

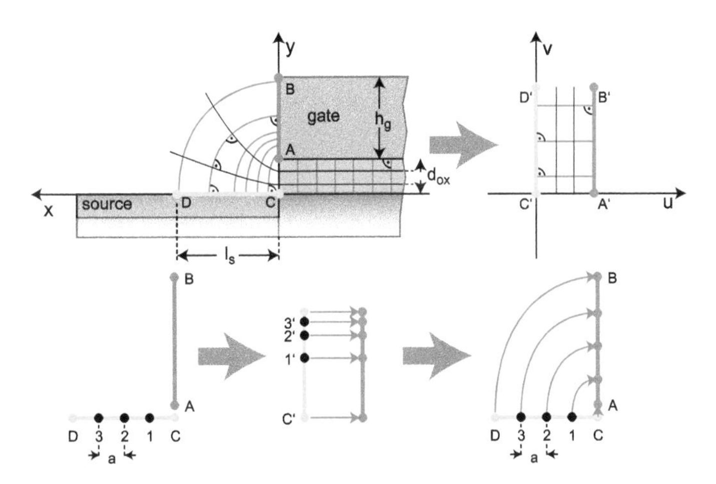

Figure 6.3: Conformal mapping allows transforming the area in the source(drain)-gate region onto a parallel plate capacitor. Mapping the horizontal field lines back yields a spatially dependent effective oxide thickness. With this, the electrostatics can be described in the entire device based on the 1D modified Poisson equation including the regions with fringing fields.

A conformal map $w = f(z)$ (with $w = u + iv$ and $z = x + iy$) maps the x, y-plane onto the u, v-plane preserving orientation and angles locally. As such, the area with the fringing fields can be mapped onto a simpler geometry that allows immediate evaluation of the value of, e. g., the capacitance since the right angles between electric field lines and equipotential lines are preserved (cf. top panel of Figure 6.3). Hence, the idea here is to map the capacitor associated with the fringing fields (in between the light blue and blue thick lines) conformally onto a parallel plate capacitor as depicted in Figure 6.3. The fringing field lines in the x, y-plane are mapped onto horizontal field lines in the u, v-coordinate system. Back-transforming the field lines allows the extraction of an effective oxide thickness given by the length of the respective field line. This effective oxide thickness will then appear in the screening length λ_{ch}, i. e., a spatially dependent $\lambda_{ch}(x)$ is obtained that enables the proper use of the modified 1D Poisson equation throughout the entire device structure.

An appropriate map $w = f(z)$ from x, y-coordinates to u, v-coordinates for the present fringing field problem is provided by the transformation $ny + ix = K\sin(u + iv)$ [171]. Writing the real and imaginary parts separately this yields

$$y = \frac{K}{n}\sin(u)\cosh(v) \quad x = K\sinh(v)\cos(u) \tag{6.5}$$

where the constants n and K are given by (see Figure 6.3 for the geometrical parameters) $n = \frac{l_s}{d_{ox}\sinh(\cosh^{-1}(\frac{d_{ox}+h_g}{d_{ox}}))}$ and $K = \frac{l_s}{\sinh(\cosh^{-1}(\frac{d_{ox}+h_g}{d_{ox}}))}$.

Let us now use the map $f(z)$ and transform the points A, B, C and D. Inserting each point into the transformation Equation (6.5) yields $A = (0, d_{ox}) \to A' = (\pi/2, 0)$, $B = (0, d_{ox} + h_g) \to B' = (\pi/2, \cosh^{-1}(\frac{d_{ox}+h_g}{d_{ox}}))$, $C = (0,0) \to C' = (0,0)$ and $D = (l_s, 0) \to D' = (0, \cosh^{-1}(\frac{d_{ox}+h_g}{d_{ox}}))$ as shown in the figure. In order to obtain a spatially dependent oxide thickness, and thus $\lambda_{ch}(x)$, one needs to compute the line integral from $= 0 \to \frac{\pi}{2}$ for a constant v, which provides the desired $d_{ox}(v(x))$,

$$d_{ox}(v) = \int_0^{\pi/2} du \left|\frac{d\vec{r}}{du}\right| = \int_0^{\pi/2} du \sqrt{\left(\frac{dx(u,v)}{du}\right)^2 + \left(\frac{dy(u,v)}{du}\right)^2}. \tag{6.6}$$

With Equation (6.5) and rewriting the result this yields

$$d_{ox}(v) = \frac{K}{n}\cosh(v) \int_0^{\pi/2} du \sqrt{1 - \sin^2(u)(n^2\tanh^2(v) + 1)}. \tag{6.7}$$

Since the discrete points on the finite difference grid along the x-direction (points denoted 1, 2 and 3 in Figure 6.3) are mapped onto the v-axis where $u = 0$, the coordinate v in the expression above can be written (using Equation (6.5)) as $v(x) = \operatorname{arsinh}(x/K)$. Inserting this into Equation (6.6) and noting that $\frac{K}{n} = d_{ox}$ yields the desired $d_{ox}(x)$:

$$d_{ox}(x) = d_{ox}\sqrt{1 - \left(\frac{x}{K}\right)^2} \int_0^{\pi/2} du \sqrt{1 - \sin^2(u)\left(\frac{x^2/d_{ox}^2}{1 - x^2/d_{ox}^2} + 1\right)} \tag{6.8}$$

In the discretized case required for our simulations, lattice points on the finite difference grid are mapped onto their counterparts in the u, v-plane and then are back-transformed in order to determine the length of the effective gate oxide at a particular site of the finite difference lattice. To give a specific example, in the case depicted in the figure the source extension of length l_s is subdivided into four equidistantly spaced parts with lattice constant $a = l_s/4$, and thus the points 1, 2 and 3 are mapped to the $1'$, $2'$ and $3'$ as shown in the lower panel of Figure 6.3. Having found the mapped v-coordinates of $1'$, $2'$ and $3'$, we can compute $d_{ox}(x_j) = d_{ox}(aj)$ shown in the bottom right part of Figure 6.3. Inserting this into the screening length of, e. g., a single-gate UTB SOI-MOSFET, we finally arrive at

$$\lambda_{ch}^{j} = \sqrt{\frac{\varepsilon_{si}}{1} d_{SOI} d_{ox}(aj)} \tag{6.9}$$

where $j = 0, \ldots, 4$ with $j = 0$ being point C and $j = 4$ being point D; note that ε_{ox} is set to 1 (air). As a result, the one-dimensional modified Poisson equation can be used throughout the entire device to describe the electrostatics within the transistor including the impact of fringing fields.

In some device concepts (such as the one discussed in Section 9.1.3.4), the source/drain contacts are as illustrated in Figure 6.4, left panel, where an extension of length l_s separates source and gate electrodes from each other (for instance, in the nanowire bundle configuration shown in Figure 5.18). In this case, two subsequent mappings from $x, y \to u, v$ and $u, v \to s, t$ coordinates have to be carried out [276]. Appropriate maps are given by

$$nu + iv = k \sin(x + iy) \quad n = \frac{CB}{CD \sinh(\cosh^{-1}(\frac{CE}{CD}))} \quad k = nCD \tag{6.10}$$

with $CB = d_{ox} + h_g$, $CD = l_s$, $CE = l_s + d_{ox} + h_g$ and

$$n'v + iu = k' \sin(s + it) \quad n' = \frac{C'E'}{A'C' \sinh(\cosh^{-1}(\frac{C'B'}{A'C'}))} \quad k' = n'A'C'. \tag{6.11}$$

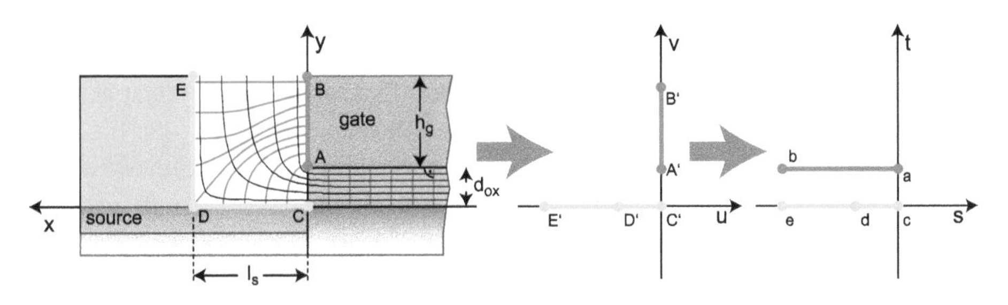

Figure 6.4: Two subsequent conformal mappings transform the source extension region illustrated in the left image onto a parallel plate capacitor [276].

6.2 Nonequilibrium Green's Function Formalism in Single-Band 1D-FETs

Having taken care of the Poisson equation and its discretization, we need to compute in a next step the carrier density and the current through the transistor. In this respect, the following questions need to be addressed. First, a transistor is basically an open quantum system since it is connected to a source and a drain contact. Numerical calculations, however, need to be carried out on a finite computational domain. Therefore, we need to know how to incorporate contacts in an appropriate way. Second, supposed we were able to solve the Schrödinger equation of the open quantum system then we would be able to calculate the local density of states (cf. Section 2.11.1), and hence the carrier density $n(x)$. However, we would only be able to do this in equilibrium. But how should the nonequilibrium situation when current flows be taken into consideration? How can the density of states be split into a part stemming from source and one part from drain? Answers to these questions can be obtained using the nonequilibrium Green's function formalism (NEGF) that has been used for most simulations in this book. In the following sections, an introduction to NEGF will be given that follows closely the approach by Datta [60]. A pragmatic approach is chosen that is certainly not rigorous but provides sufficient background to understand the derivation, and thus allows for working with NEGF.

6.2.1 The Green's Function

Consider a linear operator $\mathcal{L}$ that operates on an unknown function $\Phi(x)$ resulting in a given source $f(x)$ according to $\mathcal{L}\Phi(x) = f(x)$ (for simplicity, the current derivation will be carried out in 1D). An operator is linear if $\mathcal{L}(a\phi + b\varphi) = a\mathcal{L}\phi + b\mathcal{L}\varphi$ with ϕ and φ denoting two vectors/functions that are mapped onto another vector/function using the linear operator and a, b are complex numbers. For any linear operator, a Green's function $G(x, x')$ can be found that is the solution of $\mathcal{L}G(x, x') = \delta(x - x')$. If we write $f(x) = \int dx' f(x')\delta(x-x')$ and using the defining equation for $G(x, x')$, one obtains $f(x) = \int dx' f(x')(\mathcal{L}G(x, x'))$. Since $\mathcal{L}$ is a linear operator, $f(x) = \mathcal{L}\int dx' G(x, x')f(x') = \mathcal{L}\Phi(x)$. Comparison of the two sides of this equation yields $\Phi(x) = \int dx' G(x, x')f(x')$ (plus a boundary term). This means that once $G(x, x')$ is known, a solution can be computed for any source function $f(x)$ by integrating $G(x, x')f(x')$ over x'.

The Green's function depends on two spatial coordinates x and x' and on two different times t and t'. However, here only stationary systems will be considered, such that the Green's function merely depends on the time difference $t-t'$. Fourier transformation then yields $G(x, x', E)$, i. e., an energy-dependent Green's function. We are interested in the Green's function associated with the Hamiltonian $\mathcal{H}$. Therefore, the linear operator is $\mathcal{L} = E - \mathcal{H}$ and the defining equation for the Green's function is

$$(E - \mathcal{H})G(x, x', E) = \delta(x - x'). \tag{6.12}$$

For the following derivations, a single band, effective mass approximation is assumed. With this, it can be shown that the Green's function of the operator $E - \mathcal{H}$ is given by

$$G(x, x', E) = \sum_k \frac{\phi_k(x)\phi_k^\star(x')}{E - \epsilon_k} \tag{6.13}$$

where $\phi_k(x)$ are the eigenfunctions and ϵ_k the eigenenergies of the system under consideration. Indeed, inserting expression (6.13) into the defining equation for the Green's function results in

$$\begin{aligned}(E - \mathcal{H})\sum_k \frac{\phi_k(x)\phi_k^\star(x')}{E - \epsilon_k} &= \sum_k \frac{(E - \mathcal{H})\phi_k(x)\phi_k^\star(x')}{E - \epsilon_k} = \sum_k \frac{(E - \epsilon_k)\phi_k(x)\phi_k^\star(x')}{E - \epsilon_k} \\ &= \sum_k \phi_k(x)\phi_k^\star(x') = \delta(x - x')\end{aligned} \tag{6.14}$$

where we used the linearity of the operator in the second step. The final result is the so-called completeness relation, which is $\sum_k \phi_k(x)\phi_k^\star(x') = \delta(x - x')$, and thus Equation (6.13) is indeed a solution for the Green's function. However, what do we gain from this expression since for Equation (6.13) we need to know the eigenfunctions and eigenenergies, i. e., we need the full solution of the Schrödinger equation. This question will be answered further below. Let us first further elaborate on what we have found so far. To this end, we assume that the 1D system under consideration is at a constant potential (set equal to zero for simplicity) stretching from $-\infty$ to ∞. In this case, the eigenfunctions are known to be plane waves of the form $\frac{1}{\sqrt{L}}e^{ikx}$ and the eigenenergy is simply $\epsilon_k = \frac{\hbar^2 k^2}{2m^\star}$. The sum over all k can be transformed into an integral using the trick detailed in the info-box in Section 2.11.1, which yields

$$G(x, x', E) = \sum_k \frac{\Delta k}{2\pi/L} \frac{\frac{1}{L}e^{ikx}e^{-ikx'}}{E - \epsilon_k} \overset{\Delta k \to dk}{\longrightarrow} \int_{-\infty}^{\infty} dk \frac{L}{2\pi} \frac{\frac{1}{L}e^{ik(x-x')}}{E - \frac{\hbar^2 k^2}{2m^\star}}. \tag{6.15}$$

Obviously, the integration cannot be carried out because the denominator diverges whenever the energy E is equal to the eigenenergy ϵ_k. To avoid this issue, and thus to find a unique solution for G the denominator is written as $E - \frac{\hbar^2 k^2}{2m^\star} = -\frac{\hbar^2}{2m^\star}(k^2 - K^2)$ where K is defined as $K = \sqrt{\frac{2m^\star E}{\hbar^2}}$. This allows writing the denominator as $-\frac{\hbar^2}{2m^\star}(k + K)(k - K)$, and thus the integrand diverges at $k = \pm K$. Next, a small imaginary part $i\delta$ is added to or subtracted from K to move the divergence away from the real k-axis. Doing so, the residue theorem can be used to compute the integral. To this end, the integral over the real-values k is extended into a contour integration over a closed path in the complex κ-plane, i. e., $\to \oint d\kappa$. Obviously, the closed contour has to include the path along the real k-axis, and thus, in order to decide how to close the contour, we have to distinguish the following cases:

(1) $\delta > 0$ and $x > x'$: This means that the denominator diverges at $k = K + i\delta$ and at $k = -(K + i\delta)$. Since $x > x'$, the exponential factor in the numerator of the integrand $e^{i\kappa(x-x')}$ yields an exponentially damped term for complex κ-values in the upper half of the κ-plane as illustrated in Figure 6.5(a). Therefore, a contour as displayed in (a) is chosen and if the contour is extended to range from $k = -\infty ... + \infty$ the contribution of the upper half circle can be neglected. As a result, $\oint d\kappa \to \int_{-\infty}^{\infty} dk$. With the residue theorem,

$$\oint dz f(z) = 2\pi i \sum_a \mathrm{Res}_a f(z), \quad \mathrm{Res}_a f(z) = \lim_{z \to a} (z-a) f(z), \tag{6.16}$$

one obtains

$$\begin{aligned} G(x,x',E) &= -\frac{m^*}{\pi\hbar^2} \oint dk \frac{e^{ik(x-x')}}{(k-(K+i\delta))(k+(K+i\delta))} \\ &= -2\pi i \frac{m^*}{\pi\hbar^2} \lim_{k \to K+i\delta} (k-(K+i\delta)) \frac{e^{ik(x-x')}}{(k-(K+i\delta))(k+(K+i\delta))} \end{aligned} \tag{6.17}$$

since there is only a single residue within the contour. With $\delta \to 0$, one finally obtains

$$G(x,x',E) = -\frac{i}{\hbar v} e^{iK(x-x')} \quad \text{where } v = \frac{\hbar K}{m^*}. \tag{6.18}$$

(2) $\delta > 0$ and $x < x'$: A very similar derivation as in case (1) can be carried out. The only difference is that the contour is closed in the lower half of the complex plane (see Figure 6.5 (b)), since $x < x'$. One finally ends up with $G(x,x',E) = -\frac{i}{\hbar v} e^{iK|x-x'|}$.

(3) $\delta < 0$ and $x > x'$ or $x < x'$: Subtracting a small imaginary part from K and distinguishing again between $x > x'$ and $x < x'$ yields the solution $G(x,x',E) = \frac{i}{\hbar v} e^{-iK|x-x'|}$ using basically the same derivation as under (1) and (2).

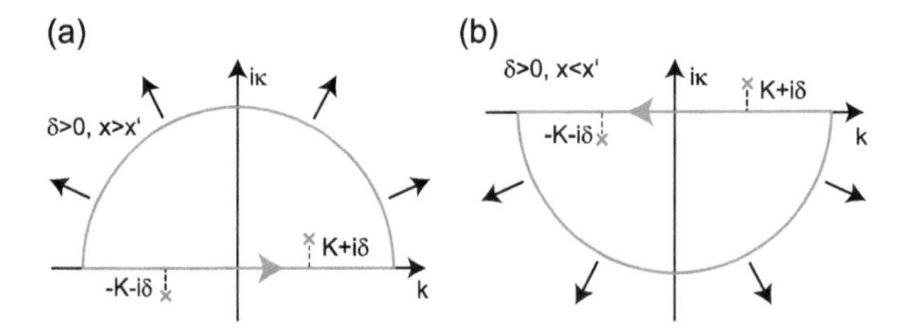

Figure 6.5: Contour integration for the two cases $\delta > 0, x > x'$ and $\delta > 0, x < x'$. The sign of $(x - x')$ determines whether the contour is closed in the upper or lower complex half-plane.

As a result, we obtain two solutions of the Green's function, the first is an outgoing wave, the second an incoming wave. If the delta function is interpreted as an impact excitation at $x = x'$, then the first solution of the Green's function would be the causal one since it represents the outgoing wave due to the excitation and is therefore called a "retarded" Green's function G^r. The incoming wave is called an advanced Green's function G^a. From the equations above, it is clear that the two solutions are not independent but can be converted into each other by exchanging the arguments x and x' and computing the conjugate complex: $G^r(x, x', E) = (G^a(x', x, E))^\star$.

Let us now come back to answer the question what all this is good for. The small imaginary part $+i\delta$ can be accounted for by a small imaginary part $+i\eta$ added to the energy such that the equation for the retarded Green's function becomes $(E + i\eta - \mathcal{H})G^r(x, x', E) = \delta(x - x')$ with the solution

$$G^r(x, x', E) = \sum_k \frac{\phi_k(x)\phi_k^\star(x')}{E - \epsilon_k + i\eta}. \tag{6.19}$$

Consider the imaginary part of $G^r(x, x, E)$ (i. e., for $x = x'$), which can be computed easily to be

$$\mathrm{Im}(G^r(x, x, E)) = -\sum_k |\phi_k(x)|^2 \underbrace{\frac{\eta}{(E - \epsilon_k + i\eta)(E - \epsilon_k - i\eta)}}_{\frac{\eta}{(E-\epsilon_k)^2+\eta^2}}. \tag{6.20}$$

Next, $\eta \to 0$, which leads to $\frac{\eta}{(E-\epsilon_k)^2+\eta^2} \xrightarrow{\eta\to 0} \pi\delta(E - \epsilon_k)$. As a result, Equation (6.20) multiplied with $-\frac{1}{\pi}$ yields the local density of states (cf. Section 2.11.1). That is a very important result since we know now how to compute the spatially-dependent carrier density (in equilibrium, though):

$$n(x) = \int dE \left(-\frac{1}{\pi}\,\mathrm{Im}\,G^r(x, x, E)\right) f(E - E_f). \tag{6.21}$$

Before taking care of contacts and addressing the question of how to simulate an open quantum system, let us first discretize the equation for G^r.

6.2.2 Discretization of the Green's Function

The Green's function is discretized on the same finite difference grid with lattice constant a as the Poisson equation (cf. Figure 6.1). To this end, we need a discrete form of the operator $E+i\eta-\mathcal{H}$, of $G^r(x, x', E)$ and of the delta function $\delta(x-x')$ (cf. Equation (6.12)), which leads to a matrix equation. In order to state these matrices explicitly, a particle-in-the-box system with infinitely high potential barriers at its edges and consisting of four finite difference lattice points is considered in the following. Let us begin with $E + i\eta$, which is straightforward to write in matrix notation as $(E + i\eta)\mathbf{1}$, i. e., as multiplication with the unit matrix $\mathbf{1}$. Next, the discretized form of $\mathcal{H}$ can also be written down immediately because it was already discussed in Section 2.4.1 and we arrived at Equation (2.14) for an arbitrary lattice point j. In the case of the PIB system with four lattice points, we therefore obtain

$$E + i\eta - \mathcal{H} \to (E + i\eta)\begin{pmatrix} 1 & 0 & 0 & 0 \\ 0 & 1 & 0 & 0 \\ 0 & 0 & 1 & 0 \\ 0 & 0 & 0 & 1 \end{pmatrix} - \begin{pmatrix} 2t + \Phi_f^1 & -t & 0 & 0 \\ -t & 2t + \Phi_f^2 & -t & 0 \\ 0 & -t & 2t + \Phi_f^3 & -t \\ 0 & 0 & -t & 2t + \Phi_f^4 \end{pmatrix} \tag{6.22}$$

where again $t = \frac{\hbar^2}{2m^* a^2}$ and Φ_f^j is the potential at lattice site $j = 1, \ldots, 4$. Note that due to the infinitely high potentials at the edges of the PIB system the wavefunctions at virtual points $j = 0$ and $j = 5$ vanish completely and Equation (6.22) is obtained. Moreover, the discrete form of $G^r(x, x, E)$ is also easy to write down, since we just need to note that $x \to x_j$ and $x' \to x_l$ and we obtain $G^r(j, l, E) = G^r_{j,l}(E)$ as

$$G^r(j, l, E) = \begin{pmatrix} G^r_{1,1} & G^r_{1,2} & G^r_{1,3} & G^r_{1,4} \\ G^r_{2,1} & G^r_{2,2} & G^r_{2,3} & G^r_{2,4} \\ G^r_{3,1} & G^r_{3,2} & G^r_{3,3} & G^r_{3,4} \\ G^r_{4,1} & G^r_{4,2} & G^r_{4,3} & G^r_{4,4} \end{pmatrix}. \tag{6.23}$$

Finally, the delta function can be cast into a discrete form by replacing it with the Kronecker delta $\delta_{j,l}$ multiplied with $\frac{1}{a}$ in order to ensure a smooth transition from the discrete form to the continuous representation when $a \to 0$. As a result, $\delta(x-x') \to \frac{1}{a}\delta_{j,l} = \frac{1}{a}\mathbf{1}$. Putting everything together, we arrive at

$$((E + i\eta) - \mathcal{H})G^r = \frac{1}{a}\mathbf{1} \to G^r_{j,l}(E) = \frac{1}{a}([E + i\eta - \mathcal{H}]^{-1})_{j,l}. \tag{6.24}$$

This equation means that the Green's function is simply the inverse of the matrix $(E+i\eta-\mathcal{H})$ (apart from the factor $1/a$). Therefore, computing the retarded Green's function and from it the density of states, and finally the local carrier density is (at least conceptually) rather simply, following the recipe:

(1) Set up a vector containing all discrete energies E_n that need to be considered. In the simplest form, this is just a vector with equidistant energies $E_n = n \cdot \Delta E$ with some appropriately chosen ΔE.
(2) Set up the matrix $E_n + i\eta - \mathcal{H}$ for each n and invert it. If equidistant energies are chosen and if there are bound states (such as in a PIB), η should be on the same order of magnitude as ΔE. Without bound states $\eta \sim 10^{-8}$ eV works well (see the video accessible through the QR code #48 for more details).

(3) Extract the imaginary part of the main diagonal elements (i. e., $j = l$ meaning $x_j = x_l$ which is $x = x'$) of the inverse and store. When multiplied with $-\frac{1}{\pi}$, this yields the local density of states $-1/\pi \,\mathrm{Im}\, G^r(j,j,E_n) = D(x_j, E_n)$.
(4) Sum over all E_n to obtain the carrier density as $n_j = \sum_n \Delta E \cdot D_j(E_n) f(E_n - E_f)$ where $f(E_n - E_f)$ is the Fermi distribution function.

Details on an example of such a calculation in a simple PIB with constant potential are provided through the QR code #48.

6.2.3 Including Contacts

In a next step, contacts need to be included in order to enable the calculation of an open quantum system. To this end, a semiinfinite system separated into a contact area (denoted with subscript C in the following) and a device (denoted with subscript D) is considered. The device may consist of an arbitrary number N of lattice sites; the contact is semiinfinite. Hence, we can write the matrix equation for determining the Green's function as a 2×2 block matrix of the form

$$\left[(E + i\eta)\begin{pmatrix} \mathbf{1}_D & 0 \\ 0 & \mathbf{1}_C \end{pmatrix} - \begin{pmatrix} \mathcal{H}_D & \tau \\ \tau^\dagger & \mathcal{H}_C \end{pmatrix}\right]\begin{pmatrix} G_D^r & G_{DC}^r \\ G_{CD}^r & G_C^r \end{pmatrix} = \frac{1}{a}\begin{pmatrix} \mathbf{1}_D & 0 \\ 0 & \mathbf{1}_C \end{pmatrix} \tag{6.25}$$

where the matrix τ is a $N \times \infty$-dimensional coupling matrix with only a single entry $-t$ on the $(N,1)$-element (and zero anywhere else), $\tau^\dagger$ is an $\infty \times N$-dimensional matrix with $-t$ only on the $(1,N)$-element and zero otherwise; $\mathcal{H}_{D,C}$ are the Hamiltonians of the isolated device and isolated contact, respectively.

Multiplying the first column of G^r with the second row of $E + i\eta - \mathcal{H}$ yields

$$\begin{aligned} &(E + i\eta)\mathbf{1}_C G_{CD}^r - \tau^\dagger G_D^r - \mathcal{H}_C G_{CD}^r = 0 \\ &\to [(E + i\eta) - \mathcal{H}_C] G_{CD}^r = \tau^\dagger G_D \to G_{CD}^r = \underbrace{[E + i\eta - \mathcal{H}_C]^{-1}}_{=G_C^r} \tau^\dagger G_D^r. \end{aligned} \tag{6.26}$$

Next, multiplying the first column of G^r with the first row of $E + i\eta - \mathcal{H}$, one obtains

$$\begin{aligned} &(E + i\eta)\mathbf{1}_D G_D^r - \mathcal{H}_D G_D^r - \tau G_{CD}^r = \frac{1}{a}\mathbf{1}_D \\ &\to [(E + i\eta)\mathbf{1}_D - \mathcal{H}_D] G_D^r - \tau G_C^r \tau^\dagger G_D^r = \frac{1}{a}\mathbf{1}_D \end{aligned} \tag{6.27}$$

where the result of Equation (6.26) has been used. Equation (6.27) can be rewritten as $[(E + i\eta)\mathbf{1}_D - \mathcal{H}_D - \Sigma^r] G_D^r = \frac{1}{a}\mathbf{1}_D$ where $\Sigma^r = \tau G_C^r \tau^\dagger$ is called the retarded self-energy function. The importance of Equation (6.27) is that an infinite, open quantum system can now be computed based on a finite number of lattice points N. Note that up to now the derivation is exact and no approximation has been used. While the matrix equation has only finite dimensions $N \times N$, we still need to know what G_C^r is and this matrix belongs to the semiinfinite contact, i. e., it has $\infty \times \infty$ dimensions. So, what did we gain by computing Equation (6.27)? Looking at the self-energy Σ^r, it becomes clear that after the matrix multiplication there is only a single matrix element different from zero, namely the (N, N)-element which is equal to $(-t)(G_C^r)_{1,1}(-t)$. Now attention has to be paid to the different indices: It should be clear that the $(1, 1)$-element of G_C^r is actually equal to the $(N + 1, N + 1)$-element of the overall matrix G^r. This means that in order to determine the self-energy function we only need this $(1, 1)$ element of the isolated contact, which is called the surface Green's function. Hence, if we find an alternative way of computing the surface Green's function of an isolated contact, we will be in a position to incorporate a semiinfinite contact in an exact way.

In order to compute the surface Green's function of an isolated contact, we consider a semiinfinite contact on a constant potential Φ_f^0 throughout all lattice points. Splitting the contact into a device consisting merely of the first point (illustrated in the equation below with the horizontal and vertical lines), we go through the same analysis as above:

$$\left[(E + i\eta)\begin{pmatrix} 1 & 0 & 0 & \cdots \\ 0 & 1 & 0 & \cdots \\ 0 & 0 & 1 & \cdots \\ 0 & 0 & \vdots & \ddots \end{pmatrix} - \begin{pmatrix} 2t + \Phi_f^0 & -t & 0 & \cdots \\ -t & 2t + \Phi_f^0 & -t & \cdots \\ 0 & -t & 2t + \Phi_f^0 & \cdots \\ 0 & 0 & \vdots & \ddots \end{pmatrix}\right] \times \begin{pmatrix} G_{1,1}^r & G_{1,2}^r & G_{1,3}^r & \cdots \\ G_{2,1}^r & G_{2,2}^r & G_{2,3}^r & \cdots \\ G_{3,1}^r & G_{3,2}^r & G_{3,3}^r & \cdots \\ \vdots & \vdots & \vdots & \ddots \end{pmatrix} = \frac{1}{a}\begin{pmatrix} 1 & 0 & 0 & \cdots \\ 0 & 1 & 0 & \cdots \\ 0 & 0 & 1 & \cdots \\ 0 & 0 & \vdots & \ddots \end{pmatrix}. \tag{6.28}$$

Since we are considering only a single point, we can immediately write down what the Green's function is (the inverse of the "matrix" is simply the inverse of a number):

$$\begin{aligned} G_{1,1}^r &= \left[(E + i\eta) - \mathcal{H}_{1,1} - \Sigma_{1,1}^r\right]^{-1} \\ \rightarrow G_{1,1}^r &= \frac{1}{E + i\eta - (2t + \Phi_f^0) - t^2 G_{2,2}^r} \end{aligned} \tag{6.29}$$

Note that $G_{2,2}^r$ is the surface Green's function we are looking for. So, it is the $(1, 1)$-element of the isolated semiinfinite contact (which is the $(N+1, N+1) = (2, 2)$-element of the overall matrix) as mentioned above. The important step now is to realize that $G_{1,1}^r$ and $G_{2,2}^r$ are actually equal: both are the surface Green's function of a semiinfinite contact with constant potential Φ_f^0. $G_{1,1}^r$ is simply the surface Green's function of the contact ranging from $1, \ldots, \infty$ where the part from $2, \ldots, \infty$ has been taken care of with an appropriate self-energy. At the same time, $G_{2,2}^r$ is the surface Green's function of the contact ranging from $2, \ldots, \infty$. But there is no difference between the gray area in Equation (6.28) and the entire matrix. As a result, the surface Green's function, called G^r_{surf} in the following, can be determined by solving the quadratic equation

$$G^r_{\text{surf}} = \frac{1}{E + i\eta - (2t + \Phi^0_f) - t^2 G^r_{\text{surf}}}. \tag{6.30}$$

Note that $i\eta \to 0$ and it will therefore be dropped in the following. Equation (6.30) yields two solutions of the form $G^r_{\text{surf}} = \frac{1}{t}(\frac{E-2t-\Phi^0_f}{2t} \pm i\sqrt{1 - (\frac{E-2t-\Phi^0_f}{2t})^2})$. The solution with "–" sign is appropriate here (the "+"-sign results in the advanced surface Green's function). Next, utilizing $e^{i\arccos(x)} = x + i\sqrt{1-x^2}$, setting $x = -\frac{E-2t-\Phi^0_f}{2t}$ and using the discrete dispersion relation $E = 2t - 2t\cos(ka) + \Phi^0_f$ (cf. Section 2.4.1) it is easy to see that $x = \cos(ka)$ and therefore we obtain for the surface Green's function the following expression:

$$G^r_{\text{surf}} = -\frac{1}{t}e^{ika}. \tag{6.31}$$

As a result, we arrived at a very simple expression for the retarded self-energy $\Sigma^r = -te^{ika}$ where k is the wave number that can be computed from the dispersion relation. Using the surface Green's function computed above, we can now immediately write down the self-energy for a semiinfinite contact connected to the right of a device: Σ^r is a matrix of the same dimension N as the device with the (N, N)-element equal to $-te^{ika}$ as the only nonzero matrix element. In fact, this result cannot be overappreciated; with an analytically computable self-energy the coupling to a semiinfinite contact, and hence the expansion of a finite device into an open quantum system is taken into account without any approximation.

In a subsequent step, the same analysis as before can now be carried out for a contact that is connected at the left end of the device in order to realize source and drain contacts. Note that this contact may be on a different potential such that the dispersion relation in the source contact is $E = 2t - 2t\cos(k_s a) + \Phi^s_f$ and in the drain $E = 2t - 2t\cos(k_d a) + \Phi^d_f$. Eventually, the following equation is obtained:

$$G^r = \frac{1}{a}[E + i\eta - \mathcal{H} - \Sigma^r_s - \Sigma^r_d]^{-1} \tag{6.32}$$

with $\Sigma^r_s = -te^{ik_s a}\mathbf{0}_{1,1}$ and $\Sigma^r_d = -te^{ik_d a}\mathbf{0}_{N,N}$ where $k_{s,d}a$ are determined from the dispersion relations given above and $\mathbf{0}_{1,1}(\mathbf{0}_{N,N})$ is a matrix with 1 at its $(1,1)((N,N))$-entry and zero anywhere else.

One final important remark needs to be made: In order to be able to describe the coupling of a contact with the simple self-energy obtained, the potential needs to be constant throughout the semiinfinite contact and equal to the potential at the point where the contact is connected to the device. Therefore, the device that we want to compute includes source and drain sections long enough such that the potential becomes constant.

For instance, the conduction band displayed in Figure 6.2 can be used and contacts with $\Phi_f^s = 0$ and $\Phi_f^d = -eV_{ds}$ can be connected at the right and left edges (see QR code #49 for a detailed calculation and local density of states plots).

Task 23.
Retarded Green's function: Consider a "system" consisting of a single finite difference lattice site (a single dot) connected to two one-dimensional, semiinfinite contacts, all at the same potential (set equal to 0) and coupled via t. Compute the density of states in this system and compare with your expectation.

6.2.4 Carrier Density in Nonequilibrium

Up to now, we discussed how to compute the carrier density as a function of x in an open quantum system with Green's functions where the contacts have been incorporated exactly through analytic self-energy functions. However, the consideration so far is only valid in equilibrium. In order to describe nonequilibrium situations, we need to split the density of states into one part for carriers from the source and a part for carriers from the drain contact. To do so, an additional function $i\Gamma = \Sigma^a - \Sigma^r$ is defined where Σ^a is the advanced self-energy function, which is $(\Sigma^r)^\dagger$.[2] This means that $\Gamma = 2\,\mathrm{Im}(\Sigma^r)$. Moreover, since we can define a retarded and advanced self-energy function for the source and drain contacts separately, Γ will be the sum $\Gamma_s + \Gamma_d$ for the source and drain contacts.

Next, we can write the equation for G^r and G^a as (the factor $1/a$ has been dropped here for a more compact notation)

$$G^r = (E - \mathcal{H} - \Sigma^r)^{-1}, \quad G^a = (E - \mathcal{H} - \Sigma^a)^{-1}. \tag{6.33}$$

Note that $\Sigma^{r,a}$ are complex-valued functions that provide the imaginary part needed in order to obtain a unique retarded or advanced Green's function. The expressions for $G^{r,a}$ in Equation (6.33) can both be inverted giving rise to

$$(G^r)^{-1} - (G^a)^{-1} = (E - \mathcal{H} - \Sigma^r) - (E - \mathcal{H} - \Sigma^a) = \Sigma^a - \Sigma^r = i\Gamma. \tag{6.34}$$

Equation (6.34) will now be multiplied from the left with G^r and from the right with G^a, which leads to

$$\begin{aligned} G^r i\Gamma G^a &= G^r[(G^r)^{-1} - (G^a)^{-1}]G^a = G^r(G^r)^{-1}G^a - G^r(G^a)^{-1}G^a \\ &= G^a - G^r = -2i\,\mathrm{Im}(G^r). \end{aligned} \tag{6.35}$$

2 Since only the first and last main diagonal elements are occupied transposing the matrix does not yield any change so Σ^a is simply the conjugate complex here.

The function $G^r\Gamma G^a$ is called spectral function A. From Equation (6.20), we know that $-\frac{1}{\pi}\operatorname{Im} G^r(x,x,E) = D(x,E)$ is equal to the density of states and as a result $A(x,x,E)$ divided by 2π yields the local density of states $D(x,E)$. The importance of writing the local DOS as the main diagonal of the spectral function is that $\Gamma = \Gamma_s + \Gamma_d$, and thus

$$A = G^r\Gamma_s G^a + G^r\Gamma_d G^a = A_s + A_d. \tag{6.36}$$

This means that the main diagonal elements of the spectral functions A_s and A_d provide the density of states for carriers injected from the respective contacts and as a result the nonequilibrium carrier density is given by

$$n_{\text{neq}}(x) = \int dE\, \frac{1}{2\pi}\big(A_s(x,x,E)f_s(E_f^s) + A_d(x,x,E)f_d(E_f^d)\big). \tag{6.37}$$

Equation (6.37) only provides the carrier density in 1D. As already mentioned above, the NEGF equations need to be solved self-consistently with the Poisson equation (see next section). To do so, a 3D carrier density is needed, which can be obtained in the following way. Let us consider a nanowire FET with a very small quadratic cross-section $d_{\text{nw}} \times d_{\text{nw}}$ of the nanowire in order to justify 1D electronic transport (i. e., only the first subband will be occupied). If the potential variations along the x-direction can be considered slow enough to justify a separation ansatz, Equation (6.19) for the retarded Green's function can be written in 3D as

$$\begin{aligned} G^r(\vec{r},\vec{r}',E) &= \sum_{\vec{k}} \frac{\phi_{\vec{k}}(\vec{r})\phi_{\vec{k}}^\star(\vec{r}')}{E-\epsilon_{\vec{k}}+i\eta} \\ &\to \sum_{k_x} \frac{\psi_1(y)\xi_1(z)\chi_{k_x}(x)\psi_1^\star(y')\xi_1^\star(z')\zeta_{k_x}^\star(x')}{E-(\epsilon_{k_x}+E_1^y+E_1^z)+i\eta} \end{aligned} \tag{6.38}$$

where we used the fact, that in the y- and z-directions the nanowire can be considered as a closed PIB system where the wavefunctions are known. Due to the strong quantization, only the first subband (therefore the energies E_1^y and E_1^z within the denominator) needs to be considered, and hence the index '1' at $\psi(y)$ and $\xi(z)$. Because of the coupling to the contacts in the x-direction we do not know what the wave-functions in x-direction are, and thus leave them as given in Equation (6.38). Ultimately, we are only interested in the main diagonal elements of G^r, and hence

$$G^r(\vec{r},\vec{r},E) \approx |\psi_1(y)|^2|\xi_1(z)|^2 \underbrace{\sum_{k_x} \frac{\chi_{k_x}(x)\zeta_{k_x}^\star(x')}{E-(\epsilon_{k_x}+E_1^y+E_1^z)+i\eta}}_{G_{1D}^r \text{ computed with NEGF}}. \tag{6.39}$$

This means that we can simply multiply our results so far with the absolute square of the wavefunctions in the y,z-directions and obtain a 3D version of the Green and spectral functions, and hence of the carrier density. Keep in mind, however, that this approximation was made on the assumption that a separation ansatz can be used.

In many cases, it is sufficient that we average the wavefunctions over y and z, meaning that we integrate $\int dydz$ and divide by the area of the nanowire d_{nw}^2. Since the wavefunctions in the y- and z-directions are normalized, the integration yields unity and as a result, the three-dimensional carrier density is obtained simply by dividing the computed 1D density by the cross-sectional area of the nanowire and by taking the quantization energy $E_1^y + E_1^z$ of the first subband into account. If you do so, it is important to keep in mind that the carrier density does not steadily decrease $\propto 1/d_{\text{nw}}^2$ since, for increasing nanowire cross-section, multiple 1D subbands may contribute to the charge in the nanowire, which need to be taken into consideration.

6.2.5 Self-Consistency

In order to compute the carrier density based on NEGF, we need to know what the potential profile $\Phi_f(x)$ along the direction of current transport is. However, $\Phi_f(x)$ is the solution of the Poisson equation and to solve Poisson's equation we need to know what the carrier density is. Therefore, the Poisson equation and the NEGF equations need to be solved self-consistently. A simple iteration between Poisson equation and NEGF will most likely not lead to a solution because of the highly nonlinear dependence of the carrier density on the exact potential profile. Therefore, we use the Newton–Raphson method to find a self-consistent solution.

The Newton–Raphson method is used to iteratively find the root of a function. In the present case of the simulations considered here, we need to find the root of

$$F(\Phi_f(x)) := \frac{\partial^2 \Phi_f(x)}{\partial x^2} - \frac{\Phi_f(x) - \Phi_g - \Phi_{\text{bi}}}{\lambda^2} + \frac{e^2 \tilde{n}(x)}{\varepsilon_0 \varepsilon_{\text{si}}} \tag{6.40}$$

where the density of ionized dopants has been incorporated into $\tilde{n}(x)$ to simplify the notation.

Before we work on finding the root of $F(\Phi_f(x))$, let us quickly go through the Newton–Raphson method in the simplest case, namely finding an approximate root of an arbitrary one-dimensional function $f(x)$ in order to understand how the method works. To do so, the function $f(x)$ is linearized around a test point x_0. The linear approximation $y_1(x) = mx + n$ has the slope $m = \frac{df}{dx}|_{x=x_0} = f'(x_0)$ in x_0. Since $y_1(x_0) = f(x_0)$ and $y_1(x_0) = f'(x_0)x_0 + n$, the constant n can be determined to be $n = f(x_0) - f'(x_0)x_0$. Consequently, as a next best guess for the actual root, we take the root of $y_1(x)$, which follows from $y_1(x_1) \stackrel{!}{=} 0$ resulting in $0 = f'(x_0)x_1 + f(x_0) - f'(x_0)x_0$. Rewriting the latter expression yields $f'(x_0)(x_1 - x_0) = -f(x_0)$ or $f'(x_0)\delta x = -f(x_0)$ with $\delta x = x_1 - x_0$. Therefore, computing $\delta x = -\frac{f(x_0)}{f'(x_0)}$ yields $x_1 = x_0 + \delta x$ as the next approximation to the actual root. x_1 will then be used for a subsequent linearization, which provides another approximation x_2 of the root with $f'(x_1)\delta x = -f(x_1)$ and $\delta x = x_2 - x_1 \rightarrow x_2 = x_1 + \delta x$. It is now easy to see that we obtain a general iteration sequence with

$$\delta x_i = \frac{-f(x_i)}{f'(x_i)} \rightarrow x_{i+1} = x_i + \delta x_i \tag{6.41}$$

where i is the index of the respective iteration step. The Newton–Raphson method is not globally convergent, and hence the zeroth iteration guess of the solution has to be close enough to the real root in order to obtain a solution. Furthermore, a simple way to improve the convergence is to use in each iteration step not the full Newton step δx but only a part of it. The exact value is a matter of the particular situation but a factor of ~0.3, …, 0.4 yields a converging iteration in most cases. Hence, $x_{i+1} = x_i + 0.3\delta x_i$.

In a next step, we need to expand the method to find the root of $F(\Phi_f(x))$. To this end, Equation (6.41) needs to be extended in the following way: instead of x, the variables are now the N potential values at the N points of the finite difference grid, and thus $x \rightarrow \Phi_f^j$ with $j = 1, \ldots, N$. Consequently, an update of the potential is obtained with

$$\underbrace{\begin{pmatrix} \Phi_f^1 \\ \Phi_f^2 \\ \vdots \\ \Phi_f^N \end{pmatrix}_{i+1}}_{\widehat{=}x_{i+1}} = \underbrace{\begin{pmatrix} \Phi_f^1 \\ \Phi_f^2 \\ \vdots \\ \Phi_f^N \end{pmatrix}_{i}}_{\widehat{=}x_i} + \underbrace{\begin{pmatrix} \delta\Phi_f^1 \\ \delta\Phi_f^2 \\ \vdots \\ \delta\Phi_f^N \end{pmatrix}}_{\widehat{=}\delta x}. \tag{6.42}$$

The function $f(x_i)$ with Neumann BC is extended into

$$f(x_i) \rightarrow \begin{pmatrix} F(\Phi_f^1) \\ F(\Phi_f^2) \\ \vdots \\ F(\Phi_f^N) \end{pmatrix}_i = \begin{pmatrix} F_1 \\ F_2 \\ \vdots \\ F_N \end{pmatrix}_i = \begin{pmatrix} \frac{-2\Phi_f^1+2\Phi_f^2}{a^2} - \frac{\Phi_f^1-\Phi_g-\Phi_{\mathrm{bi}}}{\lambda_{\mathrm{ch}}^2} + \frac{e^2\tilde{n}_1}{\varepsilon_0\varepsilon_{\mathrm{ch}}} \\ \frac{\Phi_f^1-2\Phi_f^2+\Phi_f^3}{a^2} - \frac{\Phi_f^2-\Phi_g-\Phi_{\mathrm{bi}}}{\lambda_{\mathrm{ch}}^2} + \frac{e^2\tilde{n}_2}{\varepsilon_0\varepsilon_{\mathrm{ch}}} \\ \vdots \\ \frac{-2\Phi_f^{N-1}+2\Phi_f^N}{a^2} - \frac{\Phi_f^N-\Phi_g-\Phi_{\mathrm{bi}}}{\lambda_{\mathrm{ch}}^2} + \frac{e^2\tilde{n}_N}{\varepsilon_0\varepsilon_{\mathrm{ch}}} \end{pmatrix}_i. \tag{6.43}$$

Finally, the derivative of $f'(x)$ with respect to x becomes

$$f'(x_i) \rightarrow \begin{pmatrix} \frac{\partial F_1}{\partial\Phi_f^1} & \frac{\partial F_1}{\partial\Phi_f^2} & \cdots & \frac{\partial F_1}{\partial\Phi_f^N} \\ \frac{\partial F_2}{\partial\Phi_f^1} & \frac{\partial F_2}{\partial\Phi_f^2} & \cdots & \frac{\partial F_2}{\partial\Phi_f^N} \\ \vdots & \vdots & \ddots & \\ \frac{\partial F_N}{\partial\Phi_f^1} & \frac{\partial F_N}{\partial\Phi_f^2} & \cdots & \frac{\partial F_N}{\partial\Phi_f^N} \end{pmatrix}_i = \mathbf{J}_i. \tag{6.44}$$

In all equations above, i indicates the iteration step. The matrix $\mathbf{J}$ is called the Jacobi matrix. In the case considered here (i. e., based on the modified one-dimensional Poisson equation), its elements are given by

$$\frac{\partial F_l}{\partial \Phi_f^j} = \frac{\partial}{\partial \Phi_f^j}\left(\frac{\Phi_f^{l-1} - 2\Phi_f^l + \Phi_f^{l+1}}{a^2} - \frac{\Phi_f^l - \Phi_g - \Phi_{\text{bi}}}{\lambda_{\text{ch}}^2} + \frac{e^2 \tilde{n}_l}{\varepsilon_0 \varepsilon_{\text{si}}}\right). \tag{6.45}$$

The partial derivatives of the first two terms in Equation (6.45) result in a tridiagonal Jacobi matrix very similar to the matrix of the Poisson equation (cf. Equation (6.4)). What remains to be determined is the derivative of the charge density, i. e., $\frac{\partial \tilde{n}_l}{\partial \Phi_f^j} = \frac{\partial n_l}{\partial \Phi_f^j}$ (note that the dopant density drops out here because it does not depend on Φ_f). In order to determine this factor a local, semiclassical approximation is followed and we neglect all nondiagonal terms. Hence, $\frac{\partial n_l}{\partial \Phi_f^j} \approx \delta_{lj} \cdot \frac{\partial}{\partial \Phi_f^j}(\int_{\Phi_f^j}^{\infty} dE\, D(E - \Phi_f^j) f(E_f^j))$ can be used where $D(E - \Phi_f^j)$ is the analytically computed density of states as given in Table 2.4 and E_f^j is the quasi-Fermi level extracted from $n_j \stackrel{!}{=} \int_{\Phi_f^j}^{\infty} dE\, D(E - \Phi_f^j) f(E_f^j)$ [162].

To be explicit, putting everything together we arrive at the following iterative scheme for the Newton–Raphson method:

$$\begin{pmatrix} \delta\Phi_f^1 \\ \delta\Phi_f^2 \\ \vdots \\ \delta\Phi_f^N \end{pmatrix} = -\begin{pmatrix} -\frac{2}{a^2} - \frac{1}{\lambda_{\text{ch}}^2} + \frac{e^2}{\varepsilon_0 \varepsilon_{\text{si}}} \frac{\partial n_1}{\partial \Phi_f^1} & \frac{2}{a^2} & 0 & \cdots \\ \frac{1}{a^2} & -\frac{2}{a^2} - \frac{1}{\lambda_{\text{ch}}^2} + \frac{e^2}{\varepsilon_0 \varepsilon_{\text{si}}} \frac{\partial n_2}{\partial \Phi_f^2} & \frac{1}{a^2} & 0 \\ \vdots & \vdots & \vdots & \ddots \end{pmatrix}^{-1} \times \begin{pmatrix} F_1 \\ F_2 \\ \vdots \\ F_N \end{pmatrix}. \tag{6.46}$$

Note that the vector $\vec{F}$ contains the full charge density $\tilde{n} = n - N_d$ whereas the Jacobi matrix only contains the derivative of n since N_d is considered to be constant, as mentioned above. Furthermore, note the factor of $2/a^2$ in the $(1, 2)$ and $(N, N-1)$ entries (and also in the first and last entry of $\vec{F}$) due to the Neumann BC.

The update of the potential distribution computed with the Newton–Raphson method will be used to set up the Hamiltonian to calculate an update of the charge density. With the new charge density, the Newton–Raphson method is again employed to obtain a further update of the potential. This procedure will be iterated until a specified convergence criterion is met (for instance, that the maximum change of the potential variation $\delta\vec{\Phi}_f$ is less than 1 meV).

6.2.6 Current Flow

After self-consistency has been reached, the current has to be calculated. This can be done with the so-called Fisher–Lee relation [73], which relates the Green's function with the transmission function used in the Landauer formalism. The details of the derivation of the Fisher–Lee relation (also called the Caroli expression) will be left out here and only the final result is given:

$$T(E) = \mathrm{Tr}(\Gamma_s G^r \Gamma_d G^a) \tag{6.47}$$

where Tr means taking the trace of the matrix. Let us again consider a "device" consisting of 4 finite difference points so that the matrix relations leading to $T(E)$ can be stated explicitly. In this case, one obtains for the part $\Gamma_s G^r$,

$$\begin{aligned}\Gamma_s G^r &= \begin{pmatrix} -2t\sin(k_s a) & 0 & 0 & 0 \\ 0 & 0 & 0 & 0 \\ 0 & 0 & 0 & 0 \\ 0 & 0 & 0 & 0 \end{pmatrix} \cdot \begin{pmatrix} G^r_{1,1} & G^r_{1,2} & G^r_{1,3} & G^r_{1,4} \\ G^r_{2,1} & G^r_{2,2} & G^r_{2,3} & G^r_{2,4} \\ G^r_{3,1} & G^r_{3,2} & G^r_{3,3} & G^r_{3,4} \\ G^r_{4,1} & G^r_{4,2} & G^r_{4,3} & G^r_{4,4} \end{pmatrix} \\ &= -2t\sin(k_s a) \begin{pmatrix} G^r_{1,1} & G^r_{1,2} & G^r_{1,3} & G^r_{1,4} \\ 0 & 0 & 0 & 0 \\ 0 & 0 & 0 & 0 \\ 0 & 0 & 0 & 0 \end{pmatrix} . \end{aligned} \tag{6.48}$$

A similar calculation for $\Gamma_d G^a$ yields a matrix where only the last line has entries so that

$$\Gamma_d G^a = -2t\sin(k_d a) \begin{pmatrix} 0 & 0 & 0 & 0 \\ 0 & 0 & 0 & 0 \\ 0 & 0 & 0 & 0 \\ G^a_{4,1} & G^a_{4,2} & G^a_{4,3} & G^a_{4,4} \end{pmatrix} . \tag{6.49}$$

Finally, multiplying the two matrices and taking the trace one obtains for the transmission probability

$$T(E) = 2t\sin(k_s a) G^r_{1,4} G^a_{4,1} 2t\sin(k_d a) = \Gamma_s^{1,1} |G^r_{1,4}|^2 \Gamma_d^{4,4} . \tag{6.50}$$

This equation can be interpreted in the following way: $\Gamma_s^{1,1}$ is the rate of carrier injection at the interface between the source and the computed device, $|G^r_{1,4}|^2$ is the probability for propagation from the first lattice point 1 to the last lattice point 4 (in general N) and finally $\Gamma_d^{4,4}$ is the rate of carrier extraction at the interface between the device and the semiinfinite drain contact. $T(E)$ is then inserted into the Landauer expression and the final current is computed as

$$I_d = \frac{2e}{h} \int dE\, \Gamma_s^{1,1} |G^r_{1,N}|^2 \Gamma_d^{N,N} (f_s(E_f^s) - f_d(E_f^d)). \tag{6.51}$$

6.3 Extensions of the 1D NEGF

6.3.1 Scattering with Buettiker Probes

Up to now, the electronic transport has been considered as being completely ballistic. Assuming ballistic transport is often justified in nanoscale devices, particularly in Schottky-barrier MOSFETs and band-to-band tunneling FETs where a tunneling barrier exists that usually represents the main scattering event (cf. Chapters 7 and 9). However, there are also many cases where scattering, especially inelastic scattering, plays an important role and needs to be included in the analysis.

One of the major benefits of NEGF is that scattering, for instance due to electron–electron, electron–phonon interaction etc., can be taken into consideration with additional self-energies. However, when the details of the scattering are not important, scattering—elastic as well as inelastic—can be accounted for by so-called Buettiker probes [259]. Buettiker probes are virtual contacts that are attached to the device. Each Buettiker probe is in equilibrium with its own Fermi distribution function. But in contrast to regular contacts, Buettiker probes are not connected to any voltage source so that their Fermi level is not fixed. As a result, the Fermi level will be adjusted so that each Buettiker probe carries zero net current; this is important in order to ensure current continuity. While Buettiker probes carry zero net current, their contribution is to impose a carrier distribution according to their Fermi distribution. If the Buettiker probes are connected to each site of the finite difference lattice (illustrated in Figure 6.6 (a)) and, moreover, if the Buettiker probes are connected to the various sites with a certain strength $\gamma = 0, \ldots, 1$ then the scattering is spread over several probes; the dimensionless coupling parameter γ is set to zero for ballistic transport and $\gamma \to 1$ for increasing scattering. As a result, the scattering can be adjusted in terms of its strength and its location via the Buettiker probes and the coupling factor γ.

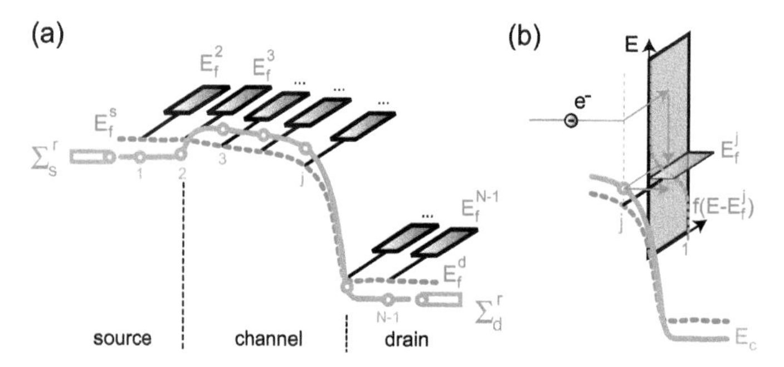

Figure 6.6: Incorporation of scattering by attaching Buettiker probes with floating Fermi levels at each grid point. The resulting Fermi levels in the probes constitute the quasi-Fermi level distribution along the current transport direction (red dashed line).

Buettiker probes can be used to describe elastic as well as inelastic scattering. Elastic scattering can be mimicked by the requirement that the current each Buettiker probe provides is zero at each energy. In this case, carriers are extracted from the channel and reinjected into the channel at the same energy, which indeed represents elastic scattering. Alternatively, a zero net current can also be obtained by requiring that the injected and extracted current components summed over all energies vanish. The latter means that a charge carrier can be extracted into a Buettiker probe at site j at a specific energy. The extracted carrier relaxes within the Buettiker probe and charge is reinjected from the Fermi distribution of the respective Buettiker probe as illustrated in Figure 6.6(b); this way of using Buettiker probes is equivalent to inelastic scattering used in some device simulations discussed in later chapters.

Although the existence of a Fermi level is strictly an equilibrium property, it is common practice to work with so-called quasi-Fermi levels. A quasi-Fermi level is the level a certain carrier density would exhibit if it was in equilibrium; in many cases, quasi-Fermi levels are even defined for the electron and hole population separately [247]. The Fermi levels of the Buettiker probes represent such a (spatially dependent) quasi-Fermi level as is shown in Figure 6.6(a) with the red dashed line. As mentioned above, the scattering rate is mediated by the dimensionless parameter γ. This parameter can be related to the mean-free path for scattering l_{mfp}. As was shown by Venugopal (see the Appendix of [259]) $l_{\text{mfp}} = 2a \times \frac{1}{\gamma}$ where a is the lattice constant of the finite difference grid. Obviously, when $\gamma \to 0$ the mean-free path becomes very large, i. e., one approaches the ballistic limit.

When working with Buettiker probes, one has to make sure during the self-consistent simulation that the Fermi levels of the Buettiker probes are always adjusted in such a way as to ensure zero net current. Since the current in each probe depends on all other probes, this requires a second self-consistency loop within the Poisson–Schrödinger (NEGF) self-consistency loop. To understand how this is done, let us consider the case where Buettiker probes are attached at each finite difference lattice site (shown in Figure 6.6(a)) via appropriate self-energy functions. Each electrode at site j (either Buettiker probe or source/drain contact) contributes to the overall self-energy Σ^r a matrix, which has zero entries everywhere except of the (j,j)-matrix element such that Σ^r becomes a diagonal matrix. To be specific, the diagonal elements of Σ^r are given by

$$\Sigma^r = \underbrace{-t\begin{pmatrix} e^{ik_s a} & 0 & 0 & \cdots & \cdots \\ 0 & 0 & 0 & \cdots & \cdots \\ \vdots & \vdots & \ddots & \ddots & \vdots \\ \vdots & \vdots & \ddots & 0 & 0 \\ \cdots & \cdots & 0 & 0 & e^{ik_d a} \end{pmatrix}}_{=\Sigma_s^r+\Sigma_d^r} \underbrace{-\gamma t\begin{pmatrix} 0 & 0 & 0 & \cdots & \cdots \\ 0 & e^{ik_2 a} & 0 & \cdots & \cdots \\ \vdots & \cdots & e^{ik_j a} & \ddots & \vdots \\ \vdots & \vdots & \ddots & e^{ik_{N-1} a} & \vdots \\ \cdots & \cdots & 0 & 0 & 0 \end{pmatrix}}_{=\Sigma_{\text{buet}}} \tag{6.52}$$

where $j = 2, \dots, (N-1)$ runs over all sites with Buettiker probes and N being the dimension of the finite difference grid. Note that the $k_j a$-terms in the self-energy for the Buettiker probes are simply given by the discrete dispersion relation with the potential energy Φ_f^j at the point j: $k_j a = \arccos(\frac{\Phi_f^j - E - 2t}{2t})$. The imaginary part of the self-energies stated above yields the matrices Γ_j where j includes all sites of the finite difference grid. Using the Fisher–Lee relation, Equation (6.47), and the quasi-Fermi levels E_f^j the current of each Buettiker probe at site j can be computed according to

$$I_j = \frac{2e}{h} \int dE \sum_i \Gamma_i |G_{i,j}^r|^2 \Gamma_j (f(E_f^i) - f(E_f^j)) \tag{6.53}$$

where the retarded Green's function is computed including the full diagonal self-energy matrix Equation (6.52) and the sum over i includes source and drain.

The (quasi-) Fermi levels $E_f^j, j = 2, \dots, N-1$ need to be found ensuring that the net currents at all Buettiker probes are zero. This has to be done iteratively using again the Newton–Raphson scheme (cf. Equation (6.46)), which leads to

$$\begin{pmatrix} \delta E_f^2 \\ \delta E_f^3 \\ \vdots \\ \delta E_f^{N-1} \end{pmatrix} = - \begin{pmatrix} \frac{\partial I_2}{\partial E_f^2} & \frac{\partial I_2}{\partial E_f^3} & \cdots & \frac{\partial I_2}{\partial E_f^{N-1}} \\ \frac{\partial I_3}{\partial E_f^2} & \frac{\partial I_3}{\partial E_f^3} & \cdots & \frac{\partial I_3}{\partial E_f^{N-1}} \\ \vdots & \vdots & \vdots & \vdots \\ \frac{\partial I_{N-1}}{\partial E_f^2} & \frac{\partial I_{N-1}}{\partial E_f^3} & \cdots & \frac{\partial I_{N-1}}{\partial E_f^{N-1}} \end{pmatrix}^{-1} \begin{pmatrix} I_2 \\ I_3 \\ \vdots \\ I_{N-1} \end{pmatrix} \tag{6.54}$$

where $\frac{\partial I_j}{\partial E_f^j} = \frac{2e}{h} \int dE \sum_i \Gamma_i |G_{i,j}^r|^2 \Gamma_j \frac{\partial f(E_f^j)}{\partial E_f^j}$ and $\frac{\partial I_j}{\partial E_f^i} = -\frac{2e}{h} \int dE\, \Gamma_i |G_{i,j}^r|^2 \Gamma_j \frac{\partial f(E_f^i)}{\partial E_f^i}$. Having determined all (quasi-) Fermi levels, the carrier density can be computed by multiplying all diagonal elements of Γ with the Fermi distribution function using the E_f^j including source and drain, i. e., $\Gamma_s f(E_f^s)$, $\Gamma_j f(E_f^j)$ for $j = 2, \dots, N-1$ and $\Gamma_d f(E_f^d)$. Multiply from left with G^r and from the right with G^a and divide by 2π. Finally, integrate over all energies. This is basically the same as was done in the ballistic limit where the spectral function was split up into one part from source multiplied with $f(E_f^s)$ and one part from drain multiplied with $f(E_f^d)$; the difference is that this has to be done for the Buettiker probes, too. With the carrier density, an update of the potential needs to be computed as has been detailed above. Finally, after a self-consistent solution has been found including (quasi-) Fermi levels in the Buettiker probes that render the current of each Buettiker probe (close to) zero, the drain current is computed as stated in Equation (6.53) setting j equal to the drain.

6.3.2 Gate Leakage with Buettiker Contacts

When the gate dielectric thickness is scaled down, gate leakage occurs because of direct tunneling of carriers through the gate insulator. As a result, current is injected from the gate electrode into the channel contributing to the overall drain current I_d; the results of gate leakage have been discussed in Section 5.9.3. In the simulations, gate leakage can be taken into account using Buettiker probes [140]. The difference between the Buettiker probes used for scattering and the Buettiker probes for gate leakage is that in the latter case the probes are all short-circuited and share a Fermi level that is determined by the gate voltage (illustrated in Figure 6.7). As such, these gate Buettiker probes may inject (or extract) carriers into (from) the channel.

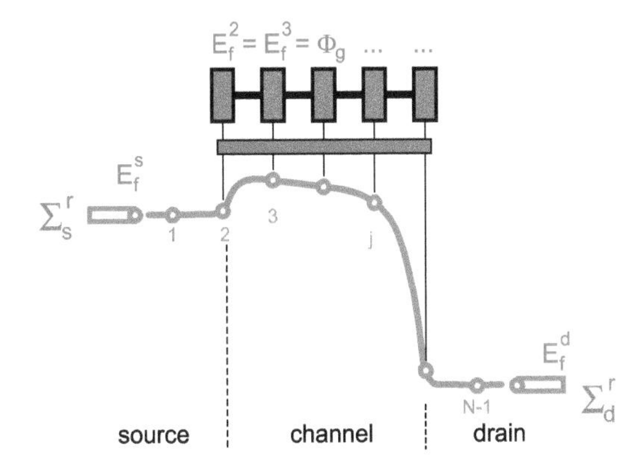

Figure 6.7: Incorporation of gate leakage with short-circuited Buettiker whose Fermi levels equal Φ_g.

In order to account for the tunneling through the gate dielectric, the gate Buettiker probes are connected to each finite difference lattice point j of the channel via a spatially dependent coupling parameter t_g^j that, for any finite thickness of the gate dielectric, is smaller than the coupling t along the source-to-drain direction. These coupling parameters are essentially determined by an exponential factor due to tunneling of carriers through the gate dielectric. However, because the electric field across the gate dielectric depends on the position (cf. Figure 5.1(b)) along the source-to-drain direction and since the electric field substantially impacts the tunneling probability, we need to compute an individual coupling parameter for every finite difference grid point j within the channel region. The coupling between each gate (Buettiker) probe and the channel is again described with a factor $\gamma_j = 0, \dots, 1$, and thus $t_g^j = \gamma_j \cdot t$ with $\gamma_j = 0$ representing a perfect insulator and $\gamma_j \neq 0$ leads to gate leakage.

The coupling parameter γ_j at point j can be related to the potential barrier of the gate insulator by analytically calculating the transmission probability $\mathcal{T}(E)$ through a potential barrier using the WKB approximation and comparing this with the transmission function $T(E)$ calculated with the Fisher–Lee relation of a single site system that is connected to the right to a semiinfinite contact and to the left to a semiinfinite contact whose coupling is mediated by the parameter γ.

The WKB approximation for the transmission $\mathcal{T}(E)$ through a potential barrier has already been discussed in Section 4.6.2. In the present case, this results in

$$\mathcal{T}(E) = \exp\left(-2\int_0^{d_{\mathrm{ox}}} dz \sqrt{\frac{2m^\star}{\hbar^2}}\sqrt{\Phi(z) - E}\right). \tag{6.55}$$

The potential $\Phi(z = 0)$ is given by $\Phi_g + \Phi_{\mathrm{bi}} + \Phi_{\mathrm{ox}}$ where Φ_g is the gate potential, Φ_{bi} the built-in potential of the channel and $\Phi_{\mathrm{ox}} = 3.1\,\mathrm{eV}$ is the conduction band off-set between silicon and SiO_2 (note that $\Phi(z)$ depends on the lattice site and, therefore, an index j will be added further below); for simplicity, flat-band conditions at $\Phi_g = 0$ are assumed here. Moreover, $\Phi(z = d_{\mathrm{ox}}) = \Phi_f^j + \Phi_{\mathrm{ox}}$ where Φ_f^j is the surface potential for the channel at position j. As a result, $\Phi^j(z) = \frac{\Phi_f^j - \Phi_g - \Phi_{\mathrm{bi}}}{d_{\mathrm{ox}}} z + \Phi_g + \Phi_{\mathrm{bi}} + \Phi_{\mathrm{ox}}$. Inserting this expression into Equation (6.55) yields

$$\mathcal{T}(E) = \exp\left(-2\sqrt{\frac{2m^\star}{\hbar^2}}\frac{2d_{\mathrm{ox}}}{3(\Phi_f^j - \Phi_g - \Phi_{\mathrm{bi}})}\left[(\Phi_f^j + \Phi_{\mathrm{ox}} - E)^{\frac{3}{2}} - (\Phi_g + \Phi_{\mathrm{bi}} + \Phi_{\mathrm{ox}} - E)^{\frac{3}{2}}\right]\right). \tag{6.56}$$

This equation can now be compared with the expression for the transmission probability obtained from NEGF. As mentioned above, a single site connected to a semiinfinite contact on the one side and to the gate electrode via $\gamma \cdot t$ on the other side is considered. In this case, the retarded Green function G^{r} is given by

$$G^{\mathrm{r}} = \left[E - 2t + te^{ika} + \gamma te^{ika}\right]^{-1} = \left[\gamma te^{ika} - te^{-ika}\right]^{-1}. \tag{6.57}$$

With $\Gamma_{\mathrm{l}} = -2\gamma t\sin(ka)$ and $\Gamma_{\mathrm{r}} = -2t\sin(ka)$, the transmission function $T(E)$ is

$$T(E) = \frac{4\gamma\sin^2(ka)}{1 + \gamma^2 - 2\gamma(\cos^2(ka) - \sin^2(ka))} \overset{ka\ll 1}{\approx} \frac{1}{1 + \frac{t(1-\gamma)^2}{4\gamma(E-\Phi_{\mathrm{bot}})}}, \tag{6.58}$$

where Φ_{bot} is considered as the conduction band bottom of the electrodes (contact and gate) attached to the single site. Since γ is very small, $(1 - \gamma)^2 \approx 1$ and, moreover, $\frac{t(1-\gamma)^2}{4\gamma(E-\Phi_{\mathrm{bot}})} \gg 1$, and hence $T(E) \approx \frac{4\gamma(E-\Phi_{\mathrm{bot}})}{t}$. Equating this with $\mathcal{T}(E)$ and solving for γ yields

$$t_g^j = \gamma_j \cdot t = \frac{t^2}{4(E - \Phi_{\mathrm{bot}})}\exp\left(-2\sqrt{\frac{2m^\star}{\hbar^2}}\frac{2d_{\mathrm{ox}}}{3(\Phi_f^j - \Phi_g - \Phi_{\mathrm{bi}})}\left[(\Phi_f^j + \Phi_{\mathrm{ox}} - E)^{\frac{3}{2}} - (\Phi_g + \Phi_{\mathrm{bi}} + \Phi_{\mathrm{ox}} - E)^{\frac{3}{2}}\right]\right). \tag{6.59}$$

Moreover, we can simply replace $E - \Phi_{\mathrm{bot}}$ with the Fermi energy of the metallic gate electrode. The effective mass in Equation (6.59) is the effective mass for tunneling through the gate dielectric and can serve here together with the Fermi energy as a fit parameter to adjust the gate leakage to experimental data. In the case considered in Section 5.9.3 (see Figure 5.20), the effective mass was set to $m^{\star} = 0.42m_0$ and the Fermi energy 1 eV [140]. The expression above gives an upper cut-off for possible gate leakage currents.

The total drain current is the sum of the current from source to drain and from each individual gate (Buettiker) probe to the drain contact. I_d is therefore explicitly given by

$$I_d = I_{sd} + \sum_{j=1}^{N} I_{gd}^{j}, \tag{6.60}$$

where $I_{gd}^{j} = \frac{2e}{h} \int dE\, \Gamma_j |G_{j,N}^{r}|^2 \Gamma_d [f(E_f^s - |e|V_{\mathrm{gs}}) - f(E_f^d)]$ with $\Gamma_j = -2\,\mathrm{Im}(\Sigma_j^r)$ of each gate contact and $\Gamma_d = -2\,\mathrm{Im}(\Sigma_d^r)$ for the drain contact.

6.3.3 Multimode Transport with Independent 1D Subbands

When the diameter of the nanowire/tube FET increases, multiple one-dimensional subbands have to be incorporated in the simulation. In a rigorous analysis, we would have to extend NEGF and the calculation of the Poisson equation to higher dimensions (see Section 6.4). However, in many cases the 1D subbands can be considered as being independent of each other and if we can still assume that a separation ansatz is appropriate, the contribution of multiple subbands can be simply computed individually and then added (both in terms of charge and current). This saves a great deal of computational burden since the time for the calculation only increases linearly according to the number of subbands. As an example, Figure 6.8(a) shows the local density of states of a conventional nanowire FET where the contribution of three 1D subbands have been accounted for. Here, a nanowire with quadratic cross-section $d_{\mathrm{nw}} \times d_{\mathrm{nw}}$ and particle-in-the-box-like quantization in the direction perpendicular to current transport is assumed. Thus, the subband energies are split off the conduction band bottom leading to the 1D subbands; the appropriate contact self-energy functions exhibit the same subband structure. For simplicity, self-consistency is not considered here.

At the end of Section 6.2.4, it was discussed that the charge in each subband decreases when the cross-sectional area of the nanowire increases according to $1/d_{\mathrm{nw}}^2$. However, it is important to keep in mind that an increasing number of subbands will be occupied with carriers in the case of larger nanowire diameters. For instance, if one dimension of the nanowire is continuously increased, an increasing number of modes will gradually lead from a 1D nanowire with a single or few contributing subbands to a two-dimensional system (as has been done with an analytic calculation in Section 5.2).

Scattering in the case of multimode transport can also be take into consideration by employing Buettiker probes. In this case, one can even distinguish between intra-

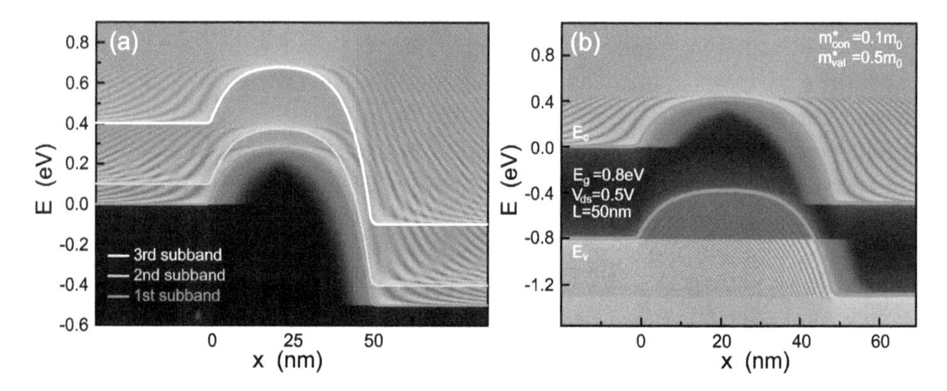

Figure 6.8: (a) Local density of states of the first three 1D subbands in a MOSFET. The subbands are considered to be independent of each other, such that the local DOS is simply obtained by adding the contributions of the subbands. (b) Local density of states in a MOSFET with conduction and valence bands exhibiting different effective masses. For simplicity, both calculations were done without self-consistency.

and inter-subband inelastic scattering. In the case of intra-subband scattering, Buettiker probes can be connected to each finite difference site for each individual subband with the requirement that the net current each Buettiker probe carries will vanish within each subband. This means that at each finite difference site there will be as many Buettiker probes connected as there are subbands. Each of the probes injects carriers only from the local subband energy on at the particular finite difference site and will have its own floating quasi-Fermi level that ensures that the net current injected by the Buettiker probes within each subband is zero. Alternatively, one Buettiker probe per finite difference site and subband can be connected as before. But now, all Buettiker probes at a specific finite difference site are short-circuited, sharing the same quasi-Fermi level. In this case, the overall net current at each finite difference site has to be zero. As a result, carriers extracted at a certain energy from one of the Buettiker probes can also be injected at another energy into a different subband. Hence, using Buettiker probes in this way mimics the effect of inter-subband scattering. An example of this will be discussed in Section 8.1.2.

6.3.4 Including Conduction and Valence Bands—Energy-Dependent Effective Mass

So far, we only considered the conduction band including multiple subbands in the effective mass approximation. This implies that for energies below the conduction band (i. e., within the band gap of the semiconductor), the quadratic dispersion relation will be continued with a purely imaginary κ (see Figure 2.23, thin black lines). However, the band gap E_g is certainly finite and (at least in a direct semiconductor) it has to be ensured that the complex band structure within the band gap connects the conduction and valence bands (cf. Figure 2.23, green dotted line). A smooth transition from conduction to valence band, while maintaining the simplicity of the effective mass approach, can be implemented with Flietner's dispersion relation [74, 101], which was already briefly discussed in Section 2.5.2. Using Equation (2.45), the energy effective mass $m_E^\star$ can be written as

$$m_E^\star = m_c^\star \left(1 + \frac{E - \Phi_f^j}{E_g}\right) \cdot \left(1 + a\frac{E - \Phi_f^j}{E_g}\right)^{-2} \tag{6.61}$$

which is valid for energies between the conduction band Φ_f^j at j and $\Phi_f^j - E_g/2$. Here, $a = 1 - \sqrt{m_c^\star/m_v^\star}$ and $m_c^\star$ and $m_v^\star$ are the conduction and valence band effective masses at the band edges, respectively. For energies in the lower half of the band gap, the energy effective mass is given by

$$m_E^\star = m_c^\star \left(1 + \frac{(\Phi_f^j - E_g) - E}{E_g}\right) \cdot \left(1 + a\frac{(\Phi_f^j - E_g) - E - E_g}{E_g}\right)^{-2}. \tag{6.62}$$

Note that in the case of a symmetric semiconductor (for instance, a semiconducting carbon nanotube) the effective masses in the conduction and valence bands are equal, and consequently, $m_c^\star = m_v^\star = m^\star$ so that $a = 0$. The energy effective masses can be used to determine the hopping parameter $t = \frac{\hbar^2}{2m_E^\star a^2}$ within the band gap. However, since $m_E^\star$ depends on the lattice site j the discretized form of the Hamiltonian has to be modified according to (compare with Equation (6.22))

$$\mathcal{H}_{jl} = \begin{cases} t^- + t^+ + \Phi_f^j & j = l, \\ -t^- & j - 1 = l, \\ -t^+ & j + 1 = l, \\ 0 & \text{otherwise}, \end{cases} \tag{6.63}$$

where j, l are indices of different points of the finite difference grid. $t^\mp = [\frac{1}{2t_j} + \frac{1}{2t_{j\mp1}}]^{-1}$ and the hopping parameters $t_j^{c(v)} = \frac{\hbar^2}{2m_{c(v)}^\star a^2}$ in the conduction (valence) band and $t_j = \frac{\hbar^2}{2m_E^\star a^2}$ within the band gap. With Equation (6.63), the equations to solve for the retarded and advanced Green function can be set up in the way described in the preceding sections. Figure 6.8(b) shows a typical local density of states of a MOSFET with conduction and valence bands exhibiting two different effective masses (see figure) computed with an energy effective mass using Flietner's dispersion relation; the local DOS plot in Figure 4.19 has also been calculated accordingly.

6.3.5 Including Conduction and Valence Bands—1D Simulation with Two-Band Tight-Binding Calculation

The use of Flietner's dispersion relation maintains the simplicity of the effective mass approximation to compute the Green's functions. This, however, implies an energy-dependent effective mass that renders the Hamiltonian non-Hermitian. As an alternative approach, which is almost as computationally efficient and can serve as starting point for NEGF simulations taking a full tight-binding calculation of the semiconductor into account, a 1D artificial solid can be defined and the band structure based on a tight-binding description (cf. Section 2.4.3 and 2.4.6) is included in the NEGF. This artificial 1D solid works particularly well for a symmetric semiconductor with equal effective masses $m_c^\star = m_v^\star$ as is the case in, e. g., a carbon nanotube.

Considering the recipe for TB calculations discussed in Section 2.4.6, it is clear that such a 1D solid needs to consist of either two atoms with a single orbital in each unit cell or a single atom with two orbitals in each unit cell. Here, we choose the former with two atoms exhibiting different nearest neighbor overlap integrals within a unit cell and in between adjacent unit cells as illustrated in Figure 6.9. Doing so we can reproduce the band structure of a carbon nanotube well within a certain energy range (see below). We have

$$\left| \begin{pmatrix} \epsilon_r & V_{ss}^1 \\ V_{ss}^1 & \epsilon_b \end{pmatrix} + \begin{pmatrix} 0 & 0 \\ -V_{ss}^2 & 0 \end{pmatrix} e^{ika} + \begin{pmatrix} 0 & -V_{ss}^2 \\ 0 & 0 \end{pmatrix} e^{-ika} - \begin{pmatrix} E & 0 \\ 0 & E \end{pmatrix} \right| = 0. \tag{6.64}$$

In order to get a symmetric band structure, red and blue atoms are equal, and hence the on-site energies $\epsilon_r = \epsilon_b$ and for simplicity we set $\epsilon_r = 0$. Solving Equation (6.64) yields

$$E(k) = \pm\sqrt{(V_{ss}^1)^2 + (V_{ss}^2)^2 - 2V_{ss}^1 V_{ss}^2 \cos(ka)}. \tag{6.65}$$

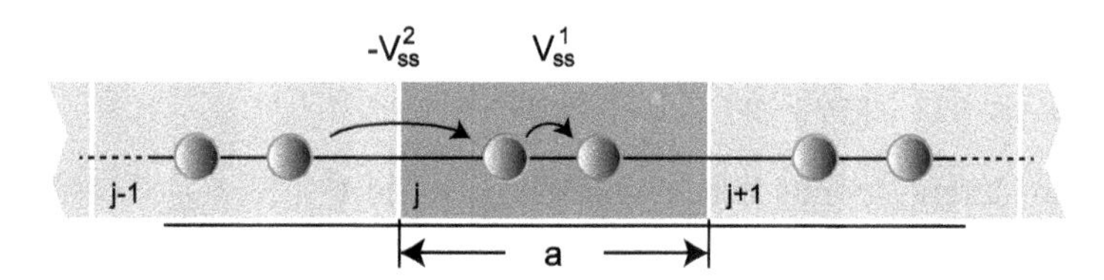

Figure 6.9: Schematic of the artificial, 1D solid with two atoms per unit cell (gray area, lattice constant *a*) each providing a single *s*-orbital. A tight-binding calculation yields a two-band structure with conduction and valence bands.

In the case $|(V_{ss}^1)^2 + (V_{ss}^2)^2| > |2V_{ss}^1 V_{ss}^2|$, the desired band structure with symmetric conduction/valence bands and direct band gap is obtained (green solid lines in Figure 6.10). Around the band gap, the computed band structure resembles a quadratic dispersion (green dashed line). However, further away from E_g the band structure becomes almost linear (compare with the blue dashed lines in Figure 6.10) before deviating from the linear behavior. Thus, the band structure of the artificial 1D solid nicely reproduces the qualitative appearance of the band structure of a carbon nanotube within the light gray energetic range. In fact, it reproduces the nanotube dispersion significantly

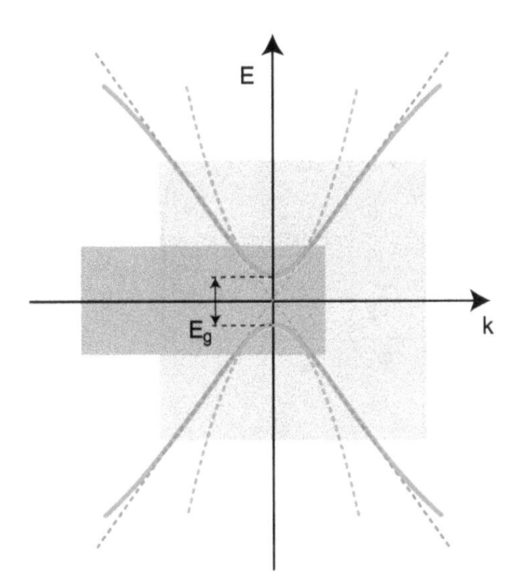

Figure 6.10: The band structure according to the two-band model approximates a quadratic dispersion relation well within the gray energy range. Within the substantially larger, light gray range, the two-band model reproduces well the band structure of a CNT.

better than an effective mass approximation (which only fits within the gray range). Note that with the artificial 1D solid no energy-dependent effective mass has to be introduced, since the appropriate complex band structure is part of the solution: indeed, for small k, the cosine-term can be Taylor expanded, and as a result one obtains the same band structure as given in Equation (2.60). This also means that replacing k with $i\kappa$ one obtains the same result as given in Section 2.9.

Depending on the particular carbon nanotube (or another semiconductor if the energy range is restricted to the dark gray energy range in Figure 6.10), one has to adjust the parameters in order to realize certain effective masses and band gaps. The band gap is $E_g = 2 \times E(k = 0) = 2 \times \sqrt{(V_{ss}^1)^2 + (V_{ss}^2)^2 - 2V_{ss}^1 V_{ss}^2} = 2|V_{ss}^1 - V_{ss}^2|$. Extracting the effective mass around $k \approx 0$, one obtains $m^\star = \hbar^2 \frac{E_g}{2a^2 V_{ss}^1 V_{ss}^2}$. If a real nanotube is to be simulated, E_g and $m^\star$ can be extracted from, e. g., a full tight-binding calculation and then the expressions above can be used to solve for V_{ss}^1 and V_{ss}^2. Before setting up the Green's function for the two-band model, solve Task 24 to recapitulate the computation of surface Green's functions (this has been used to compute the surface DOS in Section 2.12.3):

Task 24.
Surface Green's function: Explicitly compute the surface Green's function of the surface of the semiinfinite 2D crystal shown in Figure 2.21(a) with s, p_x-orbitals; set $V_{sp} = 0$.

In a next step, the NEGF matrices have to be set up in a similar way as above. From Equation (6.64) it is known how the Hamiltonian has to be constructed, which yields

$$\mathcal{H} = \left(\begin{array}{cc|cc|cc|c} \Phi_f^j & V_{ss}^1 & 0 & 0 & 0 & 0 & \cdots \\ V_{ss}^1 & \Phi_f^j & -V_{ss}^2 & 0 & 0 & 0 & \cdots \\ \hline 0 & -V_{ss}^2 & \Phi_f^{j+1} & V_{ss}^1 & 0 & 0 & \cdots \\ 0 & 0 & V_{ss}^1 & \Phi_f^{j+1} & -V_{ss}^2 & 0 & \ddots \\ \hline 0 & 0 & 0 & -V_{ss}^2 & \Phi_f^{j+2} & V_{ss}^1 & 0 \\ \vdots & \vdots & \vdots & \vdots & \vdots & \vdots & \ddots \end{array}\right). \tag{6.66}$$

Setting all potentials to the same value Φ_f^0 one can go through the same analysis as above to incorporate contacts via appropriate self-energies (cf. Section 6.2.3). The self-energy in this case will be a 2×2-matrix. However, since only one of the atoms of each unit cell is connected to the adjacent cells, the self-energy is

$$\Sigma^r = \tau G_C^r \tau^\dagger = \left(\begin{array}{cc|cc} 0 & 0 & 0 & \cdots \\ -V_{ss}^2 & 0 & 0 & \cdots \end{array}\right) G_C^r \left(\begin{array}{cc} 0 & -V_{ss}^2 \\ 0 & 0 \\ \hline 0 & 0 \\ \vdots & \vdots \end{array}\right) = \begin{pmatrix} 0 & 0 \\ 0 & (V_{ss}^2)^2 (G_C^r)_{1,1} \end{pmatrix}. \tag{6.67}$$

The surface Green's function is therefore obtained from

$$\begin{pmatrix} (G_C^r)_{1,1} & (G_C^r)_{1,2} \\ (G_C^r)_{2,1} & (G_C^r)_{2,2} \end{pmatrix} = \left[E + i\eta - \begin{pmatrix} \Phi_f^0 & V_{ss}^1 \\ V_{ss}^1 & \Phi_f^0 \end{pmatrix} - \begin{pmatrix} 0 & 0 \\ 0 & (V_{ss}^2)^2 (G_C^r)_{1,1} \end{pmatrix} \right]^{-1}. \tag{6.68}$$

The inverse of the 2×2-matrix can easily be computed (see also Task 24) so that the surface Green's function (only the $(G_C^r)_{1,1}$-element is needed) is obtained by solving

$$(G_C^r)_{1,1} = \frac{E + i\eta - (V_{ss}^2)^2 (G_C^r)_{1,1}}{(E - \Phi_f^0)(E + i\eta - \Phi_f^0 - (V_{ss}^2)^2 (G_C^r)_{1,1}) - (V_{ss}^1)^2}. \tag{6.69}$$

All other relations will then be similar to the discussion above.

6.4 Simulations of Devices in Higher Dimensions

In the preceding sections, electronic transport and the Poisson equation have been described in a 1D framework. Within this framework, the source/drain extensions can be included using conformal mapping and multiple independent subbands can be incorporated to simulate nanowire devices with larger diameter. However, a more rigorous treatment of the device behavior is sometimes needed. In particular, the electrostatics of the device plays a very important role for the functionality. For instance, the calculation of the parasitic capacitances of nanowire bundle FETs in Section 5.9.2 is based on a 1D NEGF of the electronic transport, while the electrostatics is considered in 3D (here 2D due to the (approximately) cylindrical symmetry of the device). Therefore, an extension of the numerical solution of the Poisson equation to the 2D/3D case will be provided before a brief introduction into NEGF in higher dimensions will be given.

6.4.1 Poisson Equation

As an example for setting up a numerical finite difference framework for the 2D case, Figure 6.11 shows a schematic image of a conventional bulk transistor. If the width W of the device can be considered to be very large, all quantities are independent of the y-coordinate and the calculation can be reduced to a 2D finite difference problem as illustrated.

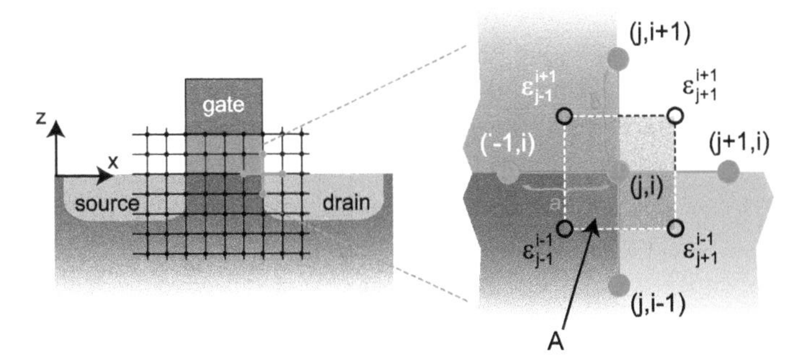

Figure 6.11: Two-dimensional cross-section of a bulk-MOSFET with finite-difference grid (left). The right panel shows a close-up of the five-point stencil of the 2D finite-difference grid centered around lattice point (j, i) at the silicon-gate dielectric interface.

Let us start by subdividing the computational domain into a regular grid with lattice constant a in the x- and b in the z-direction, as illustrated in Figure 6.11. The right panel of Figure 6.11 shows a close-up of the so-called five-point stencil for the potential $\Phi(x_j, z_i)$ centered around the point (j, i). From the simple 1D finite difference formula given above (cf. Equation (6.1)), it is easy to see that the Laplace operator in 2D is

$$\frac{\partial^2\Phi(x,z)}{\partial x^2} + \frac{\partial^2\Phi(x,z)}{\partial z^2} \to \frac{\Phi_{j+1,i} - 2\Phi_{j,i} + \Phi_{j-1,i}}{a^2} + \frac{\Phi_{j,i+1} - 2\Phi_{j,i} + \Phi_{j,i-1}}{b^2}. \tag{6.70}$$

Equation (6.70) can be written in matrix form and solved for the potential distribution once the carrier density is known. The extension from 1D to 2D is thus straightforward including boundary conditions (see Task 25); only the interfaces of regions with different dielectric constants need more attention.

Task 25.
2D Poisson equation: Write down explicitly the matrix equation for the 2D Poisson equation considering a square lattice with 3×3 lattice points and lattice constants a (in the x- and y-directions). Assume Neumann boundary conditions with zero electric field at the bottom and left boundaries and Dirichlet boundary conditions at the remaining two boundaries with constant Φ_0 at the top and Φ_1 at the right boundary. Moreover, a constant charge density is assumed.

To include dielectric boundaries in the Poisson equation, we need to consider the generalized form of it given by

$$\nabla(\varepsilon(\vec{x})\nabla\Phi(\vec{x})) = -\frac{e^2 n(\vec{x})}{\varepsilon_0}. \tag{6.71}$$

In order to find an appropriate expression for the finite difference representation of the generalized Poisson equation incorporating regions of different dielectric constants, we define an auxiliary lattice with the same lattice constants a and b but shifted with respect to the original one as illustrated with the hollow black circles in Figure 6.11, right panel. If Equation (6.71) is integrated over the gray shaded area, i. e., over one unit cell of area $A = a \cdot b$ of the auxiliary finite difference lattice, the right-hand side will lead to the charge $(-e)n_{j,i} \cdot a \cdot b$ (per length, since we do not consider the y-direction) on the site (j, i) divided by ε_0. On the other hand, Gauss' law tells us that $\int_A \nabla(\varepsilon\nabla\Phi)\, dx\, dz = \oint_C \varepsilon\nabla\Phi\, d\vec{n}$ where C is the circumference (dashed line) of A and $d\vec{n}$ is the differential unit normal vector. The contour integral $\oint$ can be subdivided into individual integrations along the four sides of A. Exemplarily, the integration along the right vertical edge is carried out. Using the coordinate system as displayed in the figure, it is clear that at this edge the scalar product of $\nabla\Phi$ and $d\vec{n}$ yields only $\frac{\partial\Phi}{\partial x} \cdot dz + \frac{\partial\Phi}{\partial x} \cdot 0$. The derivative can be approximated with the forward difference $\partial\Phi(x,z)/\partial x \approx \frac{1}{a}(\Phi_{j+1,i} - \Phi_{j,i})$. Integrating along the edge, one half is in the region with ε^{i-1}_{j+1} and the other half with ε^{i+1}_{j+1} (doped silicon and air in the example displayed in Figure 6.11). Since the partial derivative is constant along the edge, we obtain (assuming the origin is at lattice point (j, i))

$$\int_{-b/2}^{0} dz\, \varepsilon^{i-1}_{j+1} \frac{\Phi_{j+1,i} - \Phi_{j,i}}{a} + \int_{0}^{b/2} dz\, \varepsilon^{i+1}_{j+1} \frac{\Phi_{j+1,i} - \Phi_{j,i}}{a} = b\frac{\varepsilon^{i-1}_{j+1} + \varepsilon^{i+1}_{j+1}}{2}\frac{\Phi_{j+1,i} - \Phi_{j,i}}{a}. \tag{6.72}$$

Similar expressions can be set up for the remaining three edges of the contour. Putting everything together (and dividing both sides by $a \cdot b$) we obtain the following discrete Poisson equation:

$$\frac{\frac{\varepsilon^{i-1}_{j+1}+\varepsilon^{i+1}_{j+1}}{2}\Phi_{j+1,i} - \frac{\varepsilon^{i-1}_{j+1}+\varepsilon^{i+1}_{j+1}+\varepsilon^{i+1}_{j-1}+\varepsilon^{i-1}_{j-1}}{2}\Phi_{j,i} + \frac{\varepsilon^{i+1}_{j-1}+\varepsilon^{i-1}_{j-1}}{2}\Phi_{j-1,i}}{a^2}$$
$$+ \frac{\frac{\varepsilon^{i+1}_{j-1}+\varepsilon^{i+1}_{j+1}}{2}\Phi_{j,i+1} - \frac{\varepsilon^{i-1}_{j+1}+\varepsilon^{i+1}_{j+1}+\varepsilon^{i+1}_{j-1}+\varepsilon^{i-1}_{j-1}}{2}\Phi_{j,i} + \frac{\varepsilon^{i-1}_{j+1}+\varepsilon^{i-1}_{j-1}}{2}\Phi_{j,i-1}}{b^2} = -\frac{e^2 n_{j,i}}{\varepsilon_0}. \tag{6.73}$$

This equation has a simple form, namely it is basically Equation (6.70) where each potential is multiplied with the average dielectric constant at the particular point.

A further extension into 3D is conceptually straightforward. However, since this requires a third index together with possible eight different dielectric constants the notation will become somewhat messy and will not be stated here explicitly. Boundary conditions, either Dirichlet or Neumann, can be implemented in the same way as has been done in the one-dimensional case stated earlier.

Task 26.
Finite difference with cylindrical coordinates: Derive a finite difference expression for the Poisson equation in cylindrical coordinates. A regular lattice with lattice constant a should be used.

6.4.2 NEGF Devices in 2D/3D

The one-dimensional NEGF simulations presented above can be extended to the 2D or 3D case. In Section 6.3.5, the 1D model was already extended to include a conduction and a valence band. While in this case the inversion of the matrix equation (6.68) can be done analytically, in the general case of a high-dimensional system, the surface Green's functions that describe the contacts need to be computed iteratively with Equation (6.29). The iteration scheme to determine the surface Green's function converges slowly but can be sped up using, e. g., the algorithm suggested in [178]. Since in the case of ballistic

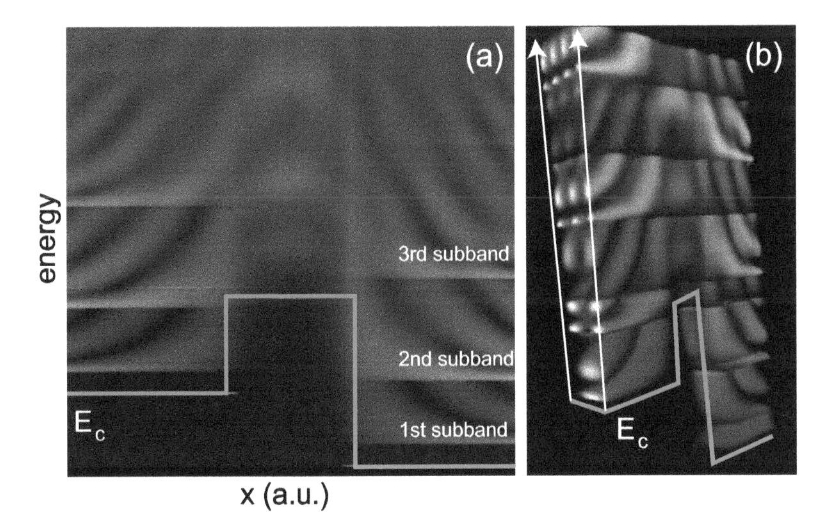

Figure 6.12: (a) Local density of states in an ultrathin-body 2D device with idealistic step-function potential distribution (green line). For simplicity, a particle-in-the-box quantum confinement was assumed in one direction and three subbands are visible in the plotted energy range. (b) 3D plot of the same local density of states.

transport only the first and the last columns of the overall Green's function are needed a recursive Green's function algorithm is usually employed. The reader is referred to the extensive literature available on the subject. Figure 6.12 shows a local density of states plot at the center of an ultrathin-body device where carrier confinement in one spatial direction leads to the formation of subbands. For simplicity, a particle-in-the-box quantization has been assumed. Furthermore, in order to illustrate the local density of states in the 3D image, a simple step-function potential distribution has been implemented. Figure 6.12(b) allows a clear observation of the three subbands shown in (a). An extension of the NEGF calculations presented here is in principle straightforward but requires clever coding in order to cope with the numerical burden of a 2D or even 3D simulation.

Exercises

Exercises together with solutions are accessible via the QR code.

7 Metal-Source-Drain Field-Effect Transistors

When scaling down transistors, one has to ensure that the parasitic resistances associated with the doped source and drain regions are always substantially smaller than the resistance of the channel in the device's on-state. The reason for this is the following: Looking at Figure 7.1, it is clear that the effective drain-source voltage $V_{\mathrm{ds}}^{\mathrm{eff}}$ and effective gate-source voltage $V_{\mathrm{gs}}^{\mathrm{eff}}$ are substantially reduced if the parasitic source/drain resistances $R_{s,d}^{\mathrm{par}}$ become significant. In this case, one obtains $V_{\mathrm{gs}}^{\mathrm{eff}} = V_{\mathrm{gs}} - I_d R_s^{\mathrm{par}}$ and $V_{\mathrm{ds}}^{\mathrm{eff}} = V_{\mathrm{ds}} - I_d(R_s^{\mathrm{par}} + R_d^{\mathrm{par}})$. Inserting this into the expression for the drain current calculated with the gradual channel approximation in Section 5.1 yields

$$I_d = \mu C_{\mathrm{ox}} \frac{W}{L}\left((V_{\mathrm{ds}} - (R_s^{\mathrm{par}} + R_d^{\mathrm{par}})I_d)(V_{\mathrm{gs}} - R_s^{\mathrm{par}} I_d - V_{\mathrm{th}}) - \frac{(V_{\mathrm{ds}} - (R_s^{\mathrm{par}} + R_d^{\mathrm{par}})I_d)^2}{2}\right), \quad (7.1)$$

which can be solved for I_d.

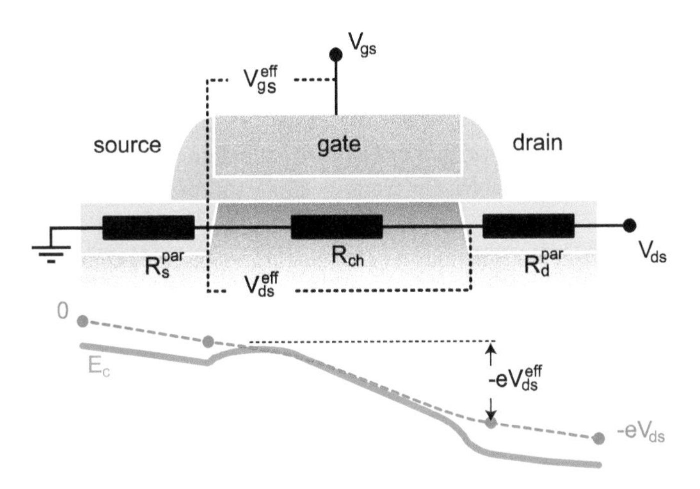

Figure 7.1: Parasitic source/drain resistances $R_{s,d}^{\mathrm{par}}$ in series with the channel resistance R_{ch} lead to a reduction of the effective gate voltage $V_{\mathrm{gs}}^{\mathrm{eff}}$ and the effective drain-source bias $V_{\mathrm{ds}}^{\mathrm{eff}}$.

However, reducing the parasitic resistances becomes increasingly difficult when scaling down the transistor dimensions. Issues associated with doping source and drain include a finite solid solubility of dopants in the semiconductor, the nanoscale dimensions of source/drain extensions in nanowire devices that lead to deactivation of dopants (see Section 4.3.2 and Chapter 5) and variability due to random dopant effects. All these issues make a replacement of the doped source/drain regions with metallic electrodes highly attractive. Obvious advantages are: (i) very low parasitic resistances, (ii) no implantation/activation is necessary, and thus, dopant diffusion is avoided, leading to (iii) well-defined, atomically abrupt interfaces between source/drain and channel holding promise for reduced variability.

https://doi.org/10.1515/9783111054421-007

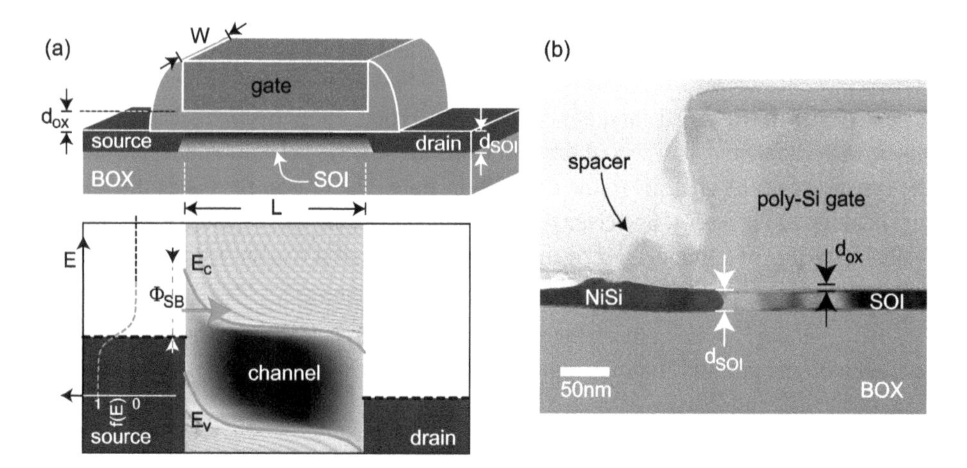

Figure 7.2: (a) Schematic of a silicon-on-insulator Schottky-barrier MOSFET with a thin channel layer $d_{ch} = d_{SOI}$, channel length L and width W; the local density of states together with the conduction and valence bands along the direction of current transport are depicted underneath. Metal-induced gap states can clearly be identified at the metal-semiconductor interfaces. (b) TEM image of the source area of an experimental SOI SB-MOSFET with NiSi contacts [149].

Figure 7.2(a) (top panel) shows a schematic of a MOSFET with metallic source/drain contacts and (b) displays a transmission electron micrograph of the source region of such a MOSFET with nickel silicide contacts (see Chapter 3) clearly showing the well-defined geometry of the source electrode [149]. However, as was elaborated on in Section 4.6, at a metal-semiconductor interface a Schottky barrier builds up due to Fermi level pinning, which leads to a substantial potential barrier at the contact channel interfaces. The lower panel of Figure 7.2(a) shows a plot of the local density-of-states (lDOS) in the channel of a MOSFET with metallic source/drain contacts. Fermi level pinning at approximately mid-gap has been assumed in the present case. Metal-induced gap states lead to a high density of interface states as apparent from the bright areas at the contact-channel interfaces. In the case of $\Phi_{SB} \gtrsim 4–5 \times k_B T$, carriers are injected into the channel mainly via tunneling through the SB as indicated with the red arrow in Figure 7.2(a). Therefore, MOSFET devices with metallic source/drain contacts are called Schottky-barrier MOSFETs (SB-MOSFETs).[1] SB-MOSFETs have attracted a great deal of interest due to the reasons mentioned above. However, the presence of the Schottky barrier has a large impact on the device's on- and off-state behavior. Hence, a great deal of research has been focused on the reduction of the Schottky-barrier height.

In the present chapter, the operating principles of SB-MOSFETs, strategies for their optimization and limitations will be discussed. Note that, even though SB-MOSFETs ex-

1 Note that there is a Schottky barrier for electron injection Φ_{SB}^{elec} and for hole injection Φ_{SB}^{hole} with $\Phi_{SB}^{elec} + \Phi_{SB}^{hole} = E_g$. Throughout the book, the term Schottky barrier refers to the barrier for electron injection (if not stated otherwise) denoted with Φ_{SB}.

hibit a number of drawbacks, there is good reason to thoroughly study them: virtually all devices based on novel materials (such as carbon nanotubes and 2D materials; see Chapters 8 and 10) are basically Schottky-barrier MOSFETs because contacting those materials by simply depositing a metal on them is the most straightforward (and sometimes the only) way for making a device. Knowledge about the operating principles of SB-MOSFETs is therefore mandatory, since wrong conclusions may be drawn from experimental data, in particular with respect to extracting carrier mobilities.

7.1 Operating Principles of SB-MOSFETs

It has been already mentioned above (cf. Section 4.6) that at metal-semiconductor interfaces a Schottky barrier builds up due to Fermi level pinning.[2] In the following we will assume that Fermi level pinning is strong enough for the Schottky barrier to be fixed and that it depends neither on the doping concentration in the semiconductor nor on the carrier density induced by applying a gate voltage or the gate voltage itself. This approximation works very well in many cases and allows for drawing a number of conclusions regarding the operation of SB-MOSFETs without complicated calculations. Hence, before further quantitatively elaborating on this let us first try to understand qualitatively how SB-MOSFETs work.

To start with, we will look at a long-channel, *n*-type SB-MOSFET and will concentrate on the injection of carriers at the source–channel interface into the conduction band of the channel. Figure 7.3 shows several conduction band profiles along the direction of current transport. At small gate voltages, the potential maximum Φ_f^0 that determines the current flow (cf. Section 5.2.1) is located energetically above the Schottky barrier. Increasing the gate voltage moves Φ_f^0 to lower energies. As long as $\Phi_f^0 \geq \Phi_{SB}$, the Schottky barrier will have no impact on the current flow and the device behaves exactly like a conventional-type transistor since it is based purely on thermal emission of carriers over Φ_f^0. As a result, an exponential increase of the current with an inverse subthreshold slope of at best 60 mV/dec is obtained. However, as soon as Φ_f^0 is moved below Φ_{SB}, the Schottky barrier is the maximum barrier in the channel, and hence will determine the injection of carriers.

How the current further increases with increasing V_{gs} depends on the details of the potential profile of the Schottky barrier (which will be a function of the device geometry) and on the material properties (effective mass, band gap, dielectric constant etc.) of the channel. In any case, it is clear that a further increase of current will be somewhere within the gray shaded area in Figure 7.3: if there was no tunneling through the Schottky barrier at all, the current would be determined by thermal emission over the

2 Even in the absence of Fermi level pinning a potential barrier may build up according to the difference in work functions of the contact metal and the semiconductor in use.

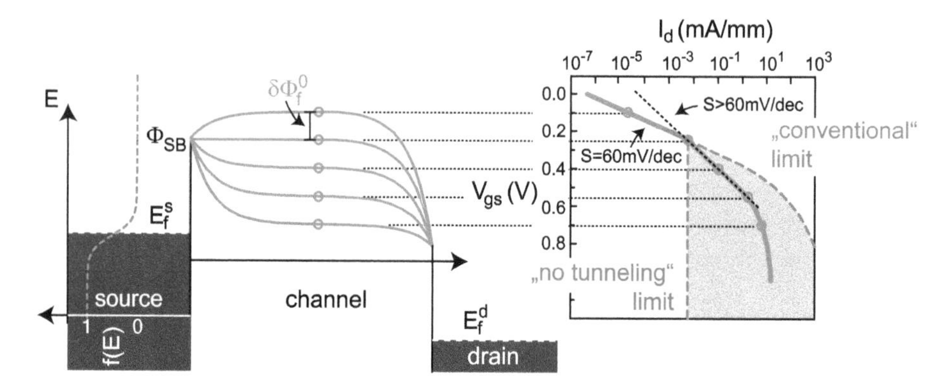

Figure 7.3: Left: Conduction band in a SB-MOSFET for different gate voltages. As long as $\Phi_f^0 \geq \Phi_{SB}$, the transistor behaves like a conventional MOSFET with an inverse subthreshold slope of at best 60 mV/dec as illustrated in the right panel. If $\Phi_f^0 < \Phi_{SB}$, the increase of current is given by the change of the tunneling probability with V_{gs} yielding $S > 60$ mV/dec and a deteriorated on-state performance.

constant Schottky barrier, and thus there would be no further change of I_d with gate voltage as illustrated with the red dashed line in Figure 7.3. If, on the other hand, the tunneling probability through the Schottky barrier approached unity, the current would further increase as in a conventional MOSFET (blue dashed line in Figure 7.3). Thus, in the case of finite tunneling the curve will lie in-between these two extremes. As a result, one observes a distinct kink in the I_d–V_{gs}-curves at the voltage where the device behavior changes from thermal emission to tunneling. The lower the tunneling probability through the SB, the more pronounced the kink.

Up to now, we only discussed the contribution of the conduction band. However, from the bottom panel of Figure 7.2(a) it is clear that because the metallic source and drain electrodes do not have a band gap, carriers can be injected into the conduction and into the valence band of the channel depending on the particular gate voltage applied. This leads to so-called ambipolar behavior, i. e., SB-MOSFETs can be operated as *n*-type and as *p*-type transistors, depending on the applied voltages.

Figure 7.4 shows I_d–V_{gs}-curves for three different Schottky-barrier heights: (a) displays the transfer characteristics and three conduction/valence band profiles in the case $\Phi_{SB} = 0$, (b) shows the same for $\Phi_{SB} = E_g/2$ and (c) in the case of $\Phi_{SB} = E_g$. The transfer characteristics in each case can be constructed using the behavior discussed in Figure 7.3 by plotting the branches for the injection of electrons and holes separately. The cross-over from electron to hole branch (or vice versa) occurs when the currents of both branches are equal. Each branch consists of a thermal emission and tunneling part with a kink separating the two regimes. The position (with respect to the gate voltage) of the kink depends on the height of the Schottky barrier and on the applied bias: In the case of $\Phi_{SB} = 0$ ($\Phi_{SB} = E_g$), the kink in the transfer characteristics of the electron branch (hole branch) lies close to the on-state of the SB-MOSFET when operated as *n*-type (*p*-type) FET, and thus can hardly be observed. On the other hand, the kink in the hole branch (elec-

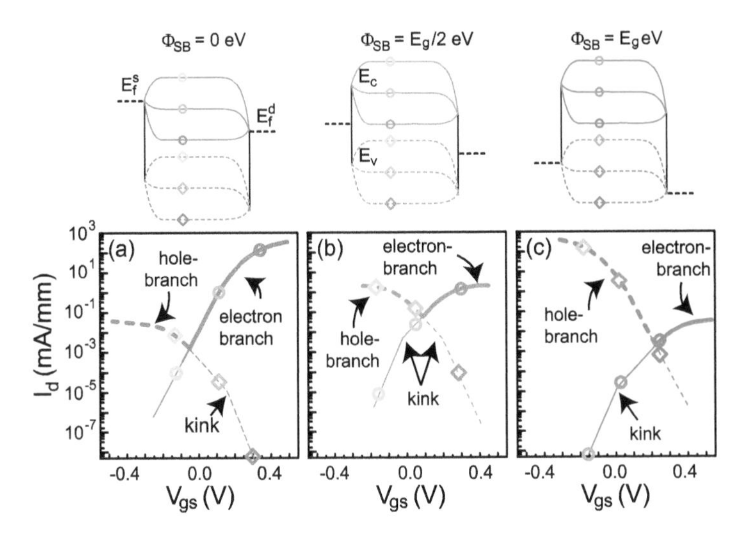

Figure 7.4: Ambipolar characteristics of SB-MOSFETs for three different Φ_{SB} explaining how electron and hole branches determine the appearance of the I_d–V_{gs}-curves in SB-MOSFETs.

tron branch) is hidden by the quickly increasing current of the electron (hole) branch. As a result, in the case of (a) and (c) one obtains a device with a switching behavior similar to a conventional FET but increased leakage due to the ambipolar operation. If the Schottky barrier is at mid-gap, the switching behavior is exclusively determined by tunneling through the Schottky barriers for the electron and hole branches. Note that in this symmetric case the kink in the $I_d - V_{gs}$-curves is hidden for both electron and hole branches, which becomes obvious when looking at the conduction/valence band profiles on top of Figure 7.4(b); if we assume a similar tunneling probability for electrons and holes, the hole branch already sets in before the electron branch enters the regime of thermal emission, and hence no kink can be observed. The inverse subthreshold slope in this case is $S > 60$ mV/dec, for both, electron and hole branches. QR code #55 provides more details on the transfer characteristics of SB-MOSFETs.

The simultaneous injection of carriers into the conduction and the valence has an important consequence for the operation of SB-MOSFETs; Figure 7.5 shows the conduction and valence bands of a SB-MOSFET for a fixed gate voltage but two different drain-source biases. Obviously, the injection of electrons does not change since the potential distribution at the source-side of the channel has not changed. However, the drain side looks different in the two cases. In fact, relative to the drain, the valence band has been "moved" upwards yielding a thinner drain-side Schottky barrier, and thus an increased injection of holes from the drain into the channel when increasing V_{ds}. Consequently, a higher V_{ds} yields a higher gate-drain voltage V_{gd} (at constant V_{gs}), and thus a higher current of the hole branch as shown in the right panel of Figure 7.5.

Figure 7.6 shows experimental transfer characteristics belonging to the silicon-on-insulator SB-MOSFET with nickel silicide source/drain electrodes shown in Figure 7.2(b)

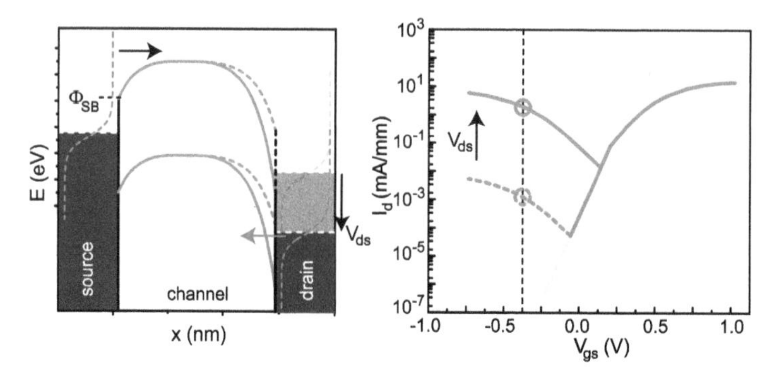

Figure 7.5: Ambipolar transfer characteristics of a SB-MOSFET for two different V_{ds}. In the case of a constant gate-source voltage V_{gs}, an increased V_{ds} appears as an increased gate-drain voltage, which leads to a larger injection of holes.

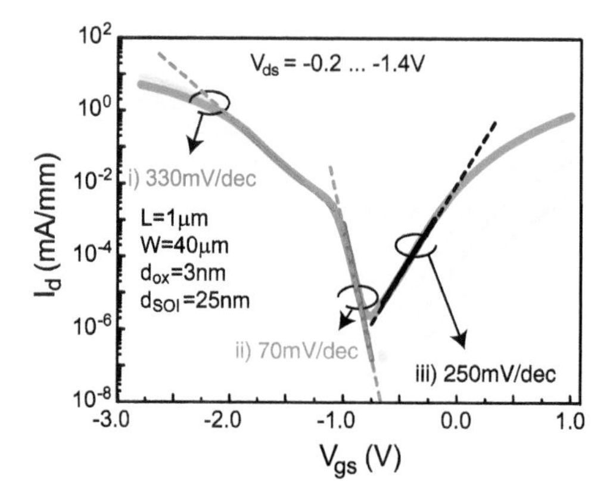

Figure 7.6: Typical transfer characteristics of a SOI Schottky-barrier MOSFET with NiSi contacts [289].

(the fabrication of such a device will be discussed in the next section) [289]; note that in the present case, the transistor is operated as a p-type device, i. e., V_{ds} is biased negatively. The data show all features that have been discussed so far: One can clearly observe ambipolar behavior with a strongly increasing current for fixed V_{gs} of the electron branch when V_{ds} decreases from -0.2 V to -1.4 V. Moreover, a distinct kink is visible in the hole branch. The clear appearance of a kink means that the Schottky barrier for hole injection Φ_{SB}^{hole} is smaller than the corresponding for electron injection. Three different inverse subthreshold slopes can be extracted from Figure 7.6; for gate voltages larger than the kink position, the current decreases with an inverse subthreshold slope of 70 mV/dec (region ii)), i. e., close to the thermal limit showing that this part of the transfer characteristics is determined by thermal emission of holes in the valence band and not tunneling through a Schottky barrier. In the case of more negative V_{gs}, the I_d–V_{gs}-curves kink and current increases with a substantially larger inverse subthreshold slope of 330 mV/dec (region i)). The latter is clearly determined by hole injection into the valence band via tunneling through the source side Schottky barrier. The inverse subthreshold slope of 250 mV/dec (region iii)) of the electron branch is a result of tunneling through the drain-side Schottky barrier due to electron injection into the conduction band. Although $\Phi_{SB}^{elec} > \Phi_{SB}^{hole}$ the inverse subthreshold slope $S_{iii)}$ is steeper for electron injection compared to hole injection $S_{i)}$. The reason for this will become clear in the next section. The point is that the height of the Schottky barrier determines the magnitude of the drain current but to first order not the *change* of current with gate voltage, i. e., the subthreshold swing. The steeper slope of the electron branch is then a consequence of the lighter effective mass of electrons in silicon compared to holes.

7.1.1 Ultrathin-Body SB-MOSFETs

After the rather qualitative discussion of Schottky-barrier MOSFETs in the preceding section, we will now introduce a model that allows computing the electrical behavior of SB-MOSFETs without the necessity of elaborate simulations. This model will also help extracting guidelines for the optimization of SB-MOSFETs.

In order to discuss electronic transport through a transistor, we need to compute the potential distribution within the device. However, particularly in long-channel SB-MOSFET (more precisely without SCE) it is the potential distribution of the source-side Schottky barrier that matters most, since the current through the device is to a large extent determined by the injection through this barrier. In order to compute the potential distribution, we will use the one-dimensional modified Poisson equation derived in Section 5.5. As a model system, a single-gate device based on a thin, fully depleted silicon-on-insulator substrate is chosen, as is illustrated in Figure 7.2(a). Note, however, that the following discussion does not depend on the particular model system and is also valid for other (ultra)thin-body FETs such as nanowire/tube transistors with multiple gates. The channel length of the device is assumed to be long enough to ensure that the drain-source bias has no impact on the potential distribution at the source contact (i. e., that short-channel effects are avoided). In this case, we can restrict the discussion to electron injection through the source-side Schottky barrier. In the next sections, we will consider the off-state and the on-state of SB-MOSFETs separately.

7.1.1.1 Off-State Behavior

In the off-state of a transistor, the density of mobile charge is negligible, i. e., $(-e) \cdot n \approx 0$. As a result, Equation (5.34) can be solved analytically leading to an exponential dependence of the potential distribution $\Phi_f(x)$ on the spatial coordinate x. With the assumption made that the SB-MOSFET under consideration has a very long-channel length L, the energetic position of the conduction band Φ_f^0 within the channel (approximately at $L/2$, i. e., far enough from the source-channel and channel-drain interfaces) is determined by the applied gate voltage, built-in voltage and work-function difference. This means that

Φ_f^0 is the same as in a conventional, long-channel transistor in the off-state, and hence is given by

$$-\frac{\Phi_f^0 - \Phi_{\text{bi}} - \Phi_g}{\lambda_{\text{ch}}^2} \approx 0. \tag{7.2}$$

Thus, $\Phi_f^0 = \Phi_{\text{bi}} - eV_{\text{gs}}$ where V_{gs} is the gate-source voltage and Φ_{bi} the built-in potential (see Section 5.5). Since a sufficiently long channel is assumed (and V_{ds} is considered large enough so that only carrier injection from source matters), Equation (5.34) can be solved with the boundary conditions $\Phi_f(x = 0) = \Phi_{\text{SB}}$ and $\Phi_f(x = L/2) \overset{L/2\to\infty}{\longrightarrow} \Phi_f^0$. As a result, the potential distribution at the source-side Schottky barrier is

$$\Phi_f(x) = (\Phi_{\text{SB}} - \Phi_f^0)\exp\left(-\frac{x}{\lambda_{\text{ch}}}\right) + \Phi_f^0. \tag{7.3}$$

An analytic expression for the current in the off-state, and hence the inverse subthreshold slope can be obtained when the potential distribution given by Equation (7.3) is replaced by an effective Schottky barrier $\Phi_{\text{SB}}^{\text{eff}}$ for thermal emission of carriers. Figure 7.7 shows the transmission probability as a function of energy (red line) computed quantum mechanically with the Fisher–Lee relation (see Section 6.47) using the potential distribution (green line) given by Equation (7.3). A sharp transition is observed with the transmission probability decreasing stronger than a simple exponential function for increasing thickness of the Schottky barrier. Therefore, a tunneling distance d_{tunnel} can be introduced that divides the Schottky barrier into two regimes; if the potential barrier is thinner than d_{tunnel}, i. e., close to the tip of the Schottky barrier, the probability for tunneling through the barrier is set to unity $T(E) = 1$. On the other hand, if the potential

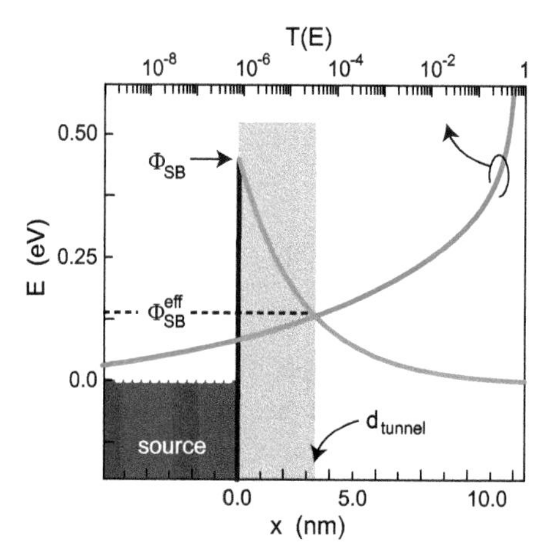

Figure 7.7: Conduction band profile (green line) of the source-side Schottky barrier together with the transmission probability $T(E)$ (red line) on a semilogarithmic plot (top-axis).

is thicker than d_{tunnel} tunneling is completely neglected by setting $T(E) = 0$ (see insets of Figure 7.8). Compared to the WKB approximation used in Section 4.6.2, this rather crude approximation is possible since in the present case the potential distribution of the source-side Schottky barrier depends exponentially (in contrast to the quadratic dependence of doped MS-contacts) on the spatial coordinate. Because $T(E)$ itself depends exponentially on the thickness of the potential barrier, it shows the sharp transition from negligible values to a probability close to one, shown in Figure 7.7.

The discussion above justifies replacing the actual Schottky barrier with a simple effective Schottky barrier for thermal emission alone $\Phi_{SB}^{eff} = \Phi_f(x = d_{tunnel})$. With $T(E) = 0$ for $E \leq \Phi_{SB}^{eff}$ and $T(E) = 1$ if $E > \Phi_{SB}^{eff}$, we can use the Landauer formalism to compute the current. Inserting d_{tunnel} into Equation (7.3) leads to the following result for the effective Schottky-barrier height Φ_{SB}^{eff}:

$$\Phi_{SB}^{eff} = (\Phi_{SB} - \Phi_f^0) \exp\left(-\frac{d_{tunnel}}{\lambda_{ch}}\right) + \Phi_f^0. \tag{7.4}$$

The exact value of d_{tunnel} does not matter for the time being. However, an estimate can be obtained using the WKB approximation showing that d_{tunnel} mainly depends on material-specific parameters in particular $m^\star$. Hence, d_{tunnel} is considered to be independent of V_{gs} in the following. In silicon SB-MOSFETs, a reasonable value for d_{tunnel} is in the range of 3.5–3.7 nm [147, 149], whereas, e. g., in carbon nanotubes, d_{tunnel} is in the range of 5 nm due to the lighter effective mass (see Section 8.1) [136].

From Equation (7.4), it is obvious that the effective Schottky barrier sensitively depends on λ_{ch}, and thus on the gate dielectric thickness d_{ox}, the channel layer thickness d_{ch}, the respective dielectric constants and on the device geometry (cf. Figure 5.12 for the λ_{ch}-values associated with different device geometries). Figure 7.8 shows the potential distribution of the source-side SB in the case of (a) a rather large λ_{ch} and (b) a much

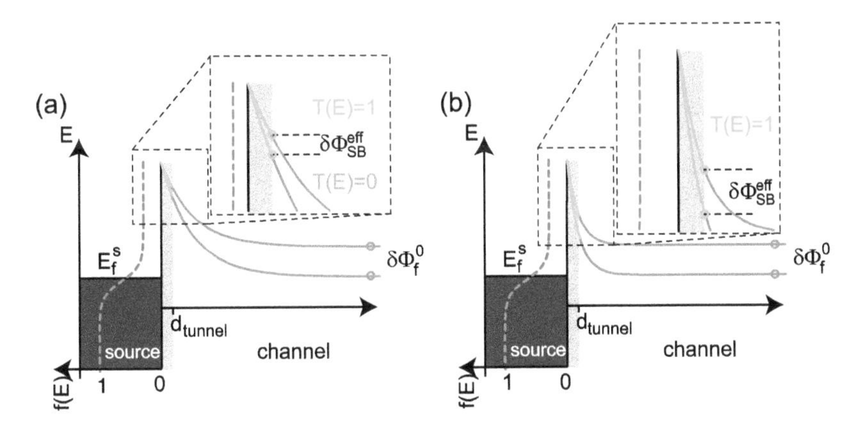

Figure 7.8: Conduction band for two different V_{gs} at the source-side Schottky barrier in the case of a large λ_{ch} (a) and a small λ_{ch} (b). The insets show close-ups illustrating the impact of V_{gs} on Φ_{SB}^{eff}.

smaller λ_{ch} for two different V_{gs}. In the case of the smaller screening length λ_{ch}, Figure 7.8(b) shows that the effective Schottky-barrier height $\Phi_{\text{SB}}^{\text{eff}}$ is changed substantially more with V_{gs} compared to the case of a larger λ_{ch}, i. e., $\delta\Phi_{\text{SB}}^{\text{eff,small}\,\lambda_{\text{ch}}} > \delta\Phi_{\text{SB}}^{\text{eff,large}\,\lambda_{\text{ch}}}$. Figure 7.8 also explains that in all cases $\delta\Phi_{\text{SB}}^{\text{eff}} < \delta\Phi_f^0$.

We can now compute the current in the off-state of SB-MOSFETs using the Landauer formula. For instance, in the two-dimensional case this yields (cf. Section 5.2):

$$I_d = \frac{2e}{h}\sum_{k_y}\frac{\Delta k_y}{2\pi/W}\int_{\Phi_{\text{SB}}^{\text{eff}}+\frac{\hbar^2k_y^2}{2m^*}}^{\infty} dE\, f_s(E_f^s) \tag{7.5}$$

where the source–drain bias was assumed large enough so that $f_d(E_f^s)$ can be neglected in Equation (7.5) but at the same time small enough so that ambipolar operation is suppressed and I_d is determined by the injection of carriers from source alone. In the off-state, the Fermi distribution function can be replaced with the Boltzmann distribution and as a result of the integration the drain current is proportional to $\exp(-\frac{\Phi_{\text{SB}}^{\text{eff}}}{k_BT})$ (compare with Equation (5.20)). The inverse subthreshold slope is now calculated in the same way as was done in Section 5.2.2:

$$S = \left(\frac{\partial\log(I_d)}{\partial V_{\text{gs}}}\right)^{-1} = \ln(10)\left(\frac{\partial I_d}{\partial\Phi_{\text{SB}}^{\text{eff}}}\frac{\partial\Phi_{\text{SB}}^{\text{eff}}}{\partial\Phi_f^0}\frac{\partial\Phi_f^0}{\partial\Phi_g}\frac{\partial\Phi_g}{\partial V_{\text{gs}}}\frac{1}{I_d}\right)^{-1}. \tag{7.6}$$

Due to the exponential dependence of the current on $\Phi_{\text{SB}}^{\text{eff}}$, one obtains $\frac{\partial I_d}{\partial\Phi_{\text{SB}}^{\text{eff}}} = -\frac{I_d}{k_BT}$. From Equation (7.4), it is clear that $\frac{\partial\Phi_{\text{SB}}^{\text{eff}}}{\partial\Phi_f^0} = 1 - e^{-d_{\text{tunnel}}/\lambda_{\text{ch}}}$. Furthermore, the change of surface potential $\delta\Phi_f^0$ with changing gate potential $\delta\Phi_g$ is the same as in a conventional MOSFET and, therefore, $\frac{\partial\Phi_f^0}{\partial\Phi_g} = 1$ in the optimum case. Finally, noting that $\frac{\partial\Phi_g}{\partial V_{\text{gs}}} = -e$ a closed expression for S is obtained:

$$S = \frac{k_BT}{e}\ln(10)\frac{1}{1-\exp(-d_{\text{tunnel}}/\lambda_{\text{ch}})} \overset{\lambda_{\text{ch}}>d_{\text{tunnel}}}{\approx} \frac{k_BT}{e}\ln(10)\left(\frac{1}{2}+\frac{\lambda_{\text{ch}}}{d_{\text{tunnel}}}\right). \tag{7.7}$$

Note that the approximation on the right side of Equation (7.7) is only valid if $\lambda_{\text{ch}} > d_{\text{tunnel}}$, which is often the case in experiments. However, $\lambda_{\text{ch}} \lesssim d_{\text{tunnel}}$ would result in $S < 60$ mV/dec, which is not possible for a SB-MOSFET at room temperature. In this case, the full expression (left part of Equation (7.7)) needs to be used.

Task 27.
SCEs in SB-MOSFETs: Compute an approximate expression for the inverse subthreshold slope of a SB-MOSFET in the case of a device that shows substantial short-channel effects. Use the simple expression for the drain capacitance C_d as given in Equation (5.19).

Task 28.
Minimal off-current of SB-MOSFET: Consider a carbon nanotube, single gate Schottky-barrier transistor with ballistic transport and mid-gap Fermi level pinning. Compute an explicit expression for the minimal off-state current as a function of bias voltage V_{ds}, channel layer thickness d_{ch}, gate dielectric thickness and constant d_{ox}, ε_{ox} as well as Schottky-barrier height Φ_{SB} and band gap E_g.

7.1.1.2 Comparison with Experiments

According to Equation (7.7), the inverse subthreshold slope S, i. e., the switching behavior, depends strongly on λ_{ch}. This means that the electrical characteristics of SB-MOSFETs can be improved substantially using thinner gate dielectrics, employing an ultrathin-body channel, gate dielectrics with high ε_k-value and a multigate device architecture. Fortunately, all these measures have to be implemented when scaling down field-effect transistor devices in order to avoid short-channel effects.

The improvement of SB-MOSFETs concerning their switching behavior can be ascribed to the exponential screening of potential variations on the length scale λ_{ch} that emerged as one of the main results of the derivation of the one-dimensional modified Poisson equation (5.34). As will be discussed in Chapter 9, the same reasoning is also important for optimization strategies of so-called band-to-band tunnel field-effect transistors. Hence, it is important to verify the assertion of increased tunneling through the Schottky barrier by comparing theoretical predictions with experiments. For such a comparison, the inverse subthreshold slope is the ideal figure of merit. First, it is a differential measure and, therefore, independent of variations of the threshold voltage in experimental devices. Second, since it is extracted in the off-state of the transistor, variations of the on-current due to channel length variations and/or parasitic source/drain resistances do not play a role.

Schottky-barrier MOSFETs with fully nickel silicided source/drain electrodes are fabricated on SOI wafers with a p-type doping of $1 \times 10^{15}\,\text{cm}^{-3}$ of varying thickness (see Chapter 3 for details on the mentioned fabrication processes). Digital etching is used to thin down the wafers to the desired thickness. Next, the SOI thickness is measured using spectroscopic ellipsometry; here, a map of the entire wafer/sample piece is recorded in order to extract a probability distribution of the measured SOI thicknesses. Subsequently, optical lithography is used to pattern the active areas of the devices. Reactive ion etching is then employed to etch through the SOI layer creating insulated mesa structures. After mesa insulation, gate oxides with different thicknesses are grown by low temperature wet thermal oxidation [6] followed by the deposition of 200 nm n-type polysilicon with LP-CVD. The gate is patterned with optical lithography and RIE and spacers are formed with conformal SiO_2-deposition and anisotropic dry etching. The last steps include a self-aligned silicidation process with nickel deposition, silicidation at 500 °C for 20 s and the removal of superficial nickel (cf. Section 3.11). The process sequence is schematically shown in Figure 7.9. Only long-channel devices with a channel length of $L = 2\,\mu\text{m}$ are fabricated in order to rule out any influence of short-channel ef-

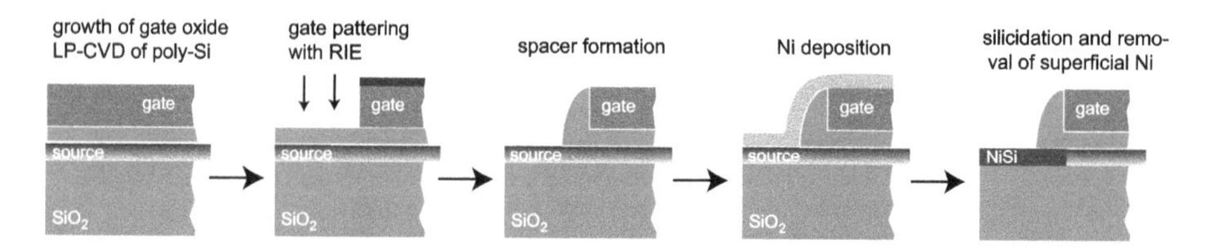

Figure 7.9: Schematics of the SOI SB-MOSFET fabrication with a self-aligned silicidation process.

fects. Figure 7.2(b) shows a TEM image of the source area of a readily fabricated device. One clearly sees the encroachment of the nickel silicide underneath the spacer and part of the gate electrode as has been discussed in Section 3.11.1. Since the SOI thickness has a rather strong impact on the NiSi encroachment, the temperature and time for the silicidation are chosen long enough to ensure that the NiSi-silicon interface lies well below the gate electrode in all cases. As already mentioned above, a possible channel length variation that may be obtained with this procedure is irrelevant here since we only use the off-state region of the transistor characteristics to carry out the analysis.

A multitude of SB-MOSFETs are measured and the inverse subthreshold slope is extracted from the transfer characteristics. Figure 7.10 shows exemplarily transfer characteristics of devices with fixed d_{ox} and varying d_{SOI} (a) and fixed d_{SOI} and three different gate oxide thicknesses (b). While the chosen gate oxide thicknesses can be realized easily by an appropriate oxidation step, adjusting different SOI thicknesses is not as simple. However, using the fact that according to Equation (7.7) the SOI thickness has a direct impact on S whereas small variations of the channel length or parasitic resistance do not, a statistical approach is chosen and the probability distribution of measured S-values is correlated with the measured probability distribution of SOI thicknesses [40, 149]; a few transmission electron microscopy images are taken to verify the correlation between

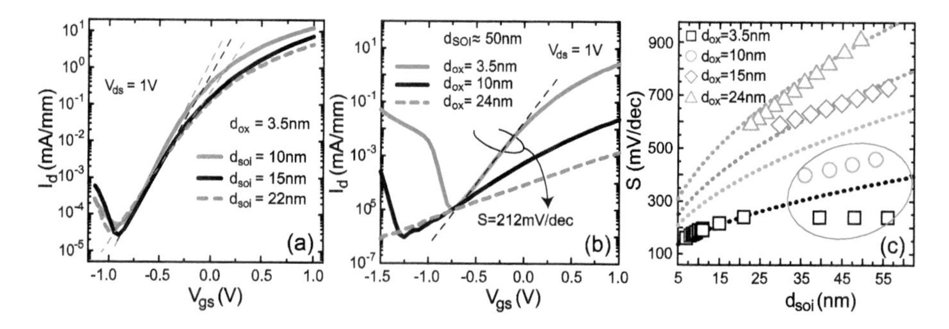

Figure 7.10: (a) I_d–V_{gs}-characteristics for SOI SB-MOSFETs with constant d_{ox} and varying d_{SOI} and (b) constant d_{SOI} thickness and different d_{ox}; the bias is 1 V in all cases [287]. (c) Inverse subthreshold slopes as a function of d_{SOI} in SOI-SB-MOSFET [149, 287]. Symbols belong to experimentally extracted S-values, the dotted lines are calculated based on Equation (7.7).

the two probability distributions. As a result, S-values from ~60 SB-MOSFETs with d_{SOI} in the range of 7–55 nm are obtained.

Figure 7.10(c) shows extracted values of the inverse subthreshold slope (symbols) together with theoretical S-values (dotted lines) based on Equation (7.7) as a function of the channel layer thickness d_{SOI} for four different d_{ox} [149, 287]. Note that for all theoretical curves only a single $d_{tunnel} = 3.4$ nm was used. Obviously, theoretical expectation and the experimental data are in very good agreement apart from the data points marked with the red ellipse. The deviation of these data points from the theoretical prediction is due to the fact that in these cases rather thick d_{SOI} is combined with thin gate dielectrics. In this case, the description of the electrostatics using the one-dimensional modified Poisson equation breaks down since the actual channel layer thickness cannot be replaced by the physical SOI thickness anymore; in this case, the theoretical model overestimates the subthreshold swings.

The conclusion of the comparison between experiments and theoretical data is that decreasing d_{SOI} and d_{ox} leads to significantly steeper inverse subthreshold slopes, and consequently, to an improved injection of carriers through the Schottky barrier. The strong impact of the device geometry on the switching behavior of ultrathin-body SB-MOSFETs is not only valid for silicon-on-insulator as channel material but it also explains why carbon nanotubes and 2D materials (cf. Chapters 8 and 10) are very well suited for high-performance SB-MOSFETs. However, as usual, there is a trade-off that needs to be taken into consideration. If the transmission probability is increased by making λ_{ch} as small as possible in order to, say, increase electron injection, this is also true for hole injection. As a result, devices based on ultrathin channels/nanowires with multigate architecture show strongly increased leakage due to ambipolar operation. This is particularly true if bias voltages on the order of the band gap are applied. Figure 7.11 shows the transfer characteristics of two nanotube devices simulated with self-consistent NEGF (cf. Chapter 6) with asymmetric Schottky barriers $\Phi_{SB}^{elec} = 0.2$ eV and $\Phi_{SB}^{hole} = 0.8$ eV for

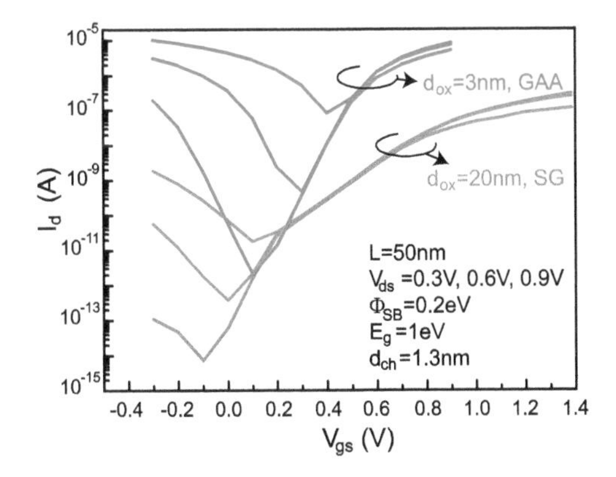

Figure 7.11: Transfer characteristics simulated with self-consistent NEGF of nanotube SB-MOSFETs with single-gate (blue) and gate-all-around (red) device architecture.

different V_{ds}; the blue curves belong to a device with $d_{ox} = 20\,\text{nm}$ in single-gate (SG) geometry and the red curves belong to a FET with gate-all-around (GAA) architecture and $d_{ox} = 3\,\text{nm}$. Obviously, the on-state and the switching are improved. Unfortunately, the extremely thin SB in the GAA case also leads to strong leakage. In fact, for $V_{ds} = 0.9\,\text{V}$ the transistor cannot be switched off anymore restricting its use to the low bias range.

7.1.1.3 On-State Behavior of SB-MOSFETs

While in the off-state of SB-MOSFETs, the tunneling of carriers through the source-side Schottky barrier can be described well by thermionic emission of carriers over an effective barrier Φ_{SB}^{eff}, the on-state of SB-MOSFETs requires further consideration.

In the device's on-state, the charge in the channel cannot be neglected anymore since it will have a significant impact on the potential distribution within the channel and in turn also on the potential landscape of the source and drain Schottky diodes. In order to account for the charge in the channel and also for scattering in the channel, we subdivide the channel into two segments. The first segment (denoted with a roman I) comprises the Schottky barrier and has a spatial extend on the order of $\sim 2\lambda_{ch}$, as illustrated in Figure 7.12(a). The second segment (denoted II) extends up to the drain contact (where the particular potential landscape at the device's drain end is disregarded for simplicity). Depending on the tunneling probability through the source Schottky barrier and the scattering within the channel, part of the drain-source voltage will drop across segment I and the remainder drops along the channel. In order to calculate an approximate expression for the on-state current, the potential $\Phi_f^0(x \approx 2\lambda_{ch})$ and with this the effective Schottky barrier Φ_{SB}^{eff} need to be computed. In contrast to the preceding section, in the on-state Φ_f^0 needs to be computed self-consistently and to do so the charge density is required. With the concept of a quasi-Fermi level (orange dashed line in Figure 7.12), the charge density can be computed by simply integrating over energy the product of the density of states at $x = 2\lambda_{ch}$ and an equilibrium Fermi distribution function $f(E_f^0)$ where E_f^0 is the quasi-Fermi level at $x = 2\lambda_{ch}$. The quasi-Fermi level E_f^0 can be calculated by equating the individual current components through segment I and II yielding[3]

$$\frac{l_{mfp}}{l_{mfp} + 2\lambda_{ch}} \int_{\Phi_{SB}^{eff}}^{\infty} dE(f(E_f^s) - f(E_f^0)) \overset{!}{=} \frac{l_{mfp}}{l_{mfp} + L} \int_{\Phi_f^0}^{\infty} dE(f(E_f^0) - f(E_f^d)) \tag{7.8}$$

where a device with one-dimensional electronic transport was assumed for simplicity; for devices in 2D or higher dimensions, the summation over appropriate $k_{y,z}$ can be carried out (as has been done before) without changing the qualitative picture of the

3 Note that this is similar to attaching a Buettiker probe at position $x = 2\lambda_{ch}$, i. e., an inelastic scattering mechanism has been assumed implicitly, which is responsible for the potential drop across the Schottky barrier.

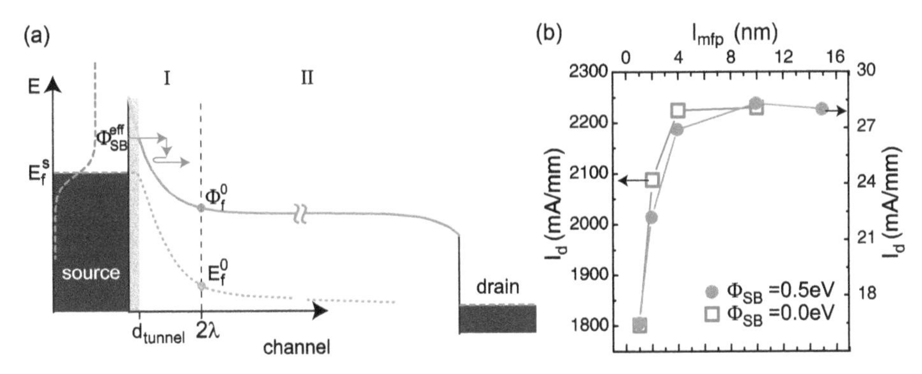

Figure 7.12: (a) Conduction band in the on-state of a SB-MOSFET. The orange dotted line illustrates the quasi-Fermi level within the device. The exponential potential distribution of the Schottky barrier suppresses effectively backscattering of carriers (red arrows) [7]. (b) Simulated on-current at constant V_{gs} and V_{ds} for two devices with high (green) and low (red) Φ_{SB} as a function of the mean free path for scattering [147].

argumentation. In Equation (7.8), scattering in the channel has been accounted for with an energy-independent transmission function $T = l_{mfp}/(l_{mfp} + L)$ where l_{mfp} is the mean free path for scattering (cf. Section 5.3 and [183]).

For carriers that scatter within the region with steep potential variation (segment I), it is unlikely to be scattered back into the source contact once they have lost $k_B T$ in energy since the Schottky diode rapidly becomes "thicker" preventing the carriers from tunneling back [181, 147]. Thus, the exact transmission function $T(E)$ can be replaced with the same step function used in the preceding section. In addition, the channel length L needs to be replaced with $2\lambda_{ch}$ as has been done in the left term of Equation (7.8) [7]. Equation (7.8) yields a transcendent equation for E_f^0 that can be solved numerically or graphically. Note that $|E_f^0/e|$ is the voltage drop across segment I. This means that, if $E_f^0 \approx 0$, then almost all of the drain-source bias drops across the channel whereas for $|E_f^0| \to |eV_{ds}|$ all voltage drops at the source Schottky diode. If the scattering mean free path l_{mfp} is not excessively small, the main dependence of the left part of Equation (7.8) on λ_{ch} is due to the exponential dependence of Φ_{SB}^{eff} on λ_{ch} (cf. Equation (7.4)) and not because of $l_{mfp}/(l_{mfp} + 2\lambda_{ch})$. This means that in many experimental situations where devices with moderate channel lengths L are studied, the current is mostly determined by the injection of carriers through the source-side Schottky barrier, which represents the main scattering event. In this case, the drain current hardly depends on L (if at all). In other words, in such a SB-MOSFET the current does not depend on the carrier mobility in the channel. Therefore, care has to be taken when analyzing and interpreting experimental data with respect to the electronic transport properties of the channel material used.

To study the impact of scattering on the on-state performance of SB-MOSFETs further, simulations are carried out using the Buettiker-probe approach presented in Section 6.3.1 [147]. Devices with a Schottky-barrier height of $\Phi_{SB} = 0.5$ eV and $\Phi_{SB} = 0$ eV, different channel lengths L and scattering mean free paths l_{mfp} are simulated. If not stated otherwise, $V_{ds} = 1.3$ V, $V_{gs} = 1.2$ V, $d_{SOI} = 5$ nm and $d_{ox} = 1$ nm in all cases. Figure 7.12(b) shows results for a constant channel length of $L = 25$ nm [147]. The green solid circles belong to devices with a high SB of $\Phi_{SB} = 0.5$ eV; the red hollow squares belong to a SB-MOSFET with $\Phi_{SB} = 0$ eV. In both cases, the on-current increases quickly and then saturates for $l_{mfp} > 4$–8 nm. This can be understood based on the model of SB-MOSFET discussed above where the transmission probability within the steep potential region of the SB (regime I) was $T \approx \frac{l_{mfp}}{l_{mfp}+2\lambda_{ch}}$. Since for the given d_{SOI} and d_{ox}, the screening length $\lambda_{ch} \approx 3.8$ nm it is clear that for $l_{mfp} > 8$ nm the transmission T plays a decreasing role. Note that although the absolute current levels of the two SB-MOSFETs displayed in Figure 7.12(b) are very different due to the different SB heights the dependence on l_{mfp} is the same. The reason for this is the rather short L considered in the simulation.

Figure 7.13(a) shows the on-state current for the same two devices and the same parameters as above as a function of the channel length L for devices with $l_{mfp} = 2$ nm, $l_{mfp} = 4$ nm and $l_{mfp} = 10$ nm. Increasing L decreases the on-current in all three cases in the device with the low SB height of $\Phi_{SB} = 0$ eV. In contrast, for the devices with $\Phi_{SB} = 0.5$ eV and the same different l_{mfp} the channel length is irrelevant (for the channel length range considered here) reconfirming the conclusions drawn from the semi-analytic consideration above, namely that in SB-MOSFETs with typical SB heights (here 0.5 eV) and where the channel length is not very large, the on-current is limited by scattering that occurs within the characteristic length scale $2\lambda_{ch}$ at the source-channel interface. In other words, for $\Phi_{SB} = 0.5$ eV, even for a thin gate oxide of 1 nm and a $d_{SOI} = 5$ nm and channel lengths at least up to ~100 nm the Schottky barrier is the main scattering

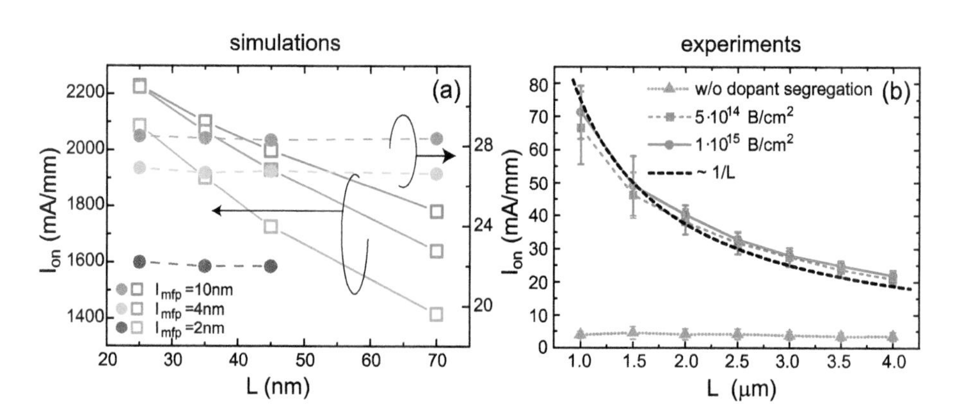

Figure 7.13: (a) Simulated on-currents for $l_{mfp} = 2$ nm, 4 nm and 10 nm for different channel lengths; $V_{ds} = 1.3$ V and $V_{gs} = 1.2$ V in all cases [147]. (b) Averaged on-currents of several p-type SB-MOSFETs with different gate lengths ranging from $L = 1.0..4.0$ μm with and without dopant segregation at $V_{gs} - V_{th} = V_{ds} = -1.8$ V [255]. Dopant segregation enables a substantial reduction of the SB such that the experimental data agrees qualitatively with the simulated data shown in (a).

event rendering scattering (and hence mobility of the channel material) irrelevant in this case. Note, however, that for very long channels the transmission through the channel $T = \frac{l_{mfp}}{l_{mfp}+L}$ eventually will dominate over the tunneling through the source Schottky barrier, and in this case, electronic transport through the device is again determined by the carrier mobility. The required minimum channel length for this depends sensitively on the SB height, the effective mass and λ_{ch} of the device (see also the discussion in Section 9.1.2.2).

Figure 7.13(b) displays experimental on-currents (at $V_{gs} - V_{th} = V_{ds} = -1.8$ V) extracted from measurements of multiple SOI SB-MOSFETs with nickel silicide (NiSi) Schottky contacts [255]. In the present case, devices with and without dopant segregation (DS) were compared. Dopant segregation, which will be discussed in detail in Section 7.2 (see also Section 3.11.2), allows one to strongly reduce the effective Schottky-barrier height. This means that the green data points (without DS) belong to a SB-MOSFET with a typical Schottky barrier of $\Phi_{SB}^{hole} \sim 0.45$ eV of the NiSi-Si interface.[4] In qualitative agreement with the results obtained from simulations, the experimental on-currents do not show any dependence on the channel length, which in turn means that the injection through the Schottky barrier is the main scattering event in the device (see also the discussion leading to Equation (8.6)). In contrast, the two other devices with substantially reduced effective Schottky-barrier height due to DS (red and blue data points) show a dependence $I_d \propto 1/L$ expected from a conventional MOSFET. Again, qualitative agreement with the simulation results shown in Figure 7.13(a) is observed.

7.1.2 Output Characteristics

So far, the on-state has been only considered at large bias. However, the particular layout of SB-MOSFETs with two metal electrodes in immediate contact with a semiconductor impacts the entire output characteristics. Especially in the case of low bias, a sublinear I_d–V_{ds} behavior is obtained that leads to a substantial degradation of voltage transfer curves of inverters and is therefore highly undesirable.

Task 29.
Inverter with SB-MOSFETs: Suppose a conventional MOSFET and a SB-MOSFET exhibit very similar output characteristics. Their only difference is that the SB-MOSFETs exhibits a pronounced sublinearity for small bias (cf. Figure 7.15(a)). Construct (graphically) a voltage transfer curve of an inverter (load free) for both device types assuming that *n*- and *p*-type FETs exhibit the same on-state performance. Exploit the fact that no current flows through the output terminal of the inverter.

4 Note that the device is operated in the *p*-branch of the ambipolar characteristics.

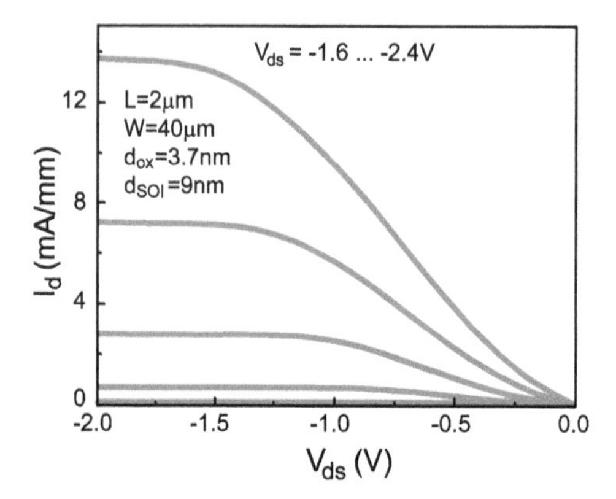

Figure 7.14: Typical experimental output characteristics of the silicon-on-insulator Schottky-barrier MOSFET with NiSi contacts shown in Figure 7.2(b) [289].

Figure 7.14 shows typical experimental output characteristics of a *p*-type SOI SB-MOSFET with NiSi source/drain contacts [289]. The sublinear behavior for small bias voltages is clearly observable. At first sight, it appears obvious that this sublinearity is due to the forward-biased Schottky junction (i. e., diode) at the drain end of the transistor and SB-MOSFET device characteristics have been described accordingly [271, 33, 119]. However, this interpretation can obviously not be correct. If the sublinearity was a pure drain property, one would not observe linear output characteristics in experimental SB-MOSFETs with very long-channel lengths where—as has been discussed above—the transport properties of the channel dominate. Rather, one would always expect a sublinear behavior for small bias.

In order to clarify the reason for the sublinearity, self-consistent NEGF simulations were carried out using the formalism presented in Chapter 6. For simplicity, a one-dimensional channel contacted by metallic source/drain electrodes with fixed SB Φ_{SB}^{s} at the source and Φ_{SB}^{d} at the drain is considered (the parameters of the device are provided in the figures below). Scattering has been incorporated via Buettiker probes (cf. Section 6.3.1) and a mean free path of $l_{mfp} = 50\,\text{nm}$ has been assumed.[5] Below, only *n*-type characteristics are simulated, *p*-type SB-MOSFETs, however, show the same qualitative behavior.

It is important to note that in the following, devices are considered that may appear unreasonable in terms of the parameters chosen. The reason for this choice is that using Equation (5.34) together with 1D electronic transport allows to adjust the charge density within the channel, and hence the feedback on the potential distribution while keeping the geometrical electrostatics (expressed through the screening length λ_{ch}) unchanged. For instance, considering a wrap-gate device with quadruple gate, λ_{ch} =

5 The chosen mean free path $l_{mfp} \sim L$ ensures that the device characteristics are not determined by the electronic transport properties of the channel as discussed in Section 7.1.1.3.

$\sqrt{\varepsilon_{si}/\varepsilon_{ox} d_{ox} d_{nw}/4}$ (see Section 5.8.3), the same screening length can be obtained either with $d_{nw} = 1\,\text{nm}$ and $d_{ox} = 20\,\text{nm}$ or $d_{nw} = 5\,\text{nm}$ and $d_{ox} = 4\,\text{nm}$. Thus, the screening length will be the same in both cases but the charge density and its impact on the potential will be substantially different.

The output characteristics of the following devices are simulated and the results for a constant V_{gs} are plotted in Figure 7.15: $L = 60\,\text{nm}$, $l_{mfp} = 50\,\text{nm}$, $d_{nw} = 1\,\text{nm}$, a band gap of $E_g = 1\,\text{eV}$ and an effective mass of $m^\star = 0.2m_0$ in a single gate geometry. Figure 7.15(a) shows the I_d–V_{ds} curves at $V_{gs} = 1.4\,\text{V}$ in the case of (i) $\Phi^s_{SB} = \Phi^d_{SB} = 0.5\,\text{eV}$, (ii) $\Phi^s_{SB} = 0.5\,\text{eV}$ and $\Phi^d_{SB} = 0.3\,\text{eV}$, (iii) $\Phi^s_{SB} = 0.5\,\text{eV}$ and $\Phi^d_{SB} = 0.1\,\text{eV}$ as well as (iv) $\Phi^s_{SB} = 0.5\,\text{eV}$ and $\Phi^d_{SB} = 0.0\,\text{eV}$. Although the sublinearity is slightly reduced with decreasing Φ^d_{SB}, in all cases a strong sublinear behavior is observed even though the Schottky barrier at the drain end vanishes in the case (iv). The reason for this behavior can be understood when looking at the conduction bands for different bias voltages as depicted in Figure 7.15(b) (note that although the simulations include electron and hole contributions only the conduction band is displayed for clarity of the illustrations). Here, the conduction band is plotted for constant V_{gs} and three different V_{ds} in the case of $\Phi^d_{SB} = 0\,\text{eV}$. A strong impact of V_{ds} on the potential distribution within the channel is observed. Indeed, $\delta\Phi^0_f/\delta\Phi_d$ approaches almost unity in the displayed case. In conventional MOSFETs, $\frac{\partial\Phi^0_f}{\partial\Phi_d} > 0$, is due to short-channel effects and is called DIBL (see Section 5.6). However, here it is not a short-channel effect[6] but it is due to the fact that the charge carrier density in the channel varies much more in SB-MOSFETs than in a conventional FET; for $V_{ds} = 0$, carriers from the source and the drain are injected and the channel is filled according to an equilibrium Fermi distribution irrelevant of the SBs. A MOS capacitor is obtained where the position of Φ^0_f is determined by the charge in the channel and the applied gate voltage.

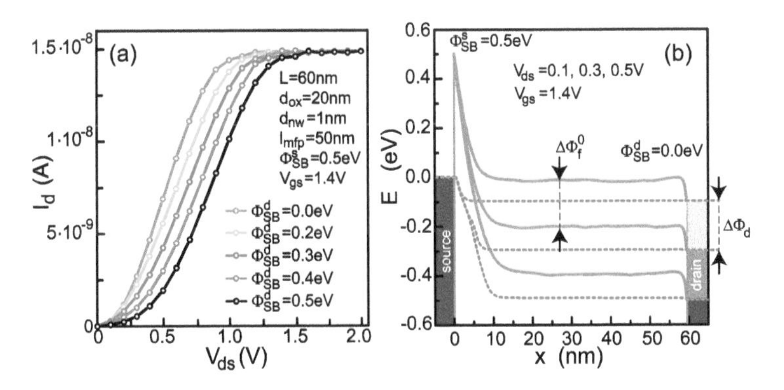

Figure 7.15: (a) Output characteristics for a fixed $V_{gs} = 1.4\,\text{V}$ of a NW SB-MOSFET (see the figure for the simulation parameters) with $\Phi^s_{SB} = 0.5\,\text{eV}$ and varying Φ^d_{SB}. (b) Band profiles (solid green lines) and quasi-Fermi level (dotted red lines) in the case of $\Phi^s_{SB} = 0.5\,\text{eV}$ and $\Phi^d_{SB} = 0\,\text{eV}$ for varying V_{ds} [146].

6 A similar behavior is also observed in tunnel FETs, as will be discussed in Chapter 9.

If V_{ds} is increased, carrier injection occurs increasingly from source only and since the carriers now need to tunnel through the source-side SB, their density drops strongly. As a result, the bands are moved downwards because of the applied V_{gs} (and thus also is Φ_f^0), thereby reducing the effective Schottky-barrier height leading to an increased carrier injection, and thus the sublinear current increase. Once V_{ds} is large enough such that the position of Φ_f^0 is only determined by V_{gs} and the charge injected from the source, current saturation is obtained.

The discussion above is supported by the behavior of the quasi-Fermi level. In Figure 7.15(b), the quasi-Fermi level (red dotted line) drops at the source side and is (almost) at the same level as E_f in drain. Hence, the sublinearity in the present case is purely due to the source-side Schottky-barrier.[7] A further important observation can be made when looking at Figure 7.15(a); the saturation current is constant in the case of varying Φ_{SB}^d, which underlines that the source-side Schottky diode is decisive for the electrical behavior of SB-MOSFETs.

The fact that in a SB-MOSFET it is the impact of the channel charge on the source-side junction that plays the dominant role leading to the sublinearity is corroborated by simulations carried out for NW SB-MOSFETs with constant $\Phi_{SB}^s = \Phi_{SB}^d = 0.5$ eV but varying d_{ox} and d_{nw}. In order to avoid an immediate impact of d_{ox} and d_{nw} on the transmission probability through the SB, the gate oxide and channel thicknesses are changed so as to leave the screening length λ_{ch} constant.[8] To be specific, the following four devices are considered: (1) $d_{ox} = 20$ nm, $d_{nw} = 1$ nm, (2) $d_{ox} = 10$ nm, $d_{nw} = 2$ nm, (3) $d_{ox} = 5$ nm, $d_{nw} = 4$ nm and (4) $d_{ox} = 2$ nm, $d_{nw} = 10$ nm. Since a single-gate device architecture is considered, $\lambda_{ch} = \sqrt{\frac{\varepsilon_{nw}}{\varepsilon_{ox}} d_{ox} d_{nw}} = 7.58$ nm in all cases. Furthermore, an increase in d_{nw} yields a reduced carrier density because purely 1D electronic transport is considered with the charge carriers confined to the NW cross-section (i. e., a particle-in-the-box approximation is used). As a result, the wavefunction spreads across the nanowire cross-section leading to a carrier density reduction (approximately according to $n(x) \propto n^{1D}(x)/d_{nw}^2$ where $n^{1D}(x)$ is the carrier density computed with NEGF) for increasing d_{nw} since contributions from higher subbands of the nanowire have deliberately not been taken into account. In addition, to carve out the impact of the channel charge more clearly, the effective mass is lowered from $m^* = 0.2m_0$ to $m^* = 0.05m_0$ (an increase of the effective band gap due to carrier confinement is neglected).

Figure 7.16 shows output characteristics for the different d_{nw} and d_{ox} mentioned above. Obviously, an increasing sublinear behavior develops if d_{nw} decreases (and d_{ox} increases). The reason for this is the increasing charge carrier density with decreasing

7 Note that the case $\Phi_{SB}^s < \Phi_{SB}^d$ is not discussed here since it obviously leads to a sublinear behavior due to the forward-biased SB diode at drain. Further information including this case is given in [146].

8 Note that in the on-state and small bias, the charge $(-e)n$ in the channel yields additional screening λ_n as discussed in Section 5.7. As a result, the transmission probability is determined by an effective screening length $\lambda_{eff}^{-2} = \lambda_{ch}^{-2} + \lambda_n^{-2}$. In the present case, $\lambda_{ch} =$ const. and λ_n strongly increases with V_{ds}.

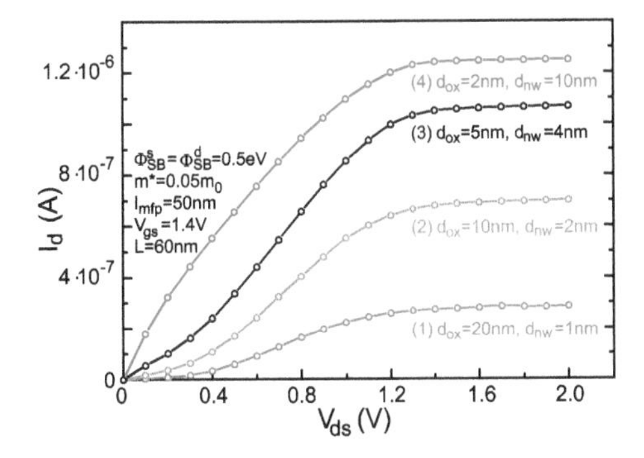

Figure 7.16: Output characteristics for a fixed V_{gs} = 1.4 V of a nanowire SB-MOSFET (see the figure for the simulation parameters) with $\Phi^s_{SB} = \Phi^d_{SB}$ = 0.5 eV and varying d_{nw} and d_{ox} resulting in the same λ_{ch} [146].

d_{nw}, leading to a stronger charge-mediated impact of the potential distribution of the source-side SB with larger V_{ds}. In contrast, the carrier density becomes rather small and the oxide capacitance large in the case (4) and, therefore, substantially less impact of the charge on the potential distribution is obtained. As a result, the decreasing impact of the charge (from device (1) toward device (4)) on the source-side SB gradually results in linear output characteristics corroborating the role of the source-side SB determining the electrical behavior of SB-MOSFETs.

7.2 Schottky-Barrier Lowering with Dopant Segregation During Silicidation

From the discussion of the preceding sections, it is clear that lowering the Schottky-barrier height is of utmost importance to improve the switching behavior and increase the on-state performance of MOSFETs with metallic source/drain electrodes. In Section 4.6.2, it was shown that a proper ohmic contact between a metal and a semiconductor is usually made by heavily doping the semiconductor, which yields a thin, highly transmissive Schottky barrier. However, doping the channel results in a large depletion capacitance, which can only be tolerated as long as $C_{ox} \gg C_{depl}$. Moreover, in order to suppress short-channel effects, ultrathin channel layers need to be used eventually leading to fully depleted SOI. While this is certainly preferable to provide optimum switching of the transistor (cf. Section 5.8.1), doping the channel in fully-depleted SOI SB-MOSFETs simply yields a shift of the threshold voltage but no improvement of the carrier injection [147]. This can be understood by looking at Equation (5.34); if the charge density due to donors (acceptors) $N_{d,a}$ is constant (as is the case for fully-depleted SOI), $N_{d,a}$ simply leads to an additional constant term that results in a shift of Φ_{bi}, and hence to a shift of the threshold voltage. The injection of carriers, however, is the same, so that the inverse subthreshold slope and the on-current at the same gate voltage overdrive remain unchanged. This scenario is shown in Figure 7.17(a), which shows conduction

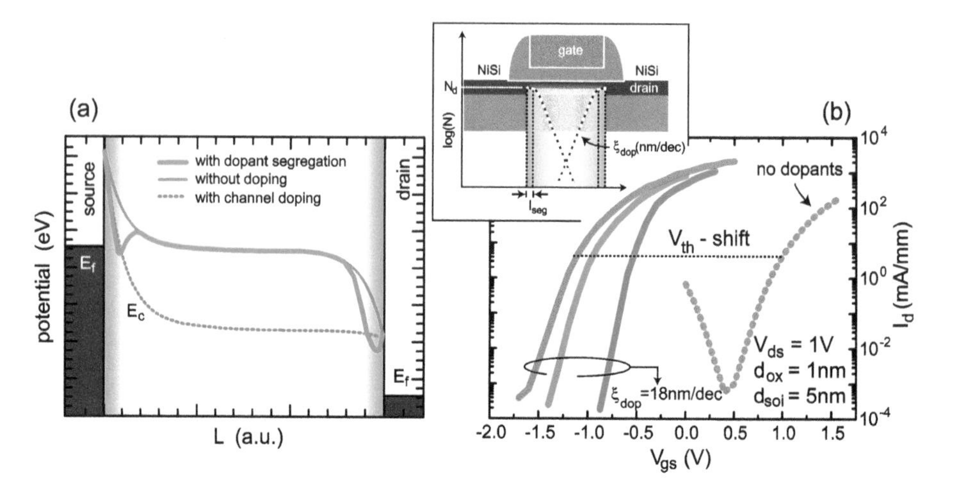

Figure 7.17: (a) Conduction band profile along the channel of a SB-MOSFET for three different doping scenarios. In the case of fully-depleted SOI, the doping results merely in a shift of the threshold voltage. An inhomogeneous doping profile (orange areas) with high doping concentration at the contact-channel interfaces leads to a reduction of Φ_{SB}^{eff}. (b) Simulated transfer characteristics of SB-MOSFETs with nonuniform dopant profile and $L = 28$ nm (blue), $L = 31$ nm (green) and $L = 38$ nm (red). In all cases, the dopant concentration drops with $\xi_{dop} = 18$ nm/dec. The blue dotted line belongs to a device without any dopants in the channel and $L = 28$ nm [147].

band profiles for three different cases: the blue curve belongs to a SB-MOSFET without any dopants within the channel, the red dashed profile depicts the conduction band for a homogeneous, high doping concentration in the channel.

What is needed in order to improve the device behavior without (merely) shifting the threshold voltage is a nonuniform doping profile with highly doped semiconductor sections only at the interfaces between the contact electrodes and the channel as illustrated with the orange areas in Figure 7.17(a). Such a nonuniform doping profile results in the conduction band profile shown with the green curve. Strong band bending is observed, which makes the Schottky barrier thinner thereby promoting the tunneling of carriers without changing the threshold voltage. As a result, a reduced effective Schottky barrier $\Phi_{SB}^{reduced}$ is obtained.

The required doping profile must have a very small lateral extent in order to avoid parasitic capacitances. Furthermore, it must exhibit the steepest possible drop of the dopant concentration to enable scaling the SB-MOSFETs to very short-channel lengths. The latter requirement is very important as the simulations displayed in Figure 7.17(b) prove. Here, the transfer characteristics of SB-MOSFETs are shown that exhibit a doping profile with a high concentration N_d at the contact-channel interfaces; the drop of the dopant concentration, however, is not steep with $\xi_{dop} = 18$ nm/dec [147]. When the channel length of the SB-MOSFET is scaled down the left and right doping profiles eventually overlap, leading to an increasing doping concentration in the channel with decreasing L. This in turn leads to a shift of V_{th} and eventually degrades the switching behavior for the shortest device with $L = 28\,\text{nm}$ (cf. main panel of Figure 7.17(b)). In this case, the overlapping dopant profiles have increased the concentration of dopants in the channel so much that the profile rather resembles a uniform doping. As a result, the switching behavior is deteriorated and now shows the same inverse subthreshold slope as a reference SB-MOSFET without any dopants in the channel (blue dotted line in Figure 7.17(b)). Therefore, the simulations reconfirm the considerations above and prove that very steep and narrow dopant profiles with high concentration are required in order to improve the performance of SB-MOSFETs.

The question now is how such dopant profiles can be realized. A very elegant way is the use of dopant segregation during nickel silicidation (NiSi), which is discussed in detail in Section 3.11.2. During the NiSi formation, dopants are collected at the silicide-silicon interface due to the substantially different solid solubilities. As a result, a narrow, ultrasteep dopant profile with concentrations beyond $10^{20}\,\text{cm}^{-3}$ can be achieved, as has been confirmed with SIMS and ECV measurements (see Figure 3.55).

In order to exploit dopant segregation in n-type and/or p-type SB-MOSFETs, a fabrication sequence as schematically shown in Figure 7.18 is carried out. The fabrication is actually the same as has been described in Section 7.1.1.2. The only difference is that after the patterning of the gate electrode, ions are implanted into the source and drain regions but no activation anneal is carried out. In the present case, arsenic is implanted at 5 keV with a dose of $5 \times 10^{14}\,\text{cm}^{-2}$, leading to an implantation depth of approximately 8 nm; boron is implanted at 2 keV with a dose of $3 \times 10^{15}\,\text{cm}^{-2}$ with an implantation depth of 10 nm. Next, spacers are generated and in order to avoid any diffusion of dopants, this is done with PE-CVD. Finally, the gate oxide is removed in the source/drain areas, nickel is deposited and a self-aligned silicidation is carried out. During silicidation, the entire source/drain areas are fully silicided. The silicidation is carried out at a temperature of 450 °C for 20–30 s. This ensures the encroachment of NiSi underneath the spacers facilitating the segregation of dopants. Thus, all dopants implanted into the region underneath the spacers are piled up at the silicide/silicon front. Afterwards, the unreacted nickel is selectively removed using Piranha (cf. Chapter 3). A transmission electron microscopy cross-section of such a SOI SB-MOSFET is shown in Figure 7.2(b) [289].

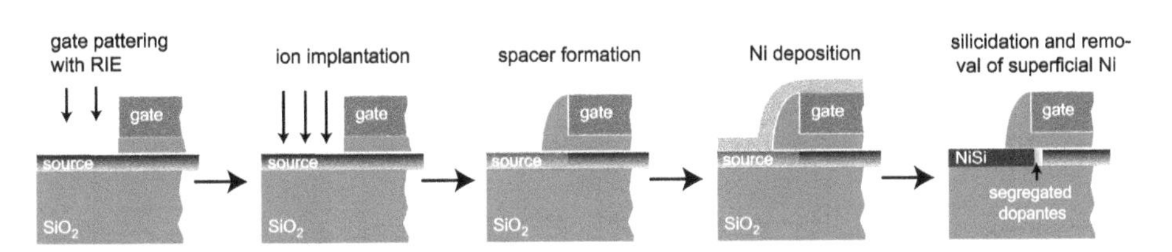

Figure 7.18: Schematics of the SOI SB-MOSFET fabrication with dopant segregation; the realization of the control samples without dopant segregation is depicted in Figure 7.9.

7.2.1 Measurements and Discussion

Typical transfer characteristics of a SOI SB-MOSFET with nickel silicide source/drain contacts but without dopant segregation have already been discussed above (see Figure 7.6). Incorporating dopant segregation into the fabrication as described in the preceding section yields the required highly doped and steep dopant profiles at the NiSi/Si interfaces.

Figure 7.19 shows transfer (main panels) and output (insets) characteristics of SOI SB-MOSFETs with boron (left) and arsenic (right) segregation [289]. Ambipolar operation can still be observed but the n-type (p-type) branch is significantly suppressed and the on-current in the p-type(n-type) branch is increased by approximately one order of magnitude in the boron (arsenic) segregation devices compared to the SB-MOSFET displayed in Figure 7.6. The most prominent difference of the devices with dopant segregation compared to the device without dopant segregation is the inverse subthreshold slope of the p-type (n-type) branch: $S = 65$ mV/dec ($S = 70$ mV/dec), which is close to the thermal limit and proves the effectiveness of dopant segregation in reducing the effective Schottky barrier height $\Phi_{SB}^{reduced}$. This reduction is also observed when plotting the on-current for a fixed gate voltage overdrive and bias as a function of the channel length L of the devices. Figure 7.13(b) exemplarily shows this in a device with boron segregation (blue and red). Two different dopant doses have been used to implant boron into the source/drain regions prior to segregation, which yields two different dopant concentrations in the segregation layer. When discussing the on-state of SB-MOSFETs this figure has already been used to show, that a strong reduction of the SB height yields a dependence of the on-current $\propto 1/L$ as in a conventional MOSFET (cf. Equation (5.4)), which is indeed observed. This underlines that dopant segregation is capable of strongly reducing $\Phi_{SB}^{reduced}$.

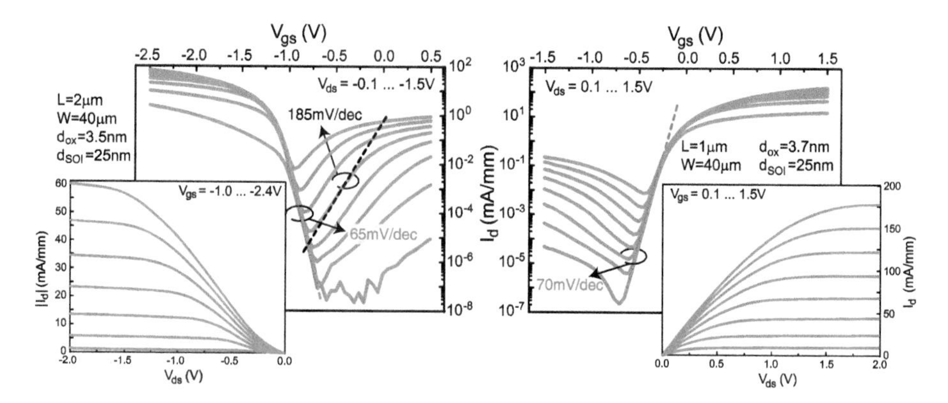

Figure 7.19: Transfer and output (insets) characteristics of SOI SB-MOSFETs with boron segregation (left) and arsenic segregation (right) during nickel silicidation [289].

Looking, however, at the output characteristics (insets of Figure 7.19), a difference between the boron and arsenic devices becomes apparent. While the arsenic device shows a linear increase of current for small bias (right inset), typical of a conventional MOSFET, the boron device exhibits the nonlinear increase of current for small drain-source voltages usually observed in SB-MOSFETs (left inset). The reason for this is that in the case of the arsenic device the dopant concentration in the segregation layer is larger than in the boron device leading to a more efficient reduction of $\Phi_{SB}^{reduced}$ [289].

7.2.2 Temperature Dependence

More insights into the operational principles of SB-MOSFETs can be obtained with temperature-dependent measurements. Although the experimental data discussed below belongs to SB-MOSFETs with dopant segregation, it is important to note that DS is only a means to modify the effective Schottky barrier height, and thus the results are also true for other ways to vary the SB height (of particular interest is, of course, how much the effective Schottky barrier can be lowered when dopant segregation during silicidation is employed).

Temperature-dependent measurements of the drain current as a function of V_{gs} are carried out and the effective potential barrier is extracted from the measurements. The potential barrier extracted this way is an effective barrier Φ^{eff} for thermal emission alone. This means, if the effective SB height Φ_{SB}^{eff} is smaller than the potential maximum in the channel Φ_f^0, $\Phi^{eff} = \Phi_f^0$. If, on the other hand, $\Phi_f^0 < \Phi_{SB}^{eff}$ then $\Phi^{eff} = \Phi_{SB}^{eff}$. Both Φ_f^0 and Φ_{SB}^{eff} show a distinctly different dependence on a change of the gate potential. In an electrostatically well-behaved FET, we have $\Delta\Phi_f^0 = \Delta\Phi_g = -e\Delta V_{gs}$, and hence $\Delta\Phi^{eff} = \Delta\Phi_f^0$ if $\Phi_{SB}^{eff} \leq \Phi_f^0$ and $\Delta\Phi^{eff} = \Delta\Phi_{SB}^{eff} = (1 - \exp(-d_{tunnel}/\lambda_{ch}))\Delta\Phi_f^0$ if $\Phi_{SB}^{eff} > \Phi_f^0$ (see Equation (7.4)). The reduced effective Schottky-barrier height $\Phi_{SB}^{reduced}$, i. e., the barrier height without the impact of the gate electric field can be deduced from a plot of the extracted Φ^{eff} as a function of V_{gs}. Such a plot is displayed in the left panel of Figure 7.20 showing the extracted values of Φ^{eff} as a function of V_{gs} at a bias of $V_{ds} = 0.1$ V. In the case of small gate voltages, Φ^{eff} exhibits an almost one-to-one dependence on V_{gs}. As discussed above, this means that $\Phi_{SB}^{eff} < \Phi_f^0$ and hence $\Delta\Phi^{eff} = \Delta\Phi_f^0$ as is the case in a conventional MOSFET. The right panel of Figure 7.20 shows a close-up of the conduction band profiles at the source-channel interface. Obviously, in the green-marked gate voltage range the strong band bending due to the segregation layer yields a substantially reduced Schottky barrier $\Phi_{SB}^{reduced}$. This reduced Schottky barrier can be determined from the point where the Φ^{eff}–V_{gs}-graph deviates from the slope –1 (one-to-one behavior). In the present case, $\Phi_{SB}^{reduced} \approx 0.1$ eV, i. e. substantially smaller than the original SB of $\Phi_{SB} = 0.67$ eV at the NiSi/Si interface. This means that in the V_{gs}-range leading to the two red potential profiles, the injection of carriers is determined by $\Phi_{SB}^{reduced}$. The magnitude of the change of the tunneling probability through this reduced barrier with changing

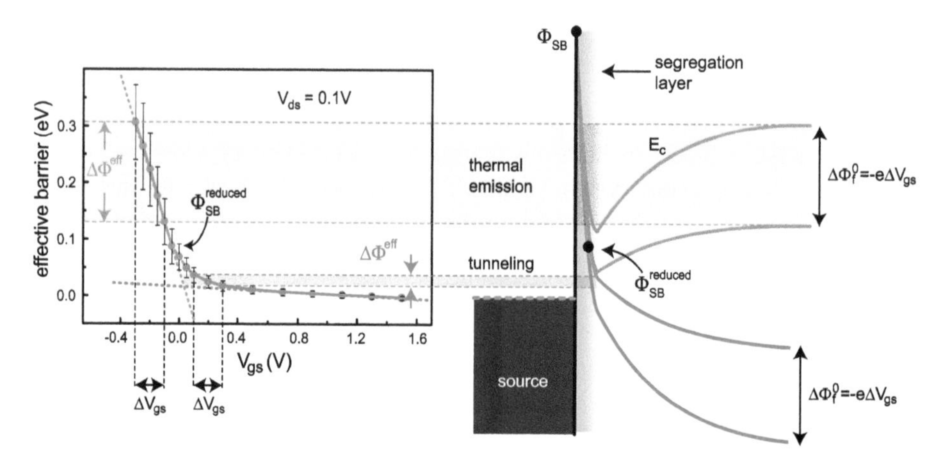

Figure 7.20: Effective barrier heights extracted from temperature-dependent measurements (left panel) as a function of V_{gs} to determine $\Phi_{SB}^{reduced}$. The right panel shows four different conduction band profiles that belong to the two green and red indicated gate voltage ranges [288].

gate voltage is much smaller than unity, meaning that $|\Delta\Phi_{SB}^{eff}(V_{gs})| \ll |e\Delta V_{gs}|$ (illustrated with the red dashed lines in Figure 7.20), which eventually limits the on-state performance.

From the temperature-dependent measurements, one can also extract the inverse subthreshold slopes of the SB-MOSFET. Figure 7.21 shows experimental values (light blue squares) extracted from the same devices with dopant segregation as discussed in the preceding section. S apparently depends almost linearly on temperature up to approximately 200 K, then remains constant and drops again with a different slope below ~125 K.

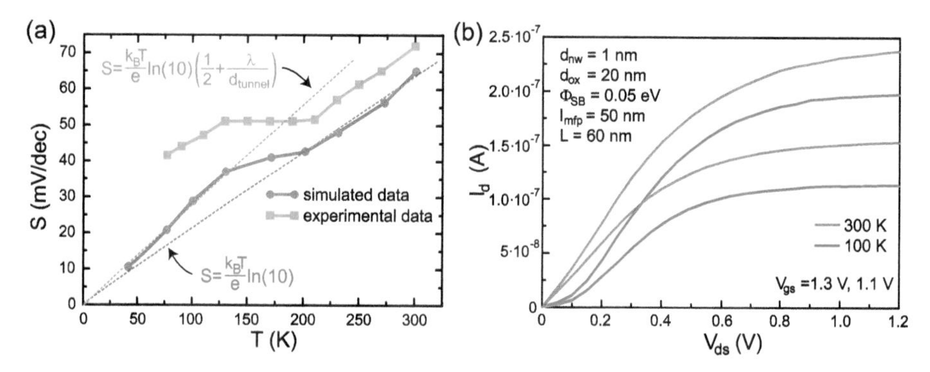

Figure 7.21: (a) Simulated (dark blue) and experimental (light blue) values of the inverse subthreshold slopes as a function of temperature for SB-MOSFETs with dopant segregation. In the temperature range where $4 \times k_BT \gtrsim \Phi_{SB}^{reduced}$, an almost ideal switching is obtained. For lower temperatures, the switching approaches the behavior of a SB-MOSFET, which is determined by the change of the tunneling probability through the Schottky barrier [148]. (b) Simulated output characteristics of a nanowire SB MOSFET with $\Phi_{SB} = 0.05$ eV at 300 K and 100 K.

This behavior can be explained with a cross-over from a transistor that behaves like a conventional MOSFET (i. e., along the red dotted line in Figure 7.21) at temperatures in the range 200–300 K and turns into a SB-MOSFETs with larger S at lower temperatures (green dotted line). This cross-over occurs when $\sim 4k_BT < \Phi_{SB}^{reduced}$, which is consistent with the value $\Phi_{SB}^{reduced} \lesssim 0.1\,\text{eV}$ found with the temperature-dependent measurements. To confirm this behavior, simulations assuming a SB-MOSFET with dopant segregation were carried out (dark blue circles in Figure 7.21(a)) that show qualitatively the same behavior with cross-over in the same temperature range as the experimental devices [148].

Finally, let us now turn to the temperature dependence of the output characteristics. Suppose $\Phi_{SB}^{reduced} \lesssim 0.1\,\text{eV}$, which leads to a carrier injection at room temperature that is dominated rather by thermionic emission than tunneling. As a result, the output characteristics show a linear $I_d - V_{ds}$ behavior for small bias as is indeed observed experimentally; see Figure 7.19 (right inset). Simulations of a SB nanowire device with $\Phi_{SB} = 0.05\,\text{eV}$ reflect this behavior and are shown in Figure 7.21(b), blue curves (simulation details are given in the figure). If the temperature is lowered, the thermal broadening of the source Fermi distribution is decreased, eventually leading to $k_BT \ll \Phi_{SB}$. As a result, carrier injection is again determined by the tunneling through the source-side SB, and hence the sublinear behavior of the output characteristics is recovered. Figure 7.21(b) displays simulated output characteristics (see Figure for details) at room temperature (blue) and at $T = 100\,\text{K}$ (red) showing the expected transition from a linear to sublinear behavior.

7.3 Interface Engineering with Depinning Layers

In the preceding section, it was discussed how dopant segregation during silicidation can be used in order to obtain a reduced Schottky-barrier height $\Phi_{SB}^{reduced}$. However, the actual Schottky barrier Φ_{SB} that is a result of Fermi level pinning is not reduced with dopant segregation. Only the tunneling probability through the barrier is increased.[9]

It was discussed in Section 4.6.3 that depinning of the Fermi level can be accomplished by inserting an ultrathin insulating layer in-between the metal and the semiconductor. The insulator needs to ensure that the density of dangling bonds at the surface of the semiconductor is reduced. More importantly, the insulator needs to be thick enough to reduce the interface density of states due to metal-induced gap states (MIGS) while at the same time being thin enough to allow for tunneling of carriers with a sufficiently high transmission probability so that overall an optimized trade-off in terms of the contact resistance can be found.

9 One could certainly argue that the experimental observations would be the same if the barrier was simply reduced by the additional positive charge of the ionized dopants within the segregation layer.

7.3.1 Metal Source/Drain Device Using Interface Engineering

As discussed in Section 4.6.3, silicon nitride is a viable option for the realization of depinned metal-semiconductor contacts. In fact, it is expected that the use of silicon nitride allows *n*-type and *p*-type contacts to be realized merely by using metallic contact electrodes with appropriate work functions. The reason for this is that on the one hand the band gap of SiN is sufficiently large to ensure that no repinning occurs (cf. Section 4.6.1 and Figure 4.21); repinning which would simply pin the Fermi level of a contact metal at a different energy level prohibits either *n*- or *p*-type devices. On the other hand, the band gap is also substantially smaller than SiO_2, allowing one to find an optimum insulator thickness that combines depinning with appropriate carrier injection and, therefore, a reduction of the contact resistance [54]. In addition, SiN can be grown very precisely on top of silicon, preventing any further oxidation, thereby providing a stable insulator (cf. Section 3.3).

In order to study the interface engineering approach with SiN experimentally, silicon-on-insulator substrates (5–10 Ωcm, *p*-type) with a SOI layer of 200 nm and 200 nm buried oxide (BOX) are used for the fabrication of so-called pseudo-MOSFETs, i. e., devices where the BOX is used as gate dielectric and the handle wafer as a large area back-gate. Circular mesa structures are patterned into the SOI using optical lithography and reactive ion etching (SF_6/O_2 plasma). After resist removal, a standard clean is carried out. Immediately before the samples are mounted in a cold-wall rapid thermal annealing furnace, the chemically grown oxide is removed with a short HF dip. The samples are then annealed in an Ar/NH_3 atmosphere for 1 min at different temperatures. Since the SiN is grown and not deposited an increase in thickness can only occur when NH_3 diffuses through the grown SiN layer to generate further nitride. But because SiN is an excellent diffusion barrier, its growth is self-terminating (cf. Section 3.3). The thickness of the grown SiN layer almost exclusively depends on the process temperature (exponential dependence). Hence, very stable and reproducible thicknesses of SiN layers are obtained by simply changing the temperature enabling the controlled generation of sub-1 nm thin nitride layers [54, 84, 72, 223].

In the present case, ~8 Å SiN was grown on top of the SOI mesa structures. In contrast to Connelly and coworkers who found the optimum SiN thickness to be approximately 4 Å [54], we deliberately chose a SiN thickness approximately twice as large, since this yields contacts that behave similar to doped semiconductor sections enabling FETs with unipolar operation as has been explained in Section 4.6.3 [84]. After the SiN growth, source/drain metal contacts are deposited. This is done either with a shadow mask or using optical lithography, electron beam evaporation and lift-off. The contact metals consist either of 200 nm aluminum or 25 nm/200 nm platinum/aluminum, thereby providing a rather low and a high work function for the realization of *n*-type and *p*-type devices.

Figure 7.22(a) shows the transfer characteristics of a pseudo-MOSFET with Al contacts (green curves) as reference and a pseudo-MOSFET with SiN interface layer in-between the silicon and the Al contact (red curves) [84]. The reference device shows

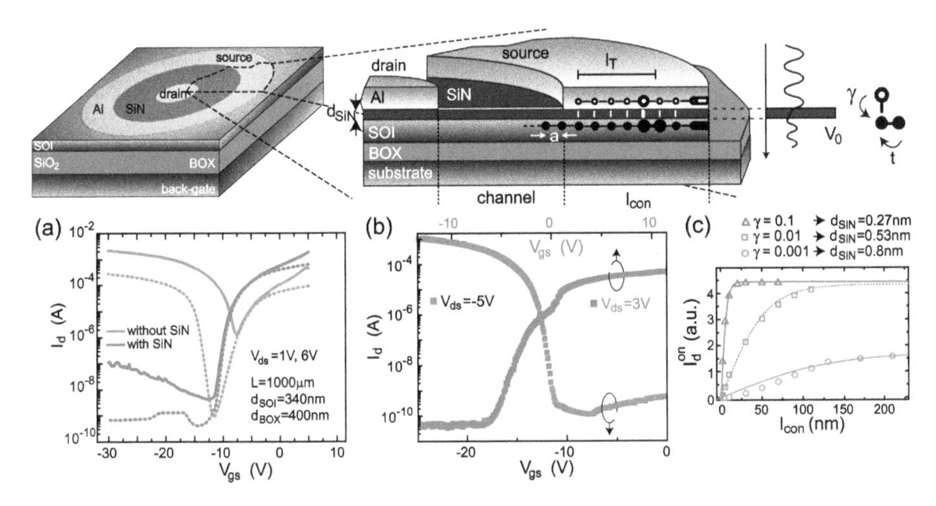

Figure 7.22: The top panel shows an image of a pseudo-MOSFET with circular mesa structure and Al contacts. The schematic image provides details on the metal-insulator-silicon contact. (a) Experimental transfer characteristics of a pseudo-MOSFET with Al contact with and without SiN interface layer [84]. (b) Transfer characteristics of a pseudo-MOSFET with SiN interface layer and Al- (red curve) and Pt-contacts (blue curve) [72]. (c) Simulated on-currents for SB-MOSFETs with horizontal contacts exhibiting three different SiN layer thicknesses as a function of the contact length.

the typical ambipolar behavior as expected from a SB-MOSFET. The device with interface layer, on the other hand, exhibits an increased on-current and a steeper inverse subthreshold slope because the depinning leads to a reduced SB. Furthermore, the characteristics of the device with SiN interface layer show unipolar behavior. As mentioned above, the reason for this is the strong suppression of MIGS with the relatively thick SiN of ~8 Å leading to contacts behaving similar to a doped semiconductor.

Figure 7.22(b) shows transfer characteristics of another pseudo-MOSFET with aluminum contacts (red curve) and platinum/aluminum contacts (blue curve); note that the reason for the large gate voltages is the thick buried oxide serving as gate dielectric. Obviously, *n*-type behavior is achieved in the case of Al contacts, whereas *p*-type behavior is obtained in the Pt case. Furthermore, unipolar device characteristics can be observed in both cases [72]. This proves that the SiN is effective in depinning (not repinning) the Fermi level because the work function of the respective contact metal acts equivalently to a gate voltage in a MIS capacitor shifting the conduction or valence bands close to the Fermi level.

7.3.2 Horizontal Metal-Insulator Contacts

Inserting a potential barrier in-between the metal contact effectively depins the Fermi level at a metal-semiconductor junction. However, the price paid for the depinning is simply another potential barrier and this raises the question of the contact resistivity and the required contact lengths. Horizontal contacts where the contact metal is deposited over a certain length l_{con} (see the schematic in the top panel of Figure 7.22) are usually described by a transmission line model consisting of a resistor network as discussed in Section 4.6.4. However, such a model does not account for the modification of the density of states in the semiconductor underneath the metal contact electrodes. Therefore, self-consistent NEGF simulations were carried out based on the formalism presented in Chapter 6. A central ingredient in the model used to study the impact of different insulator thicknesses and contact lengths is the way of incorporating the horizontal contact; the contact metal is represented by multiple Buettiker probes all sharing the same Fermi level. These Buettiker probes are connected to the semiconductor via a coupling factor γ that allows for tuning the coupling strength and basically reflects the thickness of the insulating layer in-between the metal and the semiconductor. The contact model is indeed the same as used to explain gate leakage in Chapter 6 and to study the electrical behavior of carbon nanotube field-effect transistors discussed in the next chapter. More details on the contact model can therefore be found in Section 8.1.3.

The important point is that the coupling between metal and silicon can be connected with the potential barrier and effective mass of the insulator (SiN). In the case of weak coupling (small γ), there is hardly any impact on the density of states within the band gap of the semiconductor and as a result, the metal-insulator-semiconductor system behaves like a doped contact as mentioned already above. This means that a weak coupling, i. e., a thicker SiN layer is preferable to obtain unipolar behavior. On the other hand, a weak coupling increases the contact resistance and makes longer contacts necessary. This fact is depicted in Figure 7.22(c), which shows the on-current through the simulated device for three different coupling strengths ($\gamma = 0.1, 0.01$ and 0.001) as a function of contact length computed with the NEGF model, briefly described above [84]. The chosen values for γ belong to SiN layer thicknesses of $d_{SiN} = 2.7\,\text{Å}$, $d_{SiN} = 5.3\,\text{Å}$ and $d_{SiN} = 8\,\text{Å}$ (if an effective mass of $0.35m_0$ is used). Obviously, in the case of weak coupling significantly longer contacts are required. However, if the coupling gets too weak it cannot be compensated for anymore with an increase of l_{con}. Nevertheless, in a range between $d_{SiN} = 4, \dots, 8\,\text{Å}$, which is experimentally accessible, depinned contacts can be realized and their properties can be tuned to exhibit a low contact resistance and to yield unipolar device behavior.

7.4 Reconfigurable Devices

In the preceding sections, we found that Schottky-barrier devices usually perform worse than conventional MOSFETs with doped source/drain regions. Even with processes such as dopant segregation during silicidation, it is difficult to lower the Schottky barrier sufficiently. The use of ultrathin-body channel layers such as nanowires in a wrap-gate architecture with ultrathin gate dielectrics strongly improves the device characteristics but also increases the off-state leakage current due to the ambipolar operation of SB-MOSFET. As a result, the specific advantages of SB-MOSFETs are overcompensated by their drawbacks. However, separating the gate electrode in two or three individual gate electrodes along the channel length allows realizing so-called reconfigurable MOSFETs (reconFETs). Reconfigurable transistors can be switched, i. e., configured to operate either as an *n*-type or as a *p*-type device [104, 193]. While one could argue that this is also true for a SB-MOSFET due to the ambipolar operation, what is meant here is that reconFETs can be switched from unipolar *n*-type to unipolar *p*-type operation, thereby avoiding the large off-state leakage of SB-MOSFETs. The realization of reconFETs relies on the ability to change the "doping" character of certain sections of the channel. This can be accomplished by using additional, independent gate electrodes. Obviously, additional gate electrodes increase the complexity of the fabrication process but it has been shown that reconFETs enable a reduction of the complexity of integrated circuits overcompensating the drawback of at least one additional gate in such devices [193].

In the simplest configuration, a reconFET features two independent gate electrodes as illustrated in the top panel of Figure 7.23. Since in reconFETs it is desired that the *n*- and *p*-type devices operate similarly, Fermi level pinning around mid-gap is actually required, meaning that both device types suffer from the tunneling injection through a rather high Schottky barrier. Therefore, reducing λ_{ch} as much as possible, i. e., realizing a wrap-gate architecture with a nanowire/tube with smallest possible diameter and employing ultrathin gate dielectrics with high dielectric constants, is most effective in

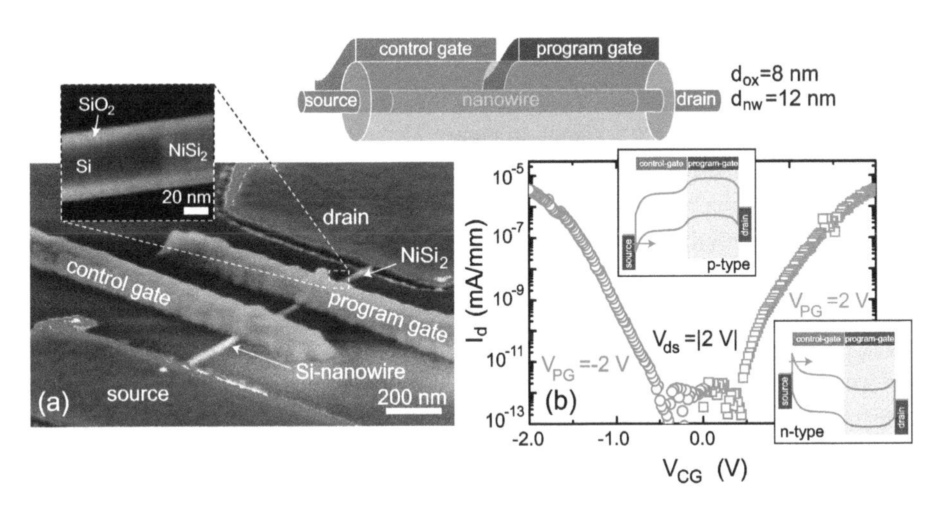

Figure 7.23: A schematic illustration of a reconFET is shown in the top panel. The two-gate electrodes are called control and program gate, respectively. (a) Scanning electron micrograph of a reconfigurable silicon nanowire FET with nickel silicide source/drain contacts based on a VLS-grown Si nanowire. The inset shows the interface between the silicon nanowire and the silicide. (b) Transfer characteristics of a reconfigurable silicon nanowire FET operating as *p*-type FET (red hollow circles) and as *n*-type FET (blue hollow squares) depending on the program voltage [103] (W. Weber, NaMLab, Dresden and TU Vienna) [103].

order to improve the electrical behavior of reconFETs. An electron micrograph of an experimental reconfigFET with nanowire channel and trigate is shown in Figure 7.23 (a). This reconFET is based on a VLS-grown silicon nanowire (cf. Section 3.13.2) that is thermally oxidized to generate an 8 nm SiO_2 gate dielectric [103]. The oxide is removed in the contact areas, nickel is deposited and a silicidation at 450 °C is carried out leading to an axial diffusion of the silicide along the nanowire axis (cf. Section 3.11.1). The inset of Figure 7.23(a) shows the $NiSi_2$-silicon interface. Finally, Ti/Al program and control gates are fabricated with electron beam lithography and lift-off.

At the drain-side (here: right) gate electrode, called the program gate (PG), a constant voltage is applied to generate a *p*- or *n*-type section, whereas the source-side (left) gate, called the control gate (CG), is used to switch the transistor. Two conduction and valence band profiles of the device configured as *n*-type (blue) and *p*-type (red) transistor are shown in the insets of Figure 7.23(b). From the band profiles, it becomes clear that the fixed program gate suppresses the leakage from drain, i. e., hole(electron) injection in the *n*-(*p*-)type device, yielding unipolar device characteristics. The control gate switches the reconFET by exploiting the ambipolar carrier injection at a metal semiconductor Schottky junction to modulate the electron (*n*-type) or hole (*p*-type) injection, respectively. Exemplarily, Figure 7.23(b) shows experimental $I_d - V_{gs}(= V_{CG})$ characteristics of both configurations of the device displayed in (a), proving that reconfiguration can indeed be accomplished. In addition, an almost perfectly symmetric *n*- and *p*-FET performance is obtained [103].

7.4.1 Program-Gate at Drain versus Program-Gate at Source

When operating the reconFET shown in Figure 7.23, the control gate that switches the transistor and controls the current flow through the device is at the source-side Schottky-junction. In order to provide a symmetric *n*- and *p*-type performance, the Fermi level at the metallic source contact needs to be aligned approximately mid-gap resulting in equal electron or hole injections through Φ^e_{SB} and Φ^h_{SB}, respectively. Therefore, the switching is always determined by tunneling, and hence worse compared to a conventional MOSFET as has been discussed in Section 7.1.1.1. The question now is: what would the device performance be if control and program gates were interchanged? In this case, we have to distinguish between the two configurations program-gate at drain (PGAD), which has been discussed so far and program-gate at source (PGAS). The answer which configuration is preferable is not straightforward and depends crucially on the semiconductor in use and the supply voltage of an envisaged circuit. To elucidate the question, self-consistent NEGF simulations have been carried out (see Chapter 6). To simplify the computation, a one-dimensional nanotube channel with symmetric band structure has been assumed. Scattering (l_{mfp} is 15 nm) has been implemented with Buettiker probes. Furthermore, $E_g = 1$ eV, $m^\star = 0.1$, $d_{ch} = 1$ nm, $d_{ox} = 5$ nm and a wrap-gate architecture have been assumed. These parameters are chosen for convenience (to obtain a converged solution of the simulation) and do not affect the general validity of the results. Finally, a symmetric Schottky barrier of $E_g/2 = 0.5$ eV is used.

Simulations have been carried out for PGAD and PGAS configurations; for clarity, only *n*-type reconfigurable FETs are studied in the present section but the results hold equally well for *p*-type reconFETs when voltages are appropriately reversed. The resulting transfer characteristics are shown in Figure 7.24. Let us first consider the PGAD configuration (illustrated in the top right panel of Figure 7.24), which is also the configuration discussed in the preceding section. In PGAD, an *n*-doped section is formed at the drain, leading to a low leakage, and hence to unipolar device behavior also for large V_{ds} because carrier injection from drain is suppressed. The control gate is at source, and thus switching of a reconFET in PGAD is the same as in a SB-MOSFET. This is exactly the behavior shown in Figure 7.24 with the blue curves and it means that the price paid for the low off-state leakage and unipolar behavior is a deteriorated switching behavior with $S_{\text{reconf}} = S_{\text{SB}} > 60$ mV/dec.

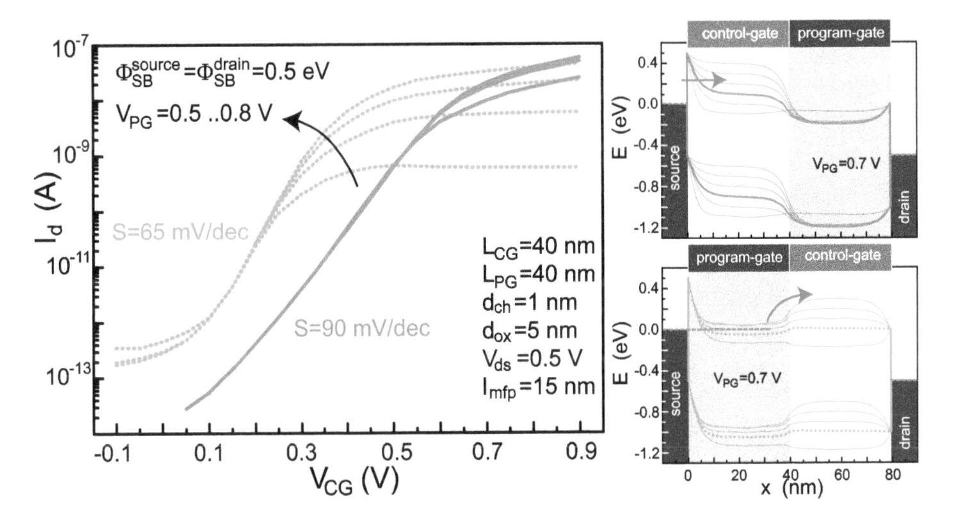

Figure 7.24: Simulated transfer characteristics of reconfigurable FETs with PGAS- (light blue dotted lines) and PGAD-configuration (blue lines); associated conduction and valence band profiles are depicted in the two panels on the right. While the PGAS yields almost ideal inverse subthreshold slopes, it shows larger off-state leakage than the PGAD-configuration.

Next, we study the PGAS configuration which is illustrated in the bottom right panel of Figure 7.24. Since the PG is now at source, significantly more leakage is expected in particular for large V_{ds} due to carrier injection from drain. However, if the drain-source bias is kept small, the leakage current due to injection of carriers at the drain end will be tolerable because the bands at the source end are sufficiently moved down. In the present case with $V_{ds} = 0.5$ eV, an effective barrier of approximately $E_g/2$ blocks the hole flow through the valence band, and as a result the off-state leakage is larger than in PGAD configuration yet smaller than in a SB-MOSFET. More importantly, the *n*-type section created with the PG is almost in equilibrium with the metallic source contact as long as the device is in the off-state. This means that the section underneath the PG acts as a virtual, *n*-doped source contact. Therefore, increasing the CG-voltage the device behaves similar to a conventional MOSFET and, correspondingly, one obtains an inverse subthreshold slope of 65 mV/dec close to the Boltzmann limit (light blue dotted curves in Figure 7.24). The on-state performance, however, is ultimately limited by the tunneling through the source Schottky barrier and hence saturates for large CG-voltages. The on-current can only be increased with a higher PG-voltage, which is indeed the case. But when considered in a circuit environment, a higher PG-voltage necessitates a higher V_{dd}, which implies a larger V_{ds} leading to an exponential increase of the off-state leakage current. Ultimately, two different voltage levels would be needed in order to exploit the PGAS configuration without suffering from increased leakage.

The findings above regarding the difference of PGAS and PGAD configurations have also been verified experimentally [244]. To this end, top-down Si nanowires are fabricated with the approach illustrated in Figure 3.58 using anisotropic silicon etching in combination with a LOCOS process [286, 244]. Figure 7.25 shows an electron micrograph of the reconFET device (see figure caption for details on the device dimensions). Transfer characteristics in both modes of operation, PGAD and PGAS, are displayed in Figure 7.26. The device can obviously be operated as *p*-type (a) and *n*-type (b) transistor and in both operation modes. In both cases, it shows the distinctly different behavior in PGAS versus PGAD that is expected from the analysis of the device simulations discussed above (compare Figure 7.26(b) with the main panel of Figure 7.24); the PGAS mode exhibits a steeper inverse subthreshold slope but also a larger off-state leakage.

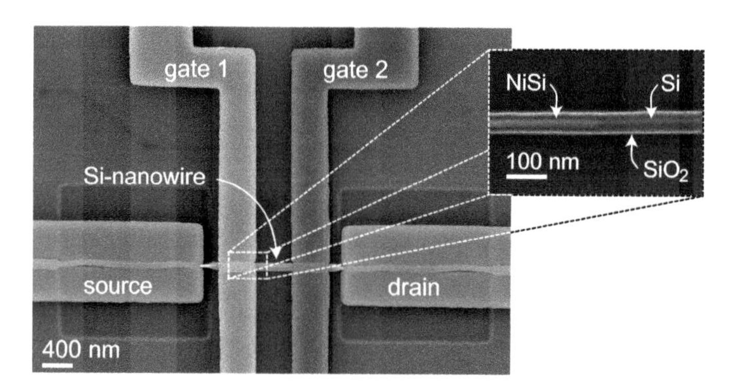

Figure 7.25: Scanning electron micrograph of a top-down fabricated nanowire reconFET with NiSi source/drain contacts. The device has an oxide thickness d_{ox} = 15 nm, a nanowire diameter d_{nw} = 25 nm and the length of the gates and their interdistance are 500 nm and 450 nm, respectively [244]. The inset shows a close-up of the NiSi-silicon nanowire interface.

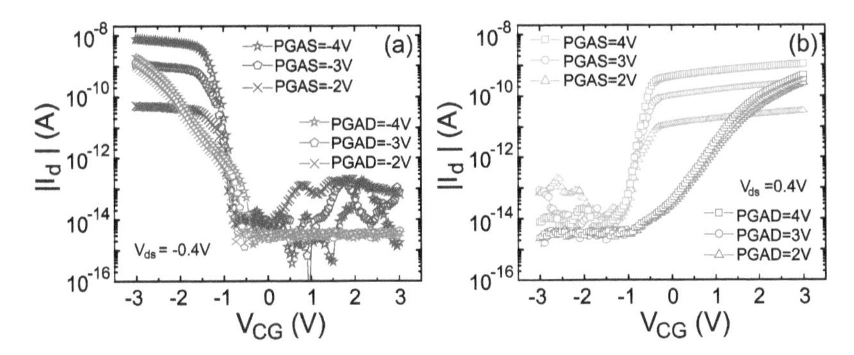

Figure 7.26: (a) ReconFET operated as *p*-type FET in PGAS (dark red) and PGAD (red) modes. (b) ReconFET operated as *n*-type FET in PGAS (light blue) and PGAD (blue) modes [244].

Looking at the transfer characteristics alone it seems that, in spite of a somewhat larger leakage, the PGAS is the preferred mode of operation. However, when taking the output characteristics into consideration things get less clear. If two different voltages are applied to the source and drain side gates, then the effective SB at the respective junctions will have different values (cf. Equation (7.4)) with the smaller being obviously at the side with the larger applied gate voltage. As a result, in PGAS there is a larger Φ_{SB}^{eff} at the drain compared to source. The current flow through the reconFET is therefore dominated by the forward biased Schottky junction at drain leading to a very pronounced sublinear behavior, which is indeed experimentally observed (see Figure 7.27(a)). However, even in PGAD, where Φ_{SB}^{eff} is larger at source, a clear sublinearity can be seen (Figure 7.27(b)) due to the charge feedback on the source-side Schottky-junction discussed in detail in Section 7.1.2 providing experimental verification of the simulation results presented there.

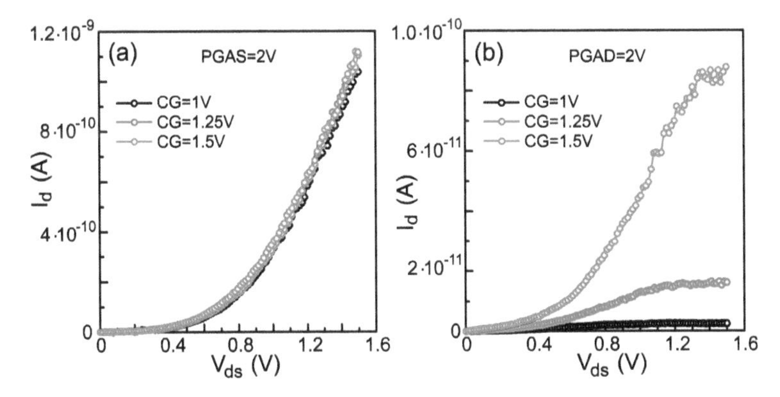

Figure 7.27: Output characteristics for different combinations of applied voltages at the source- and drain-side gate electrodes. PGAS operation mode is shown in (a), PGAD in (b) [146, 244].

In conclusion, reconfigurable MOSFETs are an attractive variant of Schottky-barrier devices that allow in principle substantially better performance than SB-MOSFETs and enable a reduction of circuit topologies despite the more complex device architecture. However, further performance improvements are necessary to exploit the potential of reconFETs. This can only be done if Fermi level pinning is removed such that the program gate is able to move conduction and valence bands with respect to the Fermi level of the metal contact, enabling carrier injection without a tunneling process through a Schottky barrier. In Section 8.1.4, an approach based on carbon nanotube FETs will be presented that potentially allows significantly higher on-currents to be obtained.

Exercises

Exercises together with solutions are accessible via the QR code.

8 Carbon Nanotube Field-Effect Transistors

In many respects, carbon nanotubes (CNTs) appear to be the best choice for ultimately scaled field-effect transistors. They exhibit an extremely small diameter, and they offer a symmetric band structure with values of the band gaps and effective masses in a range ideally suited for nanoelectronics devices (cf. Section 2.9). Recent research on carbon nanotubes resulted in significant progress regarding scalability of carbon nanotube field-effect transistors (CNTFETs), contact formation as well as the realization of integrated circuits [37, 77, 106, 217]. Therefore, the electronic properties of CNTFETs will be studied in the present chapter with experiments as well as simulations.

8.1 Carbon Nanotube FETs as SB-MOSFETs

The easiest way of realizing a CNTFET is to disperse nanotubes on an oxidized piece of silicon and contact them with metallic electrodes using electron beam lithography and lift-off. In this case, the silicon wafer serves as a large area back-gate and the oxide on top plays the role of the gate dielectric. Figure 8.1(a) shows an electron micrograph of such a simple back-gated CNTFET with two nickel source/drain electrodes that cover a larger nanotube section. The actual details of the metal-nanotube contacts are a bit more involved (see Section 8.1.3.1) and require a model for the contacts similar to the metal-SiN-Si contact studied in Section 7.3. However, in order to elucidate a number of experimental observations—in particular, the rather large dependence of the inverse subthreshold slope S and the on-current on the gate oxide thickness d_{ox}—it is sufficient to describe the metal-CNT electrodes as simple metal-semiconductor contacts with a certain Schottky-barrier (SB) height. Many experimentally observable features of CNTFETs can therefore be understood by considering them as Schottky-barrier FETs (detailed explanations on the principles of operation of SB-MOSFETs are provided in Chapter 7).

8.1.1 Impact of the Device Geometry

Figure 8.1(b) shows a typical transfer characteristic for a small drain-source bias of the CNTFET shown in (a). The device exhibits ambipolar behavior, typical of SB-MOSFETs, which is expected when the source/drain contacts are regarded as simple metal-semiconductor contacts. Extracting the inverse subthreshold slope S of a large number of CNTFETs with different thicknesses of the gate oxide d_{ox} one observes a strong dependence of S on d_{ox} shown in Figure 8.1(c) [8, 136]. As discussed in Chapter 7, this peculiar behavior is a consequence of the dependence of the carrier injection through the Schottky barrier on the screening length λ_{ch}, which in the present case is $\sqrt{\frac{\varepsilon_{CNT}}{\varepsilon_{ox}} d_{ox} d_{CNT}}$ because the electrostatics of a simple, back-gated CNTFET can approximately be described as a single-gate planar device [8]. The dotted red line in Figure 8.1(c) shows the

https://doi.org/10.1515/9783111054421-008

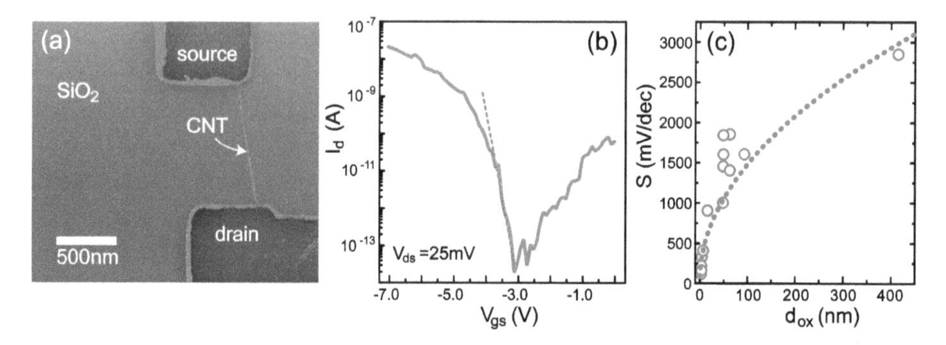

Figure 8.1: (a) Electron micrograph of a back-gated carbon nanotube field-effect transistor. The CNT is dispersed on an oxidized piece of silicon wafer that serves as large-area back-gate. Electron beam lithography and lift-off are used to form nickel source/drain contact electrodes. (b) Typical transfer characteristic of a CNTFET showing asymmetric ambipolar behavior with better *p*-type performance than *n*-type behavior. (c) S as a function of d_{ox} for CNTFETs [8]. The dotted lines are computed with Equation (7.7) and the hollow symbols are experimental data.

inverse subthreshold slope as a function of d_{ox} as computed with Equation (7.7), which overall agrees very well with the experimental data (hollow circles). The CNTFETs exhibit the same dependence of S on d_{ox} as we already obtained when discussing SOI SB-MOSFETs (see Figure 7.11). The only difference between the analytic calculation in Figure 7.11 and Figure 8.1(c) is the tunneling distance d_{tunnel}, which is ~3.4 nm in the case of the SOI SB-MOSFETs and ~5 nm in the case of the CNTFETs. With an effective mass in the CNT that is approximately a factor of two smaller than the light effective mass in silicon, the latter can be understood since the tunneling distance scales roughly as $\propto \frac{1}{\sqrt{m^*}}$ [136].

The importance of the dependence of S on d_{ox} is that this experiment and the appropriate analysis clearly shows that CNTFETs are Schottky-barrier devices and not conventional FETs as one might think when looking at, e. g., the output characteristics (not shown here) or considering the relatively steep inverse subthreshold slopes of $S \approx 100$ mV/dec that are observed in CNTFETs with thin gate oxide. Hence, extracting meaningful values for the mobility requires extra care. Furthermore, the presence of Schottky barriers at the contacts has also important implications regarding the expected multimode transport in CNTFETs as will be elaborated on in the next section.

Task 30.
CNTFET as SB-MOSFETs: Consider a (11, 7) carbon nanotube in a single-gate transistor architecture (d_{ox} = 10 nm SiO_2) and metal contacts. The Fermi level at the metal-CNT interface is at mid-gap. Compute the inverse subthreshold slope of the transistor and the minimum off-state current at V_{ds} = 0.5 V.

8.1.2 Multimode Transport in CNTFETs

In Section 2.13.3, it was discussed that in a nanowire/tube transistor where vertical quantization leads to a sufficient energetic separation of the one-dimensional subbands, the contribution of each subband may be observable as a (smeared out) step-like increase of current in the transfer characteristics of field-effect transistors. However, a step-like increase can only be observed provided that the energetic separation is significantly larger than $k_B T$ and larger than the applied bias V_{ds} (cf. Figure 2.54). In addition, electronic transport with a quantized increase of conductance according to $\frac{2e^2}{h}$ will only occur if the geometry of the source contact channel region resembles a quantum point contact where a region with many subbands (source) is gradually tapered into the nanowire/tube such that all subbands can be occupied with the same amount of charge when an appropriate gate voltage is applied.

A carbon nanotube is an inherently small, true nanoscale object and since low-index (n, m) nanotubes exhibit the necessary subband splitting it is expected that multimode transport should be observable in the electrical characteristics of CNTFETs. To be specific, let us compute what we expect in terms of subband splitting. In Section 2.9, the band structure of carbon nanotubes is computed from which the band gap as well as the subband structure can be extracted. Using Equation (2.60), the band gaps for the various subbands of a (n, m) nanotube (at $\kappa = 0$) are given as

$$E_g^{\tilde{n}} = 2\pi\sqrt{3}V_{pp\pi}\frac{|\tilde{n} - \frac{n-m}{3}|}{\sqrt{n^2 + m^2 + nm}} \tag{8.1}$$

where $\tilde{n}$ is the subband index; note that the diameter of the nanotube is $d_{\mathrm{CNT}} = \frac{|\vec{C}|}{\pi} = \frac{a\sqrt{n^2+m^2+nm}}{\pi}$. Band gaps of typical nanotubes used in the experiments discussed here are on the order of 0.6–0.7 eV. Such a band gap can be realized with a (12, 8) CNT, for instance. Inserting $n = 12$ and $m = 8$ into Equation (8.1) yields the band gap of the first subband with $\tilde{n} = 1$ (the actual band gap) to be $E_g = 0.62$ eV and $d_{\mathrm{CNT}} \approx 1.4$ nm. The band gap energies of higher subbands are obtained with other positive and negative integer values of $\tilde{n}$. Dividing these gap energies by two provides the position of the conduction band minima and the energetic separation between adjacent subbands can be determined. In the case of the (12, 8) nanotube, the minima of the conduction band subbands are given by $E_g^{\tilde{n}=1}/2 = 0.31$ eV (first subband), $E_g^{\tilde{n}=2}/2 = 0.62$ eV (second subband), $E_g^{\tilde{n}=0}/2 = 1.24$ eV (third subband), etc. (cf. Section 2.9). Exemplarily, Figure 8.2(a) shows the conduction and valence bands in a Schottky-barrier CNTFET taking into account the multiple 1D subbands of the (12, 8) nanotube with the energetic subband separations computed above.

The subband spacing in the considered (12, 8) nanotube is substantially larger than $k_B T = 25$ meV at room temperature and as a result, multimode transport with a step-function like increase of the current for increasing gate voltage is expected. Since the

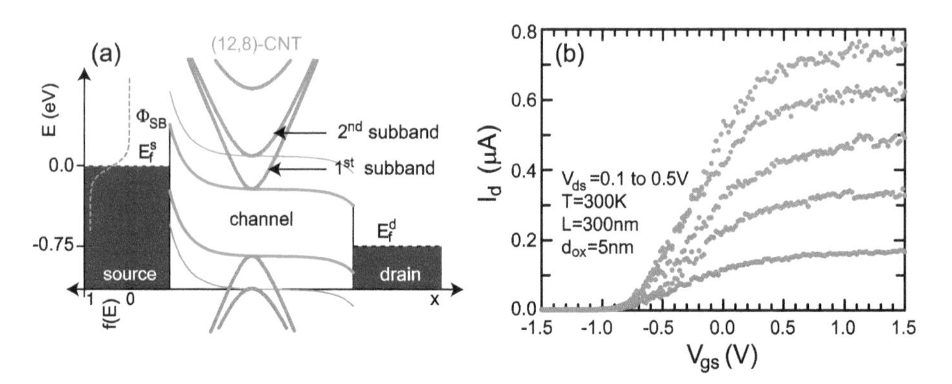

Figure 8.2: (a) Conduction and valence bands of a SB-CNTFET based on a (12, 8) nanotube with second and third 1D subband. The Fermi level is assumed to be aligned at mid-gap. (b) Transfer characteristics of a CNTFET with light potassium doping [9].

subbands are degenerate, the current steps should be proportional to $\frac{4e^2}{h}V_{\text{ds}}$. However, looking at the experimental $I_d - V_{\text{gs}}$ curves of a CNTFET (with a CNT of similar dimension as (12,8) tube) as displayed in Figure 8.2(b), no apparent subband structure can be seen, and consequently, transport in CNTFETs has been interpreted as being purely 1D. So, what is missing in the description of electronic transport? Is the energetic splitting of the subband actually far smaller than anticipated from our band-structure calculations? Or will higher subbands not be occupied because the quantum capacitance C_q of the CNT dominates, i. e., $C_q \gg C_{\text{ox}}$ (cf. Section 4.5.3) for higher gate voltages leading to $\delta\Phi_f^0 \to 0$?

Although the DOS in a carbon nanotube is somewhat different from a true 1D system in that it first drops proportional to $1/\sqrt{E}$ and then approaches a constant for higher energies (see Equation (2.104)), the inversion layer/quantum capacitance C_q nevertheless drops to a small value with increasing V_{gs}. As a result, it is possible to move the bands by applying appropriate gate voltages leading to the occupation of higher subbands. Consequently, the second question can be negated. As it turns out, it is the Schottky barrier at the source-channel interface that prevents the *observability* of multimode transport even though more than a single 1D subband contributes to the current.

Let us start the discussion by calculating the current through a carbon nanotube FET using the Landauer approach. In the case of small V_{ds}, the difference of source and drain Fermi distribution functions can be Taylor expanded yielding (cf. Equation (5.25)) $f(E_f^s) - f(E_f^d) \approx -\frac{1}{k_BT}\frac{\partial f(E_f^s)}{\partial E}(|e|V_{\text{ds}}) \approx \delta(E - E_f^s)\frac{|e|V_{\text{ds}}}{k_BT}$. Thus, the drain current is

$$I_d \approx \frac{2e^2}{h}T_{\text{SB}}^{\text{tot}}(E_f^s, \Phi_f^0)\frac{1}{1 + \exp(\frac{\Phi_f^0 - E_f^s}{k_BT})}V_{\text{ds}}, \tag{8.2}$$

where the derivative of the Fermi distribution function with respect to energy is approximated with a delta function. In addition, $T_{\text{SB}}^{\text{tot}}$ is the overall transmission probability through the Schottky barriers at the source and drain. Note that in the equation above Φ_f^0

is a function of the gate potential Φ_g, which needs to be determined taking the effect of charge self-consistently into account (see Chapters 5 and 7). Since the bias V_{ds} is considered to be small and the device as being a long channel transistor, $\Phi_f^0 - \Phi_g = -\frac{Q_{ch}}{C_{ox}}$ where the equilibrium value can be used for the channel charge Q_{ch}. Furthermore, the small V_{ds} also justifies that T_{SB}^{tot} can be assumed to consist of two equal transmission probabilities $T_{SB}^s = T_{SB}^d$, which can be approximated with the WKB approximation. The question is now how the two individual transmission probabilities $T_{SB}^{s,d}$ must be combined in order to obtain T_{SB}^{tot}. So, before proceeding, let us briefly discuss how two transmission probabilities are connected to each other.

Combining two transmission probabilities: Two transmission probabilities T_1 and T_2 (for instance, due to two potential barriers or two Schottky barriers within a device), cannot be combined by simply multiplying them, i. e., $T_{tot} \neq T_1 T_2$. If we described the resistance of a wire with length L by n pieces $\delta x = L/n$ with a transmission $T_{\delta x} = \frac{l_{mfp}}{l_{mfp}+\delta x} = \frac{l_{mfp}}{l_{mfp}+L/n} = \frac{1}{1+L/(l_{mfp}n)}$ simply by $T_{tot} = (T_{\delta x})^n$, this would result in an exponential decrease of the transmission since $\lim_{n\to\infty}\left(\frac{1}{1+L/(l_{mfp}n)}\right)^n = \exp(-L/l_{mfp})$ which is certainly in contradiction with Ohm's law. In fact, multiple reflections in between the two potential barriers need to be taken into consideration and one has to sum over all possible paths (four are explicitly shown in Figure 8.3) in order to obtain the total transmission

$$T_{tot} = T_1T_2 + T_1R_2R_1T_2 + T_1R_2R_1R_2R_1T_2 + \cdots = T_1 \underbrace{(1 + R_2R_1 + R_2R_1R_2R_1 + \cdots)}_{=\frac{1}{1-R_2R_1}} T_2 \tag{8.3}$$

and, since $R_1 = 1 - T_1$ and $R_2 = 1 - T_2$, this yields $T_{tot} = \frac{T_1T_2}{T_1+T_2-T_1T_2}$.

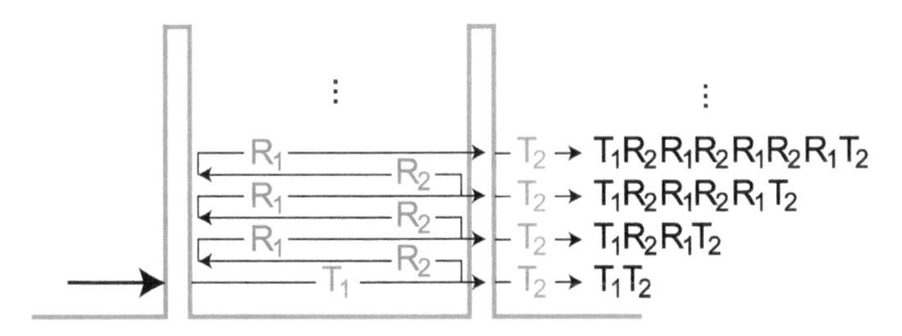

Figure 8.3: Two potential barriers as scattering centers with transmission and reflection probabilities $T_{1,2}$ and $R_{1,2}$, respectively. The overall transmission probability T_{tot} is a sum over all possible paths including multiple reflections as depicted in the figure.

According to Equation (8.3), combining the transmissions through the two Schottky barriers yield

$$T_{SB}^{tot} = \frac{T_{SB}^{WKB}}{2 - T_{SB}^{WKB}}. \tag{8.4}$$

In Section 4.33, the WKB approximation for a Schottky barrier in depletion approximation was introduced. In the present case, the potential of the Schottky barrier is determined by the screening length λ_{ch} instead of the depletion length. As a first-order estimate, we can therefore use the WKB expression with L_{depl} being replaced by λ_{ch}:

$$T_{SB}^{WKB}(E_f^s) = \exp\left(-\frac{4}{3}\lambda_{ch}\sqrt{\frac{2m^\star}{\hbar^2}}\frac{(\Phi_{SB})^{3/2}}{(\Phi_{SB}-\Phi_f^0)}\right). \tag{8.5}$$

The expression above shows that the transmission T_{SB} broadens the contributions of each subband and depending on λ_{ch}, a large gate voltage range (equivalent to a Φ_f^0-range if we assume perfect gate control) is required in order to reach a constant transmission for each subband. Indeed, in Figure 8.4 the (added) transmission computed with Equation (8.5) of three subbands in the (12, 4) nanotube considered so far is shown for three different values of λ_{ch} as a function of V_{gs}; perfect gate control was assumed, which is justified considering the small values of λ_{ch} (and thus the large geometrical oxide capacitance C_{ox}) chosen here.

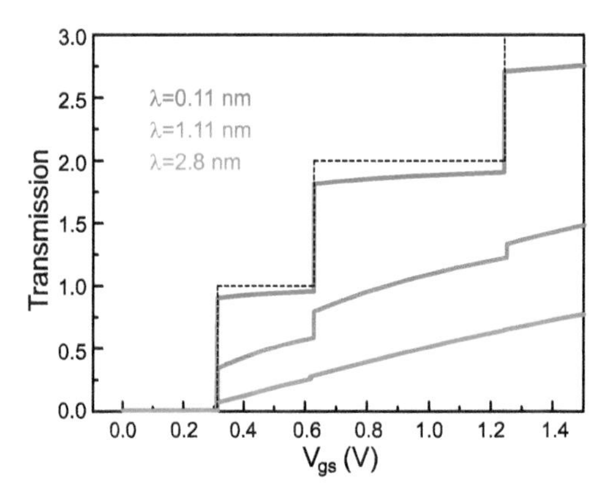

Figure 8.4: Transmission of three subbands (at 0.31 eV, 0.62 eV and 1.24 eV) as a function of gate voltage for a CNTFET with (12,4) nanotube for three different screening lengths λ_{ch}. The dashed line shows the transmission if there was no SB.

Obviously, extremely small λ_{ch} are required to obtain a step-like increase in the transmission. But even in the case of an unrealistically small $\lambda_{ch} = 0.11$ nm one would not get $T = 1$ in each of the subbands. Hence, multimode transport even in a device with such a small λ_{ch} would not be in steps of $\frac{4e^2}{h}$. For somewhat larger and more realistic values of λ_{ch}, the steps shrink, and for a $\lambda_{ch} = 2.8$ nm (which is still very small and experimentally not easy to realize), the transmission function of each subband is broadened so much that hardly any step is visible. If transport is at room temperature, no step will be visible at all, although three subbands contribute to the current at higher V_{gs}.

To be able to *observe* multimode transport, it is necessary that the current in one mode saturates with increasing gate voltage before a subsequent mode significantly contributes to the current. For a specific nanotube, this can be done in two different ways.

First, the transmission through the Schottky barriers could be strongly increased by making the screening length λ_{ch} as small as possible as explained when discussing Figure 8.4 above. However, this implies modifying the device structure employing, e. g., dielectrics with high dielectric constant, using a wrap-gate architecture, etc. which can be difficult to realize. The second approach is to limit the transmission probability through the channel independent of the Schottky barriers, so that the current in one mode saturates quickly when increasing V_{gs}. This ansatz can be followed by deliberately deteriorating the carrier mobility within the nanotube, i. e., by exploiting scattering in the channel of the devices. In Section 7.1.1.3, it was argued that in SB-MOSFETs the tunneling through the SB often dominates over the carrier mobility. Therefore, rather strong scattering in the channel is needed for the second approach.

Scattering can be incorporated into our semianalytical considerations in the following way: the transmission through the channel in the presence of scattering denoted as T_{ch} is connected to the mean free path and the channel length as $T_{ch} = l_{mfp}/(L + l_{mfp})$ (cf. Section 5.3). Hence, the overall transmission probability is

$$T_{tot} = \frac{T_{SB}^{tot} T_{ch}}{T_{SB}^{tot} + T_{ch} - T_{SB}^{tot} T_{ch}} = \frac{1}{L/l_{mfp} + 1/T_{SB}^{tot}}. \tag{8.6}$$

As a result, if the ratio $L/l_{mfp} \gg 1/T_{SB}^{tot}$ then $T_{tot} \approx l_{mfp}/L$ as in a conventional FET. Consequently, the current saturation of each subband becomes independent of the Schottky barriers and it is expected that multimode transport will be observable in this case. From Equation (8.6), it is clear that this can be achieved by making L very large and/or l_{mfp} small (and certainly also by increasing the transmission T_{SB}^{tot} for a given ratio L/l_{mfp}, i. e., realizing a wrap-gate architecture and/or decreasing the effective oxide thickness).

Experimentally, an elegant way to increase the ratio L/l_{mfp} is the introduction of scattering within the channel of an SB-CNTFET with potassium doping [31]. Potassium doping yields a large amount of Coulomb scattering sites at the surface of the nanotube reducing l_{mfp} when the K-concentration increases. Above a certain amount of doping, multimode transport becomes visible and transfer characteristics with a (broadened) stepwise increase of the drain current are obtained. The main panel of Figure 8.5(a) shows experimental transfer characteristics of K-doped CNTFETs where three modes can be identified to contribute to the current; the inset shows the same device on a larger V_{gs}-range where a larger number of modes is observed [9].

In order to support the experimental observations, self-consistent NEGF simulations have been carried out, accounting for scattering with the Buettiker probe approach explained in Section 6.3.1. A device with $d_{CNT} = 1.4$ nm, $d_{ox} = 2$ nm in a wrap-gate architecture and $L/l_{mfp} = 30$ is considered in order to guarantee a large T_{SB}^{tot} as well as L/l_{mfp}-ratio. The main panel of Figure 8.5(b) shows the result. In the present case, a constant subband spacing of 0.25 eV and a Schottky barrier of $\Phi_{SB} = 0.35$ eV are assumed. One clearly observes a step-like increase of the current and three different subbands can be identified (note that in the present case equidistant subbands were assumed in order

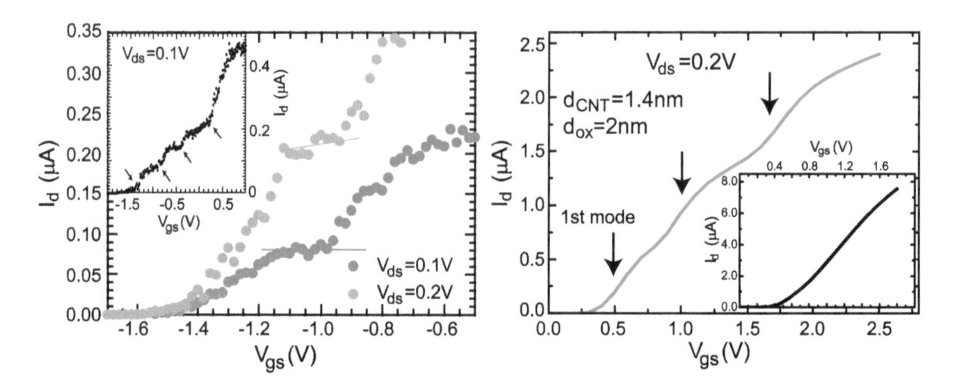

Figure 8.5: (a) Experimental transfer characteristics of a SB-CNTFET exhibiting multimode transport. The inset shows a larger gate voltage range [9]. (b) Simulated I_d–V_{gs} characteristics using NEGF and Buettiker probes to include strong scattering. The inset shows the results of the same device in the case of ballistic transport [136].

to reduce the computational burden). However, in the case of ballistic transport (shown in the inset of Figure 8.5(b)) the different modes cannot be resolved anymore although multimode transport with the first three modes contributing significantly to the current occurs in this case, too. This inability of observing multimode transport has to be taken into account not only when interpreting carbon nanotube data but also when analyzing other ultrathin-body SB-MOSFETs.

8.1.3 Contact Formation to CNTs

As discussed in the preceding section, the electrical behavior of carbon nanotube FETs is to a large extent determined by the presence of Schottky barriers at the metal-nanotube interfaces. The very small diameter of nanotubes leads to a high injection of carriers through the Schottky barriers such that the electrical behavior was at first interpreted in the framework of conventional transistors with possible false extraction of nanotube properties such as the mobility from the experimental data. Extracting the inverse subthreshold slope of a larger number of CNTFETs with different oxide thicknesses showed that the behavior of CNTFETs can indeed be described by assuming Schottky contacts at the metal-nanotube interfaces. However, a closer look at CNTFET data and comparison with the expectation from simulations show discrepancies in that the difference in the on-currents of CNTFETs with different contact electrode materials are much larger than expected (cf. Figure 8.8(a)). In particular for nanotubes with a diameter small enough to yield energy gaps in the range 0.7–1.0 eV, one observes on-currents with orders of magnitude difference for different contact metals used in otherwise equally fabricated devices. There are several aspects to this fact that cannot be explained with a metal-semiconductor contact giving rise to a Schottky barrier. First, from the MIGS model discussed in Section 4.6.1 (in particular, see Figure 4.21) a Fermi level pinning even stronger than in silicon is expected. As such, there is not much deviation in the on-currents expected when the contact metal is changed. Second, even if Fermi level pinning was not strong, the large differences of on-currents in experimental CNTs cannot be explained with a variation of the Schottky-barrier height, given the fact that the screening length, and hence the effective Schottky barrier is not large anyway. Third, the Schottky-barrier model overestimates the ambipolarity of the transfer characteristics. Figure 8.6(b) shows exemplarily a comparison between experimental (blue triangles) and simulated transfer characteristics (red curve) showing a substantial difference in the case of a band gap of 1 eV [143].

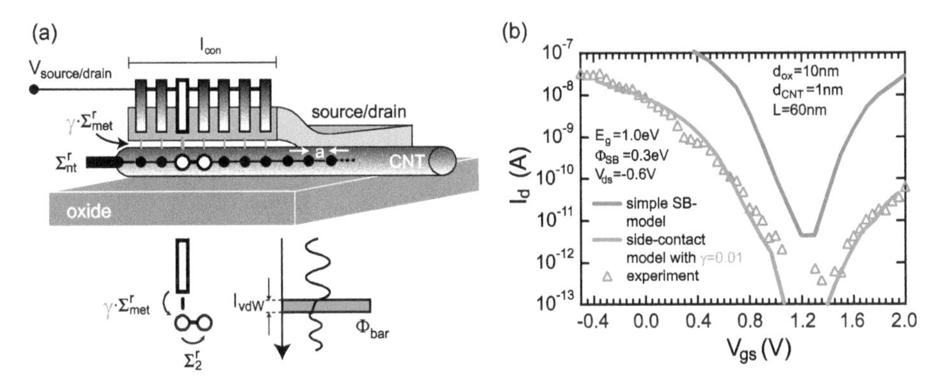

Figure 8.6: (a) Side-contact model for a metal-nanotube contact. A metal covers the nanotube over a length l_{con}. In the simulations, the metal is subdivided into a large number of individual, short-circuited Buettiker probes that are connected to each lattice site of the finite difference grid. (b) Comparison of experimental transfer characteristics (blue triangles) with a simple Schottky contact (red curve) and the side-contact model (green curve) [143].

The reason for the discrepancy between experiment and simulation is the contact geometry considered in the Schottky-contact model compared to the actual experimental situation. Instead of a metal in direct contact with the nanotube, CNTFETs are usually contacted with a metal covering a certain portion of length l_{con} of the nanotube as depicted in Figure 8.6(a). This way of contacting the nanotube makes a more elaborate contact model necessary.

It has been noted by Tersoff that a substantial potential barrier $\Phi_{bar} > 10$ eV exists at the metal-nanotube interface in the case of a metal deposited on top of the nanotube, which is a result of the rapidly decaying wavefunction perpendicular to the nanotube axis [249]. In addition, since the metal is a van der Waals distance ($l_{vdW} \approx 3$ Å) away from the nanotube, there is a very thin but rather high potential barrier in-between the metal and the nanotube. As a result, the actual contact is rather similar to the metal-ultrathin nitride-silicon contact discussed in Section 7.3.2. In fact, the high but extremely thin potential barrier between metal and nanotube can be interpreted as a depinning layer, which explains why the material of the metal contact (i. e., the work function) plays such a prominent role for CNTFETs. The potential barrier also elucidates why experimental devices exhibit less ambipolarity than expected based on the simple SB model. Since the potential barrier in between metal and nanotube is inherent to the contact it needs to be taken into account when quantitative comparisons between experiments and simulations are made.

8.1.3.1 Modeling Metal-Nanotube Contacts

Figure 8.6(a) shows schematically the geometry of what is called a horizontal or side-contact in the following: the metal is deposited on top of a nanotube and covers a section of length l_{con}. In order to incorporate such side-contacts into NEGF simulations, we use the Buettiker probe model introduced in Section 6.3.1. The metal is subdivided into individual, short-circuited contact portions. Each portion of the metal is connected with the nanotube at a specific site of the underlying finite difference grid and represented by a Buettiker probe. As in the case of gate leakage, all Buettiker probes within source (drain) share the same Fermi level fixed by the terminal voltage (e. g., ground potential in the source). The weak coupling between each Buettiker probe and the nanotube is described with the simple coupling factor $\gamma = 0, \ldots, 1$ as has already been done in Sections 6.3.1 and 6.3.2. The value $\gamma = 0$ insulates the contacts from the nanotube and $\gamma = 1$ yields the strongest possible coupling. As will be discussed below, the case $\gamma = 1$ is equivalent to the Schottky-barrier model used in Sections 8.1 and 8.1.2.

In a similar way as has already been done in the case of gate leakage (see Section 6.3.2), the coupling parameter γ can be related to the potential barrier present at the nanotube-metal interface by analytically calculating the transmission function $\mathcal{T}(E)$ through a potential barrier of height Φ_{bar} and width l_{vdW} using elementary quantum mechanics. $\mathcal{T}(E)$ is then compared with the transmission function $T(E)$ through a single site finite difference "grid" using NEGF. To this end, the single site (with potential $\Phi_{site} = 0$) is connected to the right to a semiinfinite contact and to the left to a semiinfinite contact whose coupling is mediated by the parameter γ (see lower panel of Figure 8.6(a)). The retarded Green function G^r of the one-site problem is

$$G^r = \left[E - 2t + te^{ika} + \gamma te^{ika}\right]^{-1} = \left[\gamma te^{ika} - te^{-ika}\right]^{-1}. \tag{8.7}$$

With $\Gamma_l = 2\gamma t \sin(ka)$ and $\Gamma_r = 2t \sin(ka)$, the transmission function $T(E)$ follows using the Fisher–Lee relation (see Equation (6.47)):

$$T(E) = \frac{4\gamma \sin^2(ka)}{1 + \gamma^2 - 2\gamma(\cos^2(ka) - \sin^2(ka))} \approx \frac{1}{1 + \frac{t(1-\gamma)^2}{4\gamma E}} \tag{8.8}$$

since $ka \ll 1$. This result has to be compared with the $\mathcal{T}(E)$ given by

$$\mathcal{T}(E) = \frac{1}{1 + \frac{\frac{(\Phi_{bar})^2}{\Phi_{bar}-E}\sinh^2(K l_{vdW})}{4E}} \tag{8.9}$$

where $K = \sqrt{\frac{2m^*(\Phi_{bar}-E)}{\hbar^2}}$. As was mentioned above, $\Phi_{bar} > 10$ eV and, therefore, $\Phi_{bar} - E \approx \Phi_{bar}$. As a result, $\mathcal{T}(E)$ and $T(E)$ are of the same form and a quadratic equation for γ can be identified. Since γ has to be between 0 and 1, a unique solution can be found given by

$$\gamma = \frac{2 + \frac{\Phi_{bar}}{t}\sinh^2(\sqrt{\frac{2m^*\Phi_{bar}}{\hbar^2}} l_{vdW})}{2} - \sqrt{\frac{(2 + \frac{\Phi_{bar}}{t}\sinh^2(\sqrt{\frac{2m^*\Phi_{bar}}{\hbar^2}} l_{vdW}))^2}{4} - 1}. \tag{8.10}$$

Note that γ explicitly depends on m^* and a through the hopping parameter $t = \frac{\hbar^2}{2m^* a^2}$. For typical values of m^* and a used in the simulations, γ is less than 10^{-2}. It is therefore in a similar range as in the analysis of Section 7.3, confirming that the high but extremely thin potential barrier Φ_{bar} plays the role of a depinning layer. As a result, the presence of this depinning layer enables the work-function difference between CNT and metal to determine where the Fermi level lies with respect to the conduction and valence bands. In addition, the metal-nanotube coupling yields a modified density of states (a reduction of MIGS) of the nanotube portion underneath the metal, i. e., the part of the CNT from which carriers are injected into the channel. Therefore, a strong dependence of the carrier injection, and thus of the on-current on the metal work function is expected.

Plots of the local density of states for strong and weak coupling are shown in Figure 8.7(a) and (b), respectively. The case of strong coupling would be obtained if there was no potential barrier in between metal and nanotube. In this case, the band gap would vanish completely due to metal-induced gap states within E_g originating from

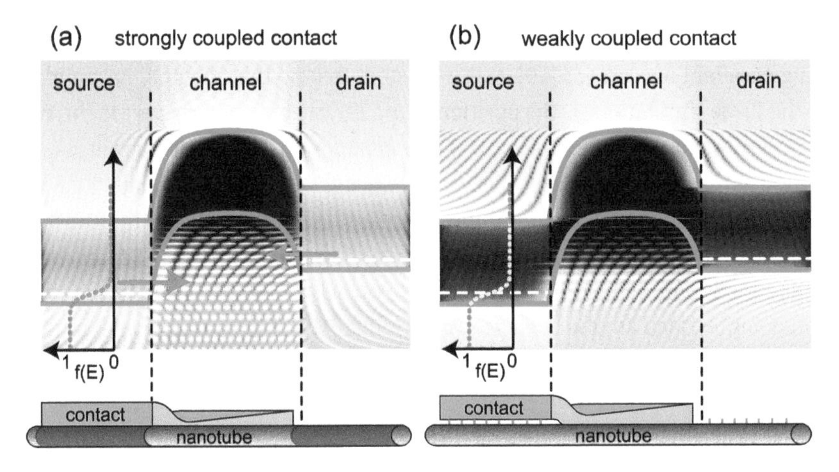

Figure 8.7: Local DOS in CNTFETs in the case of strong (a) and weak (b) metal-CNT coupling. Strong coupling yields a pronounced modification of the DOS underneath the metal resulting in a Schottky contact behavior. The weak coupling leads to a slight modification of the DOS with semiconductor-like behavior.

the intimate contact between metal and CNT. As a result, the density of states becomes "metal-like," and hence carriers can be injected from source and drain at "all" energies as is the case in a metal-semiconductor Schottky contact (Figure 8.7(a)). This would lead to pronounced ambipolar behavior with large leakage currents in the off-state (cf. Figure 7.11). A weak coupling, on the other hand, leads to a "semiconductor-like" density of states with an almost unmodified, intact band gap (cf. Figure 8.7(b)). This in turn results in a suppression of ambipolar behavior, since depending on the position of the Fermi level, either electron or hole injection into the channel dominates. For sufficiently large E_g and appropriately weak coupling, this even yields unipolar device behavior (as observed in the silicon pseudo-MOSFETs shown in Figure 7.22).

8.1.3.2 Comparison with Experiments

The side-contact model can be used to learn more about the specific metal-nanotube contact properties. To this end, simulated transfer characteristics are compared with experimental data obtained from CNTFETs fabricated with three different contact metals—aluminum, titanium and palladium. At first, experimental devices with aluminum contacts are used to find a suitable coupling parameter γ. The reason for this is that in the case of the aluminum contacts, the lowest on currents are measured, which is due to the relatively low work function (compared to the other two metals (cf. Figure 4.18)) resulting in a Fermi level that lies within the nanotube band gap. Since a coupling strength of $\gamma \lesssim 0.01$ is expected, the coupling is rather weak, and thus the density of metal-induced gap states is expected to be rather low. This means that the on-current of the CNTFET with Al contacts sensitively depends on γ, providing a rather robust determination of the coupling strength. As a result of this comparison, $\gamma = 0.007$ was found. With this coupling strength, the devices with Ti and Pd contacts were simulated.

Figure 8.8(a) displays the comparison between the simulated and the experimental data showing excellent agreement [40]. It is important to note that the only parameter changed in the three simulations is the position of the Fermi level with respect to the valence band according to the work functions of Ti and Pd, yet the different current levels and shift of the threshold voltage are nicely reproduced. Moreover, the relatively bad inverse subthreshold slope of the aluminum device and even the kink in the transfer characteristics (cf. Figure 7.4 for more explanations) of the titanium CNTFET are well replicated [40]. Deviations at very low current levels are due to the fact that in the simulations nonidealities such as gate leakage are excluded. The side-contact model is also compared with experimental data from the literature (the data is extracted from [77]). Figure 8.8(b) displays a comparison, which again shows excellent agreement between simulation and experiment. This proves that the contact model very well describes the properties of metal-nanotube contacts. It also shows that the assumption of the presence of the potential barrier Φ_{bar} is well justified. More importantly, this potential barrier Φ_{bar} indeed depins the Fermi level.

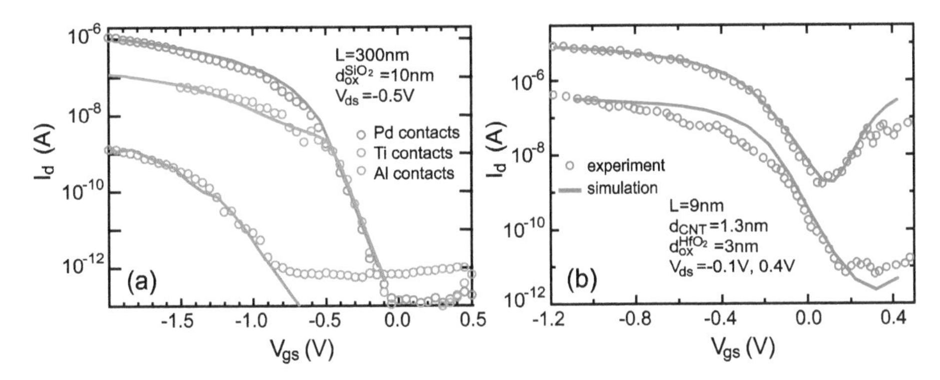

Figure 8.8: (a) Comparison of simulated (straight line) and experimental CNTFET characteristics using the side-contact model for Pd, Ti and Al as contact metals [40]. (b) Comparison of simulated [52] and experimental data extracted from [77].

Next, simulating the electrical characteristics over a wide range of metal-nanotube work-function differences with a constant coupling strength of $\gamma = 0.007$ leaves the position of the Fermi level with respect to the valence band of the nanotube $|E_v - E_f|$ as the only parameter to adjust in order to compare simulated with experimental data. In fact, the energetic difference $|E_v - E_f|$ would be the same as the Schottky-barrier height Φ_{SB}^{hole} in the simple metal-nanotube model. Therefore, comparing simulations with experiments allows for extracting the calibration plot displayed in Figure 8.9, which maps the on-currents of CNTFETs (at a constant gate voltage overdrive of $|V_{gs} - V_{th}| = 0.5$ V) to Schottky-barrier heights present at the metal-nanotube interface [40]. This calibration plot provides immediate access to the SB height in CNTFETs i. e., to $|E_v - E_f|$ from which a first estimate of the band gap, and thus the nanotube diameter in experimental devices can be extracted (when comparing different devices with the same contact metal).

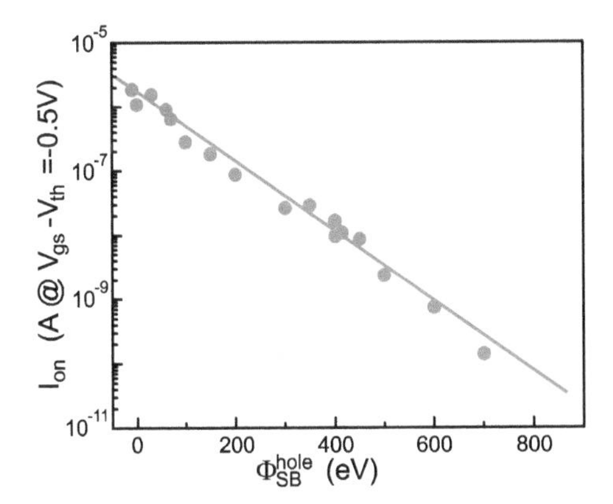

Figure 8.9: Calibration curve that maps the on-current of a CNTFET to the Schottky barrier at the metal-nanotube interface [40].

The weak metal-nanotube coupling appears to be beneficial for CNTFETs because it depins the Fermi level and suppresses ambipolar operation in the devices. On the other hand, a weak coupling increases the contact length required in order to obtain a low contact resistance. The side-contact model relies on using Buettiker probes, and thus the coupling strength γ can be related to a mean free path for scattering. This mean free path is essentially equivalent to the transfer length, and thus represents a minimum required contact length (see Section 6.3.1). Following Venugopal and coworkers [259], the mean free path can be expressed as $l_{\mathrm{mfp}} = a/\gamma \approx l_{\mathrm{con}}^{\mathrm{min}}$ where a is the lattice spacing of the finite difference grid. For the coupling strength of $\gamma = 0.007$ that enabled the excellent agreement between simulation and experiment shown above, a minimum contact length, i. e., a transfer length $l_T \approx 70$ nm is obtained. Reducing the contact length below this minimum will lead to increased contact resistivity, which has recently also been observed experimentally [76]. This would ultimately limit the scalability of CNTFETs. A viable approach to circumvent this has recently been demonstrated in [37]. Transferring molybdenum side-contacts into end-bonded contacts, low contact resistances independent of the contact length could be realized; such end-bonded contacts are illustrated in Figure 8.10 where the contact is basically transferred from a weakly to a strongly coupled contact by breaking the carbon-carbon bonds of the CNT. However, it was mentioned above that with such contacts an increased off-state leakage due to ambipolar operation is expected, which limits devices to be used with low V_{ds} only (see Figure 7.11).

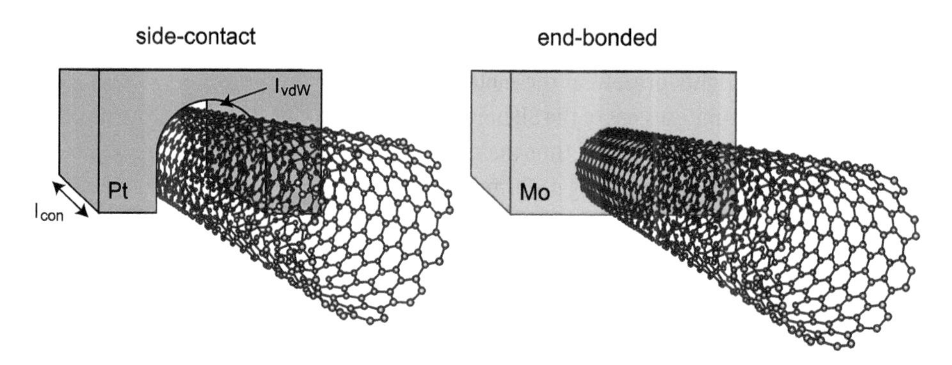

Figure 8.10: Carbon nanotube with side-contact (left) and end-bonded contact (right).

8.1.4 Reconfigurable CNTFETs

Irrespective of the contact model, it is clear from the discussion so far that CNTFETs behave similar to SB-MOSFETs. Due to the extremely small diameter of CNTs and the possibility of realizing a wrap-gate architecture, the screening length λ_{ch} can be made very small, resulting in strong improvements in the tunneling probability through the Schottky barrier. However, as discussed in Section 7.1.1, improving the tunneling probability is not exclusively beneficial: it also increases the leakage current due to the ambipolar operation (cf. Figure 7.11). Reconfigurable FETs (cf. Section 7.4), on the other hand, feature an additional gate electrode that allows for adjusting the potential distribution in the channel to block the ambipolar operation. In this respect, carbon nanotubes appear to be ideally suited for reconFETs, since they exhibit a symmetric band structure enabling symmetric injection of electrons and holes and they allow strongly increased carrier injection due to the inherently small diameter of the nanotube. The aim of the present section is, therefore, to study reconfigurable CNTFETs and to see how their performance can be improved. For convenience, some of the concepts and explanations related to the working principle of reconFETs will be repeated here (cf. Section 7.4).

Figure 8.11 shows an electron micrograph together with a schematic device cross-section of a dual-gate carbon nanotube transistor [11, 172]. The device is fabricated by depositing a thin aluminum gate on top of an oxidized silicon wafer (the thickness of the SiO_2 is 10 nm). After forming an approximately 4 nm thin Al_2O_3 (in a water rich ambient at 160 °C for 1.5 h) that serves as the actual gate dielectric, a nanotube is dispersed on top of the structure and contacted with titanium contact electrodes (for details on the device and its fabrication, see [11]). The particular device design where the aluminum gate lies on top of the oxidized silicon wafer allows using the wafer as a large area polarity gate (PG) that acts only on the source/drain extension regions since the channel area is screened by the aluminum gate. Hence, applying a positive/negative back-gate voltage induces electrons/holes in the source and drain extensions thereby creating electrostati-

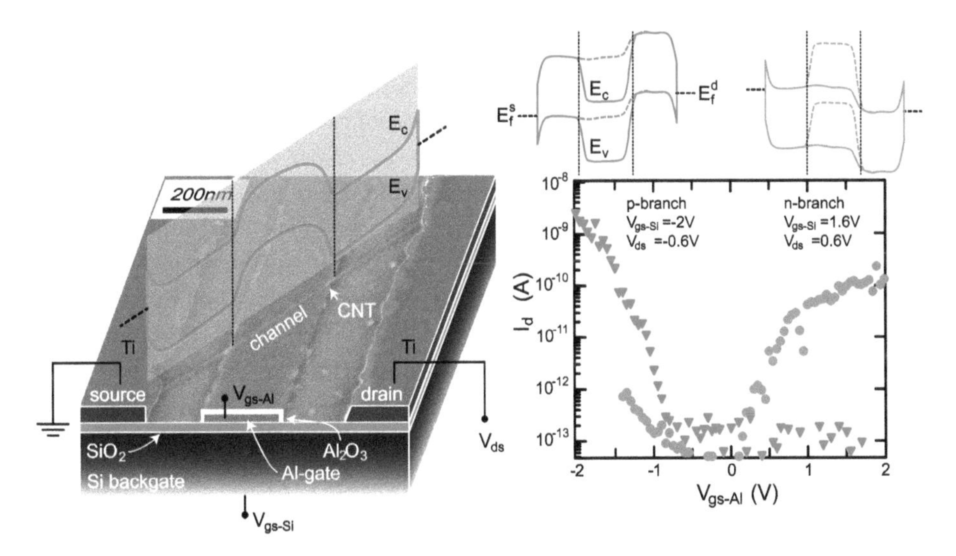

Figure 8.11: Electron micrograph and schematic illustration of a dual-gate carbon nanotube FET with a large area back-gate that serves as polarity gate (PG). A central control gate (CG) is made of aluminum and lies underneath the nanotube, thereby preventing any impact of the PG in the CG-region; exemplarily, band profiles for *p*-type and *n*-type polarity are shown on top of the respective transfer characteristics (right panel) [172].

cally *n*/*p*-type doped contact regions. This gate-controlled *n*- and *p*-type doping is shown in Figure 8.11, right panel. Here, conduction and valence bands along the direction of current transport are shown exemplarily for the two polarities. Applying an appropriate voltage at the Al control gate (CG) allows for switching the CNTFETs on and off. The lower right panel of Figure 8.11 shows transfer characteristics of such a dual-gate reconfigurable CNTFET [172] operated as *n*-FET (blue) and *p*-FET (red). Unipolar device operation is obtained as expected for a reconFET. Note, however, the difference to the reconfigurable devices presented in Section 7.4; here, the PG is a large gate acting on source *and* drain. As such, the device shown here acts similar to the PGAS configuration without leakage, since the conduction/valence band profiles resemble regular *n*-*p*-*n* or *p*-*n*-*p* MOSFETs, respectively.

In order to obtain symmetric carrier injection in reconFETs (note that the titanium contacts in the dual-gate reconFET above are not ideal since they yield better hole injection compared to electron injection), a mid-gap line up of the Fermi level with respect to the conduction and valence band is required. However, the weak metal-nanotube coupling (see preceding section) leads to strongly reduced MIGS, and thus to a strongly reduced carrier injection in the case of a Fermi level line-up close to mid-gap as was shown above in the case of Al contacts (see Figure 8.8(a)). Even the small screening length λ_{ch} that can be achieved in devices based on CNTs does not help much in this case unless the device is changed from a side-contact to an end-bonded device. But even with end-bonded contacts, the resulting SB will limit the performance of such reconFET. As a result, it would be highly desirable to realize reconfigurability without Schottky barriers; this can be accomplished exploiting the weak coupling of metal-nanotube contacts.

Consider the two device layouts for reconFETs shown in the insets of Figure 8.12. The case shown on the right side resembles the reconfigurable FETs discussed in Section 7.4 with PG and CG next to each other. In the case shown in the left inset, the polarity gate is located only underneath the metal-nanotube contact. How can this lead to reconfigurability? It was mentioned above that the metal-nanotube coupling is weak leading merely to small modifications of the density of states within the band gap (as a result of the suppressed MIGS). It is therefore expected that the conduction and valence bands of the nanotube can be moved with respect to the Fermi level of the contact electrode by applying an appropriate voltage at the PG (back-gate) underneath the nanotube. The upper left inset of Figure 8.12 displays a 2D calculation of the electrostatics of the metal-nanotube contact showing that the contact does not fully screen the impact of the bottom gate electrode. Therefore, the PG (back-gate) can be used to tune the polarity. As a result, while the device with PG and CG next to each other exhibits a Schottky barrier at the metal-nanotube interface, there will only be the extremely thin barrier Φ_{bar} associated with the depinning layer in the second case, provided that the voltages at the back-gate (PG) are sufficiently high. Hence, the latter device layout with back-gate as PG (left inset in Figure 8.12) should provide substantially larger on-currents than the other, rather conventional (i. e., with Schottky barriers), reconFET device architecture.

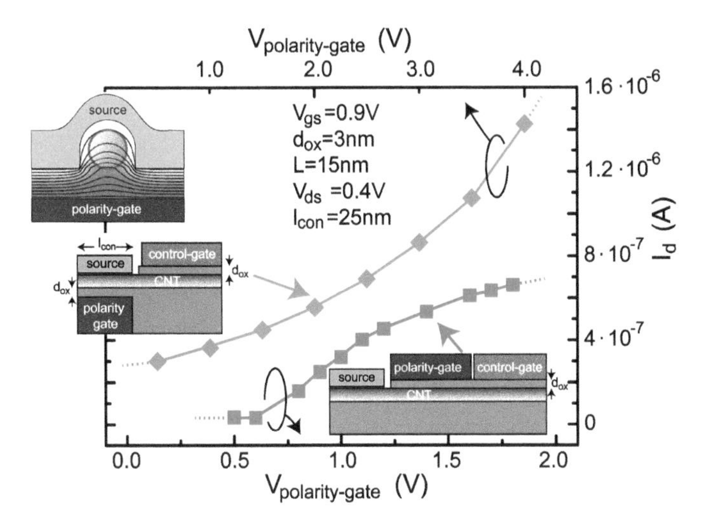

Figure 8.12: Drive current as a function of the polarity gate voltage of the two device configurations depicted in the insets. Simulations are done for devices with $L = 15$ nm, nanotube diameter 1.3 nm, $d_{ox} = 3$ nm, $V_{gs} = 0.9$ V, $V_{ds} = 0.4$ V, $E_g = 1.0$ eV and $\gamma = 0.007$ [144].

Reconfigurable MOSFETs based on the two device layouts were simulated with self-consistent NEGF calculations. Exemplarily, the drive current for the n-type polarity is computed as a function of the voltage at the PG (see the caption of Figure 8.12 for details on the device parameters) [144]. The main panel of Figure 8.12 shows the on-current extracted at a control gate voltage of $V_{gs} = 0.9$ V for the two device layouts. Obviously, the CNTFET with the polarity gate underneath the contact (green data points) shows a superior performance since for sufficiently large back-gate voltages the conduction band in the contact area is moved close to or even below the Fermi level. In contrast, the device with PG next to the contact shows a saturating current at large PG voltages. Note that Ti contacts have been assumed here; with Al contacts the difference between the two devices is expected to be even larger.

Due to the symmetric band structure of carbon nanotubes, the same behavior is expected for hole injection if an appropriate negative voltage at the polarity gate is applied. In other words, due to the weak metal-nanotube coupling, the energetic position of the Fermi level with respect to the conduction and valence bands, which is due to the work-function difference of metal and nanotube, can be changed by applying appropriate voltages at the back-gate PG. This means that the resulting potential barrier for carrier injection can be tuned such that excellent electron or hole injection is obtained. On the other hand, in the device where the polarity gate is placed next to the contact, the on-current saturates for increasing polarity-gate voltage in the case of side-contacts. Even if end-bonded contacts were realized, the current increase would be substantially weaker compared to the case of a back-gate PG since there will always be a residual effective Schottky-barrier height in the former case even for large program voltages. Therefore, gate-controlled doping in combination with the weakly coupled metal-nanotube side-contacts and back-gate PG appears ideally suited for the realization of reconfigurable FET devices.

8.2 Conventional CNTFET Devices

As mentioned above, the reconfigurable carbon nanotube FET with dual-gate architecture discussed in the preceding section allows for adjusting conduction/valence band profiles as in a conventional MOSFET. Therefore, such a dual-gate CNTFET can be used to study carbon nanotubes in a regular *p-n*/*i-p* device layout.

Figure 8.13(a) shows the transfer characteristics of a dual-gate CNTFET with a channel length of 200 nm (green curves) and 40 nm (red curves) for two different V_{ds} [12]. Unipolar, regular transfer characteristics are observed in the case of L = 200 nm. However, scaling down the channel length to 40 nm results in a severe deterioration of the off-state with high leakage currents particularly in the case of a large bias voltage. The reason for such a behavior could be short channel effects. In Section 5.6, a criterion for the appearance of SCEs was given, stating that at $(4–5) \times \lambda_{ch} \approx L$ SCEs start to appear. In the present case, the dual-gate CNTFET is equivalent to a single-gate architecture. Taking the aluminum oxide as gate dielectric and the small nanotube diameter into consideration, the screening length λ_{ch} can be estimated to be not larger than $\lambda_{ch} \lesssim 3\,\text{nm}$; this means that we do not expect any SCE in the present device, even if L = 40 nm. So, what is the reason for the degradation of the off-state in the short CNTFETs? In order to clarify the mechanism, self-consistent NEGF simulations are carried out (cf. Chapter 6). Figure 8.13(b) shows the simulated transfer characteristics of a conventional nanotube transistor. In the simulation, a channel length of 10 nm (to speed up the simulation) and a wrap-gate architecture with a gate oxide thickness d_{ox} = 3 nm are assumed [141]. Fur-

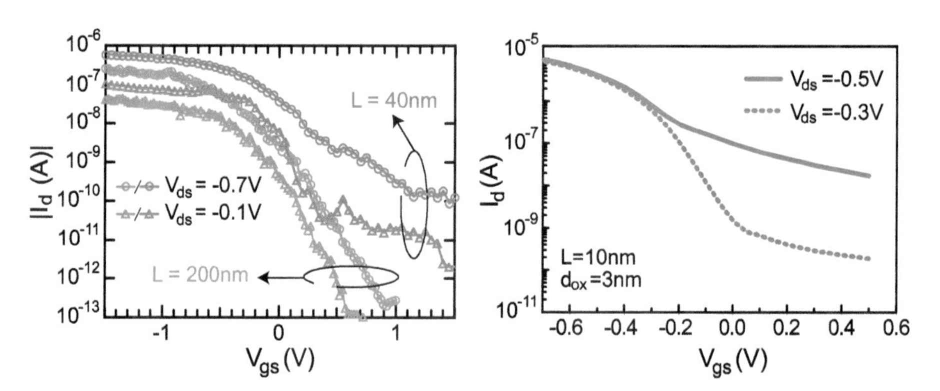

Figure 8.13: (a) Experimental transfer characteristics for a dual-gate CNTFET with L = 200 nm (green curves) and L = 40 nm (red curves) [12]. (b) Simulated transfer characteristics of a CNTFET with doped source/drain regions with L = 10 nm [141].

thermore, the Fermi energy in source and drain is considered to be 0.1 eV above the band edge and the band gap $E_g = 0.6$ eV. Although the channel length is rather short, the wrap-gate architecture yields $\lambda_{ch} \approx 1.1$ nm, and thus no SCE are expected in the simulated devices. Comparing the experimental and simulated data, it is obvious that they are qualitatively in excellent agreement although the absolute current levels do not match because of the difference in the device geometries. In particular, the deteriorated off-state is well reproduced.

The mechanism behind the deteriorated switching behavior is revealed in Figure 8.14, which shows conduction and valence band profiles in the simulated device. The dashed lines in the left panel show bands if the charge in the conduction band is completely disregarded. However, electrons are injected into the channel from the drain contact due to band-to-band tunneling (BTBT). For the chosen V_{ds}, these carriers cannot leave the channel into the source contact, because their movement is blocked by the band gap in source. Therefore, a pile-up of negative charge in equilibrium with the drain Fermi distribution occurs within the channel. This negative charge within the channel moves the conduction/valence bands upwards in energy (solid lines in the left panel of Figure 8.14 computed self-consistently) and, as a result, a larger hole current in the device's off-state will flow. The right panel of Figure 8.14 shows the conduction/valence bands for two different bias points. The larger (i. e., the more negative) V_{ds}, the larger the electron density within the channel and the more severe the effect. This way, a kind of drain induces a barrier "lowering" (in fact, the barrier moves upwards because a p-type FET is considered here) is obtained. Nevertheless, the mechanism is

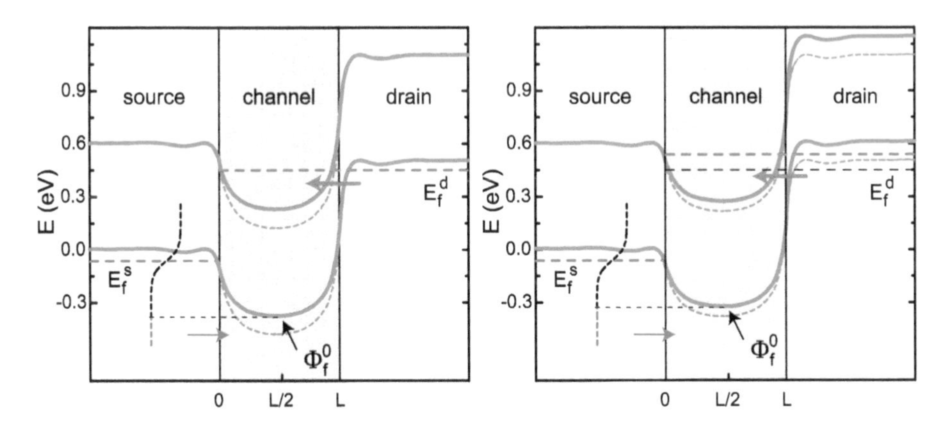

Figure 8.14: Left panel: Illustration of the impact of carriers injected into the channel due to BTBT in a conventional CNTFETs. The dashed lines show simulated conduction and valence bands at a fixed V_{ds} disregarding the charge in the conduction band. Taking the charge into consideration yields bands that are moved upwards in energy (solid lines). As a result of this, the off-state hole leakage current increases. Right panel: Bands for two different V_{ds}. The larger the bias the larger the injection charge, and thus the more severe the effect.

substantially different from DIBL[1] and relies on strong BTBT. An interesting aspect of this behavior is that it allows one in principle to measure the band movement when charge is in the channel. Moreover, the strong BTBT is certainly due to the very small values of λ_{ch} that can be realized with carbon nanotubes. In the next chapter, so-called tunnel FETs will be introduced and experimental devices will be discussed that exploit the strong BTBT in carbon nanotubes.

Exercises

Exercises together with solutions are accessible via the QR code.

1 It rather resembles the mechanism that leads to the sublinear output characteristics of SB-MOSFETs (cf. Section 7.1.2).

9 Steep Slope Transistors

In Chapter 5, performance improvements of conventional MOSFETs have been discussed. As a central result, it was found that the limitation of any conventional field-effect transistor to a minimum inverse subthreshold slope of $S_{min} = 60\,mV/dec$ at room temperature is one of the major roadblocks hindering a further reduction of the power consumption of highly integrated circuits built from such devices. Reducing the power consumption requires lower supply voltages V_{dd}. However, with devices bound to the 60 mV/dec limit, a smaller supply voltage inevitably leads to either a substantial performance loss or to a strong increase of the static power consumption due to the exponentially increasing off-state leakage. This scenario is shown in Figure 9.1 (reproduced from Chapter 5 for convenience) where the green and blue curves belong to conventional MOSFETs.

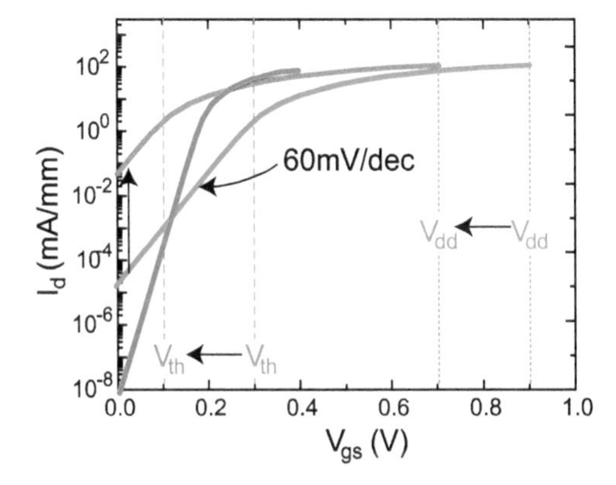

Figure 9.1: Transfer characteristics of a conventional MOSFET with optimum subthreshold swing at room temperature (green curve). Scaling down V_{dd} either leads to a loss of on-state performance or to an exponential increase of leakage (blue curve) due to $S_{min} = 60\,mV/dec$. An optimized steep-slope transistor (red curve) would allow reducing V_{dd} while maintaining the on-state performance at a lower off-state leakage.

Ideally, a device would show a behavior as illustrated by the red curve in Figure 9.1; an inverse subthreshold slope significantly steeper than 60 mV/dec in combination with a high on-state performance would allow decreasing V_{dd}, and thus reducing the dynamic power consumption (cf. Equation (5.31)). At the same time, the steep slope would enable reaching lower off-state leakage currents in a certain V_{dd}-interval. Such devices are highly desirable for the realization of ultralow power integrated circuits (ICs) needed for mobile applications, autonomous sensor devices and internet-of-things applications. Device concepts that are (at least potentially) able to provide $S < 60\,mV/dec$ are called *steep-slope transistors*. A number of different steep-slope transistor concepts have been proposed and intensively investigated in recent years. The present chapter explains the device physics and operating principles of a selection of these concepts.

https://doi.org/10.1515/9783111054421-009

9.1 Band-to-Band Tunnel Field-Effect Transistors

Field-effect transistors based on band-to-band tunneling (TFETs) are the most widespread investigated steep-slope transistor concept and have been intensively studied for more than 15 years by many groups [23, 11, 116, 179, 173, 112, 224]. The strong interest in TFETs stems from the fact that they (potentially) allow for the realization of steep-slope transistors with a device structure that is very similar to conventional MOSFETs; the only difference is that they exhibit an *n*-(*p*-)type source and a *p*-(*n*-)type drain contact instead of either *n*-type or *p*-type source/drain contacts. As an example, Figure 9.2(a) shows a schematic (left) and an experimental realization (right) of a TFET fabricated on a thin-body silicon-on-insulator substrate [229]. The fabrication of TFETs is therefore very similar to conventional MOSFETs, and thus, integrating TFET technology into existing CMOS production lines would only require small changes to be implemented. Furthermore, as will be discussed in detail below, measures that need to be taken in conventional CMOS technology to avoid short-channel effects (such as ultra-thin gate dielectrics preferably with a high dielectric constant, wrap-gate architectures, etc.) turn out to be performance boosters of TFETs. Again, very few changes would be necessary for an integration of TFETs as add-on or replacement of conventional CMOS.

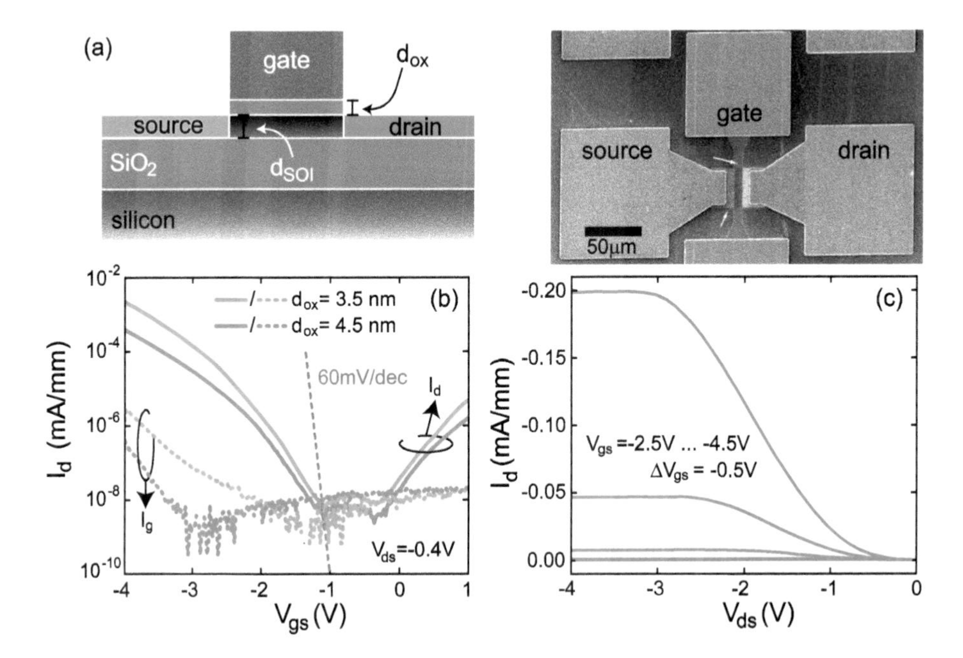

Figure 9.2: (a) Schematic and electron micrograph of a SOI TFET with *n*-doped source and *p*-doped drain [229]. (b) Transfer characteristics of two TFETs with different d_{ox} [229]. Plotted are the absolute values of the drain current I_d and gate leakage current I_g. The red dashed line illustrates an $S = 60$ mV/dec. Panel (c) shows typical output characteristics of a long-channel ($L = 2$ µm) TFET with sublinear behavior for small V_{ds} [229].

While tremendous progress regarding various TFET implementations and improvements has been made recently, experimental realizations so far still lack a satisfying performance.

TFETs exhibit a number of electrical properties that are similar to SB-MOSFETs, and hence are less favorable. Figure 9.2(b) shows transfer characteristics of the single-gate SOI TFET shown in the electron micrograph (a). Clearly observable is the fact that TFETs exhibit ambipolar operation. Moreover, the device shows an inverse subthreshold slope substantially larger than 60 mV/dec (the red dashed line shows 60 mV/dec as a reference) and the on-current is lower than in a conventional MOSFET. In addition, like SB-MOSFETs, TFETs exhibit a sublinear regime in the small bias region of their output characteristics (c), which is unfavorable for a proper functionality of inverters, for instance (see Task 29).

In order to understand the electrical behavior of TFETs and how their performance can be improved, the next sections are devoted to explaining the operating principles of TFETs. As will become clear below, many measures that at first seem to improve the performance of TFETs do have drawbacks and become rather intricate and involved when spatial dimensions are reduced to the nanoscale such as in nanowire TFETs. This will be studied at length and the various interdependencies of different TFET optimization strategies will be discussed in detail in order to elaborate on their impact and effectiveness.

9.1.1 Operating Principles of TFETs—Off-State

TFETs are gated *n-i-p*(*p-i-n*) devices with a degenerately *n*(*p*)-doped source contact, a gated intrinsic(or lightly doped) channel area and a *p*(*n*)-doped drain contact. Figure 9.3 shows the conduction band of a conventional *n*-type MOSFET (left panel) and conduction/valence bands of an *n*-type TFET (right) together with the associated transfer char-

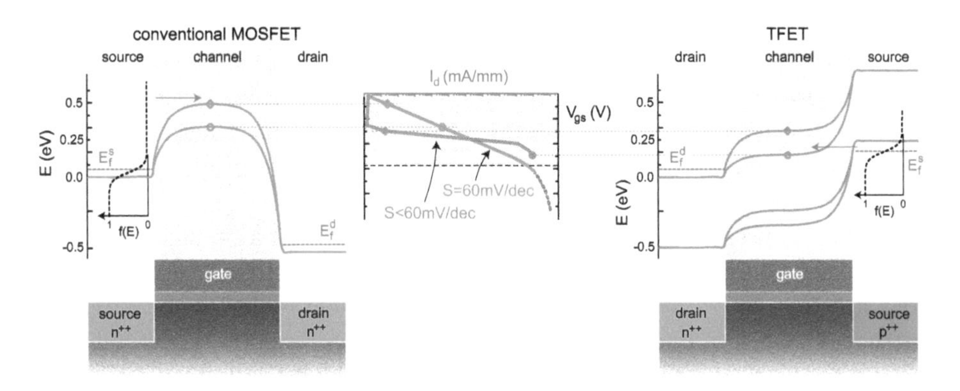

Figure 9.3: Comparison of the switching in a conventional MOSFET (left) and an optimized TFET (right panel). The center panel shows schematic I_d–V_{gs} curves of the two device types.

acteristics (middle panel). Whereas in the conventional n-MOSFET, lowering the potential maximum within the channel by applying appropriate gate voltages yields an exponential increase of the current with at best 60 mV/dec at room temperature (cf. Section 5.2.2), an energetic window for band-to-band tunneling (BTBT) is opened at the source-channel interface in the case of TFETs. As a result, the switching is not determined by modulating carrier injection from a thermally broadened Fermi distribution function but by field-effect modulated BTBT. With suitable optimizations, TFETs may therefore switch in a smaller gate voltage range than conventional MOSFETs, i. e., with an inverse subthreshold slope $S < 60$ mV/dec.

Let us start studying the operating principles of TFETs and how their performance can be optimized by introducing a semianalytical model for a TFET. To keep the discussion as simple as possible, let us consider a one-dimensional device layout, i. e., a nanowire or nanotube channel, with ballistic electronic transport. The current through such a TFET can be computed with the Landauer expression using an appropriate transmission probability $T(E)$ (cf. Equation (2.117)):

$$I_d = \frac{2e}{h} \int_{-\infty}^{\infty} dE\, T(E)(f_s(E_f^s) - f_d(E_f^d)). \tag{9.1}$$

There is an important point that needs to be clarified before we proceed. When dealing with semiconductors, we usually distinguish between electrons in the conduction band and holes in the valence band. However, here we deal with band-to-band tunneling, so how do we handle the two types of carriers and in particular the transfer from electron to hole behavior due to tunneling between the bands? First of all, remember that holes are just a construction that helps with counting charge carriers in the valence band. Instead of counting all the electrons from the top of the valence band to the bottom of the band, it is a lot easier to pretend that holes are a type of charge carriers with positive effective mass and positive charge. But holes are really only empty states in the valence band, i. e., missing electrons giving rise to a positive charge due to the positive background from the host crystal atoms. Hence, there are not two different carrier types but simply just electrons. I should rather say electrons *or* holes, but not electrons *and* holes, because instead of considering TFETs based on electrons one can equally well use holes throughout the device (i. e., within the valence and conduction bands). The bottom line is that you can either choose the electron picture or the hole picture—both do the trick.

Next, in a direct semiconductor, band-to-band tunneling at a p-n junction can be understood as charges moving through the complex band structure as illustrated in Figure 9.4(a). Here, an electron in the valence band moves toward the left, i. e., it resides on the right branch of the dispersion relation in the valence band. The carrier "follows" the complex band structure (cf. Figure 2.23) illustrated with the dark green dotted line and enters on the left side of the dispersion in the conduction band to maintain the negative carrier velocity. Hence, the complex band structure plays an important role for

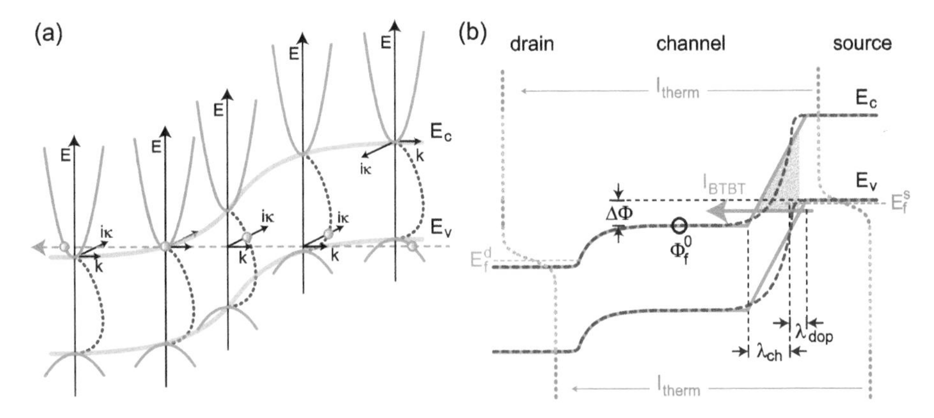

Figure 9.4: (a) Illustration of BTBT at a *n*-*p* junction. A charge carrier is transferred from the valence to the conduction band via the complex band structure (red). The carrier is transferred from the right (left) side of the dispersion in the valence band to the left (right) side of the dispersion in the conduction band to maintain the negative (positive) direction of carrier velocity. (b) Conduction and valence bands of a TFET along the direction of current transport. The BTBT barrier at the source-channel interface is replaced with a triangular shaped barrier (gray-shaded area) that allows using the WKB approximation.

the tunneling probability; it has already been dealt with in earlier chapters and can be computed, e. g., with the tight-binding method (Section 2.4.7), or taken into consideration with an energy-dependent effective mass (Section 2.5.2). Additional complications are expected in the case of an indirect semiconductor such as silicon. Here, a phonon is required in order to provide the necessary momentum change to allow carriers to tunnel from the conduction to the valence band (or vice versa). Taking all these aspects into consideration requires elaborate simulation.

For the time being, however, we will disregard these details and treat the band-to-band tunneling barrier as a regular potential barrier for carriers. With the Landauer expression for current transport, the drain current can then be written as a sum of three contributions, as depicted in Figure 9.4(b).

The first contribution is the injection of electrons from source with energies higher than the conduction band in source E_c^s, i. e., from the Boltzmann tail of the source Fermi distribution function. These electrons travel through the channel to find empty states in drain. Therefore, the current contribution is given by $I = \frac{2e}{h}\int_{E_c^s}^{\infty} dE(f_s - f_d) \approx \frac{2e}{h}\int_{E_c^s}^{\infty} dE\, f_s(E_f^s)$.

The second contribution stems from electrons injected from the valence band in source with energies below the valence band in drain E_v^d. Since the Fermi distribution of source in this energy range can be set to unity this current contribution is approximately $I \approx \frac{2e}{h}\int_{-\infty}^{E_v^d} dE(1 - f_d(E_f^d))$. The latter two current contributions are due to thermal emission and denoted accordingly I_{therm} in Figure 9.4(b).

The third contribution eventually stems from BTBT (large blue arrow in Figure 9.4(b)) and is denoted I_{BTBT}. As long as direct source-to-drain tunneling can be neglected, I_{BTBT} yields only a nonzero contribution in the energetic window $\Delta\Phi$ between the conduction band in the channel and the valence band in source. As a result, the total current is given by

$$I_d \approx \underbrace{\frac{2e}{h}\int_{E_c^s}^{\infty} dE\, f_s(E_f^s) + \frac{2e}{h}\int_{-\infty}^{E_v^d} dE\,(1 - f_d(E_f^d))}_{I_{\mathrm{therm}}=I_{\mathrm{leak}}} + \underbrace{\frac{2e}{h}\int_{\Phi_f^0}^{E_v^s} dE\, T(E)(f_s(E_f^s) - f_d(E_f^d))}_{I_{\mathrm{BTBT}}} \tag{9.2}$$

where $T(E)$ is the band-to-band tunneling probability. This means that the two current contributions due to thermal emission represent the minimum possible off-state leakage current $I_{\mathrm{therm}} = I_{\mathrm{leak}}$ of TFETs. Since I_{leak} stems from the Boltzmann tail of the respective Fermi distribution functions, it can be approximated as

$$I_{\mathrm{leak}} \approx \underbrace{\frac{2e}{h}\int_{E_c^s}^{\infty} e^{-\frac{E-E_f^s}{k_BT}}}_{=\frac{2e}{h}k_BT\exp(-\frac{E_c^s-E_f^s}{k_BT})} + \underbrace{\frac{2e}{h}\int_{-\infty}^{E_v^d} e^{-\frac{E_f^d-E}{k_BT}}}_{=\frac{2e}{h}k_BT\exp(-\frac{E_f^d-E_v^d}{k_BT})} \approx 2\frac{2e}{h}k_BTe^{-\frac{E_g}{k_BT}} \tag{9.3}$$

where we set $E_c^s - E_f^s \approx E_g$ and $E_f^d - E_v^d \approx E_g$, which is a suitable approximation since degenerately doped source/drain contacts are assumed.

An analytic expression for $T(E)$ due to BTBT can be obtained by replacing the exact conduction/valence band profiles (dark green dashed lines in Figure 9.4(b)) at the injecting contact-channel interface with straight, parallel bands as illustrated with the green lines in Figure 9.4(b). The spatial extent of the source-channel *p-n* junction is given by the sum of the screening lengths due to doping in source, λ_{dop}, and due to the electrostatics of the particular device architecture under consideration, λ_{ch} (cf. Figure 5.12). The height of the source-channel *p-n* potential barrier is equal to $E_g + \Delta\Phi$ where $\Delta\Phi$ is the energetic width of the BTBT window, i. e., the energetic difference between the valence band in the source and the conduction band in the channel. As a result, a triangular BTBT barrier is obtained (gray triangle in Figure 9.4(b)) whose tunneling probability can be computed analytically using the WKB approximation. Moreover, this triangular barrier does not change when changing the energy. This means that T_{WKB} will not depend on energy, which is a major simplification because it allows for computing a fully analytic expression for the drain current.

Let us now calculate T_{WKB} and with this the drain current through a TFET, which will then facilitate the computation of the inverse subthreshold slope and determine the device parameters that predominantly impact the TFET performance. Figure 9.5 shows the conduction/valence band profiles at the source-channel interface at two different

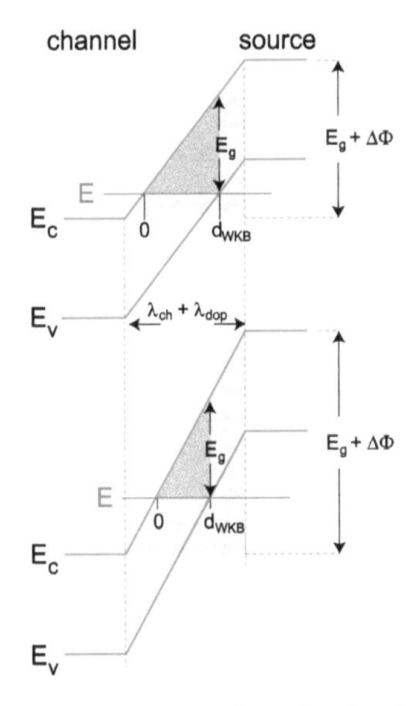

Figure 9.5: Source channel *p-n* junction at two different V_{gs}. The BTBT barrier is approximated with the (dark) gray triangle, facilitating the use of the WKB approximation.

V_{gs} where the exact potential profile has been approximated with straight lines as mentioned above. The WKB approximation yields (cf. Equation (4.33))

$$T(E) \approx T_{\mathrm{WKB}} = \exp\left(-2 \int_{0}^{d_{\mathrm{WKB}}} dx \sqrt{\frac{2m^\star(\Phi_f(x) - E)}{\hbar^2}}\right). \tag{9.4}$$

With the potential of the BTBT barrier replaced with a triangular barrier, one obtains $\Phi_f(x) - E = \frac{E_g}{d_{\mathrm{WKB}}} x$ as apparent from Figure 9.5. Inserting this into Equation (9.4) and carrying out the integration over x yields (with units $[E_g] = \mathrm{J}$ and $[\hbar] = \mathrm{Js}$)

$$T_{\mathrm{WKB}} = \exp\left(-2\frac{2\sqrt{2m^\star}}{3\hbar}\sqrt{E_g} d_{\mathrm{WKB}}\right). \tag{9.5}$$

The final thing to do is to compute d_{WKB}, i. e., the base of the triangular barrier that needs to be tunneled through. Looking at Figure 9.5 one can relate d_{WKB} to device parameters in the following way. The light and the dark gray triangles in the figure have the same angles meaning that the ratio $\frac{E_g}{d_{\mathrm{WKB}}}$ is the same as $\frac{E_g+\Delta\Phi}{\lambda_{\mathrm{dop}}+\lambda_{\mathrm{ch}}}$. Solving for d_{WKB} and inserting into Equation (9.5) finally results in

$$T_{\mathrm{WKB}} = \exp\left(-\frac{4(\lambda_{\mathrm{dop}} + \lambda_{\mathrm{ch}})\sqrt{2m^\star}(E_g)^{3/2}}{3\hbar(E_g + \Delta\Phi)}\right) \tag{9.6}$$

where the units are again $[E_g] = [E_g + \Delta\Phi] = \mathrm{J}$. As expected, T_{WKB} does not depend on energy, and thus inserting it into Equation (9.2) enables the calculation of an analytic expression for the drain current (first term), including the leakage current due to thermal emission (second term):

$$I_d \approx \frac{2e}{h}\exp\left(-\frac{4(\lambda_{\mathrm{dop}}+\lambda_{\mathrm{ch}})\sqrt{2m^*}(E_g)^{3/2}}{3\hbar(E_g+\Delta\Phi)}\right)\underbrace{\int_{\Phi_f^0}^{E_v^s} dE[f_s(E)-f_d(E)]}_{=F(\Phi_f^0,E_v^s)} + \frac{4e}{h}k_B T e^{-\frac{E_g}{k_B T}}. \tag{9.7}$$

Equation (9.7) allows us already to briefly touch on a critical point, namely the size of the band gap E_g. While the BTBT current increases proportionally to $\propto \exp(-\sqrt{E_g})$ with decreasing E_g, the leakage I_{leak} increases according to $\propto \exp(-E_g/k_B T)$. If I_{leak} becomes too large it may dominate the off-state of the TFET and as a result, the steep inverse subthreshold slope is lost. The issue of the size of the band gap is intricate and will be discussed in more detail later. For the time being, let us assume that the leakage current can be neglected. In this case, the inverse subthreshold slope is obtained from the first term of Equation (9.7). To simplify the notation, the current I_d can be written as a product of $T_{\mathrm{WKB}}(\Delta\Phi)$ and $F(\Phi_f^0, E_v^s) = F(\Delta\Phi)$ with $\Delta\Phi = E_v^s - \Phi_f^0$ (cf. Equation (9.7)). Note that the surface potential Φ_f^0 in a TFET is equivalent to the potential maximum in a conventional MOSFET; the only difference is that Φ_f^0 is the deflection point (not the maximum) of the potential distribution along the channel. Therefore, the top-of-the-barrier model can be used here to relate Φ_f^0 to the gate voltage. From Equation (5.17), we obtain for a constant V_{ds},

$$\delta\Phi_f^0 = \frac{C_{\mathrm{ox}}}{C_{\mathrm{ox}} + C_{\mathrm{depl}} + C_s + C_d + C_{\mathrm{inv}}}\delta\Phi_g. \tag{9.8}$$

If a fully depleted, long-channel TFET in the off-state is considered; this expression boils down to $\delta\Phi_f^0 = \delta\Phi_g$ and because $\Delta\Phi = E_v^s - \Phi_f^0$ it is immediately apparent that $\frac{\partial\Delta\Phi}{\partial\Phi_g} = -1$ in this case. With this, S can be computed to be

$$S = \left(\frac{\partial \log I_d}{\partial V_{\mathrm{gs}}}\right)^{-1} = \ln(10)\left(\frac{\partial I_d}{\partial\Delta\Phi}\underbrace{\frac{\partial\Delta\Phi}{\partial\Phi_g}}_{=-1}\underbrace{\frac{\partial\Phi_g}{\partial V_{\mathrm{gs}}}}_{=-e}\frac{1}{I_d}\right)^{-1}. \tag{9.9}$$

Carrying out the derivative $\frac{\partial I_d}{\partial\Delta\Phi}$, the following two contributions are obtained:

$$S = \frac{\ln(10)}{|e|}\left(\frac{T_{\mathrm{WKB}}(\Delta\Phi)\frac{\partial F(\Delta\Phi)}{\partial\Delta\Phi}}{T_{\mathrm{WKB}}(\Delta\Phi)F(\Delta\Phi)} + \frac{F(\Delta\Phi)\frac{\partial T_{\mathrm{WKB}}(\Delta\Phi)}{\partial\Delta\Phi}}{T_{\mathrm{WKB}}(\Delta\Phi)F(\Delta\Phi)}\right)^{-1}. \tag{9.10}$$

The first part is exclusively determined by the change of the incomplete Fermi integral $F(\Delta\Phi)$ whereas the second part is due to a change of the tunneling probability with changing gate potential [142].

The two contributions in Equation (9.10) allow for two different mechanisms to realize an S smaller 60 mV/dec. First, if T_{WKB} is small but rapidly changes with gate voltage, the second term in Equation (9.10) dominates. In this case, the change of current with V_{gs} is not determined by thermal emission anymore (which yields $S = 60$ mV/dec) but by the change of the tunneling probability. Using the WKB approximation for the transmission probability, S can be calculated to be

$$S \approx \frac{\ln(10)}{|e|} \frac{3\hbar(\Delta\Phi + E_g)^2}{4(\lambda_{\text{dop}} + \lambda_{\text{ch}})\sqrt{2m^\star} E_g^{3/2}} \tag{9.11}$$

where again $\Delta\Phi = E_v^s - \Phi_f^0$. This means that, for small $\Delta\Phi$, S can be made very small. Note that the dependence $\propto (\lambda_{\text{dop}} + \lambda_{\text{ch}})^{-1}$ is actually counterintuitive, since T_{WKB} drops with increasing $\lambda_{\text{dop}} + \lambda_{\text{ch}}$. However, it is important to note that S can only be made steep in a very small V_{gs}-range leading to switching in a practically irrelevant current range (this current range would actually be overcompensated by the leakage due to thermal emission). In addition, due to the quadratic dependence of S on $\Delta\Phi$ the inverse subthreshold slope rapidly increases and a small S at some gate voltage is not sufficient for a steep-slope transistor to perform better than a conventional MOSFET. An averaged $S_{\text{av}} < 60$ mV/dec over several orders of magnitude in drain current is needed to enable low off-state leakage and high on-state currents in a reduced V_{dd} voltage interval. As a result, if the second term of Equation (9.10) was dominant, TFETs would not be useful. So, let us turn our attention to the first term.

The first term in Equation (9.10) becomes dominant if T_{WKB} is close to unity and thus changes only slightly with gate voltage. In this case, the inverse subthreshold slope is determined by the change of the function $F(\Delta\Phi)$. Consequently, the inverse subthreshold slope becomes

$$S \approx \frac{\ln(10)}{|e|} \Delta\Phi. \tag{9.12}$$

This expression shows that very small inverse subthreshold slopes are feasible for $\Delta\Phi \to 0$. Moreover, S does only depend linearly on gate voltage and, as a result, $S < 60$ mV/dec can be expected over several orders of magnitude.

The reason why an inverse subthreshold slope steeper than 60 mV/dec can potentially be obtained if the second term of Equation (9.10) is dominant is the fact that the transmission T_{WKB} has become close to unity in the entire energy interval $\Delta\Phi$. Hence, the band gaps within the source and channel regions act as a band-pass filter that cuts of the high and low energy tails of the source Fermi distribution function, which is similar to cooling the source Fermi function [142]. In the video provided via QR code #62, this point is illustrated in more detail.

9.1.2 Operating Principles of TFETs—On-State

The preceding section mostly dealt with the off-state behavior of TFETs. However, based on the 1D model used above, we are also able to understand specific properties of the on-state of TFETs. Let us consider a TFET well in the on-state with a rather small bias applied. This leads to a conduction/valence band profile as illustrated in Figure 9.6. Electrons are injected from source and drain into the conduction band in the channel where they give rise to a significant charge density that needs to be taken into consideration. The charge within the channel can be computed approximately in the following way. When charge is injected from drain, carriers will have a positive group velocity and occupy the right branch of the dispersion relation in the channel (depicted in Figure 9.6), thereby occupying approximately half of the available density of states within the channel. The injected carriers move toward the BTBT barrier where a fraction T_{WKB} will be transmitted into the source contact due to BTBT and a fraction $1-T_{\mathrm{WKB}}$ will be reflected, occupying a part of the left branch (with negative group velocity) of the dispersion relation. At the same time, the left branch will also be occupied with electrons that are injected from source via BTBT (this process is depicted in Figure 9.4(a)). If we split the available DOS $D(E)$ into two equal parts for the left and right branches of the dispersion relation, the charge in the channel Q_{ch} is given by

$$Q_{\mathrm{ch}} = \underbrace{e\int_{\Phi_f^0}^{\infty} dE\frac{D(E)}{2}f_d(E_f^d)}_{\text{injected from drain}} + \underbrace{eT_{\mathrm{WKB}}\int_{\Phi_f^0}^{E_v^s} dE\frac{D(E)}{2}f_s(E_f^s)}_{\text{injected from source}} + \underbrace{e(1-T_{\mathrm{WKB}})\int_{\Phi_f^0}^{\infty} dE\frac{D(E)}{2}f_d(E_f^d)}_{\text{reflected}} \quad (9.13)$$

where in the third term the transmission is $T_{\mathrm{WKB}} = 0$ if energies are above E_v^s and thus the upper integration limit can be put to ∞, since in the on-state, the Fermi distribution

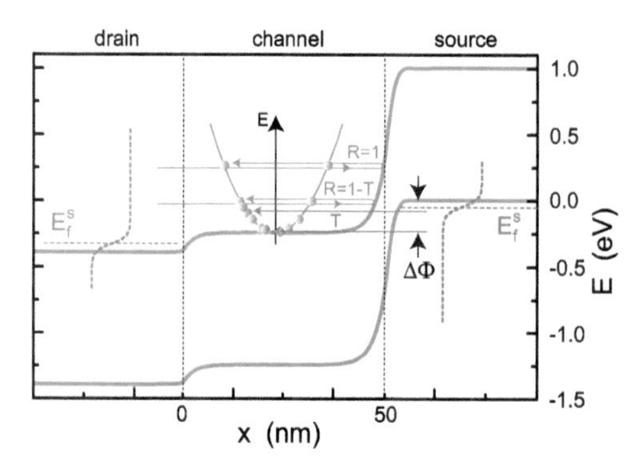

Figure 9.6: Carrier injection into the channel of a TFET. Electrons from drain are injected and occupy the right branch of the dispersion. Within the energetic window $\Delta\Phi$, the fraction $1 - T_{\mathrm{WKB}}$ of the carriers is reflected and occupies also the left branch. Carriers from source are injected with a probability T_{WKB} into the channel and occupy states on the left branch of the dispersion.

functions provide a negligible contribution for energies above E_C^s. Using Equation (9.13) together with Equation (5.12), the potential Φ_f^0 can be computed as a function of V_{ds}, gate voltage and geometrical parameters such as d_{ox} and dielectric constant (in C_{ox} and via λ_{ch} in T_{WKB}), channel thickness d_{ch}, its dielectric constant and device geometry (in λ_{ch}, and hence T_{WKB}), band gap and doping concentration (in λ_{dop}) as well as temperature. Φ_f^0 needs to be computed numerically; once known, it can be used to compute the current through the TFET using Equation (9.7).

Although a number of approximations were necessary to obtain this expression, it opens the possibility for a relatively complete (semi) analytical description of the electrical behavior of TFETs, taking most of the important device parameters into consideration; employing NEGF simulations it will become clear that some more subtle properties are not accounted for with the simple Landauer model presented here. However, the model can be used to explain an important aspect of TFET operation. Supposed T_{WKB} was rather small so that to first order the second contribution to the charge in Equation (9.13) could be neglected and $(1 - T_{WKB}) \approx 1$ in the third term. The channel would then be in equilibrium with the drain Fermi distribution and a certain surface potential Φ_f^0 is obtained. For increasing V_{gs}, the situation would actually resemble a MOS capacitor in inversion/accumulation with increasing charge density, and thus increasing inversion layer/accumulation layer capacitance $C_{inv,acc}$ (in the following, the appropriate capacitor will be called C_{inv} for simplicity). If the device is not in the quantum capacitance limit, at some gate voltage $C_{inv} > C_{ox}$, and consequently, a change of gate voltage yields $\delta\Phi_f^0 = \frac{C_{ox}}{C_{inv}+C_{ox}}\delta\Phi_g \to 0$ in the classical limit (cf. Section 5.9.1) where a constant bias and negligible $C_{s,d,depl,it}$ are assumed. If the drain-source bias is now increased the Fermi distribution of the drain that injects carriers into the channel is moved down in energy, and thus the injection of carriers drops so that only carriers from the source are injected. Since $T_{WKB} \ll 1$, the carrier density, and thus $C_{inv} \to 0$. As a result, $\delta\Phi_f^0 \approx \delta\Phi_g$ and the surface potential is moved according to the applied gate voltage. In turn, this will increase T_{WKB} giving rise to an increased drain current. This mechanism is the reason for the exponential increase of the on-current of a TFET in the small bias regime (cf. Figure 9.2(c)) [142]. Since the current increase is induced by drain, it will be called "drain-induced-barrier thinning" (DIBT) in the following. In Section 7.1.2, it was shown that a very similar effect is also responsible for the sublinearity of the output characteristics of SB-MOSFETs.

The occurrence of DIBT can be expressed in the following way. Starting from Equation (5.12), we can again compute $\delta\Phi_f^0$. However, this time the dependence of Q_{ch} on the drain-source bias (i. e., the drain potential Φ_d) needs to taken into consideration. As a result, the last term in Equation (5.12) yields two terms. Then, assuming that C_{depl} and $C_{s,d,it}$ can be neglected (i. e. $C_\Sigma = C_{ox}$), one obtains

$$\delta\Phi_f^0 = \delta\Phi_g - \underbrace{\frac{e\partial Q_{ch}/\partial\Phi_f^0}{C_{ox}}}_{=C_{inv}/C_{ox}}\delta\Phi_f^0 - \underbrace{\frac{e\partial Q_{ch}/\partial\Phi_d}{C_{ox}}}_{=-C_{inv}/C_{ox}}\delta\Phi_d \tag{9.14}$$

where $e\partial Q_{\text{ch}}/\partial\Phi_d = -C_{\text{inv}}$ because a change of Φ_d with constant potential in the channel Φ_f^0 yields the same absolute value of C_{inv} as leaving Φ_d constant and changing Φ_f^0; the only difference is the sign of the change. Equation (9.14) can be rewritten by moving the second term of the left-hand side to the right-hand side of the equation, and factor out $\delta\Phi_f^0$, which yields

$$\delta\Phi_f^0 = \frac{C_{\text{ox}}}{C_{\text{ox}} + C_{\text{inv}}}\delta\Phi_g + \frac{C_{\text{inv}}}{C_{\text{ox}} + C_{\text{inv}}}\delta\Phi_d. \tag{9.15}$$

This expression reflects the discussion above, namely that changes of Φ_g and Φ_d lead to a change of Φ_f^0, and thus to an exponential increase of the current. In fact, Equation (9.15) resembles Equation (5.17), which described the impact of short-channel effects. However, here, the dependence of Φ_d enters through C_{inv} and not C_d. Hence, DIBT is not a short-channel effect. It also appears in long-channel devices. What matters is the ratio $C_{\text{inv}}/C_{\text{ox}}$. As a result, Equation (9.15) provides a strategy to suppress DIBT: in the quantum capacitance limit (cf. Section 5.9.1), $C_{\text{inv}} \ll C_{\text{ox}}$, which would lead to the desired $\delta\Phi_f^0 = \delta\Phi_g$.

To further proceed with the analysis regarding TFET performance, self-consistent simulations of TFETs using NEGF (detailed in Chapter 6) are carried out. The quantum capacitance limit can be obtained in 1D wrap-gate structures with a material exhibiting a small effective mass. Therefore, device and material parameters are chosen accordingly (see Figure 9.7). Comparing the device in the quantum capacitance limit and classical limit is usually not trivial, because changing the device parameters such as d_{ox} or $m^\star$ will have an immediate impact on the BTBT probability. The impact of charge is always interrelated to a change in BTBT. In order to avoid this interrelation, the carrier density is multiplied with an artificial degeneracy factor during the self-consistent computation of the potential profile, that allows tuning the carrier density without major changes in the BTBT probability; for the calculation of the current, this factor is set

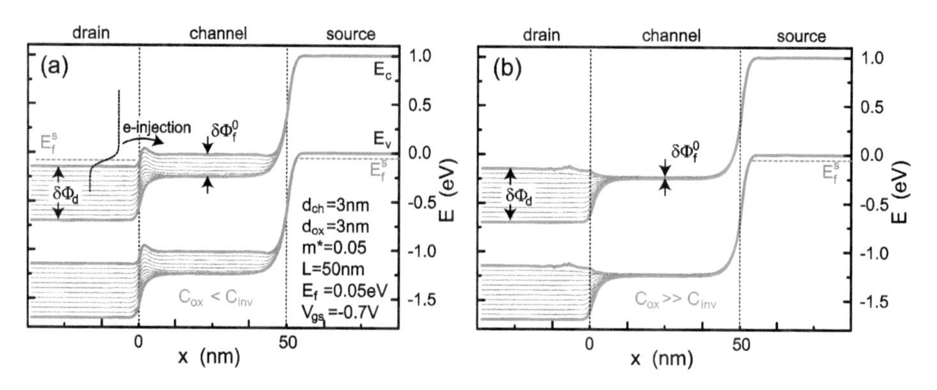

Figure 9.7: Simulated conduction and valence band profiles in the on-state of a nanowire TFET for different V_{ds} at constant V_{gs}. In (a), the carrier density was artificially increased leading to substantial DIBT because $C_{\text{ox}} < C_{\text{inv}}$ whereas in (b) $\delta\Phi_0/\delta\Phi_d \to 0$ since $C_{\text{ox}} \gg C_{\text{inv}}$.

to unity. Figure 9.7 displays the conduction and valence bands along current transport in the "classical" limit (CL) where $C_{\mathrm{ox}} < C_{\mathrm{inv}}$ (a) and in the quantum capacitance limit (QCL) with $C_{\mathrm{ox}} \gg C_{\mathrm{inv}}$ (b). The impact of the drain on the energetic position of Φ_f^0 in (a) is clearly visible reflecting the discussion above. In (b), Φ_f^0 is almost independent of Φ_d, and hence fully determined by the gate voltage.

To further illustrate the impact of charge in the channel on the output characteristics of TFETs, simulations were done with degeneracy factors (see above) of 0.1, 1, 5, 10 and 20 where 0.1 yields the QCL and 20 the CL. The corresponding output characteristics are shown in Figure 9.8. Due to the impact of drain on the channel potential as displayed in Figure 9.7(a), an exponential increase of the current in the small bias regime (blue curve in Figure 9.8) is obtained qualitatively mirroring the behavior of the experimental device shown in Figure 9.2(c). Lowering the degeneracy factor the device approaches the QCL. As a result, the sublinearity in the small bias regime becomes increasingly suppressed. Eventually, in the QCL the bands are not affected by the drain-source bias anymore as depicted in Figure 9.7(b). As a result, regular MOSFET-like output characteristics are obtained (cf. Figure 9.8, green curve). Also, note that in the QCL the on-state current increases because the gate is able to control Φ_f^0 completely, thereby increasing BTBT.

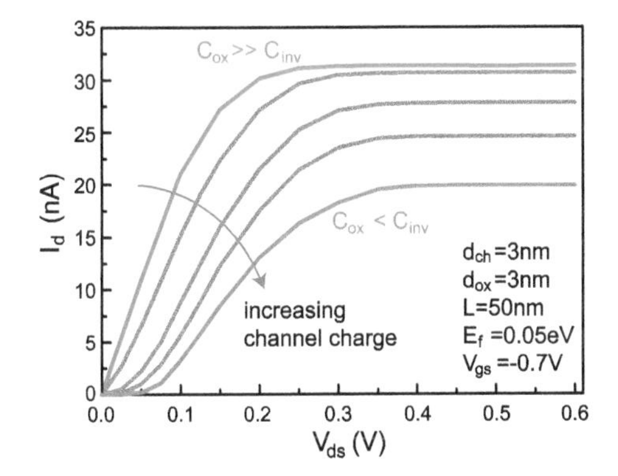

Figure 9.8: Output characteristics for constant V_{gs} of TFETs with different ratios between C_{ox} and C_{inv}. The device with $C_{\mathrm{ox}} \gg C_{\mathrm{inv}}$ (green curve) shows regular MOSFET behavior with a linear increase of I_d for low bias (cf. Figure 9.7(b)). The TFET with $C_{\mathrm{ox}} < C_{\mathrm{inv}}$ resembles a SB-MOSFET.

9.1.2.1 Channel Length Scaling—Short-Channel Effects in TFETs

In the literature, it is often stated that TFETs exhibit fewer or even no short-channel effects (recall that drain-induced barrier thinning is no SCE), when compared to conventional devices, since the tunneling process only involves the source-channel interface. However, this is not the case.[1] In Section 5.6, we saw that short-channel effects in con-

1 It is certainly a matter of the definition of SCEs. For instance, there will be no V_{th} roll-off in a short-channel TFET. But extrapolating a scaling benefit from this observation is not correct.

ventional MOSFETs appear because of overlapping source-channel and channel-drain *p-n* junctions when the channel length L is approximately of the same order as λ_{ch}. This overlap reduces the potential barrier Φ_f^0 leading to DIBL. TFETs are not too different from conventional MOSFETs in this respect. The only difference is that the potential maximum of a conventional MOSFET is replaced with the deflection point (also called Φ_f^0) as already used above. The overlapping *p-n* junctions will then impact this deflection point. In the off-state of the TFET and for sufficiently large drain-source bias, C_{inv} vanishes and we obtain $\delta\Phi_f^0 = \frac{C_{\text{ox}}}{C_{\text{ox}}+C_d}\delta\Phi_g + \frac{C_d}{C_{\text{ox}}+C_d}\delta\Phi_d$ where C_d is the drain capacitance following the same reasoning as in Section 5.6. Using this expression and assuming a constant bias, the inverse subthreshold slope can be recalculated resulting in

$$S = \frac{\ln(10)}{e}\left(\frac{\partial F/\partial\Phi_g}{F} + \frac{\partial T_{\text{WKB}}/\partial\Phi_g}{T_{\text{WKB}}}\right)\left(1 + \frac{C_d}{C_{\text{ox}}}\right) \tag{9.16}$$

i. e., we obtain the same result as above simply multiplied with the factor $1 + \frac{C_d}{C_{\text{ox}}}$ as in a conventional MOSFET. Since the ratio of C_d and C_{ox} is equal to $\frac{\lambda_{\text{ch}}^2}{L^2}$, it is obvious that severe short-channel effects will strongly degrade the switching of TFETs.

The appearance of SCE in TFETs is schematically shown in the left panel of Figure 9.9, which displays the conduction/valence bands in the case of a TFET with strong overlap of source-channel and channel-drain *p-n* junctions (light gray areas). Obviously, the potential profile in the channel is to a large extent determined by the drain potential rather than the gate. In other words, the gate potential loses control over Φ_f^0, which determines the BTBT probability T_{WKB}. Figure 9.9, right panel, shows simulated transfer characteristics of nanowire TFETs with decreasing channel lengths L [7]. Obviously, the switching of TFETs is deteriorated for smaller L. Note that in the case of the shortest device, direct source-drain tunneling occurs in addition to SCE. However, the loss of steepness of the inverse subthreshold slope is mostly due to short-channel effects [7].

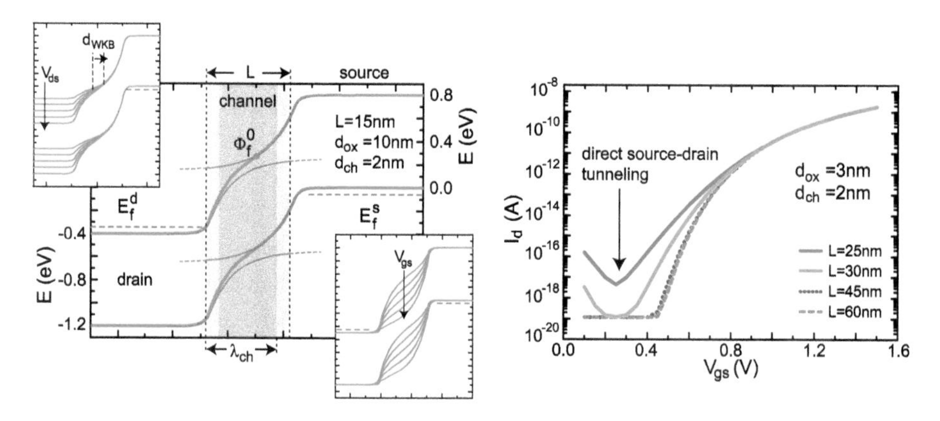

Figure 9.9: (a) Conduction and valence bands in a TFET with strong overlap of the *p-n* junctions ($L \sim \lambda_{\text{ch}}$). (b) Simulated transfer characteristics of short-channel TFETs with decreasing L [7].

Clearly, the overlapping source/drain *p-n* junctions in TFETs do not lead to an exponential increase of the leakage current as observed in a conventional MOSFET due to DIBL. Yet, TFETs that are not properly scaled lose their major benefit, namely potentially enabling $S < 60$ mV/dec.

9.1.2.2 Channel Length Scaling—The Impact of Carrier Mobility

In conventional MOSFETs, the carrier mobility μ is one of the most important figures of merit. Increasing the mobility improves the performance of MOSFETs without scaling. But what about the carrier mobility in TFETs? In order to assess the impact of carrier mobility in TFETs, the transmission probability for scattering $T_{\text{mfp}} = l_{\text{mfp}}/(l_{\text{mfp}} + L)$ (where l_{mfp} is again the scattering mean free path; cf. Equation (5.29)) can be combined with the BTBT probability T_{WKB}. In the same way as in Section 8.1.2 (see Equation (8.3)), we obtain

$$T_{\text{tot}} = \frac{T_{\text{WKB}} T_{\text{mfp}}}{T_{\text{WKB}} + T_{\text{mfp}} - T_{\text{WKB}} T_{\text{mfp}}} = \frac{1}{\exp(\frac{4\lambda\sqrt{2m^*}E_g^{3/2}}{3\hbar(E_g+\Delta\Phi)}) + \frac{L}{l_{\text{mfp}}}}. \tag{9.17}$$

From the expression above, it is clear that $T_{\text{tot}} \approx T_{\text{WKB}}$ if $T_{\text{WKB}} \ll T_{\text{mfp}}$, which is the case in most TFETs unless the channel length is excessively long or the mean free path extremely short. This makes sense because a TFET is a contact-switching device very similar to a SB-MOSFET for which we have already discussed that the transport properties of the channel material are often irrelevant (cf. Section 7.1.1.3). This means that the mobility is expected to play an insignificant role if any at all.

Figure 9.10(a) displays experimental on-state currents of SOI TFETs similar to the one depicted in Figure 9.2(a) for various gate voltages (i. e., increasing T_{WKB}) as a function of channel length, showing complete independence of L in all cases [229]. The carrier mobility μ is thus irrelevant for the TFETs. Furthermore, (b) shows simulated on-currents (discrete data points) as a function of channel length L for four different d_{ox} together with analytical current calculations (straight lines) based on the Landauer formalism and Equation (9.17). The simulations[2] and analytical calculations are done for double-gate silicon TFETs and are in excellent agreement. This reconfirms the usefulness of the WKB approximation (including scattering) to describe the performance of TFETs. As expected, I_d is almost independent of L and drops only for rather long-channel lengths (note the logarithmic scale of the axis). Based on Equation (9.17) a minimum channel length can be computed for μ to become significant. This minimum channel length is given by $L_{\text{min}} \geq l_{\text{mfp}} \exp(4\lambda\sqrt{2m^*}E_g^{3/2}/(3\hbar(E_g + \Delta\Phi)))$ and provides a design rule for TFETs. As long as the channel length of a targeted TFET is well below L_{min} one can choose the material and the geometry of the TFET without taking care of the carrier

2 Simulations were carried out with Centaurus TCAD by K. Boucart in A. Ionescu's group at EPFL, Switzerland.

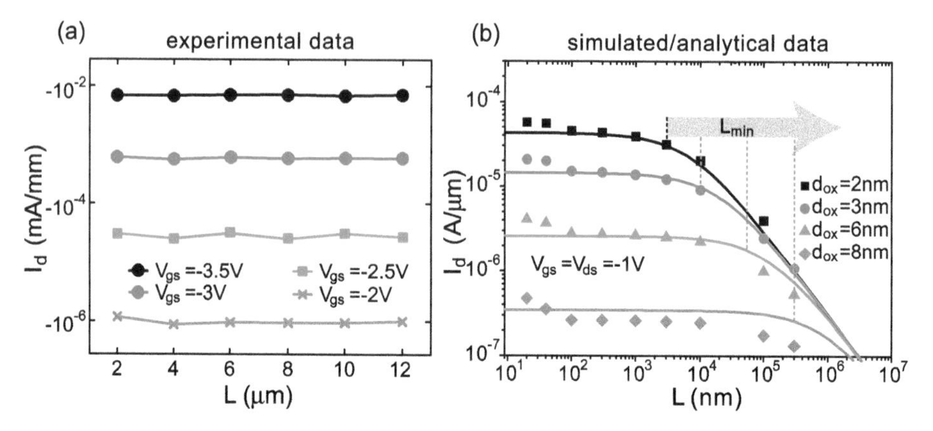

Figure 9.10: (a) Experimental on-state currents of SOI-TFETs for different channel lengths L and at four different V_{gs} [229]. The data is extracted from devices as shown in Figure 9.2(a). (b) Simulated and analytic data of the on-state current as a function of L for four different TFETs.

mobility. For instance, a possible loss of μ, likely to occur if extremely thin SOI is used, is overcompensated by the performance benefit due to increased BTBT.

9.1.3 TFET Optimization

In the preceding sections, a model for TFET performance in the off- and on-state has been developed that provides guidelines for the optimization and for achieving a steep slope over several orders of magnitude. It was already found that in a TFET a band-pass filter behavior needs to be created that enables an effective "cooling" of the source Fermi distribution. To this end, the BTB tunneling probability must be increased as much as possible, which, according to Equation (9.6), can be done by decreasing d_{ox}, decreasing E_g, using a material with a small effective mass, decreasing the channel thickness d_{ch}, increasing the dielectric constant of the gate dielectric and increasing the dopant concentration in the source contact. Furthermore, the quantum capacitance limit appears to be effective in avoiding the sublinearity of the output characteristics. This looks like a clear plan how to get rid of all TFET performance issues. Unfortunately, things are more intricate and not as straightforward as the model developed so far suggests. Therefore, each of the mentioned performance boosters will be discussed in greater depth in the following sections.

9.1.3.1 Dependence on the Gate Oxide Thickness

Scaling down the effective oxide thickness EOT (i. e., reducing the physical gate dielectric thickness and/or increasing the dielectric constant of the gate oxide) is the most effective performance booster for TFETs. The reason for this is that as long as gate leakage is avoided there is no trade-off, no accompanying interrelation that would deteriorate the performance of TFETs when reducing EOT. A small EOT reduces λ_{ch}, thereby increasing the BTBT probability. In addition, decreasing EOT increases C_{ox} and therefore drives the device more toward the QCL with its benefit of realizing linear output characteristics in the low bias regime (cf. Figure 9.8) [135].

In real experiments, gate leakage can of course occur, particularly if an ultrathin gate dielectric is used. In this case, it is very important to carefully examine the results before a clear statement concerning the transport mechanism can be made. Gate leakage can be deceiving and TFET characteristics may appear like devices that show a sub-60 mV/dec behavior. The impact of gate leakage has been discussed in Section 5.9.3 and it was found that depending on the bias conditions and the tunneling probability through the gate insulator, the drain current can exhibit a regime where $S < 60$ mV/dec seemingly appears. The smaller V_{ds}, the more likely is the observation of such a regime in the transfer characteristics. The reason for this is of course only the change of the direction of the gate-drain bias and if this occurs in the off-state of the device, a steep turn-off behavior of I_d seemingly appears. In TFETs, which are supposed to operate with small supply voltages (and hence small V_{ds}) and that require extremely thin gate dielectrics to increase the BTBT probability, one needs to be very careful when interpreting the measured data and always observe source, drain and gate currents simultaneously.

9.1.3.2 Dependence on the Channel Layer Thickness

Equation (9.6) shows the strong dependence of the BTBT probability on λ_{ch}. Therefore, employing a nanowire wrap-gate device architecture with very small nanowire diameter appears very promising for improving the performance of TFETs. However, this is only partly true. If a nanowire consisting of a material with low effective mass is used, quantization due to carrier confinement will lead to an effective increase of the band gap. If, for simplicity, a simple particle-in-the-box quantization is considered (see Section 2.2.2), the effective band gap scales as $E_g^{\text{eff}} = E_g + \frac{\hbar^2\pi^2}{m_c^\star d_{ch}^2} + \frac{\hbar^2\pi^2}{m_v^\star d_{ch}^2}$, where $m_{c,v}^\star$ are the effective masses in the conduction and valence bands. As a result, the benefit of a smaller λ_{ch} when reducing d_{ch} will be overcompensated at some diameter by the increasing E_g^{eff} that deteriorates TFET performance.

Equation (9.6) can be used to assess the impact of diameter scaling by replacing E_g with $E_g^{\text{eff}} = E_g + 2\frac{\hbar^2\pi^2}{m^\star d_{ch}^2}$ where a single effective mass $2/m^\star = 1/m_c^\star + 1/m_v^\star$ is used. If T_{WKB} is computed at the gate voltage when the TFET starts to switch (yielding $\Delta\Phi = 0$) and if we assume that the screening in the source contact due to doping results in $\lambda_{dop} \ll \lambda_{ch}$, the BTBT probability is given by

$$T_{\text{WKB}} \approx \exp\left(-\frac{4}{3\hbar}\sqrt{\frac{\varepsilon_{ch}}{\varepsilon_{ox}}\frac{d_{ox}d_{ch}}{4}}\sqrt{2m^\star}\sqrt{E_g + 2\frac{\hbar^2\pi^2}{m^\star d_{ch}^2}}\right) \tag{9.18}$$

where the screening due to the wrap-gate architecture was approximated with a quadruple gate structure to simplify the discussion. It is obvious that in the case of larger diameters the quantization energy vanishes and $T_{\text{WKB}} \propto \exp(-\sqrt{d_{\text{ch}}})$, whereas decreasing the diameter such that the quantization energy dominates (this may occur in nanowires made of a material with small m^* and small bulk E_g) the term containing the band gap becomes $\sqrt{E_g + 2\frac{\hbar^2\pi^2}{m^* d_{\text{ch}}^2}} \approx \sqrt{2\frac{\hbar^2\pi^2}{m^* d_{\text{ch}}^2}}$ and as a result, $T_{\text{WKB}} \propto \exp(-1/\sqrt{d_{\text{ch}}})$. Therefore, a material-specific optimum diameter for nanowires used in TFETs exists. A material such as carbon nanotubes is preferable in this respect; extremely small diameters on the order of $1, \dots, 1.5$ nm are required for a nanotube to exhibit a band gap suitable for a TFET (see Section 9.1.3.5).

9.1.3.3 Dependence on the Density of States

So far, the optimization of TFET performance has been concentrating on the screening length λ_{ch} in the channel. However, in order to obtain steep *p-n* junctions the screening length in the source contact λ_{dop} also needs to be as small as possible. λ_{dop} is related to the density of states within the source contact and, therefore, a more elaborate model needs to be used. In order to obtain a complete picture of the operation and optimization of TFETs, self-consistent NEGF simulations have been carried out based on the formalism presented in Chapter 6.

An essential ingredient in the simulation is the proper incorporation of the electrostatics of the device. Therefore, a TFET in a nanowire bundle configuration as depicted in Figure 9.11(a) is considered that serves as a reference device.[3] The bundle configuration allows truncating the electrostatics to the cylindrical device shown in the lower panel of (a) due to the zero electric field-condition at the boundaries of each device (cf. Figure 9.11(b)). Therefore, the full three-dimensional electrostatics of the device is taken

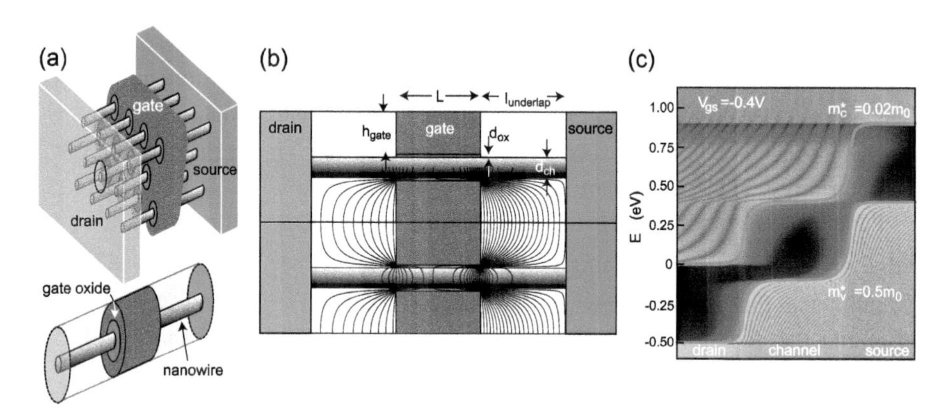

Figure 9.11: (a) TFET nanowire bundle architecture with wrap-gate that allows reducing the electrostatics to the displayed cylindrical device. (b) Equipotential lines in two adjacent nanowire TFETs. (c) Local density of states in a nanowire TFET device [135].

3 Note that this is the same nanowire bundle configuration as was used in Section 5.9.2.

into account. The nanowire is sufficiently thin, so that one-dimensional electronic transport is appropriate. The conduction and valence bands are considered to have different effective masses and the complex band structure within the direct band gap is accounted for using Flietner's dispersion relation (see Chapter 6 and Section 2.5.2). Figure 9.11(c) shows the local DOS in a TFET with different effective masses in the conduction and valence bands. All relevant device parameters are given in the figure captions showing the simulation results.

It is important to note that the nanowire bundle TFET considered here represents a best case scenario in terms of the electrostatics, since a wrap-gate architecture is studied providing the smallest λ_{ch} and the parasitic impact of the gate on the source/drain extensions is minimized due to the particular bundle configuration. Moreover, a direct semiconductor and ballistic transport are considered, thereby providing an upper limit of possible TFET performance.

Before proceeding with the analysis a new figure of merit for the off-state is introduced that is particularly useful for TFETs. The reason for this is that, in contrast to conventional MOSFETs, the inverse subthreshold slope S in TFETs depends on V_{gs} (cf. Section 9.1.1). As a result, the I_d–V_{gs}-curve exhibits a steep slope usually only in a small V_{gs}-range. Consequently, the minimum inverse subthreshold slope S_{min} of the entire I_d–V_{gs}-curve (called point slope) is not necessarily a meaningful number. Instead, an average inverse subthreshold slope S_{av} between the off-state and the threshold voltage could be used. However, V_{th} is not well-defined in TFETs making it difficult to compare different TFETs based on S_{av}. To provide a proper measure for comparison of TFETs and study how well they approach a steep-slope transistor, the current level I_{60} can be used [256], i. e., the current where $S = 60$ mV/dec is reached. The higher I_{60}, the better the TFET because up to the V_{gs} reaching I_{60}, the device operates with an average slope smaller than 60 mV/dec. The difference between point slope S_{min}, average slope S_{av} and the I_{60}-figure of merit is shown in Figure 9.12. In the following, S_{min} and I_{60} are used as relevant figures of merit.

It was discussed above that a small effective mass leads to a higher BTBT probability, provided that vertical quantization does not result in a substantial increase of the effective band gap. Furthermore, a small effective mass yields a small DOS within the channel such that the quantum capacitance limit could in principle be reached that allows avoiding DIBT. However, it is important to note that when computing the TFET behavior displayed in Figure 9.7, the source and drain contact regions were considered to have been fortified with additional gate electrodes, providing sufficient screening. Yet, in a real device a certain underlap region is required to avoid short circuiting contact and gate electrodes. The electrostatics in these source/drain extensions turns out to be problematic, once taken into account.

The left panel of Figure 9.13 shows self-consistently computed conduction and valence bands of the source-extension region in the case of four different effective masses for the nanowire bundle TFET depicted in Figure 9.11. In the present case, a constant

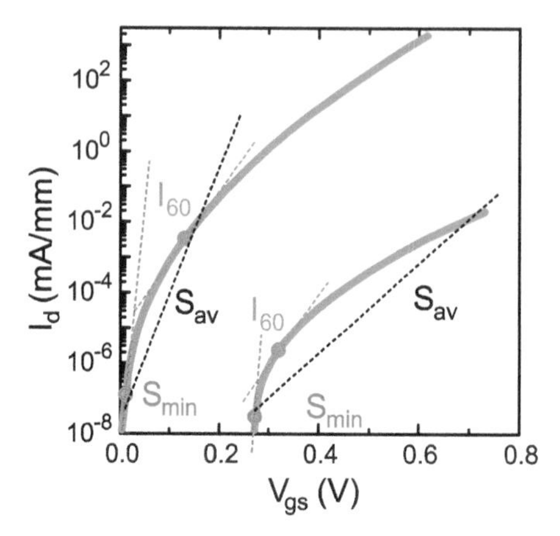

Figure 9.12: Illustration of the different figures of merit to characterize the switching of a TFET: point slope S_{min}, average slope S_{av} and I_{60}.

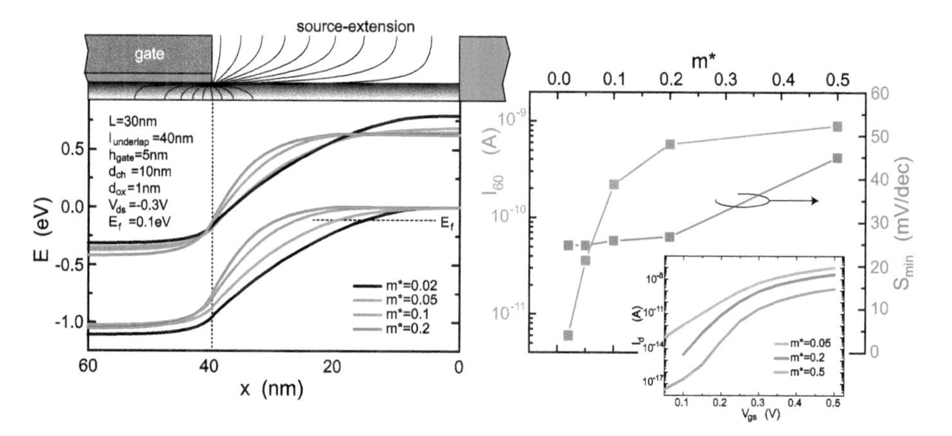

Figure 9.13: Left: Conduction and valence band profile within the source-extension region and the channel of the injecting contact for several effective masses. The top image shows a schematics of the wrap-gate nanowire device with equipotential lines computed with self-consistent NEGF simulations. The right panel displays S_{min} and I_{60} for nanowire TFETs shown on the left. The inset depicts transfer characteristics for different m^* [135].

Fermi energy of $E_f^S = 0.1$ eV is assumed within the source extension; all other parameters are given in the figure. Vertical quantization has been accounted for with a simple particle-in-the-box model, which is the reason why the band gap increases with decreasing effective mass. Decreasing m^* leads to a smaller DOS in the nanowire, and consequently, to a lower carrier density in the source extension. The reduced carrier density in turn leads to a reduced screening, which results in a broadening of the *n-p* junction within the source extension. In other words, λ_{dop} increases when m^* decreases, which ultimately limits the achievable BTBT probability even if everything else (EOT, d_{ch}, E_g, etc.) has been optimized.

The right panel of Figure 9.13 shows S_{min} and I_{60} for TFETs as a function of $m^{\star}$. Interestingly, while the smallest masses result in the steepest S_{min} due to improved BTBT, they also yield the smallest I_{60} due to the largest λ_{dop}. Vice versa, the largest $m^{\star}$ leads to the highest I_{60} but also the largest S_{min}. The inset of the right panel displays the transfer characteristics (the same parameters as given in the left panel), clearly showing that the highest on-currents are actually obtained with the largest effective masses. Again, it is the improved screening of the gate action on the source extension with increasing $m^{\star}$ that leads to this counterintuitive situation that a large effective mass yields the better TFET.

The scenario discussed so far is certainly very unsatisfactory. While it is clear that a large effective mass will degrade the BTBT probability, it is nevertheless the best choice. However, there is an important factor that was left out in the discussion regarding the screening. In the computation of the impact of $m^{\star}$, a constant Fermi energy of 0.1 eV was assumed. But if a reduced effective mass leads to a degradation due to the reduced carrier density in the source extension, what about increasing E_f? Figure 9.14(a) shows S_{min} and I_{60} as a function of E_f in the source; all other parameters of the simulated device are given in the figure. As expected, I_{60} increases with E_f because of the improved screening. However, S_{min} also increases approaching 50 mV/dec for larger E_f. The reason for this behavior is shown in (b). When the Fermi energy increases, the band-pass filter provided by the *p*-*n* junction at the source-channel interface sits energetically at the wrong position in that carriers from the Boltzmann tail of the source Fermi distribution function are injected when the device switches on, similar to a conventional MOSFET. Indeed, if the BTBT probability was close to unity, a high E_f would lead to $S \approx 60$ mV/dec [142]. As a result, the trade-off between the requirement of a low effective mass to improve BTBT and a high $m^{\star}$ to allow for optimized screening at a low E_f cannot be resolved easily.

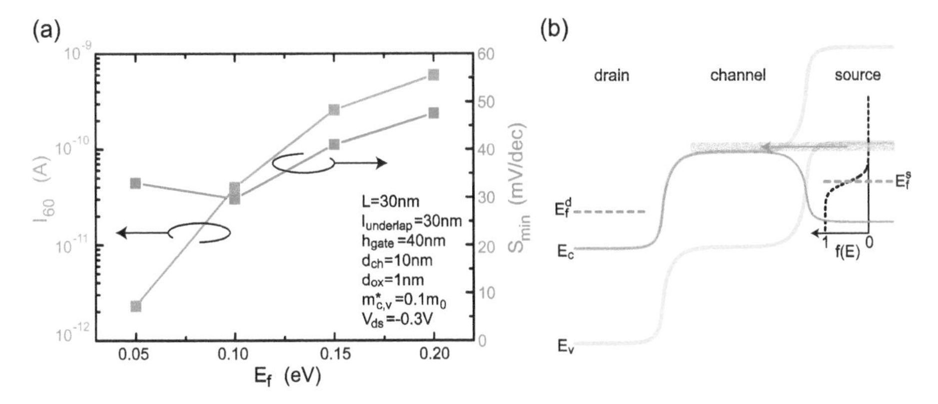

Figure 9.14: (a) S_{min} (red) and I_{60} (blue) as a function of E_f in source [135]. (b) Conduction and valence bands in a TFET (light green) when switching into the on-state. In the case of a large Fermi energy, the band-pass filter resides in the Boltzmann tail of the source Fermi distribution and the device behavior is similar to a conventional *n*-type MOSFET (blue line).

Task 31.

Source contact with large Fermi energy: Consider a TFET with ideally abrupt source-channel and channel-drain junctions leading to a step-function appearance of the potential profile. To realize such steep *p*-*n* junctions, a very high doping concentration in the source contact is necessary (apart from an extremely small screening length in the channel). Such a high doping concentration may lead to a rather large Fermi energy in source. Compute the drain current with Equation (9.1) (neglect the parts due to thermal emission) of a 1D TFET with step-function-like potential profile and a Fermi energy of $E_f^S = 0.25$ eV. You may assume that the transmission probability is large enough to justify $T = 1$ within the energy range of BTBT. Furthermore, any contribution from the drain can be neglected and perfect gate control with $\delta\Phi_f^0 = \delta\Phi_g$ is considered.

9.1.3.4 Dependence on the Band Gap—Heterostructure TFETs

It was mentioned above that the size of the band gap E_g determines to a large extent the minimum off-state leakage, since it blocks the thermally activated carrier flow (cf. Equation (9.3)). On the other hand, decreasing E_g increases the BTBT, leading to an improved TFET performance. Therefore, a trade-off between boosting BTBT while keeping the leakage low needs to be found, which is again unsatisfactory. One way out of this dilemma is the use of an axial heterostructure [22, 231, 137, 173]. Two scenarios are possible. In the first scenario, a material with substantially smaller band gap is inserted at the source-channel interface. Doing so provides a small band gap for BTBT but a larger one in the remainder of the device resulting in a low leakage current. The second approach employs a type-II heterointerface.

In Section 4.7, the three types of heterostructures have already been discussed briefly. A type-II heterointerface with broken band alignment appears ideally suited to increase TFET performance since it promises a very high BTBT probability while the band gaps in source, channel and drain can be much larger. As an example, the heterosystem $InAs/Al_xGa_{1-x}Sb$ is considered with self-consistent NEGF simulations. Using this ternary III–V compound, broken or staggered band line-ups at the source-channel heterojunction can be realized by varying the Al mole fraction x.

Figure 9.15(b) shows the device layout under consideration. A nanowire of diameter d_{ch} consisting of $Al_xGa_{1-x}Sb$ is grown onto a highly n-doped InAs substrate with doping concentration N_{d} and a metallic gate of length L is wrapped around the NW insulated by a dielectric of thickness d_{ox}. The type II heterointerface is at the nanowire-substrate interface. While the position is therefore well known, to avoid short-circuits between gate and source an underlap with length l_{ext} is required. Hence, the gate action in the underlap region must be accounted for appropriately. This is done by approximating the wrap-gate architecture with four gates (quadruple gate architecture) placed around the channel, which leads to a slight underestimation of the gate impact in the underlap region. The benefit is that the area bound by the points 1, 2, 3 and a, b can now be conformally mapped (see Section 6.1.2) onto a parallel plate capacitor (Figure 9.15(b)). The lengths of the back-transformed electric field lines yield a spatially dependent $d_{\mathrm{ox}}(x)$,

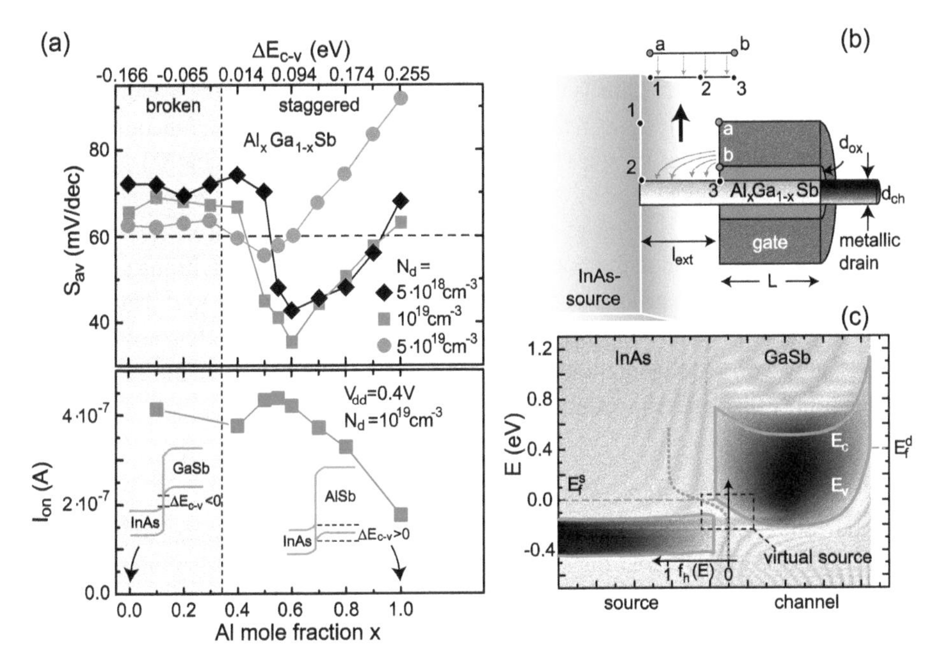

Figure 9.15: (a) Extracted S-values (top) and on-current (bottom) as a function of composition x of the InAs/Al_xGa_{1-x}SB-heterostructure for different doping concentrations. (b) Schematics of the wrap-gate nanowire TFET layout. Conformal mapping is used to take the source underlap region into account. (c) Grayscale plot of the local DOS close to the on-state of the TFET [137].

which is inserted into $\lambda_{\text{ch}}(x) = \sqrt{\frac{\varepsilon_{\text{III-V}}}{4\varepsilon_{\text{ox}}} d_{\text{ox}}(x) d_{\text{ch}}}$. Thus, the one-dimensional modified Poisson equation (5.34) derived in Section 5.5 can be used in the entire device. Since the drain contact has little impact on the TFET performance as long as it is sufficiently transmissive, we assume a metallic drain contact with the drain Fermi level always being aligned with the valence band.

Self-consistent NEGF simulations have been carried out for the nanowire heterojunction TFET as a function of the aluminum mole fraction x for three different doping concentrations in the source (InAs). Scattering with a mean free path of $l_{\text{mfp}} = 30\,\text{nm}$ is taken into consideration with Buettiker probes. This ensures that the potential notch in the virtual source (see Figure 9.15(c)) will be occupied with carriers. Inverse subthreshold slopes averaged over three orders of magnitude change in I_d and the on-currents are extracted from the simulations and plotted in Figure 9.15(a) (for details see [137]). Interestingly, an inverse subthreshold slope steeper than 60 mV/dec is only obtained in the case of low doping and staggered band gaps (with $0.5 < x < 0.8$) with a minimum of $S_{\text{av}} = 35\,\text{mV/dec}$ at $x = 0.6$ and $N_d = 10^{19}\,\text{cm}^{-3}$. This may appear to be counterintuitive since a broken band line-up seems to be best suited for TFETs. However, in the case of a broken line-up, a virtual source (black dashed box in Figure 9.15(c)) builds up from which thermal emission occurs.

For $N_d = 5 \cdot 10^{19}$ cm^{-3}, S is slightly below 60 mV/dec at $x = 0.5$ and strongly increases for larger x. The reason for this is the large Fermi energy in the source, yielding a carrier injection from the Boltzmann tail of the source Fermi distribution function (cf. Figure 9.14). The lower panel of Figure 9.15(a) shows I_{on} evaluated by placing $2/3 \cdot V_{dd}$ above and $V_{dd}/3$ below V_{th} (evaluated by extrapolating linear I_d–V_{gs} plots to zero I_d) [137]. The optimum performance in terms of on/off-current ratio is obtained at $x = 0.6$ where S occurs at a minimum and at the same time the on-current is almost at its maximum. In summary, type II heterostructures are indeed useful to improve the performance if a staggered, close to broken band line-up is chosen. The major concern of heterostructure TFETs is, however, the alignment of the heterointerface with respect to the gate, which is difficult to realize experimentally.

9.1.3.5 Electrostatic Doping in TFETs

An important result so far is that optimized TFET performance requires appropriate screening in the channel *and* the source contact. The issue is that a high doping concentration in 1D nanowire devices with low effective mass leads to a high Fermi energy, which in turn results in S approaching 60 mV/dec. Moreover, since doping is a random process, a high doping concentration leads to disorder, and thus to smearing out the band edge of the source contact with a DOS exponentially decaying into the band gap (cf. Section 2.12.2) shown in Figure 4.2. In addition, diffused dopants and/or the implantation process may result in trap states within the band gap at the source/channel interface that allow for trap-assisted tunneling even when the device is in its off-state. Both effects, a smearing of the band edge and traps, lead to a deterioration of the steepness of the switching behavior of TFETs [228]. To illustrate this effect, a one-dimensional nanotube TFET (the parameters are given in Figure 9.16) is simulated using NEGF. A smeared distribution of dopants centered at 50 meV below the conduction band was considered adopting the approach outlined in [130]. Inelastic scattering due to phonon absorption or emission, must be present in order to induce vertical transport, which is taken into consideration with Buettiker probes.

Figure 9.16 shows transfer characteristics in the case of a TFET with (red curve) and without (blue curve) trap-assisted tunneling (TAT). The inset shows a gray-scale plot of the local density of states; the trap-/dopant-related band below the conduction band is clearly visible. Vertical transport due to TAT (illustrated with the red arrow in the inset of Figure 9.16) results in a substantial degradation of S. The reason for this is that the band-pass filter functionality of the source-channel n-p junction is deteriorated and thus current can already flow before the gate voltage moves the valence band in the channel on the same energy as the conduction band is source.

In Section 7.4, reconfigurable FETs were studied where gate electrodes have been used to implement an effective n- and p-type "doping" concentration by applying appropriate gate voltages. The same approach can be used in TFETs in order to resolve the trade-off between the necessity for screening in source and the requirement for a low

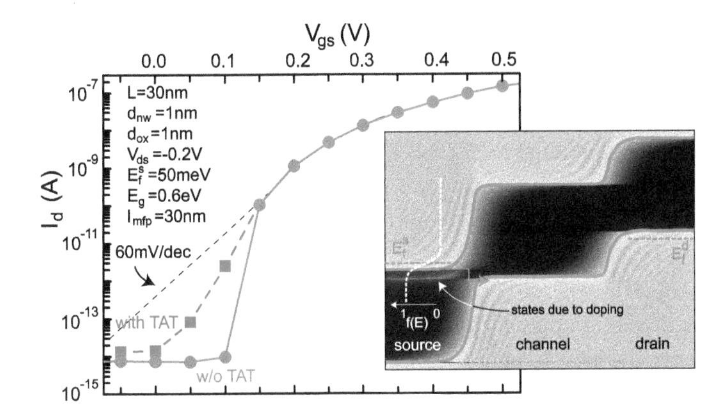

Figure 9.16: Simulated $I_d - V_{gs}$-curve for a nanotube TFET. The red curve shows the transfer characteristic with trap states present at an energy of 50 meV below the conduction band edge. The blue curve belongs to the device without TAT. The inset shows the local density of states with trap states. The presence of traps together with inelastic scattering yields a substantial deterioration of the steepness of S.

E_f and an undisturbed band gap with sharp band edge. In fact, the high carrier density within the metallic gate electrode provides the necessary screening, if the effective oxide thickness is made small enough (cf. Section 4.5.5). The latter point is very important and worth elaborating on: In Section 5.7, it has been discussed how the screening lengths due to, e. g., a gate electrode in source λ_s and mobile charge λ_n ought to be combined to obtain an overall screening length $\lambda_{\text{tot}}^{-1} = \sqrt{\frac{1}{\lambda_s^2} + \frac{1}{\lambda_n^2}}$. In order to avoid the adverse effect of a high carrier density in a 1D nanowire with low DOS (leading to S approaching 60 mV/dec) one has to ensure that λ_s is always the dominant, i. e., the smallest screening length. In other words, $\lambda_s \ll \lambda_n$ since otherwise (by inducing more carriers with applying higher gate voltages at the source gate) a large Fermi energy is eventually obtained that injects carriers from the Boltzmann tail. As a result, for optimum TFET performance the EOT of (at least) the source and channel electrodes must be as small as possible to ensure a high BTBT probability. This requirement is actually difficult to meet. When fabricating triple-gate devices in order to implement the appropriate n- and p-regions with electrostatic doping, the three gate electrodes need to be insulated from each other but in the best case without leading to an ungated underlap region in between two adjacent gates (cf. Figure 9.18).

The underlap can be avoided with a vertical arrangement of the gates with partially overlapping gates. However, this implies that the EOT in either the source or channel is larger than in the channel or source (unless an elaborate spacer process is utilized). In the best case, the channel has to be placed in between bottom- and top-gates in order to have the same small d_{ox} and no gate underlap, which may be challenging technologically. In this respect, carbon nanotubes are an attractive material since their ultra small diameters allow small λ_s even if EOT is not extremely small. Concluding, electrostatic instead of impurity doping allows disentangling the necessity for strong screening from the magnitude of the carrier density in source if properly made.

As an attempt to realize a TFET based on electrostatic doping, the reconfigurable, dual-gate carbon nanotube FET studied in Section 8.1.4 is used [11]. In this device, the nanotube is manipulated by two gate electrodes with an EOT of 10 nm and ~2.5 nm, which yields, together with the very small diameter $d_{\mathrm{CNT}} = d_{\mathrm{ch}}$ of the nanotube, screening lengths λ_s and λ_{ch} of a few nanometers only. Applying negative voltages to the large area back gate (cf. Figure 8.11) creates *p*-type sections in-between the Ti-contacts and the actual gate. If negative voltages are applied to the center gate electrode, a conventional MOSFET is obtained where the switching of the device is due to field-effect modulated thermal emission of holes. Indeed, the left branch of the transfer characteristic displayed in Figure 9.17 (gray shaded area) shows an inverse subthreshold slope of 65 mV/dec, close to the thermal limit. On the other hand, if sufficiently large positive voltages are applied at the center gate electrode, a window for BTBT opens up as is illustrated in the two right panels in Figure 9.17. Here, the conduction and valence bands for two different V_{gs} are depicted, showing that in the current situation BTBT occurs at the source-channel and channel-drain interfaces. In the present case, BTBT leads to a minimum inverse subthreshold slope of $S_{\min} = 40$ mV/dec, steeper than the thermal limit. Note that there are only few data points in the gate voltage range where the device shows a steep switching behavior. Therefore, in order to reconfirm the observation of a steep current increase, self-consistent NEGF simulations were carried out (red and blue lines in Figure 9.17), which are in excellent agreement with the experimental data. A further important point

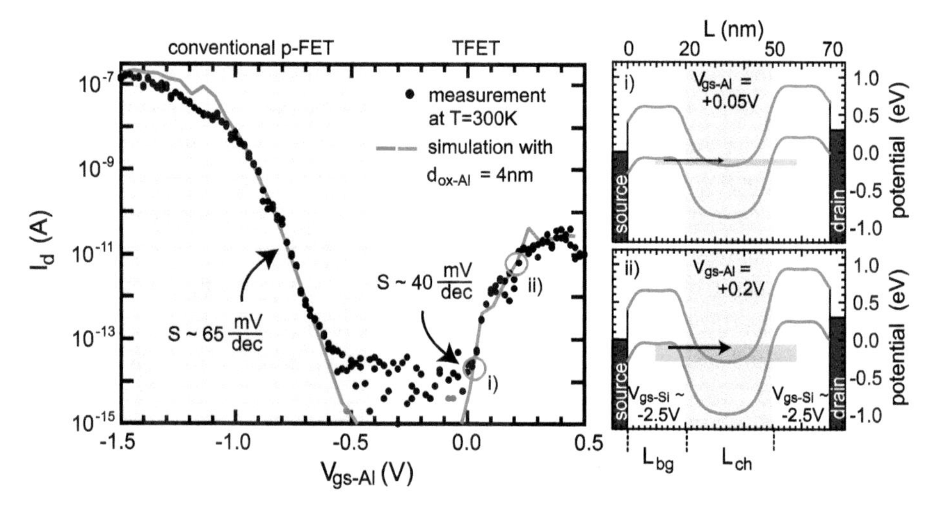

Figure 9.17: Transfer characteristic of a reconfigurable, dual-gate CNTFET operated as a *p*-type device. For negative V_{gs}, the device acts as a conventional MOSFET, for positive gate voltages as a TFET. Conduction/valence band profiles at two V_{gs} in the TFET-operation regime are shown in the two panels on the right. An inverse subthreshold slope of 40 mV/dec is achieved, reconfirmed with self-consistent NEGF simulations (straight red and blue line) [11].

is that the reason for the rather low on-state current in the right (BTBT) branch of the device is the existence of two BTBT barriers that carriers need to tunnel through.

In order to improve the CNT TFET discussed above, the bottom gate electrode needs to be split up into two separate gate electrodes in order to realize the *n-i-p* band structure of a TFET with only a single BTBT barrier (see also Section 10.2.1). Such a CNT TFET has recently been demonstrated [209]. Figure 9.18(a) shows an electron micrograph and a schematics of a TFET that exhibits three separate gates that can be used as source-side program-gate (PG1), control gate (CG) and as drain-side program-gate (PG2). The gate dielectrics consist of a single layer of HfO_2 (6 nm) in the source-/drain-side gate regions and a double layer of HfO_2 (i. e., 12 nm) in the control gate region.[4] Figure 9.18(b) displays transfer characteristics for two different source-side program voltages (PG1). A minimum inverse subthreshold slope of S_{min} = 41 mV/dec is obtained, similar to the dual-gate TFET shown above. However, as expected, the on-current is more than one order of magnitude larger.

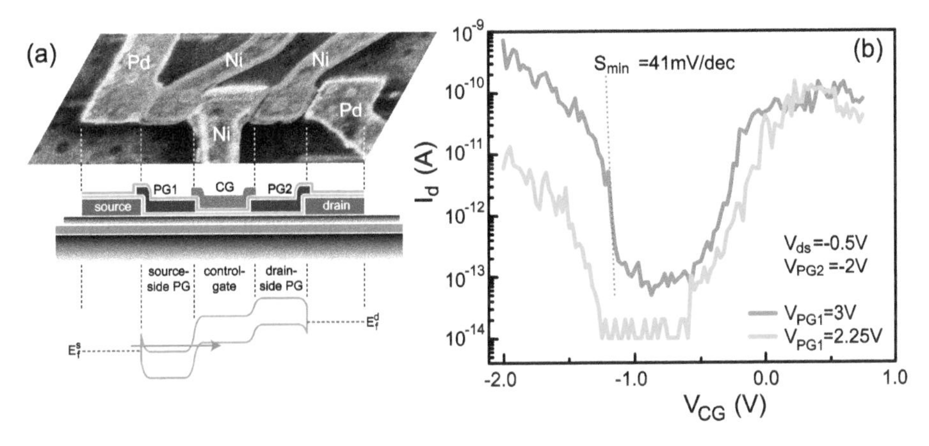

Figure 9.18: (a) Colored electron micrograph and schematics of a CNT TFET with separate source- (PG1) and drain-side program gates (PG2) (Z. Chen, Purdue University). (b) Transfer characteristics for two different source-side program gate voltages [209]. A minimum inverse subthreshold slope of 41 mv/dec is obtained.

9.2 Alternative Steep-Slope Transistor Concepts

The optimization of TFETs requires a number of trade-offs to be made due to the interdependencies of the various performance boosters. Moreover, even if a steep slope can be realized, the presence of the BTBT barrier will always deteriorate the on-state performance, when compared to conventional transistors. Therefore, modifications of

4 The fabrication leads to a $\lambda_s < \lambda_{ch}$ and a small gate underlap at the source-channel and channel-drain interfaces.

TFETs that particularly address the issue of on-state performance have been proposed. In addition, completely novel steep-slope transistor concepts have been intensively investigated. In the present section, selected alternative steep-slope transistor concepts are briefly discussed.

9.2.1 TFETs with Line-Tunneling

In TFETs as discussed so far, band-to-band tunneling occurs at the interface between source and channel, i. e., at a particular location within the device. Therefore, this tunneling is sometimes called point tunneling. In order to improve the on-state current of a TFET the idea is now to spread the BTB tunneling over a larger area, which is called line tunneling. The idea behind line tunneling is basically the same as, e. g., in horizontal metal-SiN-semiconductor contacts discussed in Section 7.3.2; increasing the length of the contact beyond the transfer length allows decreasing the contact resistance. For a TFET, this enables increasing the on-state performance.

There are two (similar) ways how line tunneling is realized that are illustrated in Figure 9.19. In (a), a buried *p*-doped pocket is realized underneath a part of the gate electrode. Applying a positive gate voltage moves the bands in the intrinsic section (thickness d_n) in between the gate dielectric and the *p*-pocket downwards in energy, so that BTBT can occur. As a result, BTBT is distributed along the length of the pocket L_n so that the overall transmission probability due to BTBT increases [110]. A very similar concept is the so-called electron-hole bilayer TFET (EHB-TFET) [4, 221] as depicted in Figure 9.19(b). Here, instead of a doped pocket, two individual gate electrodes (gate 1 and 2) are used to induce two opposing sheets of electrons and holes. For sufficiently large gate

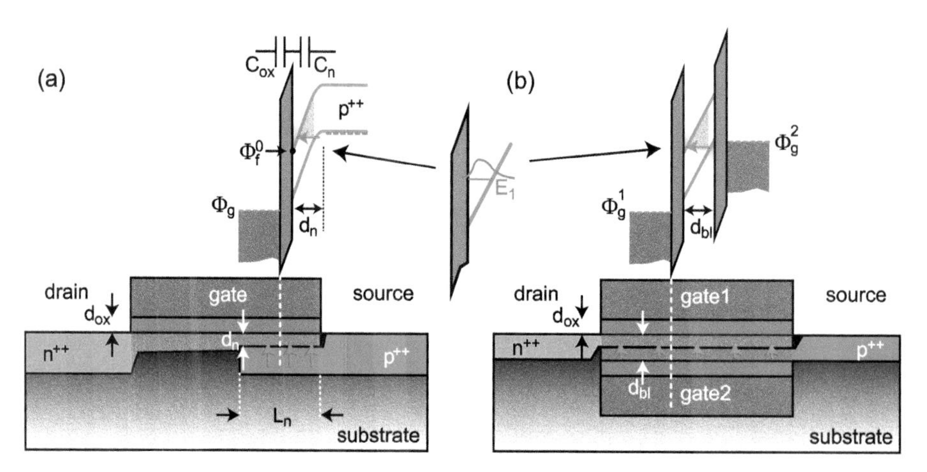

Figure 9.19: (a) TFET with *p*-doped pocket and (b) electron-hole-bilayer TFET. The schematics on top show a cross-section of the MOS capacitor at the position of the dashed line. Carrier confinement in the triangular potential well leads to an increase of the effective band gap.

voltages, BTBT occurs, distributed across the entire length of the overlap of gate 1 and gate 2, thereby increasing the BTBT probability.

Compared to a TFET with point tunneling, there is an important difference in the two device concepts. In order to obtain a steep slope the thickness d_n in the concept shown in (a) (called pocket TFET in the following) and the thickness of the electron-hole bilayer d_{bl} in (b) need to be extremely thin. The reason for this is that the potential can be thought of as being fixed by the highly p-doped section in the pocket TFET and by the two gate voltages in the EHB-TFET. The cross-sections of the BTBT region in Figure 9.19 show the conduction/valence band situation. Let us first focus on the pocket TFET. If the depletion region in the p-doped pocket and the effect of mobile charge are neglected, the bands can be approximated with straight lines (see Figure 9.19), and hence the WKB approximation elaborated in Section 9.1.1 can be used again. This time Φ_f^0 is determined by noting that between the p-doped pocket and the gate electrode there is a series combination of $C_{\mathrm{ox}} = \varepsilon_{\mathrm{ox}}/d_{\mathrm{ox}}$ and $C_{\mathrm{n}} = \varepsilon_{\mathrm{si}}/d_n$. Hence, $\Phi_f^0 = \frac{C_{\mathrm{ox}}}{C_{\mathrm{ox}}+C_n}\Phi_g$ + const. where the constant contains the work-function difference (set to zero for simplicity). Inserting into T_{WKB} yields

$$T_{\mathrm{WKB}} = \exp\left(-\frac{4d_n\sqrt{2m^*}E_g^{3/2}}{3\hbar(E_g + \Phi_g\frac{\varepsilon_{\mathrm{ox}}d_n}{\varepsilon_{\mathrm{ox}}d_n+\varepsilon_{\mathrm{si}}d_{\mathrm{ox}}})}\right). \tag{9.19}$$

A similar relation can be derived for the EHB-TFET (d_n needs to be replaced with $d_{\mathrm{bl}}/2$ if a symmetric device layout and $\Phi_g^1 = -\Phi_g^2$ are assumed). As a result, d_n and d_{bl} need to be made extremely thin in order to increase T_{WKB}. This, however, decreases the gate control over Φ_f^0, and thus a simultaneous and strong reduction of C_{ox} is mandatory, since $C_{\mathrm{ox}} \gg C_n$ is required for optimum gate control. While in TFETs with point tunneling, a reduction of the effective oxide thickness (EOT) leads to a continuous improvement, in pocket- and EHB-TFETs, once perfect gate control over Φ_f^0 is achieved, no further improvement due to EOT scaling is possible. The latter argument is certainly only true for a constant d_n and d_{bl}. However, decreasing d_n and d_{bl} below a certain thickness leads to quantization (illustrated with the inset in between Figure 9.19(a) and (b)), which increases the effective band gap. An increased band gap in turn deteriorates T_{WKB}. Moreover, in EHB-TFETs the simultaneous quantization in the conduction and valence bands may prevent BTBT to occur at all. As a result, d_n- and d_{bl}-scaling is limited and, therefore, also EOT scaling and with this the achievable BTBT probability. Obviously, optimizing pocket- or EHB-TFETs is also intricate [221] and requires materials with rather low band gaps to start with. In this respect, III–V heterostructures such as InAs/GaSb with a broken type II-heterointerface may be a solution [208]. Promising is the use of 2D materials that are one monolayer in thickness and, therefore, seem to be ideally suited for EHB-TFETs [169]. However, one major concerns remains; the approach is not very scalable, since reducing the device size also decreases the current through the device due to reduction of the length over which line tunneling occurs.

9.2.2 Energy-Filtering Devices

A major drawback of TFETs (including pocket- and EHB-TFETs) is the band-to-band tunneling barrier at the source-channel interface that limits the achievable on-currents and—if not done properly—also yields $S_{av} > 60$ mV/dec. However, as explained in Section 9.1.1 (see also the video provided by QR code #62), the ideal functionality of TFETs is obtained when the BTBT barrier plays the role of a band-pass filter cutting off the high and low energy tails of the source Fermi distribution function. Hence, a band-pass filter is what is required for a steep-slope transistor, not a tunneling barrier. Inserting an appropriate energy filter in-between source and channel that acts as a band-pass filter yields a so-called energy-filtering device (EF-FET). EF-FETs potentially allow for combining a steep slope with a high on-state current [26, 86].

Figure 9.20 illustrates an energy-filtering device. A band-pass filter has been inserted in between the source contact and the channel, and if a gate voltage is applied, the bands within the regular MOSFET channel are moved downwards in energy. Since the energy filter blocks current injection from the Boltzmann tail of the source Fermi distribution function, the off-state leakage is suppressed. Only when the potential maximum in the channel is moved below the (energetic) upper edge of the energy filter a current will flow through the device. If the Fermi level in the metallic source electrode is at an appropriate energy (illustrated in Figure 9.20), the current rises abruptly into the on-state of the transistor yielding a MOSFET-like on-state performance.

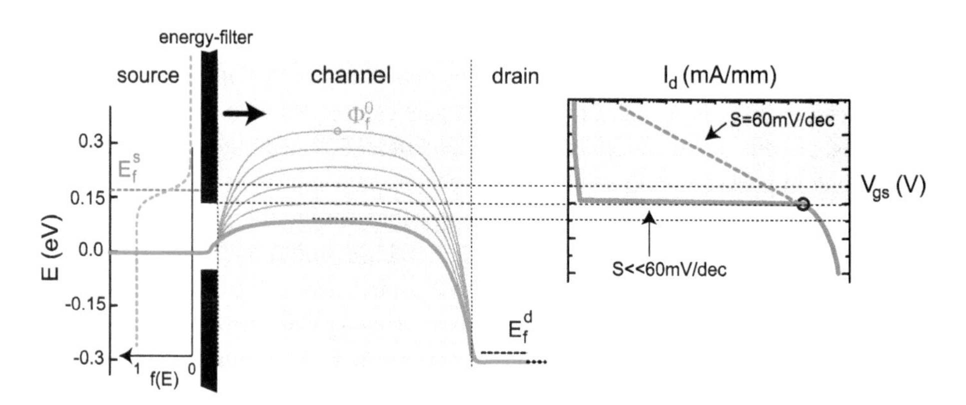

Figure 9.20: Illustration of an energy-filtering device: A band-pass filter (black) is inserted in between the source and channel of a regular MOSFET blocking carrier injection from the Boltzmann tail of the source Fermi distribution function. Once Φ_f^0 is moved below the upper energetic edge of the band-pass filter, the device abruptly switches into its on-state (right panel).

The question now is how such an energy filter can be realized. One suitable way to create an energy filter is the use of a superlattice as is illustrated in Figure 9.21. Here, a metallic contact electrode is connected to a superlattice consisting of a number (seven

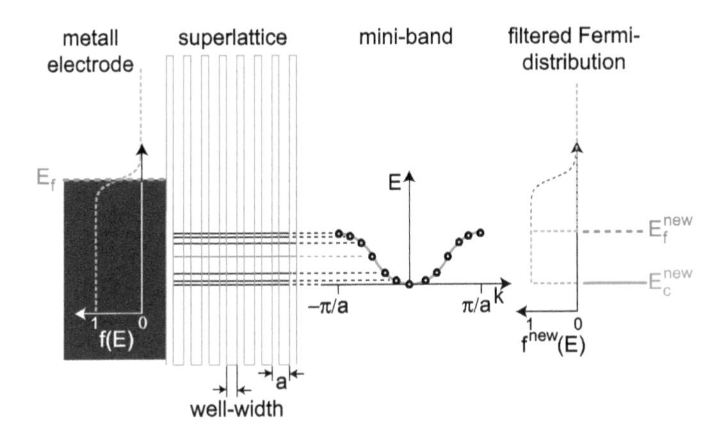

Figure 9.21: Energy filter in front of a metallic contact realized with a superlattice consisting of coupled particle-in-the-box systems in a lattice with lattice constant a. The filtered Fermi function can be used to realize steep-slope devices.

in the present case) of particle-in-a-box systems (green lines) coupled to each other. This results in a one-dimensional artificial solid state with lattice constant a giving rise to the formation of a cosine-shaped miniband as shown in Figure 9.21 (see also Section 2.4.1). If we assume that the potential barriers and well widths can be realized with great precision, the miniband has sharp band edges and represents the desired band-pass filter. Driving a current from the metallic electrode through the miniband will basically result in a filtered Fermi distribution function. If the original Fermi level lies energetically well above the top edge of the miniband, the filtered Fermi distribution function resembles a step function, i. e., a Fermi distribution function at low temperatures where the upper edge of the miniband can be thought of as an effective, new Fermi level (cf. Figure 9.21). As we know from our considerations in Chapter 2, there will be a second (and third, etc.) miniband developing, which stems from the second eigenstate of the particle-in-the-box system. As a result, the original Fermi level cannot be placed energetically too high above the upper edge of the miniband, since otherwise a substantial part of the Boltzmann tail of the Fermi function may inject carriers into the channel through the second superlattice miniband.

A superlattice as required for an energy filter can technologically be realized with heteroepitaxial growth of compound semiconductors. The GaAs/AlGaAs [86] or the InGaAs/InAlAs [177] heterosystems have been investigated where sufficiently large conduction band offsets can be realized. Figure 9.22 shows a simulation of an energy-filtering FET. In (a), the local density of states is displayed together with the conduction band [26]. Due to the superlattice, a miniband appears that can clearly be identified. Figure 9.22(b) shows the transfer characteristics of the device, simulated with NEGF (cf. Chapter 6). A steep turn-on behavior with an inverse subthreshold slope of 25 mV/dec over several orders of magnitude is visible. Furthermore, the device (a 1D nanowire FET) shows the expected MOSFET-like on-state currents.

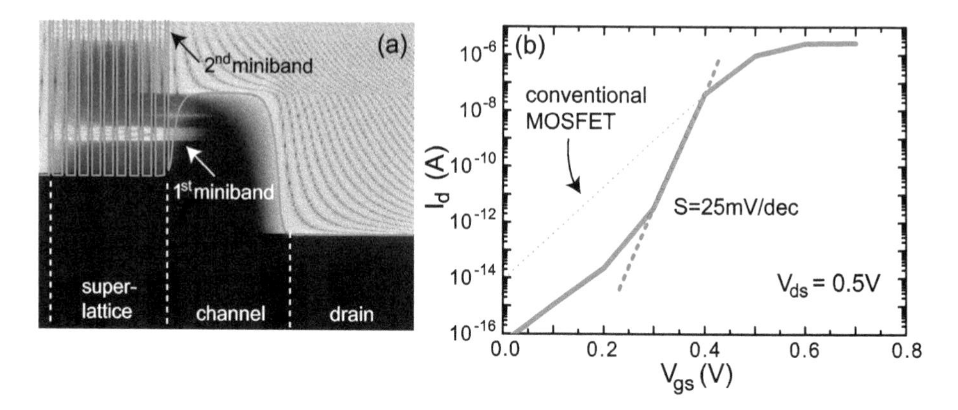

Figure 9.22: (a) Local DOS in a 1D energy filtering device computed with NEGF. A miniband forms in the source contact that filters the Fermi distribution. (b) Transfer characteristics of an energy-filtering device showing S = 25 mV/dec and MOSFET-like on-currents.

In the present case, ballistic transport has been assumed and, furthermore, any impact of the gate on the superlattice has been neglected. It is clear that taking both into consideration is likely to yield a degradation of the turn-on behavior. However, recent self-consistent simulations showed that even considering the full electrostatics of a nanowire energy-filtering device [85], a steep-slope behavior is still observed. Realizations of such a device concept have so far not been convincing, though.

9.2.3 Impact Ionization Field-Effect Transistors

Impact-ionization (II) MOSFETs (IMOS) exploit an avalanche breakdown due to II in order to amplify the current within the device and thereby circumvent the 60 mV/dec limit of conventional MOSFETs. Similar to TFETs, IMOS devices consist of a gated, reversed biased *p-i-n* structure. However, in contrast to TFETs, the gate covers only a part of the intrinsic (*i*) region. Applying sufficiently large gate-source and drain-source voltages, a high electric field builds up within the intrinsic region leading to impact ionization and eventually avalanche breakdown. The avalanche multiplication of carriers results in extremely steep turn-on characteristics with inverse subthreshold slopes significantly smaller than 60 mV/dec. At the same time, rather large on-state currents are obtained due to the internal gain [88].

For avalanche breakdown to occur, the carriers need to be accelerated to at least the kinetic energy equivalent to the band gap of the semiconducting material. Therefore, a reduction of the supply voltage is only possible using low band-gap semiconductors. A major reliability concern of IMOS devices is that the hot carriers generated by the large electric fields in the II-region lead to hot carrier injection into the gate dielectric resulting in a loss of the steep switching behavior already after a few switching cycles [88] and eventually to a dielectric breakdown. However, the amount of hot carrier injection

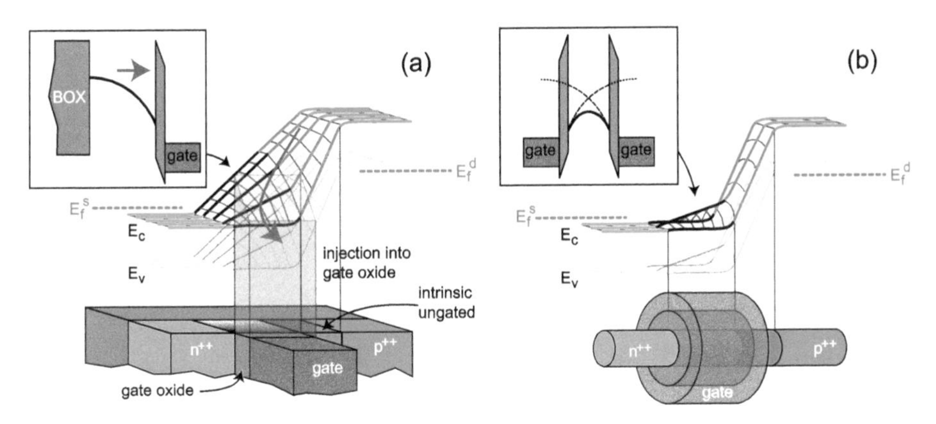

Figure 9.23: Schematics of an SOI IMOS (a) and a wrap-gate nanowire IMOS (b) together with the potential distribution in the device.

depends on the device architecture and can thus be engineered in order to reduce or even avoid hot carrier degradation. To illustrate this, let us compare Figure 9.23(a) and (b); (a) shows a schematic of a silicon-on-insulator, planar IMOS and the inset displays the potential profile perpendicular to the direction of current transport. Obviously, the large electric field induced by the positive gate-source voltage and the negative drain-source bias required for II results in a substantial field component, accelerating the hot carriers toward the gate dielectric. On the other hand, a gate-all-around nanowire device architecture exhibits a strongly reduced hot carrier injection. The potential profile in such a wrap-gate nanowire IMOS is depicted in Figure 9.23(b); reducing the nanowire diameter yields an overlap of the potential distribution perpendicular to the nanowire axis (cf. inset), and thus reduces the electric field at the nanowire surface. Since this electric field is responsible for hot carrier injection, it is expected that a nanowire IMOS shows strongly improved reliability.

Finally, carrier confinement can be employed to avoid hot carrier injection altogether. Utilizing a III–V nanowire with small bulk band gap E_g (such as InSb), the rather different conduction and valence band effective masses ($m_c^\star$ and $m_v^\star$) can be exploited. Reducing the diameter d_{nw} of the nanowire (assuming a square cross-section and a particle-in-the box quantization), the energetic difference between first and second subband in the conduction band can be made larger than the effective band gap, i. e., $\frac{\hbar^2\pi^2}{2m_c^\star d_{\mathrm{nw}}^2}((2^2+1-2)) > E_g + 2\frac{\hbar^2\pi^2}{2m_c^\star d_{\mathrm{nw}}^2} + 2\frac{\hbar^2\pi^2}{2m_v^\star d_{\mathrm{nw}}^2}$ (see Section 2.2.2). In this case, the hot carriers will remain in the first subband, and thus hot carrier degradation is avoided [24].

Task 32.
Suppression of hot carrier injection: Consider an InSb nanowire with square cross-section and assume particle-in-the-box quantization. Plot the effective band gap and the energy of the first and second subband in the conduction band and compute a criterion for avoiding hot carrier injection if the InSb nanowire is used in an IMOS device.

The benefit of a nanowire IMOS architecture is shown experimentally by realizing vertical, wrap-gate IMOS devices based on undoped Si-nanowires epitaxially grown on *n*-type (111) Si wafers (see Section 3.13.2). After nanowire growth and removal of the Au catalyst, SiO_2 as a gate dielectric is deposited via PECVD. A wrap-gate is realized by depositing Al and planarization with photoresist. After RIE etching of the resist, the aluminum covering the top of the nanowire can be removed. This gate recess defines the gate length of the device. Final steps include planarization with polyimide and the formation of a nickel top contact [25]. The center panel of Figure 9.24 shows an electron micrograph of a fabricated device without the top contact.

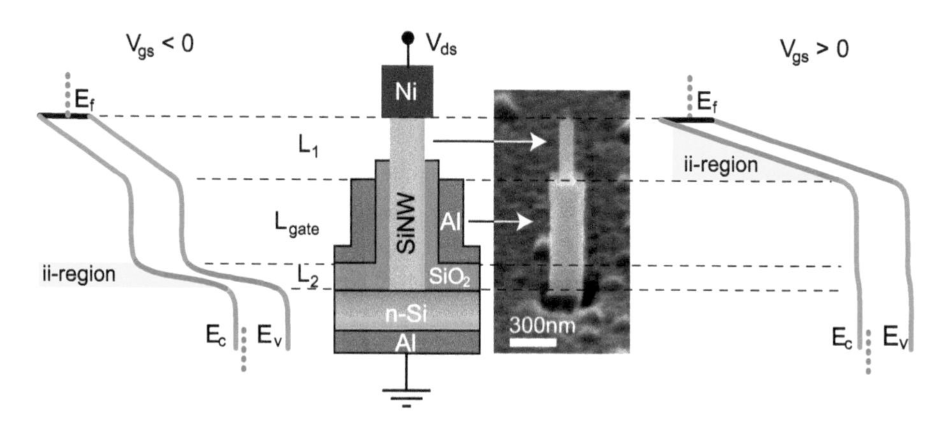

Figure 9.24: Schematic of a vertical NW-IMOS together with a scanning electron micrograph of the experimental device (center); note that the top contact has been removed. The conduction/valence band profiles in the on-state of the IMOS transistor for negative (left) and positive (right) gate voltages are displayed, too, showing the two different II regions within the device.

Figure 9.25(a) shows transfer characteristics of the IMOS device for bias voltages −0.5 V, −2.5 V and −4.5 V. In the present case, the bottom contact is grounded and V_{ds} is applied at the nickel top contact. A slight shift of the threshold voltage with changing V_{ds} can be observed, however, more importantly subthreshold swings as low as 5 mV/dec and 14 mV/dec are measured when V_{ds} is increased to −4.5 V (red curve). The reason for the ambipolar behavior is the presence of two possible impact-ionization regions in the device, which are schematically shown in Figure 9.24. Impact ionization occurs in either of the two depending on the applied voltages giving rise to the experimentally observed ambipolar behavior.

It was mentioned above that a major disadvantage of IMOS devices is that they are prone to degradation due to hot carrier injection into the gate dielectric. Figure 9.25(b) shows transfer characteristics of the vertical nanowire IMOS device for several (>100) sweeps of gate voltage. A hysteresis is observed, which is mainly due to the use of a low temperature PE-CVD gate oxide giving rise to a rather large number of defect/in-

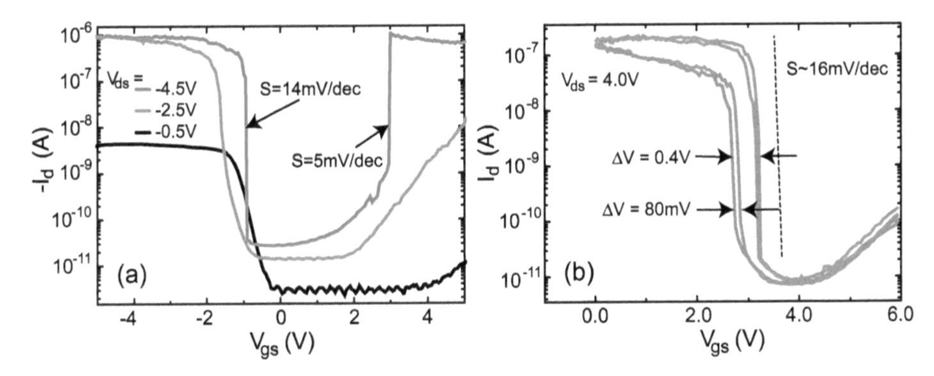

Figure 9.25: (a) Transfer characteristics of a wrap-gate Si nanowire IMOS for three different bias conditions [25]. (b) Repeated measurements (>100 voltage sweeps) of the vertical wrap-gate Si nanowire-IMOS. Note that the observed hysteresis is due to the PE-CVD-deposited gate dielectric. A shift of the threshold voltage is not observed [27].

terface states [27]. The important point, however, is that the threshold voltage (of each forward/backward sweep) remains almost unaltered while sweeping the gate voltage. Reasons for the suppression of hot electron degradation are the lower operating voltages compared to more conventional, planar IMOS FETs and—as already mentioned above—the smaller vertical fields within the channel due to the use of a nanowire with small diameter in a wrap-gate configuration [27].

9.2.4 FETs with Feedback Mechanism

An interesting approach to realize devices with an extremely steep inverse subthreshold slope is exploiting a charge feedback mechanism which can, for instance, be achieved in forward biased n-p_e-n_e-p devices where the subscript "e" indicates that these parts of the device are fortified with gate electrodes and the "doping" is realized by applying appropriate gate voltages (see top panel of Figure 9.26(a)). Such a device, called Z^2-FET has recently been demonstrated and intensively studied (see, e. g., [264]). The device works in the following way: constant gate voltages (with respect to source) are applied at gate 1 and 2 such that at small drain-source bias a n-p-n-p potential profile results where electron and hole currents are blocked. On increasing V_{ds}, the current increases exponentially since holes are injected from drain into the channel. At a certain threshold bias (V_{ds}), the amount of holes injected is so large that they pile up in the p_e-part (gate 1) of the channel where they lower the potential barrier (similar to the charge pile up discussed in Section 8.2). As a result, the injection of electrons over the p_e-barrier increases such that electrons are piled up in the n_e-region (gate 2) where they lift up the potential barrier leading to more hole injection. This feedback continues until there is a balance of electrons and holes in the p_e and n_e sections and the device switches from a low current in the off-state with basically zero S into the on-state. The main panel of Figure 9.26(a)

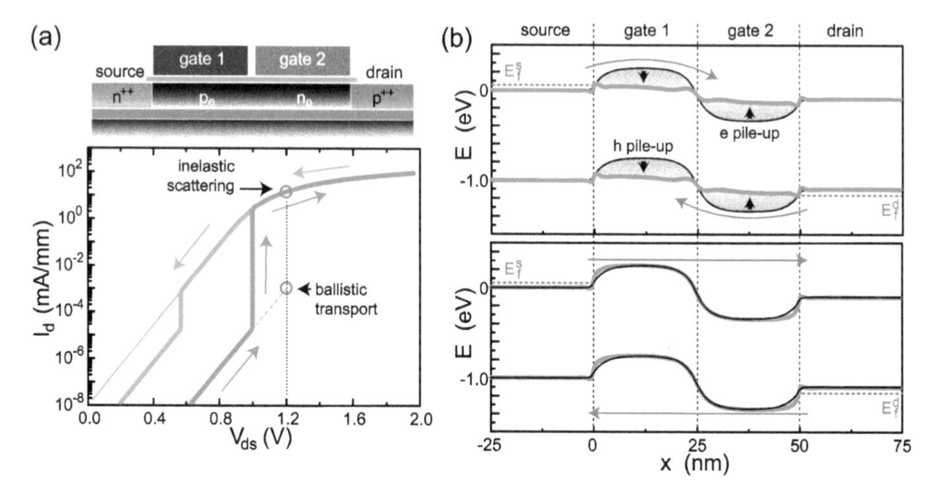

Figure 9.26: (a) Schematic of a Z^2-FET (top) and an illustration of the output characteristics with steep turn-on and hysteresis. (b) Conduction and valence bands (initial profiles in black) after a self-consistent computation (green lines). The top panel is with inelastic scattering, the lower panel shows the case of ballistic transport.

shows this behavior schematically. Switching the device off, requires a lower V_{ds} resulting in a hysteretic behavior (A similar behavior can also be obtained as a function of gate voltage (either gate 1 or 2) if appropriate voltage combinations are chosen.). The IV characteristics are therefore distinctly different when compared to a regular MOSFET but are suitable for, e. g., the realization of a 1T dynamic random access memory [264].

The functionality of the device relies on the charge feedback of the negative (positive) charges collected in the n_e (gate 2)- and p_e regions (gate 1) as illustrated in the top panel of Figure 9.26(b). Therefore, the device does only work if the following two conditions are fulfilled: (i) Inelastic scattering must be present ensuring that the electrons (holes) that are injected above (under) the potential barrier in the p_e-region (n_e-region) lose their energy and are piled-up in the n_e-region (p_e-region). The top panel of Figure 9.26(b) shows exemplary conduction and valence band profiles computed self-consistently with NEGF; here, the black curve is the initial potential profile and the green curve is the self-consistent result. Indeed, electron and hole pile-up yield a strong modification of the potential barriers in the gated regions leading to a large on-state current for gate voltages where the device would still be off without the feedback. The lower panel of Figure 9.26(b) shows the same calculation at the same voltages in the case of ballistic transport. Apparently, no feedback occurs in this case, as expected. (ii) While at room temperature, there will likely be a sufficient amount of inelastic scattering the second prerequisite for the realization of a working transistor is that charge pile-up leads to a band movement. To this end, the device must not be in the quantum capacitance limit. It was already discussed in Section 5.9.1 that when $C_{ox} \gg C_{inv}$ the charge in the channel has no impact on the energetic position of the bands anymore. For the device considered here, this means that the feedback, and hence the steep switching is completely lost. This point may become problematic if scaled, GAA nanowire representations of a Z^2-FET are considered.

9.2.5 Negative Capacitance FETs

In recent years, a lot of research has been devoted to the exploration of so-called negative capacitance field-effect transistors (NC-FETs). The idea behind NC-FETs is the following. In Section 5.2.2, the off-state of a conventional MOSFET has been introduced, the inverse subthreshold slope was calculated and Equation (5.23) has been obtained, which is reproduced below for convenience:

$$S = \frac{k_B T}{|e|} \ln(10) \cdot \frac{\partial \Phi_g}{\partial \Phi_f^0} = \frac{k_B T}{|e|} \ln(10) \left(\frac{C_{ox} + C_{semi}}{C_{ox}} \right), \tag{9.20}$$

where all capacitors related to charges in the semiconductor are summarized in C_{semi}. As a result, the term in parentheses is $1 + \frac{C_{semi}}{C_{ox}}$ and since C_{ox} and C_{semi} are both larger than zero, the parentheses is ≥ 1, and hence $S \geq 60$ mV/dec at room temperature. However, if one was able to make C_{ox} negative then $S < 60$ mV/dec would become possible. Such negative capacitances can be realized with ferroelectric gate dielectrics; the reader is referred to the large number of studies on the topic available in the literature.

Admittedly, I have neither sufficiently thought about nor worked on negative capacitance FETs in order to present a more comprehensive and thorough analysis. But if one translates the concept of negative capacitance FETs into a mechanical analogon, one obtains the schematics shown in Figure 9.27: the conduction band of a conventional (n-type) MOSFET where neither depletion charge nor interfaces charges play a role can be represented by a long rod (green line in the left panel) that is rotatably mounted (black circle) deep within the "substrate." Switching the transistor on and off requires

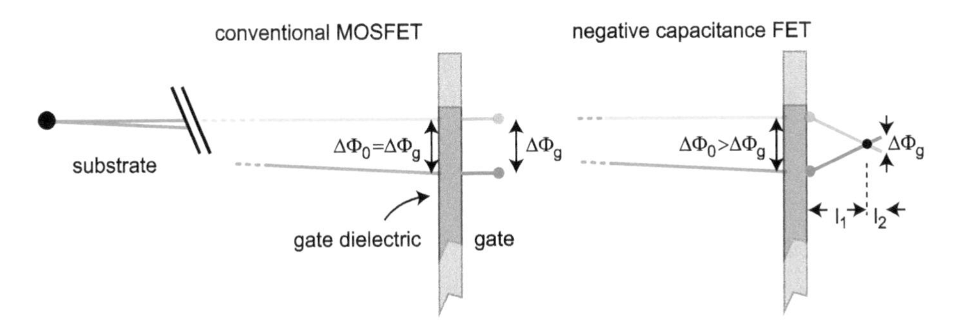

Figure 9.27: Mechanical analogons of a conventional MOSFET (left panel) and a negative capacitance FET (right panel).

a certain interval $\Delta\Phi_0$. If the gate dielectric is very thin, $\Delta\Phi_0 = \Delta\Phi_g$, i. e., the voltage interval (supply voltage) applied to the gate. On the other hand, $\Delta\Phi_0 > \Delta\Phi_g$ can be realized by inserting a joint in the red lever (black circle, right panel) ensuring that $l_1 > l_2$. While indeed a smaller gate voltage interval is required, there is more force needed to switch between on- and off-state. The bottom line is that in terms of energy needed to do cyclic switching of the transistors nothing is gained. NC-FETs are often realized with an additional, external capacitor with ferroelectric dielectric connected to the gate (or with a large area back-gate leading to a substantially larger area of the capacitance associated with the ferroelectric compared to the channel of the device). The charging and discharging of this capacitor provides the energy needed for the steep switching. In other embodiments of NC-FETs, the ferroelectric capacitor is of the same geometrical size and placed directly on top of the regular gate dielectric. These devices show hysteretic transfer characteristic (again, I do have to admit that I checked only parts of the literature) with a large hysteresis when a steeper slope is observed (and an increasing S toward 60 mV/dec for decreasing hysteresis). This means when cycling between on- and off-state between two fixed endpoints (in terms of off-state leakage and on-current) the energy consumed is the same for the hysteretic NC-FET device and a conventional MOSFET. This scenario is schematically shown in Figure 9.28.

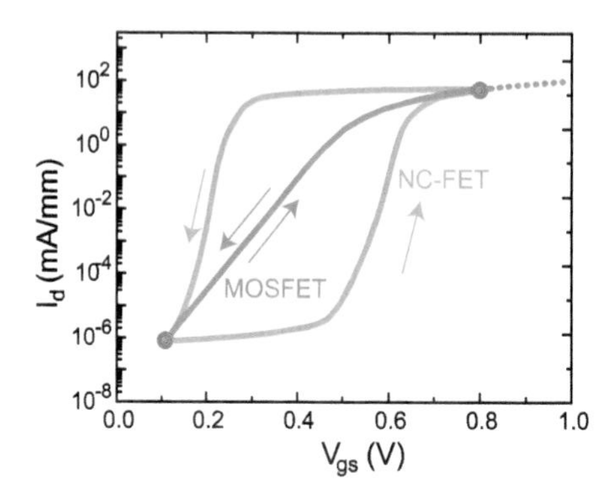

Figure 9.28: Illustration of transfer characteristics of a conventional MOSFET (green curve) and a NC-FET with hysteretic behavior. Cyclic switching between, e. g., the red points leads to the same energy consumption of the two devices.

9.2.6 Field-Effect Controlled Modulation of the Band Gap

In a regular FET, current modulation is obtained by shifting the band gap energetically in order to increase/decrease the injection of carriers into the channel. With perfect gate control, this leads to the 60 mV/dec limit, frequently discussed in this book. Now suppose we were able to shift the band gap energetically and at the same time modulate its magnitude, then a steep-slope transistor would be obtained showing MOSFET-like on-state behavior with $S < 60$ mV/dec.

In Section 2.8.2, it was shown that applying a vertical electric field allows creating and tuning a band gap in bilayer graphene in a certain energy range. Such a field-effect-controlled band gap appears highly attractive for a steep-slope transistor that combines band movement with band gap tuning. In the present section, we will therefore briefly study the suitability of bilayer graphene for steep-slope transistors via a tunable band gap. To this end, the electric field-dependence of the band gap calculated with tight-binding will be used that resulted in Equation (2.54); experimental investigations on bilayer graphene can be found in Sections 10.1.3 and 11.3.2.

In order to assess whether a steep slope, FET can be realized with bilayer graphene the most ideal case will be considered: It is assumed that the top and bottom gate capacitors to realize the vertical electric field (cf. Figure 2.33) have very high dielectric constants and are thin enough so that the potentials of the two graphene layers are the same as the potentials of the two gate electrodes. This means that $\Delta = \Phi_g^1 - \Phi_g^2$ and the band gap becomes $E_g = \frac{V_\perp(\Phi_g^1 - \Phi_g^2)}{\sqrt{V_\perp^2 + (\Phi_g^1 - \Phi_g^2)^2}}$. Now, the following two limiting cases are possible. First, Φ_g^1 is kept constant while Φ_g^2 is changed (shown in Figure 9.29(a)) and second, Φ_g^1 and Φ_g^2 are changed simultaneously with opposite sign (Figure 9.29(b)); any other case will be a mixture of these two.

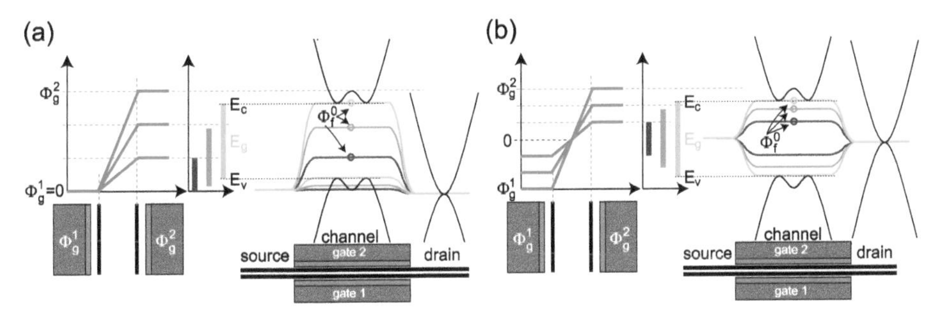

Figure 9.29: Modulation of E_g in bilayer graphene. The potential across the bilayer graphene is shown in the left part of each panel. E_g and its energetic position is shown in the center and right. In (a), the bottom gate remains constant (here $\Phi_g^1 = 0$), while gate 2 is used to move the bands and tune E_g. In (b), opposite gate voltages are applied to gates 1 and 2.

Let us consider an n-type FET, and thus the energetic position of the conduction band maximum Φ_f^0 (see Figure 9.29) determines the carrier injection. In both cases, (a) and (b), the maximum Φ_f^0 is determined by Φ_g^2. With $\Delta = \Phi_g^2 - \Phi_g^1 = \Phi_g^2$ and $\Phi_f^0 = E_g/2 + \Delta/2$ one can compute $\partial\Phi_f^0/\partial\Phi_g^2$ needed to calculate the inverse subthreshold slope of such a transistor (cf. Equation (5.23)). With Equation (2.54), the limit for small gate voltages becomes $\partial\Phi_f^0/\partial\Phi_g^2 \to 1/2 + 1/2$, whereas for large gate voltages $\partial\Phi_f^0/\partial\Phi_g^2 \to 1/2$.

As a result, the minimum inverse subthreshold slope one obtains with a tunable band gap in bilayer graphene in the device configuration displayed in Figure 9.29 is $S_{\mathrm{min}} =$ 60 mV/dec approaching 120 mV/dec for larger Φ_g^2. The reason is that E_g does not increase beyond the potential Φ_g^2, as illustrated in Figure 9.29(a) with the (dark/light)green bars. The same is true in the case depicted in (b) and any other combination of the two cases. What has been neglected so far is the contribution from holes that are injected into the valence band. In case (a), a rather high hole leakage is expected prohibiting a proper off-state behavior. However, even if we neglect this leakage component, or if the device is switched as shown in Figure 9.29(b), it is apparent that a FET based on tuning the band gap of bilayer graphene cannot exhibit an inverse subthreshold slope steeper than 60 mV/dec.

An alternative way to exploit the tunable band gap of bilayer graphene is to use mechanical strain in order to modify $V_\perp$ instead of Δ (see Equation (2.54)). $V_\perp$ stems from the van der Waals interaction and as such depends strongly on the distance between the two layers. Increasing $V_\perp$ allows larger band gaps to occur, and thus potentially a steep-slope device to be realized [58].

Exercises

Exercises together with solutions are accessible via the QR code.

10 Device Based on Two-Dimensional Materials

In Chapter 5, it was discussed that an ultrathin channel layer is required in order to preserve electrostatic integrity in MOSFET devices that are downscaled to very short channel lengths. In principle, gate-all-around nanowire/tube devices enable the realization of the smallest screening lengths λ_{ch}. In particular, carbon nanotubes are ideally suited for this purpose since they do not only represent nanostructures with a diameter in the $1, \ldots, 2$ nm range with suitable band gaps. The electronic transport in any nanostructure of similar dimensions etched out of a volume material would be significantly impaired with strong variability and degradation of mobility. A nanotube, on the other hand, is a close to perfect object with periodic boundary conditions along the circumference and a weak interaction with neighboring materials due to a radially quickly decaying wavefunction. However, although there has recently been tremendous progress and a 16bit carbon nanotube processor has been demonstrated [106], from a technological point of view, fabricating billions of nanotube devices with the same dimensions, well-defined electrical properties and at predefined locations is extremely challenging.

Initiated by the first demonstration and investigation of monolayers of graphene, the so-called 2D materials have gained tremendous attention since they offer, down to a single atomic layer, the thinnest conceivable electronic material featuring carrier mobilities not found in other two-dimensional systems (such as a two-dimensional electron gas in III–V heterostructures, for instance) of similar dimension and a great variety of different electronic properties including metallic, semiconducting, insulating and even superconducting behavior. Furthermore, the two-dimensionality facilitates the use of highly sophisticated top-down fabrication techniques for the realization of appropriate devices and circuits. From a nanoelectronics point of view, the mentioned properties make 2D materials highly attractive. Especially the combination of different 2D materials into so-called van der Waals heterostructures holds promise to realize ultracompact devices and to engineer novel materials by appropriate combinations of 2D materials. The latter is particularly appealing since van der Waals heterostructures are generated by stacking crystalline materials on top of each other without the need for epitaxy. This holds great promise for, e. g., next generation flexible electronics, multifunctional nanoelectronics, 3D integration, etc.

The present chapter provides an introduction to 2D materials for nanoelectronics and discusses some of the properties of and devices based on 2D materials.

10.1 Graphene FETs

Graphene was the first 2D material to be intensively studied [203, 82] and caused a downright hype with worldwide interest in the material. With a thickness of ~3 Å, it allows for electrostatic integrity of field-effect transistors down to smallest dimensions. In addition, graphene exhibits potentially a very large carrier mobility facilitating field-effect

https://doi.org/10.1515/9783111054421-010

transistors whose performance can be optimized through decreasing L *and* increasing μ (cf. Equation (5.31)). First MOSFET devices were fabricated and characterized [168]. However, severe ambipolar behavior and very low on/off-current ratios were observed. This is not surprising considering the band structure of graphene. As already discussed in Section 2.8.1, a linear dispersion is obtained that exhibits two cones for the conduction and valence bands that meet at the Dirac point (see Figure 2.32). Due to the honeycomb lattice, the density of states vanishes at the Dirac point (Figure 2.50), which is why graphene is called a zero-gap semiconductor. This means that the current modulation as a function of V_{gs} observed in graphene field-effect transistors (provided that a very small bias is applied) stems merely from the linear energy-dependence of the density of states. Figure 10.1 shows a typical transfer characteristic of a back-gated graphene field-effect transistor. Schematics of the conduction and valence band cones with respect to the Fermi level are depicted in the insets. When the gate voltage moves the Dirac point through the source/drain Fermi levels, the current becomes minimal, because of the decreased density of states. Due to the thermal broadening of the Fermi distribution and nonideal features such as substantial potential fluctuations (see below), a very small on/off-current ratio is obtained.

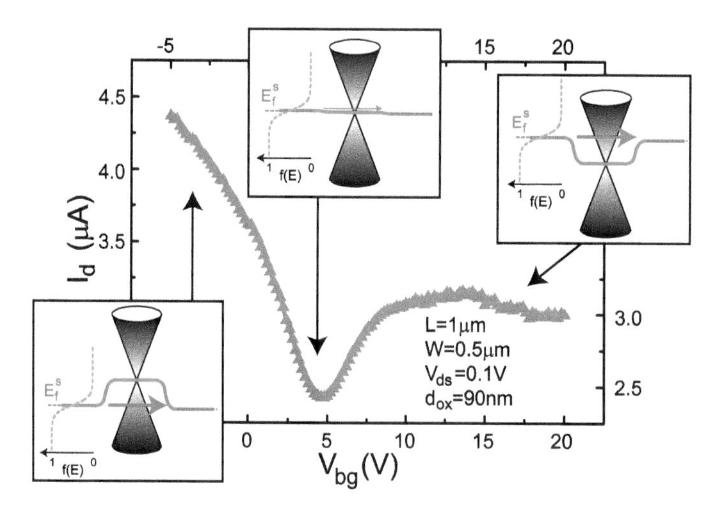

Figure 10.1: Transfer characteristics of a back-gated graphene FET with Ti/Au contacts. The insets show the cones of the graphene band structure at the respective back-gate voltage.

The lack of a band gap is one of the major issues of graphene considering nanoelectronics applications. A band gap can in principle be created either with a nanoribbon or with bilayer graphene (cf. Sections 2.8.5 and 2.8.2), which will be discussed further below. Looking at the I_d–V_{gs}-characteristics in Figure 10.1 an asymmetry is obvious, which is not necessarily expected from the idea of carrier injection into the band structure as depicted in the insets of Figure 10.1. The origin of this asymmetry is the way the contacts are made in the device. Due to the importance of contacts, this point will be discussed first.

10.1.1 Graphene FETs—Contacts

Contacts to 2D materials are often made by simply depositing an appropriate metal on top of a portion of the material. Figure 10.2(a) shows an electron micrograph of a typical graphene field-effect transistor (GFET) [139]. The contacts resemble the side contacts studied in carbon nanotube transistors in Chapter 8. Indeed, as will be discussed below, deposited metal-graphene contacts behave in a very similar way as metal-nanotube contacts and can therefore be described in the same fashion. One of the central results in metal-nanotube contacts is the existence of an extremely thin but high potential barrier in between metal and nanotube that leads to a weak coupling (and hence depins the Fermi level) between the two materials.

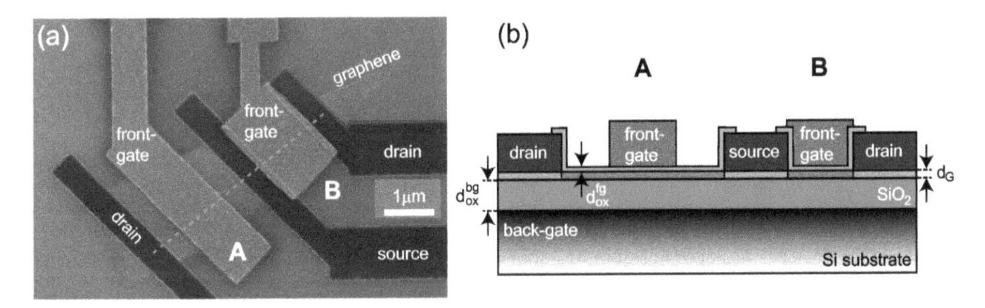

Figure 10.2: (a) Electron micrograph of the graphene device structure under investigation. A cross-section of the devices of type A and B is shown in (b) [39, 139].

The coupling strength between nanotube and metal could be studied by simulating CNTFETs with different contact metals and coupling strengths and comparing this with experimental transfer characteristics. The strong variations in on-state current particularly for metals such as aluminum facilitated a robust comparison with excellent agreement between experiment and theory (cf. Section 8.1.3). In a GFET, such a scheme cannot be employed and it is much less obvious how relevant information about the metal-graphene coupling can be obtained. However, a dual-gate graphene field-effect transistor with front- and back-gate electrodes can be employed to answer this question. Such a device is displayed in Figure 10.2: (a) shows a colored electron micrograph and (b) a schematic cross-section of such a GFET. The GFETs are realized by exfoliation of a monolayer graphene onto a heavily doped silicon substrate with 300 nm thermally grown SiO_2 (facilitating proper visibility with optical microscopy; see Section 3.13.3). The graphene is contacted using Ti/Pd/Au (0.5 nm/20 nm/20 nm) electrodes followed by the deposition of a 10 nm Al_2O_3 employing ALD. The devices are finalized with the formation of a Ti/Au(1 nm/40 nm) front-gate electrode (see [39] for details). Two different types of devices are fabricated, called A and B (Figure 10.2). In the following, only the device of type B will be considered with the source, gate and drain electrodes being as close to each other as possible [39, 139].

Information on the metal-graphene interface can be obtained by operating the device in the following way: A constant front-gate voltage V_{fg} is applied and the current is recorded as a function of the back-gate voltage V_{bg}. A rather small bias (here $V_{ds} = 0.01$ V, which is smaller than $k_B T/|e|$ at room temperature) is applied such that the device behavior can be described assuming a single Fermi level. Figure 10.3, left panel, shows the measured resistance versus back-gate sweeps for several front-gate voltages.

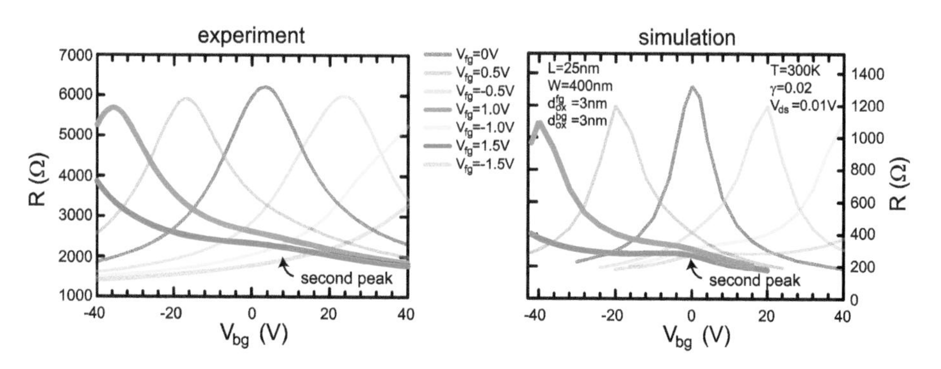

Figure 10.3: Resistance versus back-gate voltage for several V_{fg} of a dual-gate GFET. Left panel: Experimental data of the device B displayed in Figure 10.2. Right panel: Theoretical data obtained with self-consistent NEGF simulations with appropriate side-contacts [139].

For small V_{fg}, the dependence of I_d on V_{bg} shows the resistance peak (i. e., drop in current) discussed above (e. g., gray curve in Figure 10.3, left). However, for larger front-gate voltages, a second, shallow resistance peak is observed (green and red curves in Figure 10.3, left) that allows for investigating the metal contact properties. The reason for the second resistance peak is that the front-gate voltage enables separating the Dirac points in the channel area and underneath the source/drain contacts. Since the conduction and valence bands in graphene are not pinned by the metal, sweeping the back-gate voltage (at constant V_{fg}) moves the cones in source/drain and in the channel through the Fermi level. For sufficiently large (energetic) separation of the Dirac points, i. e., for sufficiently large V_{fg}, the Dirac points of the channel and of source/drain cross the Fermi level at distinctly different back-gate voltages, leading to two observable resistance peaks. While the large resistance peak stems from the graphene band structure in the channel, the second, shallow peak is due to the band structure in source and drain. Figure 10.4 shows this situation schematically: in (a) the two resistance peaks overlap (black and red dashed lines), while in (b) the large V_{fg} separates them. The fact that the second peak is substantially broader and the peak height significantly less pronounced compared to the first resistance peak suggests a metal-induced modification of the density of states in the graphene.

To study the contact properties in more detail, self-consistent NEGF simulations of the dual-gate GFET (type B in Figure 10.2) using the same simulation tool already applied

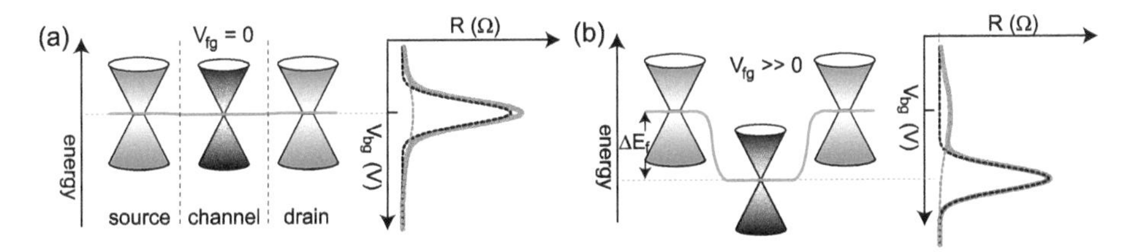

Figure 10.4: Schematic illustration of resistance versus back-gate voltage characteristics for $V_{fg} = 0$ (a) and a large front-gate voltage (b). The overall resistance peak is a superposition of the first peak (black dashed line) and the second peak (red dashed line).

successfully to describe metal-CNT contacts (see Section 8.1.3 and Chapter 6) are carried out. Due to the dual-gate device structure, the 1D modified Poisson equation (5.34) is ideally suited to describe the electrostatics of the GFET. An independent mode approach is employed as illustrated in Figure 10.6 (cf. Section 5.2) in order to account for the 2D graphene flake. An energy-dependent effective mass (see Sections 2.5.2 and 6.3.4) is used that reproduces the band structure of each 1D subband (including linearity of the dispersion for higher energies and the complex band structure within the energy gap) appropriately. The metal-graphene contacts are described with Buettiker probes within the contact area. All Buettiker probes within one contact have a common Fermi level given by the terminal voltage and are coupled to each subband. The metal-graphene coupling strength is again mediated by the coupling constant $\gamma = 0,\ldots,1$ as described in Section 8.1.3. In order to keep the computational burden as small as possible, we simulate GFETs with a channel length of $L = 25\,\text{nm}$, equal front- and back-gate dielectric (SiO_2) thicknesses of $d_{ox} = 3\,\text{nm}$ and a width of the device of $W = 400\,\text{nm}$ resulting in two-hundred 1D subbands that are considered in the simulations. In addition, the thickness of the graphene layer d_G and the metal-graphene separation are both taken to be 3 Å; finally, room-temperature conditions and ballistic transport are assumed in all simulations.

Figure 10.5 shows two color plots of the local density of states along the device. Because of the metal-graphene coupling (in the present case $\gamma = 0.05$ was chosen to better illustrate the effect) in the contact areas the density of states is modified and does not vanish anymore at the Fermi level (red line in Figure 10.5, left). In contrast, in the channel the DOS is unperturbed exhibiting the well-known linear behavior (black line in Figure 10.5, right). The simulations therefore reconfirm the assumption made when explaining the appearance of the two resistance peaks.

Simulations of the resistance versus back-gate for several front-gate voltages were carried out as a function of the coupling strength γ [139]. A strength of $\gamma = 0.01,\ldots,0.02$ was found to reproduce best the experimental observations. Figure 10.3(b) displays the simulated data showing excellent qualitative agreement with the experimental data in (a). The analysis shows that the metal-graphene coupling is rather weak resulting merely in a slight modification of the DOS of graphene underneath the contact (compare with

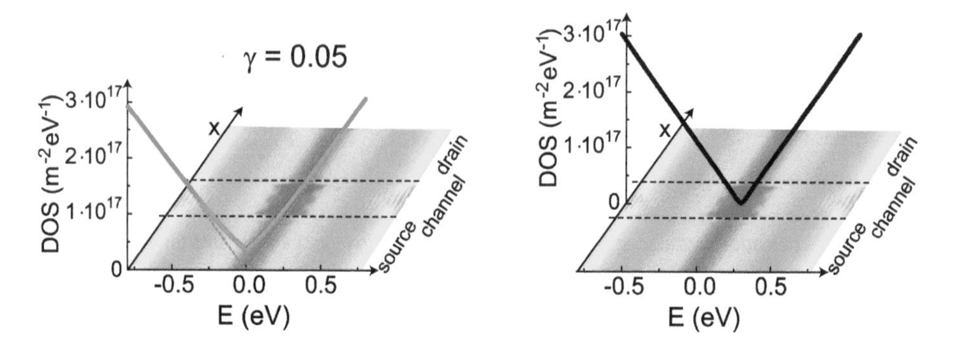

Figure 10.5: Local density of states plots in the contact and channel region. The metal-graphene coupling yields a modified density of states within the contacts that is responsible for the second resistance peak.

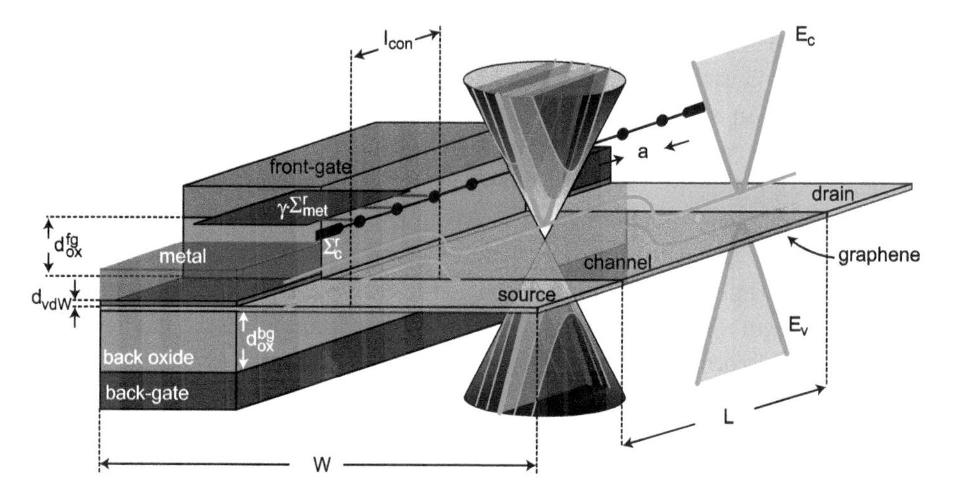

Figure 10.6: Schematics of the simulated device structure. An independent 1D-mode approach with an energy-dependent effective mass is employed as illustrated by the five intersections between the graphene cone and the planes of constant k-values quantized along the direction of W [139].

Figure 10.5). This is expected since there are hardly any surface sites the metal can bond to. While this is the reason why graphene shows excellent electronic transport properties, it limits the achievable contact resistivities at the same time. The latter can be circumvented by using end-bonded or edge contacts (similar to carbon nanotubes) [267]. Such contacts are realized by encapsulating the graphene (for instance in hexagonal boron nitride) and then etch through the entire material stack; Figure 10.7 shows a schematic of the two contact geometries. Doing so carbon–carbon bonds are cut, the bonding orbitals are exposed and form contacts to a metal along the exposed edge. Excellent contact properties have been demonstrated with this method [267]. While in carbon nanotubes, side-contacts definitely have their merits since they allow one to suppress an ambipolar operation, which may lead to severe leakage (cf. Figure 7.11), this argument is irrelevant in graphene.

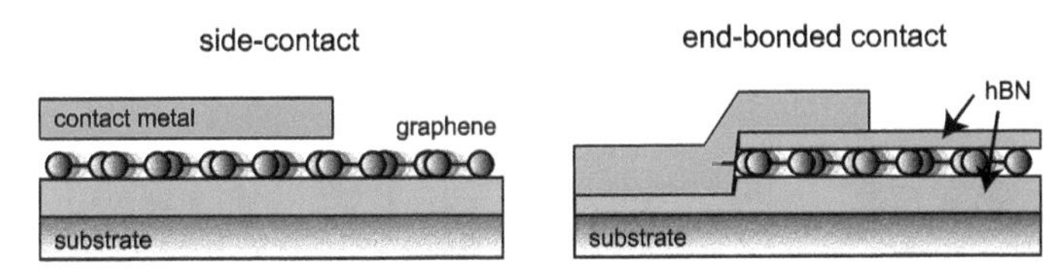

Figure 10.7: Side- and end-bonded contact geometries in metal-graphene contacts.

10.1.2 Graphene Nanoribbon FETs

One way to generate a band gap in graphene is to exploit quantization due to carrier confinement by forming a nanoribbon of appropriate width. In Section 2.8.5, we already computed the band structure of armchair graphene nanoribbons (AGNRs) and found that an extremely small width of approximately $1, \ldots, 2$ nm is required to obtain an E_g in a useful range (cf. Figure 2.38). Manufacturing such an AGNR with lithography and etching is close to impossible. It is not only the extremely small width that is problematic but one also needs to align the etch mask perfectly with respect to the graphene lattice. Moreover, the slightest line edge roughness at these dimensions will have a strong impact on the nanoribbon, resulting in significant variability due to fluctuations of the band gap. In a carbon nanotube, rolling up a nanoribbon yields perfect periodic boundary conditions along the circumference of the tube. In contrast, in a graphene nanoribbon carbon atoms at the edges exhibit dangling bonds resulting in edge scattering [280] and conductivity fluctuations [195, 41]. And even if one was able to saturate the edge dangling bonds with hydrogen, a line edge roughness of the etch mask would lead to variations of the band gap [70]. On the other hand, chemically derived nanoribbons [170] showed semiconducting behavior with appropriate band gaps that enabled the fabrication of excellently performing transistors device [268]. However, these ribbons are processed from a solution and thus the benefit over carbon nanotubes, namely the possibility of a deterministic fabrication technology is lost.

In addition to fluctuations due to the patterning of nanoribbons, the substrate plays a very important role (see also Section 11.3.2 on cryogenic bilayer graphene FETs). Substrate-induced structural distortions or charged impurities in the substrate and chemical doping due to resist residues lead to strong potential fluctuations. These fluctuations result in the formation of charge "puddles" in graphene, i. e., separated sections with positive or negative charge that result in a strong reduction of the carrier mobility. As a result, a mobility gap forms that becomes apparent when low-temperature measurements with such a graphene sample are carried out [79, 238].

As an experimental example of a graphene nanoribbon device, Figure 10.8(a) shows an electron micrograph of a nanoribbon with ~30 nm width on top of a substrate featuring three independently addressable gate electrodes. The three gate electrodes offer individual control over the electronic properties of the respective sections of the nanoribbon via gate-controlled doping. Such buried triple-gate (BTG) substrates are fabricated

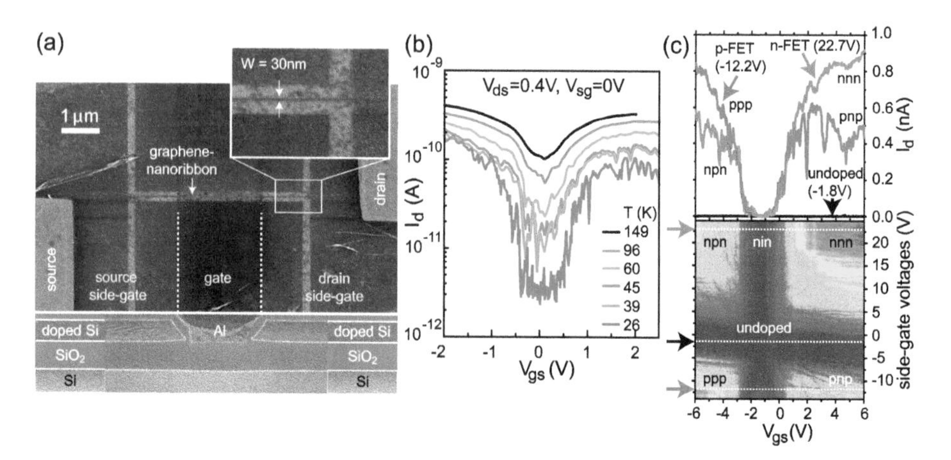

Figure 10.8: (a) Combination of top view and cross-sectional electron micrograph of a graphene nanoribbon with a width of ~30 nm on top of a buried triple-gate substrate. (b) Temperature-dependent measurements at side-gate voltages of $V_{\text{side-gate}}^{\text{source/drain}} = 0$ V. (c) Transfer characteristics of the graphene nanoribbon FET shown in (a) for various, equal side-gate voltages (bottom panel). The top panel shows exemplarily the transfer characteristics for $V_{\text{side-gate}}^{\text{source/drain}} = -12.2$ V (red), −1.8 V (black) and 22.7 V (blue) [197].

in the following way (for details on the fabrication techniques see Chapter 3): a highly doped SOI layer is etched anisotropically using TMAH to form two electrically insulated SOI regions that serve as source and drain side gates. Subsequently, an SiO_2 layer (100 nm in thickness) is grown by wet thermal oxidation followed by a sputter deposition of aluminum. After chemical-mechanical polishing (see Figure 3.51), the aluminum is oxidized to form ~5 nm of Al_2O_3 as gate dielectric. Finally, the three gates are contacted [197]. A graphene flake is transferred onto the BTG substrate using the PVA/PMMA method as described in Section 3.13.4. E-beam lithography and etching in an oxygen/argon plasma are then used to form the nanoribbon, which is displayed in the inset of Figure 10.8(a).

Temperature-dependent measurements of graphene nanoribbon devices on BTG substrates are carried out as displayed in Figure 10.8(b). Decreasing the temperature leads to substantial current fluctuations, which are associated to potential fluctuations. This is consistent with an on/off-current ratio of only ~100 at $T = 26$ K, while the off-state regime is found in a V_{gs}-range between −0.5 and 0.5 V. This range is too large for the small on-/off-ratio observed and also does not fit with the size of the semiconducting gap expected at a nanoribbon width of 30 nm. Therefore, the off-current range clearly shows a transport gap [99]. The lower panel of Figure 10.8(c) shows a two-dimensional plot of measurements at 25 K applying different side-gate voltages; exemplarily, the top panel of Figure 10.8(c) shows transfer characteristics for $V_{\text{side-gate}}^{\text{source/drain}} = -12.2$ V (red line), −1.8 V (black line) and 22.7 V (blue line), respectively. One observes that by applying appropriate side-gate voltages a device with rather n-type character (blue line), p-type character (red line) and a device without significant current flow due to the source/drain regions being within the transport gap (black line) can be realized. While fluctuations due to roughness and dangling bonds at the nanoribbon edge are difficult to mitigate, substrate-induced potential fluctuations can to a large extent be avoided by encapsulating the graphene into hexagonal boron nitride and by using graphitic gate electrodes (see Section 11.3.2) [63, 113].

10.1.3 Bilayer Graphene

A band gap in graphene can also be generated using bilayer graphene. If the bilayer is placed in-between two gate electrodes (see Figure 2.33) and a vertical electric field is applied the Dirac points of the two graphene layers are shifted by Δ with respect to each other and a band gap of size $E_g = \frac{\Delta V_\perp}{\sqrt{\Delta^2+V_\perp^2}}$ is generated where $V_\perp \approx 0.3\,\text{eV}$ is the overlap integral of the van der Waals interaction (cf. Section 2.8.2).

Figure 10.9(a) shows transfer characteristics of a dual-gate bilayer graphene FET [48]; the inset shows an electron micrograph of the device. To fabricate such a device, a bilayer is transferred to an oxidized silicon wafer that serves as large area back-gate. The SiO_2 has a thickness of 90 nm ensuring good visibility of the flakes with optical microscopy (cf. Section 3.13.3). On top of the bilayer, a 10 nm thick HfO_2 serves as the gate dielectric for the top-gate electrode [48]. Applying various constant back-gate voltages, the drain current is measured as a function of top-gate voltage at a low bias of $V_{ds} = 10\,\text{mV}$ (see Figure 10.9(a)). Obviously, the off-current, and hence the on/off-current ratio increases with increasing back-gate voltage. With a constant back-gate voltage, the gap will slightly change during the front-gate sweep. Nevertheless, the minimum current is obtained when the created band gap is shifted such that $E_g/2 - |e|V_{ds}/2$ blocks the injection of electrons and holes (see inset of Figure 10.9(b)).

Task 33.
Band gap in bilayer graphene: Based on a simple Landauer approach of current transport in a 1D channel compute the ambipolar current flow due to electron and hole injection over the band gap E_g in bilayer graphene. To this end, you may assume that the Fermi levels in source and drain $E_f^{s,d}$ are fixed (and separated by a bias of $V_{ds} = 10\,\text{mV}$) and the band gap is moved by applying appropriate gate voltages with respect to E_f^s and E_f^d. Next, compute the on/off-current ratio as a function of band gap and compare the calculation with the on/off-current ratios obtained in bilayer graphene.

In order to determine the band gap that has been created with the vertical field, a simple 1D Landauer model can be used as illustrated in the inset of Figure 10.9(b). Although the model is one-dimensional, comparing only the on/off-current ratios with the experimental data, it allows determining the band gap for electronic transport; see Task 33 for details. This band gap E_g is plotted in Figure 10.9(b) as a function of the difference between top- and back-gate voltages showing that band gaps of up to approximately 250 mV can be generated. The green line in (b) depicts the expectation from the ana-

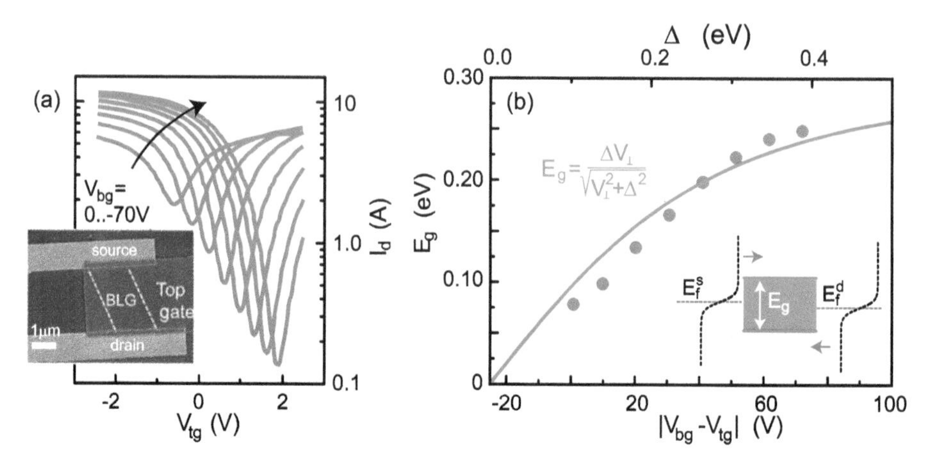

Figure 10.9: (a) Transfer characteristics at V_{ds} = 10 mV of a dual-gate bilayer graphene FET (shown in the inset) as a function of the top-gate voltage for different back-gate voltages (Z. Chen, Purdue University) [48]. (b) Extracted band gaps (blue data points) obtained from comparing the on-/off-current ratio with expectations from a simple Landauer model (illustrated in the inset). The green line belongs to the analytic expectation of E_g as a function of Δ, the energetic separation of the Dirac points of the bilayer.

lytic tight-binding calculation (cf. Section 2.8.2). Comparing the blue data points and the green line, enables an extraction of Δ and with this of the internal electric field generated between the two graphene layers. In any case, the gaps that can be created with the vertical electric field in bilayer graphene are at best ~250 meV and, therefore, too small for logic applications. Nevertheless, the ability of a field-effect tunable band gap is appealing, in particular in the framework of novel device functionalities (see the discussion in Section 9.2.6).

10.2 Transition Metal Dichalcogenides

In the preceding section, it became clear that creating a band gap sufficiently large to ensure proper switching in logic devices is difficult to achieve in graphene. Luckily, there are many other 2D materials that exhibit band gaps in a suitable range. While there is actually a whole bunch of different 2D materials with promising properties, the present section will only deal with the most widely investigated materials, namely transition metal dichalcogenides (TMDCs). The material properties and the band structure of some selected TMDCs have already been discussed in Section 2.8.3. Obviously, with monolayer thicknesses of ~6 Å and band gaps in the range 1.2–1.8 eV they appear to be ideally suited for ultimately small scaled transistor devices. Moreover, TMDCs exhibit relatively large effective masses, which at first sight may appear as a drawback. However, this is not necessarily the case. As was discussed in Section 5.9.4, direct source tunneling strongly increases with decreasing effective mass, thereby severely limiting the minimum possible channel lengths necessary for proper device functionality. Therefore, the relatively large effective masses in TMDCs are actually beneficial in terms of scalability.

One of the main issues of TMDC devices is the realization of appropriate, low-ohmic contacts [175]. Usually, strong Fermi level pinning is observed at metal-TMDC interfaces giving rise to a substantial Schottky barrier at the interface between metal contact and TMDC, which severely limits the performance of devices. Considering the fact that there are no dangling bonds at the surface of a TMDC this might be surprising at first sight.

Figure 10.10 shows different ways of contacting TMDCs. In most cases, contacts are actually realized by simply depositing them on top of the TMDC (Figure 10.10(a) and (b)), similar to contacting carbon nanotubes or graphene. Recently, also end-bonded (c) and edge-bonded (d) contacts with graphene have been demonstrated [47, 95]. While this looks like many different ways of contacting, also considering the details of the metal-TMDC interface [124], it basically boils down to two different contacts and these are exactly the same ones that were already discussed in earlier chapters. First, when covalent bonding between the metal and the TMDC occurs (as is the case for Ti and Mo, see (a)) one basically obtains an intimate metal-semiconductor contact. Since the bond length between metal and chalcogen is very short, no real difference between contact scheme (a) and the end-bonded contacts in (c) is expected [96]. The strong Fermi level pinning is simply due to metal-induced gap states as has been discussed in Section 4.6.1. Indeed, in Figure 4.21 the pinning factor S_{MIGS} is plotted as a function of $1/E_g^2$ for several different semiconductors. The red marked points belong to MoS_2 and $MoTe_2$ and they fit very well into the MIGS picture as has recently been noted in [96]. This case is basically what has been termed "strong coupling" in Figure 8.7(a) and although the model presented in Section 8.1.3.1 may appear to be too simplistic for the metal-TMDC case, it reflects the actual situation: in both cases, end-bonded and (deposited) covalently bonded contacts, Fermi level pinning within the band gap is obtained giving rise to a Schottky bar-

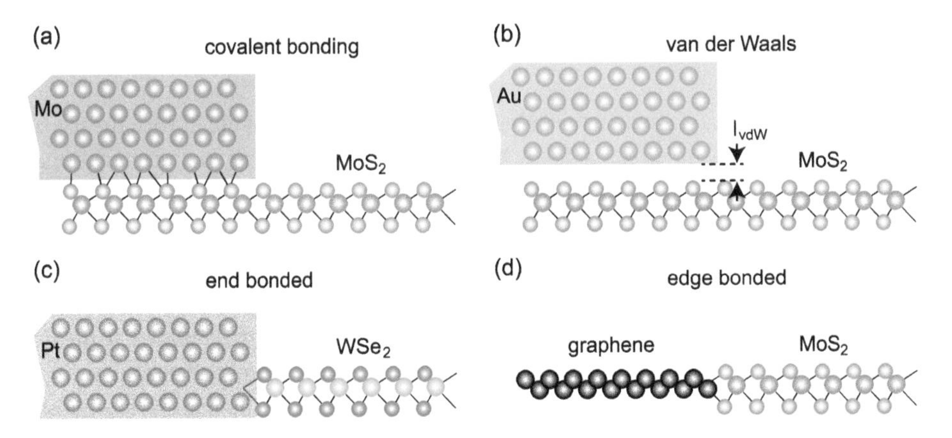

Figure 10.10: Different ways of contacting a TMDC monolayer. (a) Top contact with covalent bonding (Mo to MoS_2), (b) side-contact with a metal a van der Waals distance away from the TMDC, (c) end-bonded contact (Pt to WSe_2) and (d) edge-contact between graphene and MoS_2 [232, 47, 95].

rier through which carriers are injected into the TMDC in a device. The second contact scheme is the one depicted in Figure 10.10(b) where there is a van der Waals distance between metal and TMDC. This contact is equivalent to the side-contact model discussed in Section 8.1.3.1 in the case of weak coupling. Due to the van der Waals distance, there is a substantial potential barrier in between metal and TMDC and this potential barrier leads to a strong suppression of MIGS. In addition, since there is no bonding between metal and TMDC and the TMDC has no dangling bonds it is expected that Fermi level pinning is suppressed (as has been the case in carbon nanotubes; see Section 8.1.3.1). Why is this usually not observed in experiments? Because the deposition of the metal leads to defects [176] within the TMDC. These defects in turn lead to an additional density of defect states that results in strong Fermi level pinning. Interestingly, by laminating atomically flat metal electrodes onto a TMDC instead of depositing, a recent study was able to show that Fermi level pinning in van der Waals metal-TMDC contacts can be avoided [176]. Due to the lamination process, the defect density of states as well as the density of metal-induced gap states could be strongly suppressed. As a result, the Schottky–Mott limit was obtained, i. e., Fermi level pinning was suppressed. However, depinned van der Waals contacts are only possible due to the potential barrier in between metal and TMDC and this does not necessarily yield lower contact resistances.

In real experimental situations, a mixture of the two limiting contact situations including defects will be present. Furthermore, contact is often made to a multilayer of TMDC and in this case the injected current is distributed among the different layers because of the weak coupling in between them [275, 232]. The bottom line is that one usually has to deal with substantial Schottky barriers at the metal-TMDC interface. Devices built from TMDCs therefore show SB-MOSFET behavior with the features discussed in detail in Chapter 7.

10.2.1 Reconfigurable TMDC-FETs with Triple-Gate Structures

The impact of contact details on the behavior of a device can be avoided with the use of electrostatic doping, which has already been used in Section 8.2, to study conventional *p-i-p* CNTFETs. In addition, reconfigurable devices can be realized with additional gate electrodes. Reconfigurable devices have already been discussed in Section 7.4 (based on silicon nanowires) and 8.1.4 (using carbon nanotubes). In these two cases, the investigated transistors exhibit only two independently addressable gate electrodes, which allows adjusting *n*- and *p*-type device behavior. It would certainly be highly attractive if one was able to switch a device not only between *n*- and *p*-behavior but also to work as a band-to-band tunnel FET. This would allow using the high on-state performance of conventional transistor operation and combine this—whenever possible and/or necessary—with the low power operation of TFETs. The drawback is of course that a third gate electrode is necessary in order to realize this.

In Chapter 9, it was discussed that TFET performance can be strongly improved when the screening lengths λ_{ch} and λ_{dop} are made small in the source contact and the channel. Furthermore, the trade-off between the required high doping concentration in the source and a loss of the steepness of the inverse subthreshold slope was investigated, too. One result was that gate electrodes instead of conventional doping may be used to disentangle the screening from the necessity to dope the source region (provided the screening length λ_s in source can be made small enough without inducing a large carrier density, see p. 374). Moreover, for a reconfigurable device, a symmetric injection of electrons and holes is necessary which is obtained when the Fermi level at the metal-semiconductor interface is pinned around mid-gap (cf. Section 7.4). For these reasons, a device architecture featuring three independent gate electrodes with thin gate dielectric and using WSe_2 as ultrathin channel layer, with Fermi level pinning approximately at mid-gap, is fabricated and studied. To this end, buried triple-gate substrates are manufactured as illustrated in Figure 10.11(a) with a slightly modified approach compared to the one presented in Section 10.1.2.

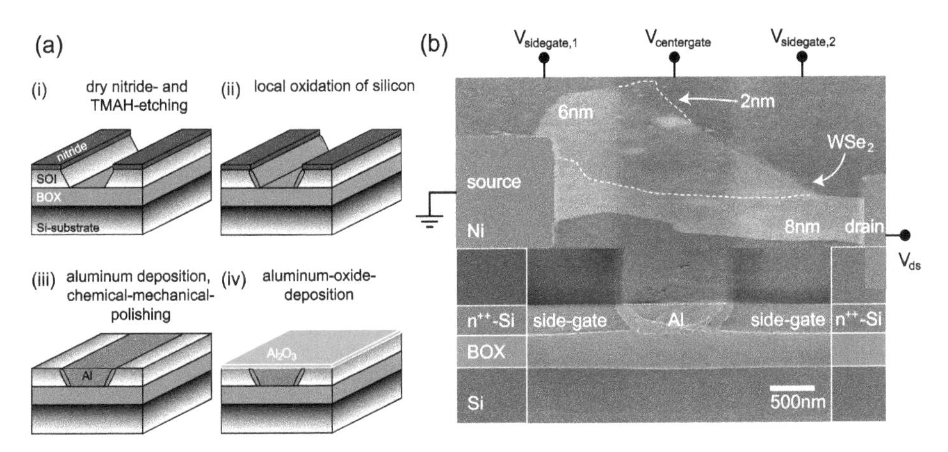

Figure 10.11: (a) Fabrication sequence for buried triple-gate substrates using anisotropic Si etching, local oxidation of silicon and an Al damascene process. (b) Combined cross-sectional SEM and AFM image of a WSe_2-flake on a BTG substrate [198].

Silicon-on-insulator with a 340 nm (100) top silicon layer is degenerately doped with phosphorous. Subsequently, 200 nm of Si_3N_4 is deposited with PE-CVD. Optical lithography and reactive ion etching in a CHF_3/O_2 plasma are employed to etch line patterns into the nitride layer. After photoresist removal and a short dip in buffered oxide etch, anisotropic silicon etching with TMAH is carried out to form a V-groove in the SOI layer. Next, the wafer is oxidized with wet thermal oxidation. Since the wafer surface is covered with nitride, a local-oxidation of silicon occurs only at the Si (111)-planes, exposed during the anisotropic silicon etching (see Figure 3.4(a)). After the removal of the Si_3N_4 with hot phosphoric/sulfuric acid, aluminum is sputtered onto the sample followed by chemical-mechanical polishing. Finally, a thin Al_2O_3-layer (~7 nm) is deposited with ALD serving as a gate dielectric [198]; details on all process steps can be found in Chapter 3.

WSe_2 is transferred to the BTG substrates by simple exfoliation with blue adhesive tape. Optical microscopy can then be used to identify thin flakes with a thickness of less than 10 nm (cf. Figure 3.62) lying across the aluminum gate. Afterwards, electron-beam lithography is used to pattern source/drain contact structures into PMMA. Subsequently, deposition of 90 nm nickel and lift-off are employed to obtain the contact structures and complete the devices. Figure 10.2(b) displays a combined scanning electron and atomic force microscopy image showing a thin (~6–8 nm) WSe_2 on top of the BTG substrates. All three gates can be used to manipulate the WSe_2.

Figure 10.12 displays transfer characteristics of a WSe_2 with a thickness of d_{WSe_2} = 3 nm determined by AFM for three operating modes. In the case of the *p*- and *n*-type FETs, a constant drain-source bias of V_{ds} = 1 V is applied and transfer characteristics are recorded for several side-gate voltages (in the case of *p*- and *n*-type devices, both side-gate voltages are equal, positive or negative). Note that due a positive V_{ds} = 1 V in the case of a *p*-type FET, holes are injected from drain (i. e., right contact in Figure 10.11(b)), which now plays the role of the source contact. One clearly observes the expected unipolar device behavior and a strong increase of the on-current with increas-

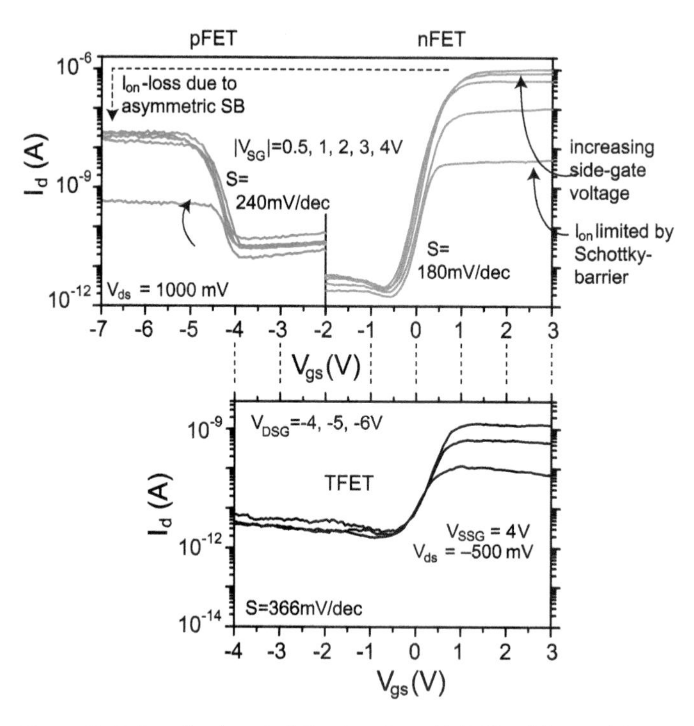

Figure 10.12: Transfer characteristics of a ~3 nm thick WSe_2-flake. Applying appropriate voltages at the gate electrodes of the triple-gate substrates allows for operation of the device as *n*-type, *p*-type and as TFET [198].

ing (decreasing) side-gate voltage in the n-type (p-type) devices. The reason for this behavior is that the Schottky barrier at the nickel-WSe_2 interface limits the current flow and this barrier becomes thinner when the side-gate voltage is increased (decreased) as has already been discussed in detail in Chapter 7. The maximum on-currents of n- and p-type devices differ substantially, which is a clear signature that the Fermi level is pinned closer to the conduction band. Together with the lower effective mass in the conduction band (cf. Table 2.2), this yields a significantly different carrier injection for electrons and holes.

To operate the device as a TFET, side-gate voltages are applied with different polarities. An n-type source with a positive source side-gate voltage $V_{SSG} = 4\,V$ and a p-type drain with the drain side-gate voltage $V_{DSG} = -4, -5, -6\,V$ are created; the drain-source bias is constant at $V_{ds} = -0.5\,V$.[1] The lower panel of Figure 10.12 shows transfer characteristics in the TFET device operation mode. First, it is important to note that the gate leakage (not shown here) is at least 30× lower than the drain current in the entire V_{gs}-range (cf. the discussion in Section 5.9.3). Together with the fact that n- and p-type devices can be realized with appropriate side-gate voltages, the exponential increase of the current around $V_{gs} \approx 0\,V$ is due to band-to-band tunneling. This is also consistent with the fact that I_d increases with increasingly negative drain side-gate voltages since the tunneling probability increases at the channel-drain interface, while at the same time thermionic transport over the potential barrier in the drain is exponentially reduced.

The I_{on}/I_{off} ratio is ~4.5×10^2 at $V_{DSG} = -6\,V$ and the inverse subthreshold slope is $S \approx 366\,mV/dec$. This rather large value is due to the fact that overpolishing during the CMP fabrication process led to an insulation (i. e., gate underlap) between the Al-gate and the highly doped silicon side gates (realized with the LOCOS process) of approximately 90 nm (see Figure 10.2(b)) that limits the performance. From the mere consideration of the band gap of WSe_2, we would expect a larger on-/off-current ratio with significantly smaller off-state currents. However, as stated above, the leakage is not due to gate leakage. An explanation of this phenomenon requires a detailed simulation of the device structure including the fact that in the present device the distribution of the injected current among the various WSe_2-layers depends on the different (side-) gate voltages [59]. Therefore, it is likely that a combination of band-to-band tunneling and interlayer coupling is responsible for the electrical behavior of the device in TFET configuration.

---

1 Note that the nomenclature of source and drain may be confusing: the left contact is denoted with source and the right contact with drain throughout the discussion as illustrated in Figure 10.2(b). However, in Chapter 9 the electrode where band-to-band tunneling occurs was called source. Here, BTBT occurs at the drain-channel interface; the nomenclature has not been interchanged in order to give the two electrodes unambiguous names in the three operating modes.

10.3 Van der Waals Heterostructures

One of the most interesting opportunities that 2D materials offer is the ability to create heterostructures that allow novel materials functionalities to be obtained and also the realization of ultracompact devices. Using 2D material transfer processes such as described in Section 3.13.4, one can stack 2D materials with different properties on top of each other enabling all-2D-material FETs, for instance. As an example, such an all-2D-material FET (see also Section 11.3.2) is shown in Figure 10.13. For the realization of this device, a trilayer graphene was transferred onto an oxidized silicon wafer and patterned into stripe-shaped electrodes with electron-beam lithography and argon/oxygen plasma etching. After the removal of the resist, a 4 nm thick (equivalent to 13 layers) multilayer of hexagonal boron nitride is transferred on top of the graphene. Subsequently, a 2 nm thin MoS_2 flake is transferred and contacted with nickel contacts using again electron-beam lithography and lift-off.

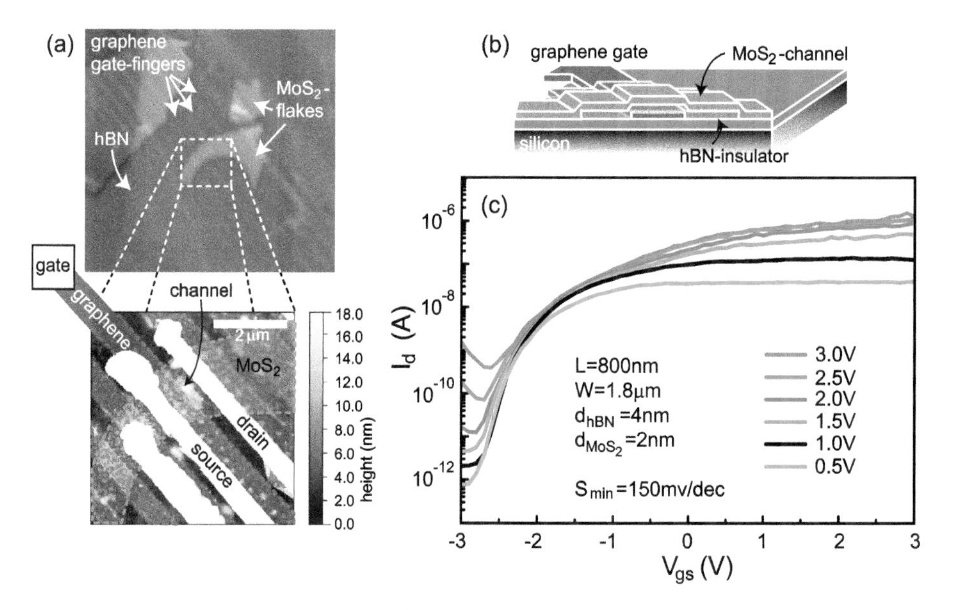

Figure 10.13: (a) Optical microscopy image of an all-2D-material FET consisting of a graphene gate, hBN gate insulator and MoS_2-channel as illustrated in (b). The lower panel of (a) shows an AFM image of the device. (c) Transfer characteristics of the all 2D-material FET.

Figure 10.13(a) shows an optical microscopy image of the device (top panel); the lower panel displays an AFM image of the active area with the different sections of the device marked with dashed lines. A cross-section schematic of the all-2D-material FET is shown in (b). The transistor has a channel length of ~800 nm and a width of ~1.8 µm. Transfer characteristics for different drain-source bias are displayed in Figure 10.13(c) showing regular switching behavior with on/off current ratios of up to 10^6, an on-current of $I_{on} \approx 1$ mA/mm and a minimum inverse subthreshold slope of 150 mV/dec. The rather strong dependence of I_{on} on V_{ds} is due to the presence of the Schottky barrier, which ultimately also limits the on-currents. A slight kink is observable as expected in the case of a Schottky-barrier MOSFET (cf. Section 7.1).

While the electrical figures of merit found in the all-2D-material FET are not too impressive one has to keep in mind that this is a device that can be fabricated merely with two electron-beam lithography steps and one etching step. No high temperature thermal annealing, no epitaxial growth, etc. are necessary. The entire device layer has a thickness of ~7 nm. This demonstrates the great potential 2D materials may offer in the future, namely the realization of material stacks with multiple active device layers and tailored properties for ultradense, three-dimensional integrated circuits.

Exercises

Exercises together with solutions are accessible via the QR code.

11 Cryogenic Electronics

Cryogenic electronics has recently attracted a renewed interest. The reason for this twofold. The first reason is related to von Neumann computing architectures: the power consumption due to the increased number of devices has become so large that it actually is more energy efficient to cool down the chips, and hence to exploit the strong reduction of power consumption that comes with the reduction of the supply voltage that the steeper inverse subthreshold slope allows at low temperatures. The second reason is related to the tremendous progress in quantum computation. Silicon-based spin qubits are a very attractive choice for the realization of complex qubit chips [282, 188, 240, 241, 283, 213, 202, 277, 186, 194] since the material system allows to combine large coherence times (particularly in nuclear spin-free, isotopically cleaned ^{28}Si) with the extremely sophisticated silicon CMOS fabrication technology. However, in order to enable an up-scaling of quantum information processors a classical electronics that controls the qubits is required, which needs to be located in the vicinity of the chip hosting the qubits [257, 34]. This means that the classical control electronics must be operable at cryogenic temperatures [80, 93, 108]. In the present chapter, we will therefore discuss the operation and optimization of cryogenic field-effect transistors.

11.1 MOSFETs at Cryogenic Temperatures

For the realization of spin qubits, gate electrodes are used in order to define electrostatically a quantum dot with two distinct energy levels to represent a two-state qubit. To properly set-up, tune, manipulate and couple qubits, a multitude of gate electrodes is required. As an example, Figure 11.1 shows an electron micrograph of an 8-qubit device based on a $Si_{0.7}Ge_{0.3}$/Si heterostructure [163]. Here, screening gates (brown) are used to define a 1D channel in-between two single-electron transistors (SET). Additional top-gates (orange, red) facilitate the formation of qubits with tunable coupling. Moreover, applying appropriate gate voltages a certain spin state can be moved along the 1D channel from one end to the opposite thereby potentially enabling coherent shuttling. While controlling a low number of gates with room temperature electronics is feasible, up-scaling the number of qubits ultimately requires a cryogenic control electronics located in immediate vicinity of the qubits as illustrated in Figure 11.1 [257]. Due to the limited cooling power of fridges at cryogenic temperatures, the control electronics needs to work at ultralow power levels.[1] From the discussion in Section 5.4 (see Equation (5.31)) and Chapter 9, it is clear that ultralow power levels require a strong reduc-

1 Although the exact operating temperature depends on, e. g., the highest possible temperature for qubit operation and how well the electronics can be thermally insulated from the qubit chip, it is fair to say that the lower the power consumption the larger the complexity of the cryogenic control electronics and hence the larger the overall number of addressable qubits.

https://doi.org/10.1515/9783111054421-011

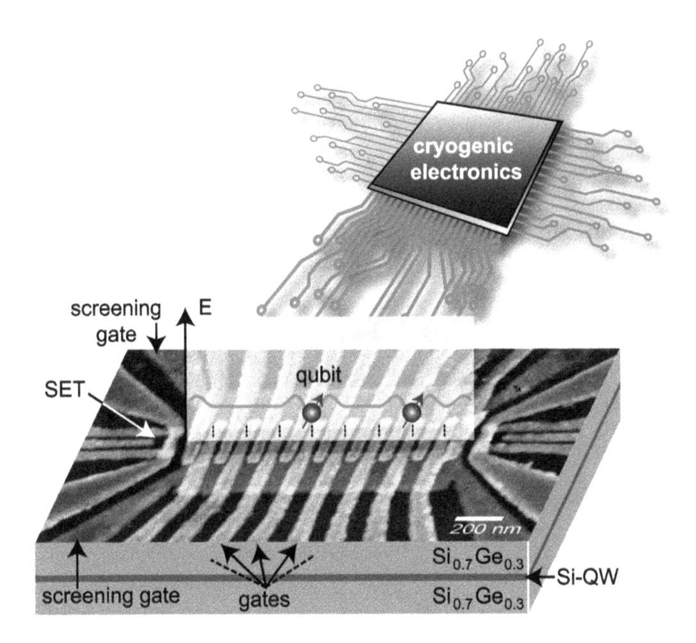

Figure 11.1: Electron micrograph of an 8-qubit device based on a Si/SiGe heterostructure with strained Si quantum well (L.R. Schreiber, RWTH Aachen University). The gate electrodes are used to manipulate, couple and move qubits along a 1D channel [163]. A classical cryogenic electronics must be located in the vicinity to enable an up-scaling of the number of qubits.

tion of the supply voltage V_{dd}. In fact, a back-on-the-envelop calculation shows that the required V_{dd} should be in the few tens of millivolts range. In turn, this means that cryogenic field-effect transistors must exhibit extremely steep inverse subthreshold slopes and require a very tight threshold voltage control. Furthermore, they must be operable with very small drain-source bias. In the following sections, the implications of these requirements will be elaborated on in detail.

11.1.1 Switching Behavior of Cryogenic MOSFETs

When discussing the switching behavior of MOSFET devices in Section 5.2.2, we obtained an ideal inverse subthreshold slope of

$$S_{\min} = \frac{k_B T}{|e|} \ln(10). \tag{11.1}$$

Hence, it appears that obtaining the required steep inverse subthreshold slopes at cryogenic temperatures comes "for free." Indeed, at a temperature of 1 K, S is expected to be as low as 0.2 mV/dec, which would allow switching of five orders of magnitude in as little as 1 mV. Then, using twice this gate voltage range as gate overdrive, a supply voltage around 3–4 mV appears feasible. However, when measuring the temperature dependence of S of a number of different field-effect transistors (bulk Si, UTB SOI, FinFETs and nanowire FETs) experimentally, a saturation of the inverse subthreshold slope below a certain temperature is observed resulting in a minimum subthreshold swing S_{min} well above the expected Boltzmann limit [20, 21, 32]. Figure 11.2 displays exemplary data of S as a function of T taken from literature (see figure). While S_{min} shows a somewhat different behavior depending on the device architecture, a saturation of S below a critical temperature T^* is observed in all devices. The saturation values of S are at least one order of magnitude larger than theoretically expected; in bulk devices, S even saturates at ~10 mV/dec.

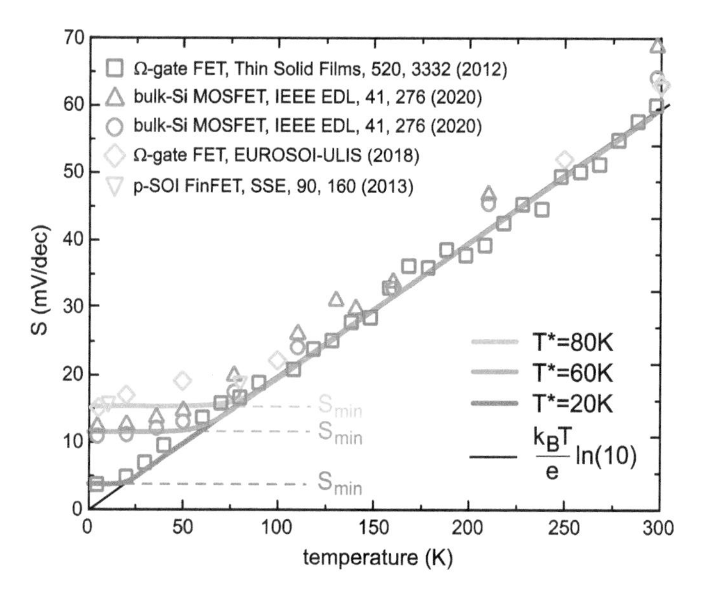

Figure 11.2: Experimental data of $S(T)$ extracted from literature (see figure) showing a saturation of the inverse subthreshold slope. The straight lines show S as a function of temperature for different T^* computed with Equation (11.6) assuming $C_{ox} \gg C_{it}^{loc}$ and $C_{ox} \gg C_{depl}$.

We already discussed in earlier Sections (see, for instance, Section 2.7, 2.12.3 and 4.5) that the metal-oxide-semiconductor interface of a MOSFET based on a three-dimensional material (e. g., silicon or silicon-on-insulator) represents a very strong disturbance of the lattice periodicity of the underlying crystal structure. Dangling bonds at the MOS surface lead to a large density of interface states that strongly impacts the functionality of a MOSFET. While the fabrication of an appropriate gate dielectric leads to orders of magnitude reduction of the interface density of states, the growth of an oxide at the silicon surface leads to oxygen diffusion resulting in a microroughness. In addition, traps due to unsaturated dangling bonds, charged defects, lattice deformations, electron phonon and electron-electron interactions all lead to disorder at the

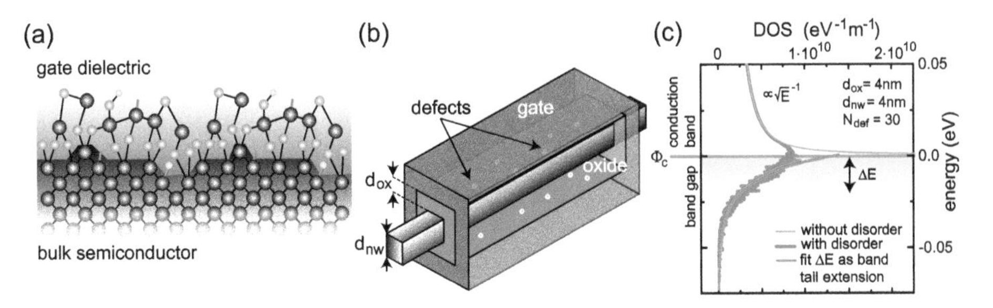

Figure 11.3: (a) Potential disorder at a MOS interface based on bulk material. (b) Nanowire with quadratic cross-section and wrap-gate. Charged defects are randomly distributed within the gate dielectric. (c) Density of states without (orange thin line) and with (red) disorder due to randomly distributed charged defects in the nanowire FET shown in (b). Band-tailing with an extension ΔE is obtained in the latter case.

silicon surface (illustrated in Figure 11.3(a)). Using a simple 1D model, the impact of disorder was already studied in Section 2.12.2 and led to a smearing of the band edges called band-tailing with a density of states that exponentially decays into the band gap (see Figure 2.48). Hence, the DOS due to band-tailing is approximately given by

$$D_{\mathrm{bt}}(E) \propto \exp\left(-\frac{\Phi_c - E}{\Delta E}\right) \tag{11.2}$$

where ΔE is the band-tail extension into the band gap around the band edge Φ_c (exemplarily, the conduction band is considered). The band tail extension ΔE is in the range of a few (tens of) meV and (energetically) located at the respective band edge (see below). Because band-tailing affects the device only in the transition region around the threshold voltage, at room temperature it can usually be neglected in conventional MOSFETs since $k_B T > \Delta E$. At cryogenic temperatures, $k_B T < \Delta E$ and, therefore, the lower the temperature the more the device behavior will be affected by band-tailing. The important point here is that D_{bt} is considered to consist of mobile states, i. e., states that carry current from source to drain. This is certainly a reasonable assumption since these states lie close to the band edge, and hence may contribute to the current through hopping transport, for instance.

In order to put the discussion on a more formal basis, let us consider a 1D nanowire with a wrap-gate. To simplify the following computation, a nanowire with a $4 \times 4\,\mathrm{nm}^2$ quadratic cross-section, surrounded by 4 nm SiO_2 is assumed as depicted in Figure 11.3(b). A configuration of charged defects randomly distributed within the gate dielectric affects the potential within the nanowire; the resulting local density of states is computed in the nanowire using NEGF (see Chapter 6). While the nanowire is considered to be one-dimensional to simplify the NEGF calculations, the Poisson equation, and hence the effect of the charged defects on the conduction band and resulting DOS is taken into consideration with a 3D finite difference approximation (Section 6.4.1). After the computation of 1000 random defect configurations with a total number of defects $N_{\mathrm{def}} = 30$

along a 100 nm long section of the wrap-gate device, the local density of states is averaged and displayed in Figure 11.3(c). A clear exponential band tail is observed; for comparison, the orange thin line shows the 1D DOS without disorder. An exponential fit (blue line) can then be used to extract the band tail extension ΔE, which is 18 meV in the present case consistent with values that have been used in a drift-diffusion simulation of cryogenic FETs [123].

Let us consider a 1D nanowire FET and compute the current at cryogenic temperatures incorporating an exponentially decaying DOS. When we derived the current in a 1D system using the Landauer approach (cf. Section 2.13.2), it was shown that with the top-of-the-barrier model I_d is given by (Equation (2.117))

$$I_d = 2e \int_{\Phi_0}^{\infty} dE \frac{D_{1D}(E)}{2} v(E)(f_s - f_d). \tag{11.3}$$

If we assume that V_{ds} is large enough such that there is no carrier injection from drain, the contribution from drain can be neglected. At cryogenic temperatures ($T \to 0$ K), we can replace the Fermi distribution with a step function, and consequently, the current is determined by carriers with the highest, i. e., the Fermi velocity v_f^s. Furthermore, when E_f^s lies below the conduction band maximum (i. e., when the device is in the off-state) we have to replace $D_{1D}(E)$ with $D_{bt}(E)$ and obtain

$$I_d \propto \int_{-\infty}^{E_f^s} dE\, v_f^s\, D_{bt} = v_f^s \Delta E \exp\left(-\frac{\Phi_c - E_f^s}{\Delta E}\right). \tag{11.4}$$

With this expression of the current, we can compute the inverse subthreshold slope based on Equation (5.21). In the present case, the derivative of the current with respect to the gate voltage yields

$$\frac{\partial I_d}{\partial V_{gs}} = \frac{\partial I_d}{\partial \Phi_c} \cdot \frac{\partial \Phi_c}{\partial \Phi_g} \cdot \frac{\partial \Phi_g}{\partial V_{gs}} = -\frac{I_d}{\Delta E} \cdot \frac{C_{ox}}{C_{ox} + C_\Sigma} \cdot (-e). \tag{11.5}$$

Note that in addition to the band-tails there may be localized interface states that do not contribute to the current but give rise to a capacitance. Therefore, C_Σ includes all capacitances in series with C_{ox} such as the depletion capacitance, localized interface state capacitance, etc. This means that we obtain the same expression as Equation (5.21) with $k_B T$ being replaced with ΔE. Thus, when cooling down MOSFETs, the inverse subthreshold slope is proportional to $k_B T$ as long as $k_B T > \Delta E$ and saturates when T drops below the critical temperature $T^\star = \Delta E / k_B$ [21, 32, 83].

It was already mentioned above that in addition to D_{bt}, there may also be a density of localized states, called D_{it}^{loc} in the following, centered around the band edge (e. g., due to defects within the gate dielectric). Such states do not carry current,[2] and thus do not contribute to I_{d} directly but they degrade the switching since they result in an additional contribution to the capacitance C_{Σ}, which is approximately given by $C_{\mathrm{it}}^{\mathrm{loc}} \approx e^2 D_{\mathrm{it}}^{\mathrm{loc}}(\Phi_c)$ (cf. Equation (4.19)) leading to a gate voltage dependent S. The difference between C_{it} considered so far and $C_{\mathrm{it}}^{\mathrm{loc}}$ is the energy(gate voltage)-range where the capacitors matter: While C_{it} stems from interface states within the band gap and deteriorates the switching of room temperature devices, $C_{\mathrm{it}}^{\mathrm{loc}}$ is due to the density of localized states around the band edge; this is an important point and will be elaborated on further below. The localized states at the band edge result in the so-called inflection phenomenon, i. e., a degradation of the inverse subthreshold slope in the transition region around the threshold voltage. Due to the thermal broadening, this region is not considered to be a part of the off-state in room temperature devices.

The discussion so far can be summarized in an empirical, closed-form expression for S over the entire temperature range (suggested in [83]):

$$S = \frac{\Delta E}{e}\ln(10)\left(1 + \frac{C_{\mathrm{it}}^{\mathrm{loc}} + C_{\mathrm{depl}}}{C_{\mathrm{ox}}}\right)\left[1 + \alpha \ln\left(1 + \exp\left(\frac{k_B T - \Delta E}{\alpha \Delta E}\right)\right)\right], \tag{11.6}$$

where $\alpha \approx 0.1$ is a smoothening parameter. Figure 11.2 displays S as a function of temperature based on Equation (11.6) for three different $T^{\star}$ clearly showing the saturation for $T < T^{\star}$.

The impact of band-tailing (i. e., a density of mobile states D_{bt}) and inflection (i. e., a density of localized states $D_{\mathrm{it}}^{\mathrm{loc}}$) on the transfer characteristic of a cryogenic MOSFET is summarized in Figure 11.4. Without disorder and without an additional $D_{\mathrm{it}}^{\mathrm{loc}}$, a sharp band edge yields a minimum inverse subthreshold slope according to the Boltzmann limit. Disorder results in band-tailing with a density of mobile states exponentially decaying into the band gap. This D_{bt} degrades S since current flows through these mobile states at a gate voltage where there is orders of magnitude less current in the ideal device (Figure 11.4(b)). Finally, with an additional $D_{\mathrm{it}}^{\mathrm{loc}}$ inflection occurs with a deterioration of the switching around V_{th}. Since inflection degrades the switching and renders S dependent on V_{gs} the inverse subthreshold slope is often plotted as a function of I_d. This way the switching behavior of cryogenic MOSFETs can be properly compared and qualified instead of giving only either the minimum inverse subthreshold slope or an average over some orders of magnitude; in TFETs, a similar issue led to the definition of the current I_{60} as a figure of merit (see Figure 9.12). Figure 11.5 shows S as a function of I_d schematically in the ideal case (green line), with band-tailing (blue line), and finally also including inflection (red line).

From the discussion so far, it is clear that in order to improve the switching of cryogenic MOSFETs, the impact of band-tailing, i. e., D_{bt}, and the density of localized

2 But they lead to a disordered potential resulting in a density of mobile states as has been used when computing Figure 11.3(c).

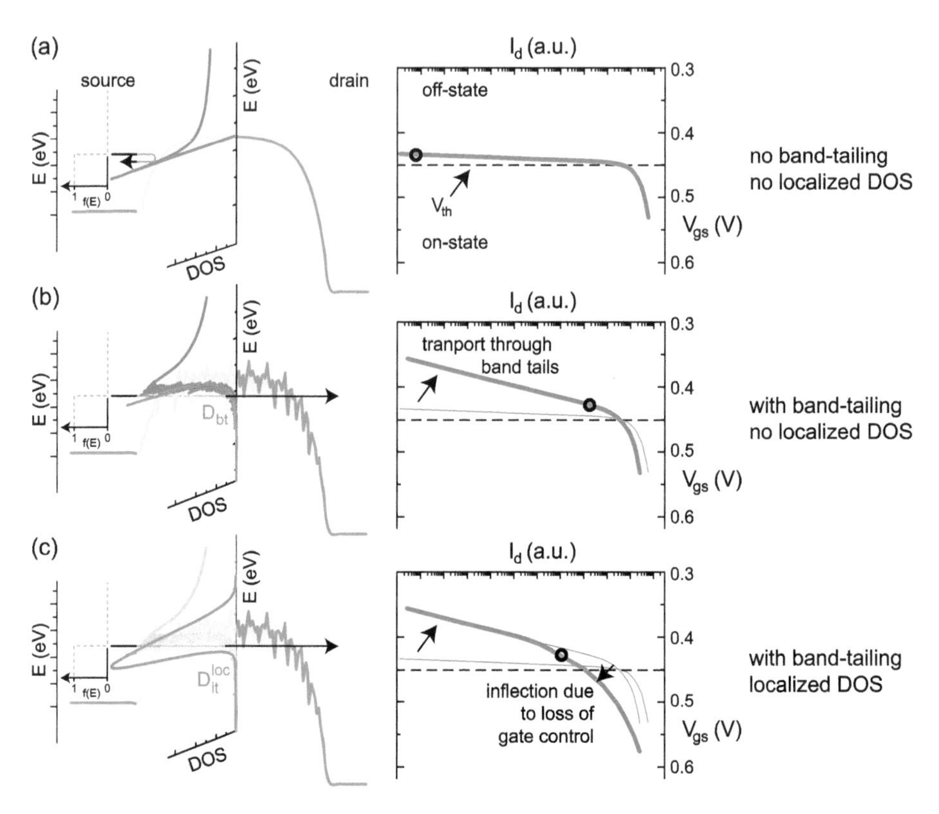

Figure 11.4: (a) Conduction band along current transport direction (green line), density of states (red line) and transfer characteristics of an ideal 1D cryogenic MOSFET, (b) 1D cryogenic MOSFET with band-tailing and (c) a 1D device with band-tailing and a density of localized states centered at the conduction band.

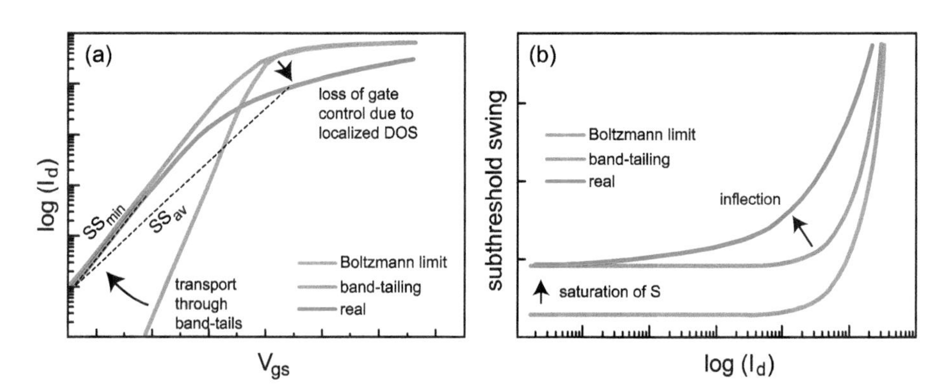

Figure 11.5: (a) Ideal transfer characteristics (green line), the impact of band-tailing (blue line) and a real curve with inflection (red). (b) In general, S depends on the gate voltage and is therefore plotted as a function of $\log(I_d)$ to show the entire off-state behavior.

states D_{it}^{loc} need to be reduced as much as possible. The strategy to reduce the impact of D_{it}^{loc}, and hence to avoid inflection is straightforward; one simply has to ensure that $C_{ox} \gg C_{it}^{loc}$, which can be accomplished by reducing the gate dielectric thickness and/or increasing the dielectric constant of the gate insulator; an optimized processing of the MOS interface is certainly required. However, how do we reduce the impact of disorder? This second point is more involved and requires an in-depth discussion. Two ways how to reduce band-tailing and hence optimize the switching of cryogenic MOSFETs will be presented in the following two sections.

11.2 Optimizing the Switching of Cryogenic MOSFETs—Improved Screening

In Chapter 5, a one-dimensional modified Poisson equation (Equation (5.34)) has been derived that has proven to be an excellent description of the electrostatics of ultrathin-body field-effect transistors. This equation shows that the presence of the gate electrode yields an exponential screening of potential variations on the length scale λ_{ch}. Since at cryogenic temperatures band-tailing is dominated by static disorder due to charged impurities, defects, etc. (phonons freeze out at those temperatures), it is expected that band-tailing can be reduced by making the screening length λ_{ch} as small as possible. This implies a very small effective oxide thickness (cf. Section 4.5.5), i. e., a very thin gate dielectric d_{ox} and a gate insulator with high dielectric constant ϵ_{ox} as well as a reduction of λ_{ch} with a wrap-gate nanowire device exhibiting a small nanowire diameter d_{nw}. In the following sections, the impact of λ_{ch}-scaling on cryogenic MOSFETs will be explored.

11.2.1 Suppression of Band-Tailing with Effective Oxide Scaling

In order to assess in how far band-tailing can be suppressed with scaling of the effective oxide thickness (EOT, cf. Equation (4.29)), the dependence of ΔE on d_{ox} and ε_{ox} in the field-effect transistor configuration displayed in Figure 11.3(b) is computed. The device structure to do this, has already been presented above but will be repeated here for convenience: A piece of a silicon nanowire with a $4 \times 4\,\text{nm}^2$ quadratic cross-section and surrounded by a gate dielectric of thickness d_{ox} with dielectric constant ε_{ox} and GAA-electrode is considered. Furthermore, a gate voltage in the device's off-state is assumed such that the effect of mobile charge can be neglected. Static disorder is simulated by randomly distributing a number N_{def} of positively charged defects within the gate dielectric as depicted in Figure 11.3(b). The 3D Poisson equation is solved using the finite difference method presented in Section 6.4.1, which yields the impact of the random distribution of defects on the conduction band of the nanowire. The local density of states is then computed using NEGF (cf. Chapter 6). Semiinfinite contacts at the same potential as the nanowire are attached with appropriate self-energies in order to avoid quantization effects along the nanowire axis. Since the diameter of the nanowire is assumed to be in the few nanometer range, carrier confinement yields one-dimensional electronic transport. The DOS is therefore only computed in the center of the nanowire. This greatly speeds up the simulations such that the local DOS of a large number (1000 in the present case) of random defect configurations can be computed and averaged to mimic static disorder. From such an averaged DOS, the band tail parameter ΔE is extracted by fitting an exponential decay to it.

The averaged DOS is computed for varying d_{ox}, dielectric constant ε_{ox} and for different numbers of defects N_{def}. Figure 11.6 shows exemplarily plots of the averaged DOS (red lines) together with the fits (blue lines) to extract ΔE for the cases stated in the different panels. Figure 11.7 shows ΔE as a function of ε_{ox} in (a) and its dependence on d_{ox} in (b). As is clearly visible from the graphs, the band-tailing parameter is proportional to d_{ox} and inversely proportional to ε_{ox}. Since in the quadruple gate case $\lambda_{\text{ch}} = \sqrt{\frac{\varepsilon_{\text{si}}}{\varepsilon_{\text{ox}}} \frac{d_{\text{ox}}}{4} d_{\text{ch}}}$, it is obvious that $\Delta E \propto \lambda_{\text{ch}}^2 \propto C_{\text{ox}}^{-1}$ where C_{ox} is the geometrical oxide capacitance per area (in 2D) or per length as in the case of the quadruple nanowire considered here. There is an important thing to mention: the dependence $\Delta E \propto \lambda_{\text{ch}}^2$ is only valid as long as $\Delta E/k_B > T$; in the case of very large oxide capacitances C_{ox} (i. e., very small λ_{ch}, and hence ΔE), Equation (11.6) approaches the ideal Boltzmann limit with $S \propto k_B T$.

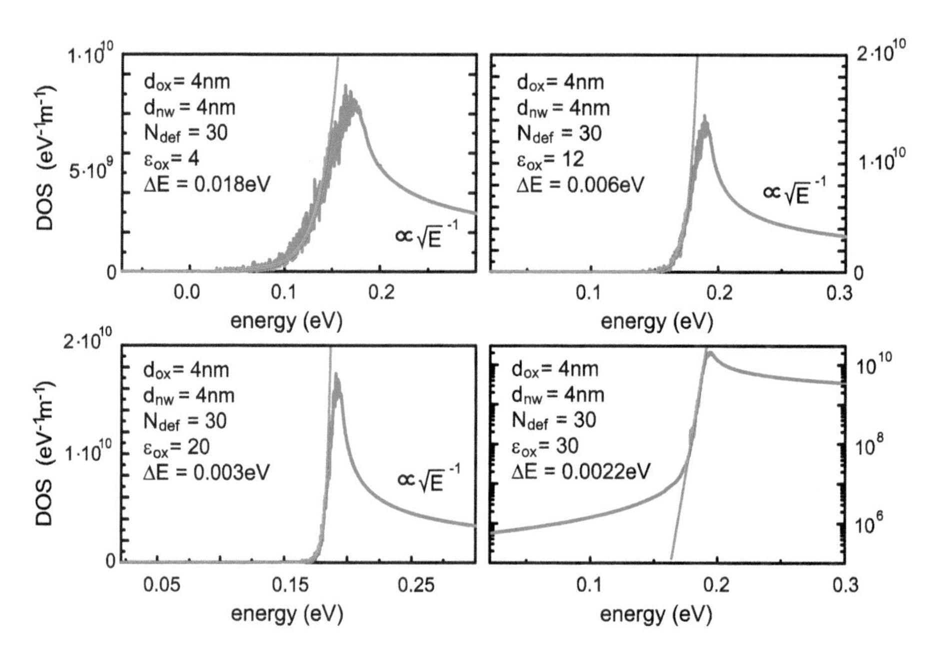

Figure 11.6: Averaged density of states (red lines) in a 100 nm nanowire section equipped with a quadruple wrap gate. The blue lines show exponential fits for the cases stated in the respective panels. The averaged DOS is based on 1,000 random defect configurations with N_{def} distributed across the gate dielectric.

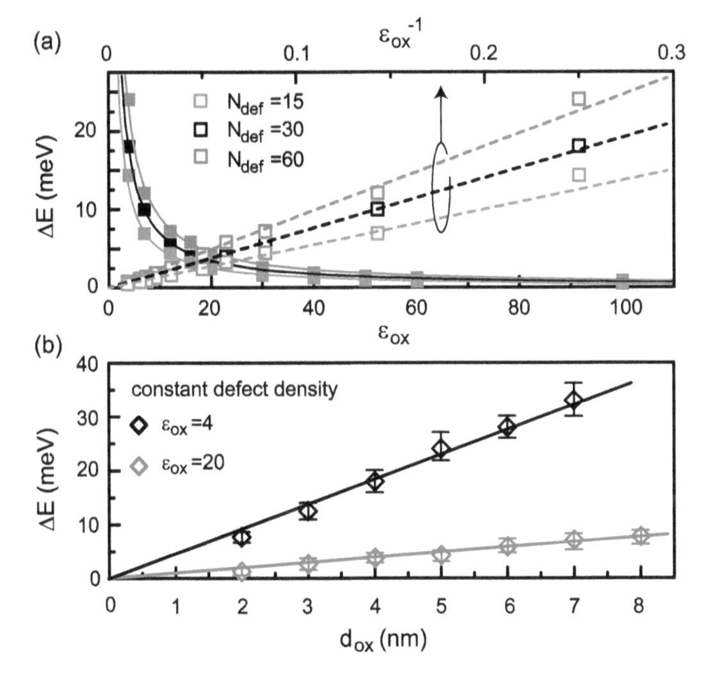

Figure 11.7: Dependence of the band-tailing parameter ΔE on the dielectric constant ε_{ox} of the gate dielectric for three different numbers of charged defects N_{def} (a), and (b) as a function of d_{ox} for two different ε_{ox} [145].

To show experimentally that the scaling of EOT is suitable to suppress band-tailing, long-channel silicon-on-insulator (SOI) MOSFETs with different effective oxide thicknesses (depicted in the inset of Figure 11.8, left panel) are fabricated. For the fabrication, standard processes such as optical lithography, reactive ion etching, etc. are used (details on the fabrication can be found in [222] and on the processes involved in Chapter 3). The important point here is that three different gate stacks with three different effective oxide thicknesses are realized: (i) 0.8 nm SiN grown with rapid thermal nitridation and 18 nm SiO_2 deposited with remote plasma enhanced CVD, (ii) chemically grown (with standard clean 2) SiO_x and 18 nm ALD-deposited HfO_2 and (iii) 18 nm HfO_2 deposited immediately on the silicon surface after the removal of the native oxide with a HF dip. The left panel of Figure 11.8 (also stating the approximate effective oxide thicknesses) shows the extracted S as a function of $\log(I_d)$ at $T = 4.2$ K. Obviously, the smaller EOT the better the switching behavior with smaller minimum values of S (i. e., reduced band-tailing) as well as reduced inflection.

Since $\Delta E \propto C_{ox}^{-1}$ and because the critical temperature $T^{\star} = \Delta E/k_B$ is substantially larger than the measurement temperature of $T = 4.2$ K, $S \propto \frac{\Delta E}{e}\ln(10)$ can be assumed. Therefore, the ratios of the minimum inverse subthreshold slopes of the different devices, i. e., the S-values for the smallest I_d where the impact of inflection is negligible, should be inversely proportional to the ratio of the oxide capacitors of the different de-

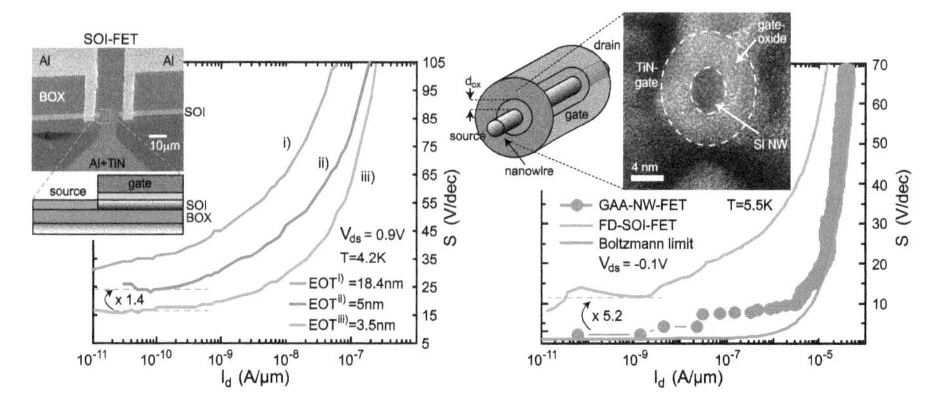

Figure 11.8: Left: Extracted inverse subthreshold slope S as a function of $\log(I_d)$ for SOI-MOSFETs with three different effective oxide thicknesses [222]. The inset shows an optical microscope image and a schematic of the fabricated SOI-MOSFETs. Right: Inverse subthreshold slope as a function of $\log(I_d)$ for a wrap-gate nanowire (blue dots) and a FD-SOI device (orange line) compared to the Boltzmann limit (green line) FET [100]. The inset shows a schematic illustration of the wrap-gate nanowire FET and a transmission electron micrograph of the fabricated device (Q. T. Zhao, FZ Jülich).

vices. However, the mentioned gate stacks do not only exhibit a different EOT but also the treatment of the interface is different. It is therefore not expected that the ratio of the inverse subthreshold slopes fits the expected behavior. Yet, comparing stack (ii) and (iii), both having the same 18 nm HfO_2 the ratio $S_{min}^{(ii)}/S_{min}^{(iii)} \approx EOT^{(ii)}/EOT^{(iii)}$ (see Figure 11.8); the ratio $S_{min}^{(i)}/S_{min}^{(ii)}$ is approximately 1.46, and thus substantially smaller than $EOT^{(i)}/EOT^{(ii)} \approx 3.68$. The different interface treatment as well as the different oxide material make a direct comparison questionable. In fact, in Section 11.3.1 the role of the treatment of the MOS interface will be further elaborated on and it will be shown that a proper interface treatment can be more effective than EOT scaling. Nevertheless, the experiments show that scaling the effective oxide is a suitable method to reduce band-tailing yielding (approximately) the expected magnitude of lowering the inverse subthreshold slope.

11.2.2 Suppression of Band-Tailing with λ_{ch}-Reduction

In the preceding section, it was demonstrated that C_{ox}-scaling is effective in reducing band-tailing. While in the experimental SOI-MOSFETs, the scaling of λ_{ch} is limited to a reduction of EOT, smaller screening lengths λ_{ch}, and hence further improvements of the switching are expected if a wrap-gate nanowire device architecture (see Figure 5.12 for the screening lengths of different device architectures) is chosen. To investigate this, Si nanowire wrap-gate transistors are fabricated with a top-down approach using electron-beam lithography and digital etching to form nanowires with diameters down to ~5 nm. The nanowires are fortified with a HfO_2 gate dielectric with a thickness of $d_{ox} = 4\,nm$ and a TiN gate electrode. Further details of the fabrication can be found in [100] and details of the processes involved are given in Chapter 3. The source and drain contacts are realized with $NiSi_2$; ion implantation and subsequent annealing yield the dopants to segregate to the $NiSi_2$/Si interface where they allow reducing the Schottky-barrier height and form (near) Ohmic contacts (see Section 7.2).

Simultaneously with the wrap-gate NW devices, SOI MOSFETs are fabricated in a mostly similar way, in order to enable a comparison of a single-gate SOI device architecture with a wrap-gate nanowire layout [100]. Again, $NiSi_2$ source/drain contacts with dopant segregation are realized and the same HfO_2 gate dielectric is used as in the case of the nanowire device.

The right panel of Figure 11.8 displays S as a function of $\log(I_d)$ (normalized to the width of the SOI device and the circumference of the nanowire) of the SOI MOSFET (orange line) and the GAA nanowire FET (blue data points) at a temperature of 5.5 K; for reference, the ideal Boltzmann limit is also shown (green line). Comparing the two experimental results clearly shows that the inverse subthreshold slope of the nanowire FET follows almost the ideally expected behavior whereas the SOI device exhibits significantly stronger band-tailing (larger minimum S-value) and inflection. Moreover, the minimum inverse subthreshold slope of the GAA nanowire device is 2.3 mV/dec, which is only a factor of 2 larger than the expected $S = 1.1\,mV/dec$ at $T = 5.5\,K$ [100]. Furthermore, comparing the minimum inverse subthreshold slopes of the GAA nanowire FET with the SOI-MOSFET, a ratio $S_{min}^{SOI}/S_{min}^{GAA} \approx 5.2$ is obtained, which fits well the ratio of $\lambda_{ch,SOI}^2/\lambda_{ch,GAA}^2$ (where again $S \propto \lambda_{ch}^2$, is used), which is between 4 and 6.7 depending on whether a quadruple or cylindrical GAA device architecture is considered. Consequently, nanowire FETs with wrap-gate device structure and high-k gate dielectric are the preferred embodiment for cryogenic MOSFETs due to the superior screening that suppresses band-tailing and reduces inflection due to the excellent electrostatics.

11.3 Optimizing the Switching of Cryogenic MOSFETs—Reduction of Disorder

In the preceding sections, the impact of band-tailing has been suppressed by appropriate screening making λ_{ch} as small as possible. However, reducing disorder in the first place is certainly a better way to avoid band-tailing and its detrimental effects on the switching behavior. In the following, two such approaches are discussed.

11.3.1 Suppression of Disorder with Ultrathin SiN Layers

The first approach to realize a reduced disorder in Si MOSFETs is using an alternative gate dielectric or dielectric stack. Here, the use of ultrathin SiN layers grown with rapid thermal nitridation (see Section 3.3) is studied. The reason for using SiN is twofold: first, it can be grown in a self-terminating way, and hence ultrathin layers can be generated reliably and reproducibly. Moreover, with an appropriate RTP tool, an in situ process of hydrogen anneal to smoothen the silicon surface, followed by the growth of the SiN layer is conceivable. Second, SiN is an excellent diffusion barrier and as such prevents silicon from oxidizing during, e. g., the deposition of a high-k gate dielectric. Since its thickness can be sub-1 nm and its dielectric constant is larger then SiO_2 it allows in principle very low effective oxide thicknesses. In the following, we employ a stack consisting of an ultrathin SiN layer followed by HfO_2 deposited with ALD.

The only drawback of SiN is that it leads to an increased density of interface states within the band gap and, therefore, deteriorates the off-state behavior of room temperature MOSFETs. Indeed, the ultrathin SiN layers used here show an increased density of interface states within the band gap. However, it has already been discussed above that at cryogenic temperatures the DOS within the band gap is to a large extent irrelevant.[3] The reason for this is the fact that switching at room temperature requires moving the band gap in the channel over a much larger energetic range (larger than $E_g/2$) whereas in a cryogenic FET only the few (tens of) meV close to the band edge matter.

Figure 11.9 shows S versus $\log(I_d)$ of SOI MOSFETs with two different HfO_2 thicknesses, with and without a 1.8 nm thin SiN interlayer (see figure for details) [222]. In both cases, the SiN interlayer helps reducing inflection although the effective oxide thickness is increased. Interestingly, there is hardly any difference for the minimum inverse subthreshold slope in the case without interlayer although EOT, and hence S_{min} should be

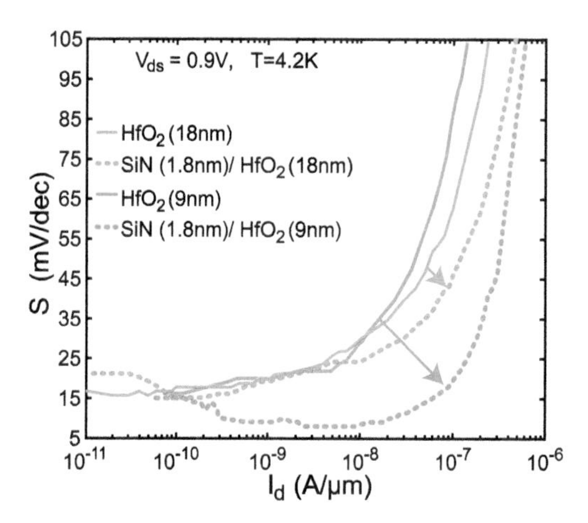

Figure 11.9: Inverse subthreshold slope S as a function of $\log(I_d)$ for planar, long-channel SOI-MOSFETs with the gate stacks given in the figure [222].

3 It is certainly true that states within the band gap will be charged by driving the device into its on-state and this may lead to a slight increase of static disorder due to charged defects counteracting partly the expected gain of using SiN.

halved when the HfO_2 thickness is reduced from 18 nm to 9 nm. The reason for this is likely to be interface states that are generated, when during HfO_2 deposition an interfacial oxide grows. Furthermore, the combination of a 1.8 nm SiN layer with 9 nm of HfO_2 (orange dashed curve in Figure 11.9) yields best results with a steep S over the largest I_d-range. Note that the increase of S at low I_d is due to increasing gate leakage; similarly fabricated devices with less gate leakage (not shown) showed a steep S also at lower currents.

The observed behavior of the cryogenic MOSFETs shows that alternative gate dielectrics, possibly unsuited for room temperature electronics, may be useful for cryogenic electronics. In the case presented here, a SiN layer reduces band-tailing and inflection. Combined with a gate-all-around nanowire device structure and a high-k gate dielectric, this holds promise to enable cryogenic MOSFETs that can be operated at extremely low operational voltages.

11.3.2 Suppression of Disorder in All-2D Bilayer Graphene Devices

In MOSFETs based on bulk materials, band-tailing can only be reduced up to a certain level due to the inherent disorder of the surface/interface of a bulk material with, e. g., unsaturated dangling bonds, a smoothness of the surface not better than ± one atomic layer, etc. (see illustration in Figure 11.3(a)). In addition, regular MOSFETs are equipped with a (usually amorphous) "bulk" dielectric containing charged defects, which cannot be completely eliminated even with the most sophisticated process technologies. On the other hand, pristine 2D materials do not exhibit dangling bonds at their surface and only interact via van der Waals interaction with their environment. Yet, when placed on regular substrates and combined with "bulk"-dielectrics, 2D materials devices suffer from the disorder induced by the substrate/dielectric resulting in low carrier mobility and in the case of graphene nanoribbons, a mobility instead a real band gap (see Figure 10.8). Recently, it was shown that in bilayer graphene encapsulated into hexagonal boron nitride (hBN) with atomically smooth graphitic top and bottom gate electrodes, a very clean band gap E_g can be induced by applying appropriate (opposite) voltages at top and bottom gates [113, 114]. The devices are fabricated by a dry van der Waals transfer using a polycarbonate membrane to sequentially stack flakes of hBN, graphite and bilayer graphene fabricated by mechanical exfoliation (see [114] for details). Applying a symmetric voltage bias $\pm V_{ds}/2$ at source and drain and changing the top and bottom gate voltages simultaneously in order to keep the displacement field (here, $D = 0.46$ V/nm), and hence the induced band gap $E_g = \frac{\Delta(D)V_\perp}{\sqrt{V_\perp^2 + \Delta(D)^2}}$ (Equation (2.54)) constant allows obtaining regular field-effect transistor device characteristics. Figure 11.10(a) illustrates the device layout and shows the Mexican-hat shaped band structure (see Section 2.8.2) that results in the double-gate area when an appropriate displacement field is implemented.

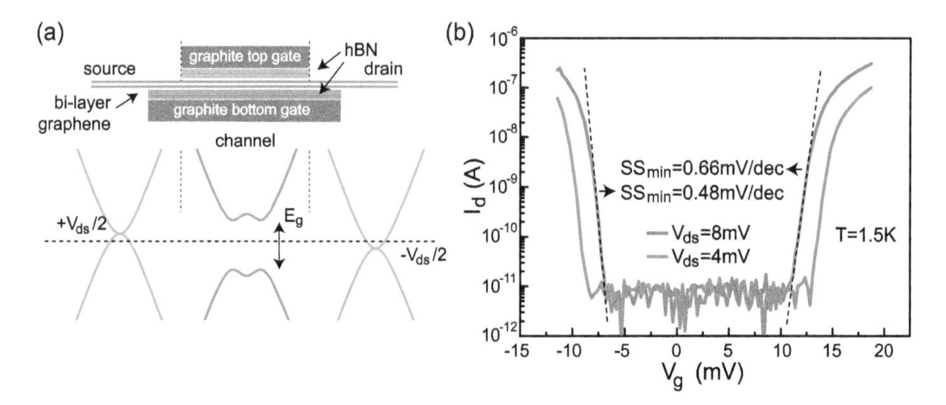

Figure 11.10: (a) Schematic of the bilayer graphene device with conduction/valence bands in source, channel and drain. (b) Transfer characteristics of a bilayer graphene device encapsulated in hBN as gate dielectric and equipped with graphite gate electrodes. The extracted minimum inverse subthreshold slope is below 0.5 mV/dec at T = 1.5 K [114].

Figure 11.10(b) shows transfer characteristics of the fabricated bilayer graphene devices for two different bias voltages at a temperature of 1.5 K. Due to the device layout (see Figure 11.10(a)), the transistor exhibits ambipolar operation. Both branches, electron and hole branch show an excellent switching behavior with an S_{min} less than 0.5 mV/dec at $T = 1.5$ K. In fact, lowering the temperature to 100 mK reduces the inverse subthreshold slope to ~200 μV/dec (not shown here) [114]. Note that these very low values have been achieved in spite of the low dielectric constant of hBN which leads to little screening of disorder (much lower than in the devices shown in the preceding section featuring HfO_2 as gate dielectric). This is therefore a clear signature of a very strong reduction of disorder in the all-2D materials FET.

11.4 Scalability of Cryogenic MOSFETs

The preceding section demonstrated that an all-2D material FET would be an ideal candidate for cryogenic electronics. However, 2D materials lack the maturity of silicon CMOS technology and in the example above with bilayer graphene, even two gate voltages needed to be applied and modified appropriately in order to maintain a proper band gap. As a result, the bilayer graphene cryogenic FET is an idealistic device demonstrating the importance of reducing disorder at the MOS interface in order to realize steep slope cryogenic MOSFETs. For real technological applications, it is rather likely that the approach discussed in Section 11.2, i. e., suppressing band-tailing by screening (smallest EOT, GAA nanowire architecture) is the most viable approach. Therefore, let us consider in the following a GAA nanowire FET with a nanowire diameter sufficiently thin to guarantee 1D electronic transport. To reduce band-tailing as much as possible, the use of a gate dielectric thickness d_{ox} as thin as possible and a high-k material with very large ε_{ox}-

value is needed since $\Delta E \propto C_{ox}^{-1}$. The smaller ΔE, the more does the inverse subthreshold slope approach the Boltzmann limit at cryogenic temperatures (cf. Equation (11.6)).

Let us assume that the cryogenic transistor should switch X orders of magnitude (e. g., $X = 5$) from the off- to the on-state and further assume a gate overdrive twice as large as the off-state V_{gs}-range. Thus, the supply voltage V_{dd} of a cryogenic circuit is $V_{dd} = 3 \cdot X \cdot S$ (because the gate overdrive of $2 \cdot X \cdot S$). As already discussed in Section 5.4, in order to reduce the power consumption as much as possible V_{dd} should be as small as possible since the dynamic power consumption $P_{dyn} \propto C_{tot} \cdot V_{dd}^2 \cdot f$ (Equation (5.31)) to comply with the small available cooling power of refrigerators.

Since the transition from the Boltzmann limit to the saturation of the inverse subthreshold slope (Equation (11.6) and Figure 11.2) occurs at $T = T^{\star} = \Delta E / k_B \propto C_{ox}^{-1}$, one should increase C_{ox} (decrease EOT) so much that $T^{\star}$ equals the desired operation temperature T_{op} of the cryogenic electronics. A further increase of C_{ox}, however, will not allow a further reduction of V_{dd}. For example, consider the wrap gate nanowire device presented in Section 11.2.2 with a minimum $S = 2.3$ mV/dec and let us assume that this is the saturation value of this device (which is not really the case). Then the critical temperature is $T^{\star} = 300\,\text{K} \cdot 2.3/60 = 11.5\,\text{K}$, and thus the band-tail extension is $\Delta E = k_B T^{\star} \approx 0.96$ meV. At a desired operation temperature of, say, $T_{op} = 1$ K the saturation of S to a value of 2.3 mV/dec requires a minimal $V_{dd} = 3 \cdot X \cdot 2.3\,\text{mV} = 34.5$ mV (where $X = 5$ decades is assumed). Now, in order to obtain the optimum low supply voltage, C_{ox} must be increased in order to ensure $T_{op} = \Delta E / k_B$. This means C_{ox} has to be increased by a factor of $T^{\star}/T_{op} = 11.5$. An increase by a bit more than one order of magnitude is rather difficult to realize considering the fact that the nanowire wrap-gate device already had a 4 nm thin HfO_2 high-k gate dielectric. This underlines the fact that from a practical point of view a further reduction of band-tailing can only be achieved by combining the wrap-gate nanowire with high-k gate dielectric with interface engineering (such as an ultrathin SiN layer) to reduce band-tailing further. Alternatively, one could possibly think about using high-k gate dielectrics with very high ε_{ox}-value. As these materials usually exhibit a decreasing band gap with increasing ε_{ox} (see Figure 4.16), they have been disregarded for room temperature devices. However, at cryogenic temperatures the small potential barrier height associated with the small band gap may be sufficient to realize the necessary insulating properties of the material. It therefore makes sense to consider the use of high-k material with very high ε_{ox}. In the example above, a reduction of $T^{\star}$ by one order of magnitude implies a reduction of V_{dd} by the same factor. And since the dynamic power consumption $P_{dyn} \propto (V_{dd})^2$, a two orders of magnitude reduction in power consumption can be obtained. However, the reduction of band-tailing, and hence V_{dd} comes with a price, namely a strong increase in C_{ox}.

According to classical scaling [65], the increase of C_{ox} due to a reduction of the effective oxide thickness is compensated by downscaling the channel length L (and width but we are considering a nanowire FET here). This ensures that the power delay product $\text{PDP} = C_g (V_{dd})^2 \approx C_{ox} (V_{dd})^2$ remains constant. However, channel length scaling in cryogenic MOSFETs is different from classical MOSFETs. The reason for this is the fact that at cryogenic temperatures, the potential barrier in the channel that blocks the current flow, is rather small since switching must not require more then a few mV of gate voltage interval. As a result, the electronic wave function of source and drain (an example is shown in Figure 11.11(a) with the white lines) penetrate rather deeply into the band gap since tunneling depends exponentially on the potential barrier height, giving rise to a density of states that exponentially decays into the band gap. This means, that even without disorder, band-tailing is obtained in short channel devices that leads to an increasing degradation of S with decreasing channel length [126, 98]. Figure 11.11 shows this scenario: (a) depicts a local density of states in a $L = 25\,\text{nm}$ GAA nanowire FET (the same parameters have been used as displayed in Figure 11.12) close to V_{th}, showing how the penetration of the electronic wave function from source and drain (bright areas) leads to a nonzero DOS within the band gap (i. e., band-tailing). (b) shows the DOS evaluated in the middle of the channel (dashed white line in (a)) in three devices with decreasing channel length. Obviously, short channels lead to a band-tailing very similar to disorder (compare Figure 11.6). Hence, direct source-to-drain tunneling affects cryogenic devices much more than devices at room temperature (see Section 5.9.4).

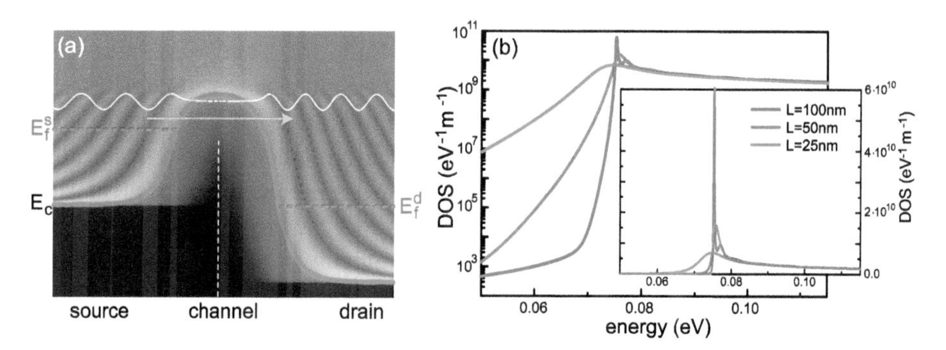

Figure 11.11: (a) Simulated local density of states in a GAA nanowire FET close to the threshold voltage; $L = 25\,\text{nm}$, all other parameters are given in Figure 11.12. The penetration of the wave function (as an example shown with the white line) leads to band-tailing. (b) Density of states (at the position of the white dashed line in (a)) for three different channel lengths showing clearly how band-tailing due to quantum mechanical tunneling increases with decreasing L. Fitting the DOS with an exponential function yields $\Delta E = 0.007\,\text{eV}$ if $L = 15\,\text{nm}$, $\Delta E = 0.003\,\text{eV}$ for $L = 25\,\text{nm}$ and $\Delta E = 0.001\,\text{eV}$ in the case $L = 50\,\text{nm}$.

In order to assess the scalability with respect to the channel length, self-consistent simulations of GAA nanowire FETs are carried out with NEGF at a temperature of $T = 4.2\,\text{K}$. The resulting transfer characteristics are plotted as a function of channel length L in Figure 11.12(a) for five different channel lengths. From such curves, S is extracted for various channel lengths and two different gate dielectrics in (b). The slightly better behavior in the case of $\varepsilon_{\text{ox}} = 20$ is due to steeper p-n junctions giving rise to a slightly longer effective channel length compared to the case $\varepsilon_{\text{ox}} = 4$. Note that in the present case band-tailing due to disorder has been dropped such that ideal switching can in principle be

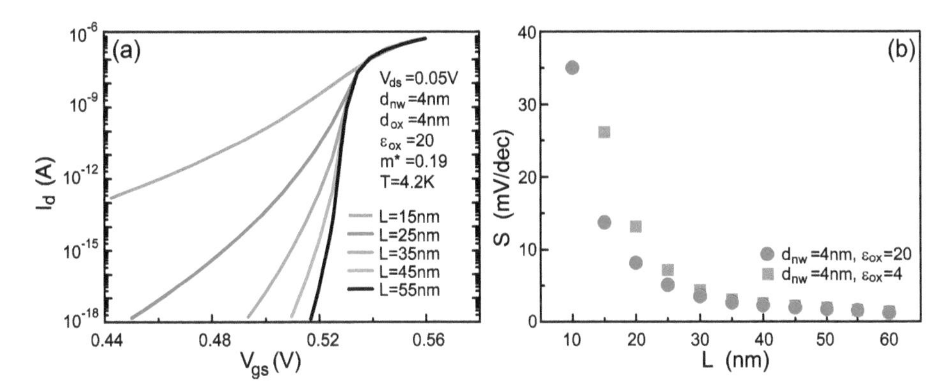

Figure 11.12: (a) Simulated $I_d - V_{gs}$ curves for five different L. (b) Inverse subthreshold slope as a function of channel length L of GAA nanowire FETs at $T = 4.2$ K for two different gate dielectrics.

expected. Obviously, the switching is increasingly deteriorated when scaling down the channel length due to band-tailing induced by direct source-to-drain tunneling. It follows that in order to avoid a degradation of S the channel length should not be scaled below ~50 nm.

According to the above finding, the channel length L is considered being constant in the following discussion (and sufficiently long to avoid a contribution to ΔE due to direct source-to-drain tunneling). To study the impact of scaling C_{ox}, the dependence of the PDP on the effective oxide thickness is computed based on the following assumptions: (i) $C_{\text{ox}} \gg C_{\text{depl}}$, since the channel is considered to be undoped, (ii) the density of localized interface states is small enough and/or C_{ox} is always large enough to guarantee $C_{\text{ox}} \gg C_{\text{it}}^{\text{loc}}$ and (iii) as above, it is assumed that an on/off current ratio of X orders of magnitude is required (resulting in $V_{\text{dd}} = 3 \cdot X \cdot S$). As a result, the PDP can be written as

$$\text{PDP} = C_g V_{\text{dd}}^2 \approx \frac{C_{\text{ox}} C_q}{C_{\text{ox}} + C_q} \left(3 \cdot X \frac{1}{C_{\text{ox}}} \frac{\ln(10)}{e} \left[1 + \alpha \ln\left(1 + \exp\left(\frac{k_B T - \Delta E}{\alpha \Delta E} \right) \right) \right] \right)^2 \tag{11.7}$$

where Equation (11.6) has been used and the gate capacitance C_g has been replaced with the series combination of the density-of-states or quantum capacitance C_q (called C_{inv} in Section 5.9.1) and the geometrical oxide capacitance C_{ox}. Moreover, $\Delta E \propto C_{\text{ox}}^{-1}$ has already been inserted in Equation (11.7). The following two cases can now be distinguished: if the device is in what was called classical capacitance limit in Section 5.9.1, $C_{\text{ox}} \ll C_q$, we have $C_g \approx C_{\text{ox}}$, and thus the PDP is being reduced according to PDP $\propto 1/C_{\text{ox}}$. This is true until $\Delta E \propto C_{\text{ox}}^{-1}$ becomes equal to $k_B T_{\text{op}}$. In this case, S approaches the Boltzmann limit, and hence does not decrease anymore. As a result, the power-delay-product will increase if C_{ox} is further increased. This means that in this case one would "over-scale" C_{ox} leading to an unnecessary increase of the capacitance, and hence of the PDP. The optimum C_{ox} depends of course on the operation temperature T_{op} of the cryogenic electronics, which is exactly reflected in Figure 11.13.

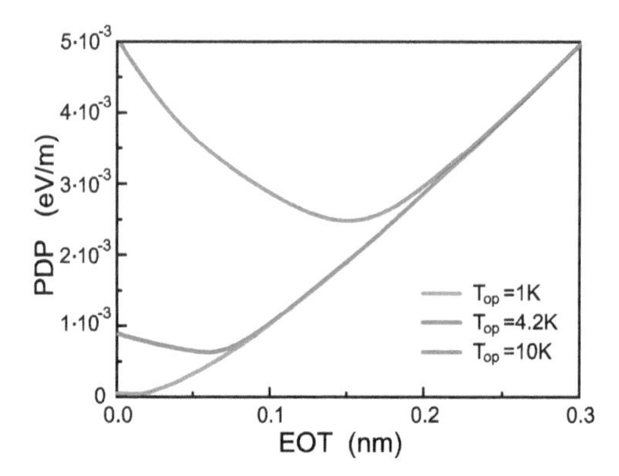

Figure 11.13: Power delay product (in eV/m) for three different operation temperatures T_{op} computed with Equation (11.7). A silicon GAA nanowire FET with quadruple gate is assumed. All device parameters are as stated in Figure 11.12.

On the other hand, if we consider the quantum capacitance limit where $C_{\mathrm{ox}} \ll C_q$, and hence $C_g \approx C_q$, a reduction of PDP is also expected. But once $\Delta E = k_B T_{\mathrm{op}}$ and the Boltzmann limit for S is reached, the power delay product saturates at the lowest level, even if C_{ox} is further reduced.

Figure 11.13 shows the PDP as a function of EOT, which has been computed with Equation (11.6) using the extracted $d_{\mathrm{ox}}/\varepsilon_{\mathrm{ox}}$-dependence displayed in Figure 11.6. In order to determine the capacitors, we consider again a gate-all-around nanowire FET with 1D electronic transport; the effective mass is assumed to be $m^\star = 0.19$, i. e., the light effective mass of silicon. Hence, the quantum capacitance (per length) is approximated as $C_q \approx e^2 D_{\mathrm{1D}}(E_f^s - \Phi_c) = e^2 \frac{2}{h}\sqrt{\frac{2m^\star}{E_f^s - \Phi_c}}$. As an upper limit for C_q, we use $E_f^s - \Phi_c \approx 2k_B T$, and hence $C_q \approx 9.24 \cdot \frac{1}{\sqrt{T}}$ nF/m with T in Kelvin. Assuming a quadruple gate-all-around electrode [53] and a silicon nanowire with $d_{\mathrm{ch}} = 4$ nm, $C_{\mathrm{ox}} \approx 0.553/\mathrm{EOT}$ nF/m where EOT is in nanometers. Hence, in the targeted temperature range for cryogenic electronics (approximately 1 K) $C_{\mathrm{ox}} < C_q$.

In summary, for the targeted low operating temperatures of cryogenic electronics, scaling C_{ox} even by utilizing dielectrics with very high dielectric constants appears to be a viable approach to reduce the overall power consumption.

11.5 Dopants in Cryogenic MOSFETs

The ability to manipulate the conductivity of semiconductors via appropriate impurity doping is one of the key techniques that enables the functionality of virtually all semiconductor devices. The behavior of dopants in cryogenic devices deserves therefore a separate consideration.

11.5.1 Dopants in Source/Drain in Nanoscale Cryogenic FETs

In nondegenerately *n*-(*p*-)doped source and drain regions, the Fermi level will move in between the conduction (valence) band and the dopant levels (see Figure 4.1) when cooled to cryogenic temperatures. As a result, the semiconductor freezes out since there is not enough thermal energy available to excite an electron into the conduction band (to occupy an acceptor level). This yields high source/drain resistances prohibiting a low power operation. Hence, cryogenic MOSFETs require degenerately doped source and drain contacts that prevent carrier freeze-out due to the formation of a dopant band that overlaps with the conduction (valence) band (see Figure 4.2). However, at nanoscale dimensions, required to ensure an effective suppression of short-channel effects and to yield appropriately small λ_{ch} to reduce band-tailing, the ionization energy E_{ion} of dopants increases due to dielectric mismatch as well as due to carrier confinement (discussed in detail in Section 4.3.2). In turn, the increase of E_{ion}[4] leads to a deactivation of the dopants already at room temperature. At cryogenic temperatures, this effect becomes more severe. Thus, an even higher doping concentration is required to ensure that the dopant band straddles the energy range between dopant level and conduction(valence) band to prevent freeze-out (see the discussion with the simple model of a dopant band in Section 4.1.1). Ultimately, this may become impossible and, therefore, alternatives for regular doping need to be employed.

One approach to circumvent issues with dopants is to use $NiSi_2$ together with dopant segregation in order to combine metallic source/drain regions with an appropriate reduction of the Schottky barrier that builds up at the $NiSi_2$-Si interface. In fact, the GAA nanowire FET shown in Figure 11.8, left panel, used this approach. As apparent from the experimental results, the dopants at the $NiSi_2$/Si interface do not freeze out. The reason for this is the fact that the $NiSi_2$ is diffused so far into the nanowire that there is no ungated underlap region anymore. As a result, the doped area is gated and it has been discussed in Section 4.3.1 that the presence of a gate provides the necessary screening in order to avoid deactivation.

A second alternative may be to exploit the NESSIAS effect discussed in Section 4.4. Since this effect does not depend on temperature, it is expected that the electronic structure shift observed in ultrathin silicon nanostructures will persist at cryogenic temperatures. Because *n*- as well as *p*-type doping can be realized with appropriate SiO_2 and SiN coatings, this may be an interesting approach for future nanoscale cryogenic MOSFETs.

4 Remember that in the case of *n*-type doping (the same is true for *p*-type doping considering the acceptor level E_a and the valence band E_v) $E_{ion} = E_c - E_d$ with E_c being the conduction band edge and E_d the donor level. As a result, a dielectric mismatch and carrier confinement increase $E_c - E_d$, i. e., they move E_d further toward midgap.

11.5.2 Dopants in the Channel of Nanoscale Cryogenic FETs

Dopants in the channel behave differently from dopants in source and drain. In a regular MOSFET device, one usually considers *p*-type dopants in the channel of an *n*-type transistor. While the *p*-type concentration is low enough and will lead to a freeze-out of the bulk, applying an appropriately high gate voltage will certainly activate the dopants as illustrated in Figure 11.14. Therefore, a depletion capacitance is present and one needs

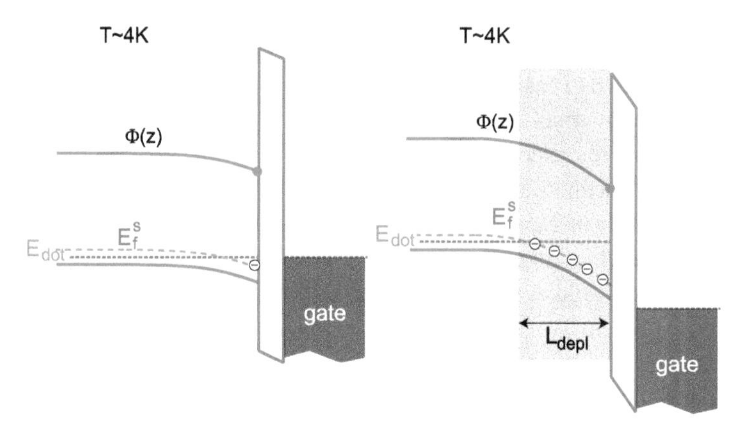

Figure 11.14: MOS capacitor with *p*-type silicon substrate at $T = 4$ K for two different gate voltages. The Fermi level is shifted in between the valence band and the dopant (acceptor) level within the Si bulk in case of cooling to cryogenic temperatures. For larger gate voltages (right panel), field-induced activation of acceptors occurs even though the bulk freezes out at cryogenic temperatures.

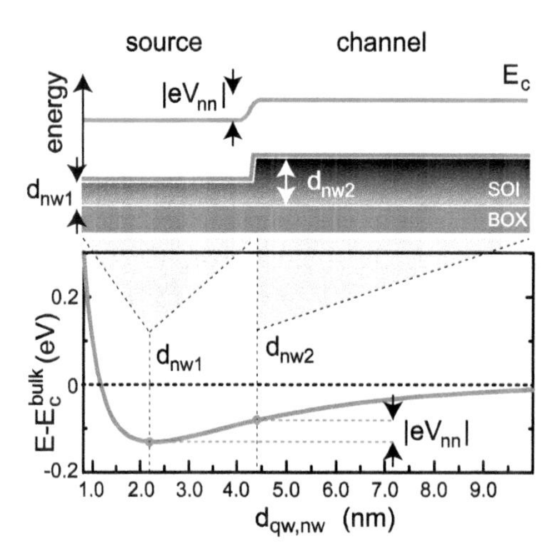

Figure 11.15: Realization of a *n-n* region with potential barrier $|eV_{nn}|$ in the meV-range exploiting the NESSIAS effect at different SOI thicknesses. The achievable small potential barriers can be used to realize the required low threshold voltages of cryogenic MOSFETs.

to ensure that $C_{ox} \gg C_{depl}$ to not degrade the inverse subthreshold slope. However, even if this is the case a strong potential variation within the channel is obtained that lead to Coulomb oscillations and strong device-to-device variability.

As a consequence of the discussion above, dopants should be avoided altogether, i. e., in the source/drain contacts as well as within the channel. In principle, the NESSIAS effect could be used in the entire device structure. If the thickness of the ultrathin silicon slab (silicon nanowire) is made slightly thicker in the channel region compared to the source/drain regions, a very shallow potential barrier of size $|eV_{nn}|$ can be realized as depicted in Figure 11.15, which can be used to achieve the required low threshold voltages of cryogenic MOSFETs. Obviously, in this case, an interface layer consisting of SiN can only be used for the realization of *p*-type cryogenic FETs.

Exercises

Exercises together with solutions are accessible via the QR code.

A Color Map for 2D Materials

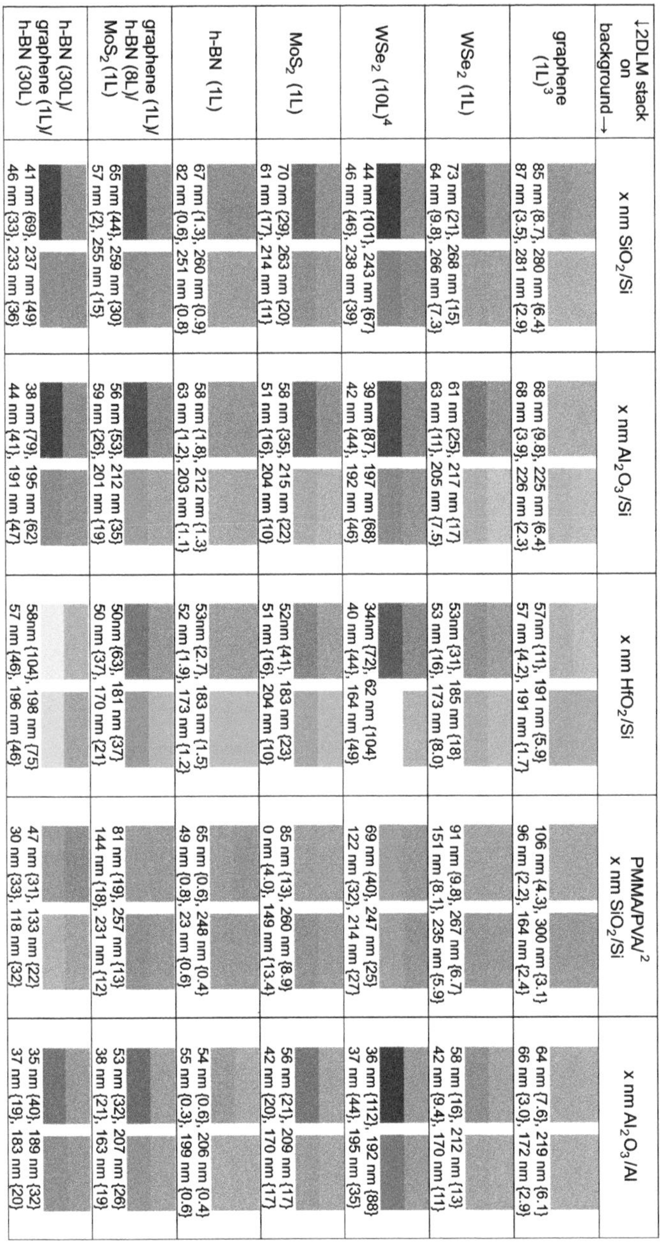

↓2DLM stack on background→	x nm SiO_2/Si	x nm Al_2O_3/Si	x nm HfO_2/Si	PMMA/PVA/[2] x nm SiO_2/Si	x nm Al_2O_3/Al
graphene (1L)[3]	85 nm {8.7}, 280 nm {6.4} 87 nm {3.5}, 281 nm {2.9}	68 nm {9.8}, 225 nm {6.4} 68 nm {3.9}, 226 nm {2.3}	57nm {11}, 191 nm {5.9} 57 nm {4.2}, 191 nm {1.7}	106 nm {4.3}, 300 nm {3.1} 96 nm {2.2}, 164 nm {2.4}	64 nm {7.6}, 219 nm {6.1} 66 nm {3.0}, 172 nm {2.9}
WSe_2 (1L)	73 nm {21}, 268 nm {15} 64 nm {9.8}, 266 nm {7.3}	61 nm {25}, 217 nm {17} 63 nm {11}, 205 nm {7.5}	53nm {31}, 185 nm {18} 53 nm {16}, 173 nm {8.0}	91 nm {9.8}, 267 nm {6.7} 151 nm {8.1}, 235 nm {5.9}	58 nm {16}, 212 nm {13} 42 nm {9.4}, 170 nm {11}
WSe_2 (10L)[4]	44 nm {101}, 243 nm {67} 46 nm {46}, 238 nm {39}	39 nm {87}, 197 nm {68} 42 nm {44}, 192 nm {46}	34nm {72}, 62 nm {104} 40 nm {44}, 164 nm {49}	69 nm {40}, 247 nm {25} 122 nm {32}, 214 nm {27}	36 nm {112}, 192 nm {88} 37 nm {44}, 195 nm {35}
MoS_2 (1L)	70 nm {29}, 263 nm {20} 61 nm {17}, 214 nm {11}	58 nm {35}, 215 nm {22} 51 nm {16}, 204 nm {10}	52nm {41}, 183 nm {23} 51 nm {16}, 204 nm {10}	85 nm {13}, 260 nm {8.9} 0 nm {4.0}, 149 nm {13.4}	56 nm {21}, 209 nm {17} 42 nm {20}, 170 nm {17}
h-BN (1L)	67 nm {1.3}, 260 nm {0.9} 82 nm {0.6}, 251 nm {0.8}	58 nm {1.8}, 212 nm {1.3} 63 nm {1.2}, 203 nm {1.1}	53nm {2.7}, 183 nm {1.5} 52 nm {1.9}, 173 nm {1.2}	65 nm {0.6}, 248 nm {0.4} 49 nm {0.8}, 23 nm {0.6}	54 nm {0.6}, 206 nm {0.4} 55 nm {0.3}, 199 nm {0.6}
graphene (1L)/ h-BN (8L)/ MoS_2 (1L)	65 nm {44}, 259 nm {30} 57 nm {2}, 255 nm {15}	56 nm {53}, 212 nm {35} 59 nm {26}, 201 nm {19}	50nm {63}, 181 nm {37} 50 nm {37}, 170 nm {21}	81 nm {19}, 257 nm {13} 144 nm {18}, 231 nm {12}	53 nm {32}, 207 nm {26} 38 nm {21}, 163 nm {19}
h-BN (30L)/ graphene (1L)/ h-BN (30L)	41 nm {69}, 237 nm {49} 46 nm {33}, 233 nm {36}	38 nm {79}, 195 nm {62} 44 nm {41}, 191 nm {47}	58nm {104}, 198 nm {75} 57 nm {46}, 196 nm {46}	47 nm {31}, 133 nm {22} 30 nm {33}, 118 nm {32}	35 nm {40}, 189 nm {32} 37 nm {19}, 183 nm {20}

Figure A.1: Color maps for identifying monolayers of different 2D materials on various layered dielectric substrates [196].

https://doi.org/10.1515/9783111054421-012

Bibliography

[1] A. Agarwal and M. J. Kushner. Plasma atomic layer etching using conventional plasma equipment. *J. Vac. Sci. Technol., A*, 27(1):37–50, 2009.

[2] B. S. Aldinger and M. A. Hines. Si(100) etching in aqueous fluoride solutions: Parallel etching reactions lead to ph-dependent nanohillock formation or atomically flat surfaces. *J. Phys. Chem. C*, 116(40):21499–21507, 2012.

[3] P. Allongue, C. Henry de Villeneuve, S. Morin, R. Boukherroub, and D. D. M. Wayner. The preparation of flat H–Si(111) surfaces in 40. *Electrochim. Acta*, 45(28):4591–4598, 2000.

[4] C. Alper, P. Palestri, J. L. Padilla, and A. M. Ionescu. The electron-hole bilayer TFET: Dimensionality effects and optimization. *IEEE Trans. Electron Devices*, 63(6):2603–2609, 2016.

[5] M. Andersen, C. Panosetti, and K. Reuter. A practical guide to surface kinetic Monte Carlo simulations. *Front. Chem.*, 7, 2019.

[6] J. Appenzeller, J. A. del Alamo, R. Martel, K. Chan, and P. Solomon. Ultrathin 600°c wet thermal silicon dioxide. *Electrochem. Solid-State Lett.*, 3(2):84, 1999.

[7] J. Appenzeller, J. Knoch, M. T. Bjork, H. Riel, H. Schmid, and W. Riess. Toward nanowire electronics. *IEEE Trans. Electron Devices*, 55(11):2827–2845, 2008.

[8] J. Appenzeller, J. Knoch, V. Derycke, R. Martel, S. Wind, and Ph. Avouris. Field-modulated carrier transport in carbon nanotube transistors. *Phys. Rev. Lett.*, 89:126801, Aug 2002.

[9] J. Appenzeller, J. Knoch, M. Radosavljevic, and Ph. Avouris. Multimode transport in Schottky-barrier carbon-nanotube field-effect transistors. *Phys. Rev. Lett.*, 92:226802, Jun 2004.

[10] J. Appenzeller, J. Knoch, E. Tutuc, M. Reuter, and S. Guha. Dual-gate silicon nanowire transistors with nickel silicide contacts. In *2006 International Electron Devices Meeting*, pages 1–4, Dec 2006.

[11] J. Appenzeller, Y.-M. Lin, J. Knoch, and Ph. Avouris. Band-to-band tunneling in carbon nanotube field-effect transistors. *Phys. Rev. Lett.*, 93:196805, 2004.

[12] J. Appenzeller, Y.-M. Lin, J. Knoch, Z. Chen, and Ph. Avouris. Comparing carbon nanotube transistors - the ideal choice: a novel tunneling device design. *IEEE Trans. Electron Devices*, 52:2568, 2005.

[13] J. Appenzeller, R. Martel, Ph. Avouris, J. Knoch, J. Scholvin, J. A. del Alamo, P. Rice, and P. Solomon. Sub-40 nm SOI v-groove n-MOSFET. *IEEE Electron Device Lett.*, 23:100–102, 2002.

[14] J. Appenzeller, R. Martel, J. Knoch, K. Chan, Ph. Avouris, J. Benedict, M. Tanner, S. Thomas, K. L. Wang, J. A. del Alamo, and P. Solomon. Scheme for the fabrication of ultrashort channel MOSFETs. *Appl. Phys. Lett.*, 77:298–300, 2000.

[15] V. Ariel and A. Natan. Electron effective mass in graphene. In *2013 International Conference on Electromagnetics in Advanced Applications (ICEAA)*, pages 696–698, 2013.

[16] N. Arjmandi, L. Lagae, and G. Borghs. Enhanced resolution of poly(methyl methacrylate) electron resist by thermal processing. *J. Vac. Sci. Technol., B Microelectron. Nanometer Struct. Process. Meas. Phenom.*, 27(4):1915–1918, 2009.

[17] S. D. Athavale and D. J. Economou. Molecular dynamics simulation of atomic layer etching of silicon. *J. Vac. Sci. Technol., A*, 13(3):966–971, 1995.

[18] S. D. Athavale and D. J. Economou. Realization of atomic layer etching of silicon. *J. Vac. Sci. Technol., B Microelectron. Nanometer Struct. Process. Meas. Phenom.*, 14(6):3702–3705, 1996.

[19] C. P. Auth and J. D. Plummer. Scaling theory of cylindrical, fully-depleted, surrounding-gate MOSFETs. *IEEE Electron Device Lett.*, 18:74–76, 1997.

[20] A. Beckers, F. Jazaeri, and C. Enz. Inflection phenomenon in cryogenic MOSFET behavior. *IEEE Trans. Electron Devices*, 67(3):1357–1360, 2020.

[21] A. Beckers, P. Jazaeri, and C. Enz. Theoretical limit of low temperature subthreshold swing in field-effect transistors. *IEEE Electron Device Lett.*, 41(2):276–279, 2020.

[22] K. K. Bhuwalka, J. Schulze, and I. Eisele. Performance enhancement of vertical tunnel field-effect transistor with SiGe in the δp^+ layer. *Jpn. J. Appl. Phys.*, 43(7A):4073–4078, Jul 2004.

https://doi.org/10.1515/9783111054421-013

[23] K. K. Bhuwalka, S. Sedlmaier, A. K. Ludsteck, C. Tolksdorf, J. Schulze, and I. Eisele. Vertical tunnel field-effect transistor. *IEEE Trans. Electron Devices*, 51(2):279–282, 2004.
[24] M. T. Bjoerk, O. Hayden, J. Knoch, E. Loertscher, H. E. Riel, W. H. Riess, and H. Schmid. Impact ionization field-effect transistor. *US 2011/0049476 A1*, 2011.
[25] M. T. Bjoerk, O. Hayden, H. Schmid, H. Riel, and W. Riess. Vertical surround-gated silicon nanowire impact ionization field-effect transistors. *Appl. Phys. Lett.*, 90(14):142110, 2007.
[26] M. T. Bjoerk, S. F. Karg, J. Knoch, H. E. Riel, W. H. Riess, and H. Schmid. Metal-oxide-semiconductor device including a multiple-layer energy filter. *US8129763B2*, 2012.
[27] M. T. Bjork, O. Hayden, J. Knoch, H. Riel, H. Schmid, and W. Riess. Impact ionization FETs based on silicon nanowires. In *2007 65th Annual Device Research Conference*, pages 171–172, 2007.
[28] M. T. Björk, H. Schmid, J. Knoch, H. Riel, and W. Riess. Donor deactivation in silicon nanostructures. *Nat. Nanotechnol.*, 4:103–107, 2009.
[29] P. E. Blöchl, O. Jepsen, and O. K. Andersen. Improved tetrahedron method for brillouin-zone integrations. *Phys. Rev. B*, 49:16223–16233, Jun 1994.
[30] Ch. Blömers, J. G. Lu, L. Huang, C. Witte, D. Grützmacher, H. Lüth, and Th. Schäpers. Electronic transport with dielectric confinement in degenerate inn nanowires. *Nano Lett.*, 12(6):2768–2772, 2012. PMID: 22494319.
[31] M. Bockrath, J. Hone, A. Zettl, P. L. McEuen, A. G. Rinzler, and R. E. Smalley. Chemical doping of individual semiconducting carbon-nanotube ropes. *Phys. Rev. B*, 61:R10606–R10608, Apr 2000.
[32] H. Bohuslavskyi, A. G. M. Jansen, S. Barraud, V. Barral, M. Cassé, L. Le Guevel, X. Jehl, L. Hutin, B. Bertrand, G. Billiot, G. Pillonnet, F. Arnaud, P. Galy, S. De Franceschi, M. Vinet, and M. Sanquer. Cryogenic subthreshold swing saturation in FD-SOI MOSFETs described with band broadening. *IEEE Electron Device Lett.*, 40(5):784–787, 2019.
[33] A. Bolognesi, A. Di Carlo, and P. Lugli. Influence of carrier mobility and contact barrier height on the electrical characteristics of organic transistors. *Appl. Phys. Lett.*, 81(24):4646–4648, 2002.
[34] J. M. Boter, J. P. Dehollain, J. P. G. van Dijk, Y. Xu, T. Hensgens, R. Versluis, H. W. L. Naus, J. S. Clarke, M. Veldhorst, F. Sebastiano, and L. M. K. Vandersypen. Spiderweb array: A sparse spin-qubit array. *Phys. Rev. Appl.*, 18:024053, Aug 2022.
[35] T. B. Boykin, G. Klimeck, R. C. Bowen, and R. Lake. Effective-mass reproducibility of the nearest-neighbor sp^3s^* models: Analytic results. *Phys. Rev. B*, 56:4102–4107, Aug 1997.
[36] E. Buitrago, M. Fernández-Bolaños, S. Rigante, C. F. Zilch, N. S. Schröter, A. M. Nightingale, and A. M. Ionescu. The top-down fabrication of a 3d-integrated, fully CMOS-compatible FET biosensor based on vertically stacked SiNWs and FinFETs. *Sens. Actuators B, Chem.*, 193:400–412, 2014.
[37] Q. Cao, S.-J. Han, J. Tersoff, A. D. Franklin, Y. Zhu, Z. Zhang, G. S. Tulevski, J. Tang, and W. Haensch. End-bonded contacts for carbon nanotube transistors with low, size-independent resistance. *Science*, 350:68–72, 2015.
[38] C. A. Chavarin, A. A. Sagade, D. Neumaier, G. Bacher, and W. Mertin. On the origin of contact resistances in graphene devices fabricated by optical lithography. *Appl. Phys. A*, 122:58, 2016.
[39] Z. Chen and J. Appenzeller. Gate modulation of graphene contacts - on the scaling of graphene FETs. In *2009 Symposium on VLSI Technology*, pages 128–129, 2009.
[40] Z. Chen, J. Appenzeller, J. Knoch, Y.-M. Lin, and P. Avouris. The role of metal-nanotube contact in the performance of carbon nanotube field-effect transistors. *Nano Lett.*, 5(7):1497–1502, 2005. PMID: 16178264.
[41] Z. Chen, Y.-M. Lin, M. J. Rooks, and P. Avouris. Graphene nano-ribbon electronics. *Physica E, Low-Dimens. Syst. Nanostruct.*, 40(2):228–232, 2007.
[42] T. Chiang. A novel scaling theory for fully depleted, multiple-gate MOSFET, including effective number of gates (engs). *IEEE Trans. Electron Devices*, 61(2):631–633, 2014.
[43] W.-J. Cho, W.-K. Chin, and C.-T. Kuo. Effects of alcoholic moderators on anisotropic etching of silicon in aqueous potassium hydroxide solutions. *Sens. Actuators A, Phys.*, 116(2):357–368, 2004.

[44] Y. Cho and N. W. Cheung. Low temperature Si direct bonding by plasma activation. *MRS Online Proceedings Library (OPL)*, 657:EE9.2, 2000.
[45] Y.-K. Choi, T.-J. King, and C. Hu. A spacer patterning technology for nanoscale CMOS. *IEEE Trans. Electron Devices*, 49(3):436–441, 2002.
[46] K. K. Christenson, J. W. Butterbaugh, T. J. Wagener, N. P. Lee, B. Schwab, M. Fussy, and J. Diedrick. All wet stripping of implanted photoresist. *Solid State Phenom.*, 134:109–112, 2008.
[47] C.-H. Chu, H.-C. Lin, C.-H. Yeh, Z.-Y. Liang, M.-Y. Chou, and P.-W. Chiu. End-bonded metal contacts on WSe_2 field-effect transistors. *ACS Nano*, 13(7):8146–8154, 2019. PMID: 31244047.
[48] T. Chu and Z. Chen. Achieving large transport bandgaps in bilayer graphene. *Nano Res.*, 8:3228–3236, 2015.
[49] D. S. Chuu, C. M. Hsiao, and W. N. Mei. Hydrogenic impurity states in quantum dots and quantum wires. *Phys. Rev. B*, 46:3898–3905, Aug 1992.
[50] I. T. Clark, B. S. Aldinger, A. Gupta, and M. A. Hines. Aqueous etching produces Si(100) surfaces of near-atomic flatness: Strain minimization does not predict surface morphology. *J. Phys. Chem. C*, 114(1):423–428, 2010.
[51] K. P. Clark, W. P. Kirk, and A. C. Seabaugh. Nonparabolicity effects in the bipolar quantum-well resonant-tunneling transistor. *Phys. Rev. B*, 55:7068–7072, Mar 1997.
[52] M. Claus, A. Fediai, S. Mothes, J. Knoch, D. Ryndyk, S. Blawid, G. Cuniberti, and M. Schröter. Towards a multiscale modeling framework for metal-CNT interfaces. In *2014 International Workshop on Computational Electronics (IWCE)*, pages 1–3, 2014.
[53] J.-P. Colinge. *Silicon-On-Insulator Technology: Materials to VLSI*. Kluwer Academic Publishers, 1991.
[54] D. Connelly, C. Faulker, P. A. Clifton, and D. E. Grupp. Fermi-level depinning for low-barrier Schottky source/drain transistors. *Appl. Phys. Lett.*, 88:012105, 2006.
[55] B. Cord, J. Lutkenhaus, and K. K. Berggren. Optimal temperature for development of poly(methylmethacrylate). *J. Vac. Sci. Technol., B Microelectron. Nanometer Struct. Process. Meas. Phenom.*, 25(6):2013–2016, 2007.
[56] A. M. Cowley and S. M. Sze. Surface states and barrier height of metal-semiconductor systems. *J. Appl. Phys.*, 36(10):3212–3220, 1965.
[57] T. J. Dalton. Microtrench formation in polysilicon plasma etching over thin gate oxide. *J. Electrochem. Soc.*, 140(8):2395, 1993.
[58] S. Das. Two dimensional electrostrictive field effect transistor (2d-efet): A sub-60mv/decade steep slope device with high on current. *Sci. Rep.*, 6:34811, 2016.
[59] S. Das and J. Appenzeller. Where does the current flow in two-dimensional layered systems? *Nano Lett.*, 13(7):3396–3402, 2013. PMID: 23802773.
[60] S. Datta. *Electronic Transport in Mesoscopic Systems*. Cambridge University Press, 1995.
[61] J. Dauber, A. A. Sagade, M. Oellers, K. Watanabe, T. Taniguchi, D. Neumaier, and C. Stampfer. Ultra-sensitive hall sensors based on graphene encapsulated in hexagonal boron nitride. *Appl. Phys. Lett.*, 106(19):193501, 2015.
[62] B. E. Deal and A. S. Grove. General relationship for the thermal oxidation of silicon. *J. Appl. Phys.*, 36(12):3770–3778, 1965.
[63] C. R. Dean, A. F. Young, I. Meric1, C. Lee, L. Wang, S. Sorgenfrei, K. Watanabe, T. Taniguchi, P. Kim, K. L. Shepard, and J. Hone. Boron nitride substrates for high-quality graphene electronics. *Nat. Nanotechnol.*, 5:722–726, 2010.
[64] A. A. Demkov, L. R. C. Fonseca, E. Verret, J. Tomfohr, and O. F. Sankey. Complex band structure and the band alignment problem at the Si–high-k dielectric interface. *Phys. Rev. B*, 71:195306, May 2005.
[65] R. H. Dennard, F. H. Gaensslen, H.-N. Yu, V. L. Rideout, E. Bassous, and A. R. LeBlanc. Design of ion-implanted MOSFET's with very small physical dimensions. *IEEE J. Solid-State Circuits*, 9(5):256–268, 1974.
[66] M. Diarra, Y.-M. Niquet, C. Delerue, and G. Allan. Ionization energy of donor and acceptor impurities in semiconductor nanowires: Importance of dielectric confinement. *Phys. Rev. B*, 75:045301, Jan 2007.

[67] E. Dornel, T. Ernst, J. C. Barbé, J. M. Hartmann, V. Delaye, F. Aussenac, C. Vizioz, S. Borel, V. Maffini-Alvaro, C. Isheden, and J. Foucher. Hydrogen annealing of arrays of planar and vertically stacked Si nanowires. *Appl. Phys. Lett.*, 91(23):233502, 12 2007.

[68] D. J. Economou. Modeling and simulation of plasma etching reactors for microelectronics. *Thin Solid Films*, 365(2):348–367, 2000.

[69] M. Elwenspoek and H. Jansen. *Silicon Micromachining*. Cambridge University Press, 2008.

[70] M. Evaldsson, I. V. Zozoulenko, H. Xu, and T. Heinzel. Edge-disorder-induced Anderson localization and conduction gap in graphene nanoribbons. *Phys. Rev. B*, 78:161407, Oct 2008.

[71] D. B. Fenner, D. K. Biegelsen, and R. D. Bringans. Silicon surface passivation by hydrogen termination: A comparative study of preparation methods. *J. Appl. Phys.*, 66(1):419–424, 1989.

[72] S. Fischer, H. I. Kremer, B. Berghoff, T. Maß, T. Taubner, and J. Knoch. Dopant-free complementary metal oxide silicon field effect transistors. *Phys. Status Solidi A*, 213(6):1494–1499, 2016.

[73] D. S. Fisher and P. A. Lee. Relation between conductivity and transmission matrix. *Phys. Rev. B*, 23:6851–6854, 1981.

[74] H. Flietner. The e(k) relation for a two-band scheme of semiconductors and the application to the metal-semiconductor contact. *Phys. Status Solidi B*, 54(1):201–208, 1972.

[75] H. Flietner. Spectrum and nature of surface states. *Surf. Sci.*, 46(1):251–264, 1974.

[76] A. Franklin and Z. Chen. Length scaling of carbon nanotube transistors. *Nat. Nanotechnol.*, 5:858–862, 2010.

[77] A. D. Franklin, M. Luisier, S.-J. Han, G. Tulevski, C. M. Breslin, L. Gignac, M. S. Lundstrom, and W. Haensch. Sub-10 nm carbon nanotube transistor. *Nano Lett.*, 12(2):758–762, 2012. PMID: 22260387.

[78] M. Frentzen, M. Michailow, K. Ran, N. Wilck, J. Mayer, S. C. Smith, D. König, and J. Knoch. Fabrication of ultrasmall Si encapsulated in silicon dioxide and silicon nitride as alternative to impurity doping. *Phys. Status Solidi A*, 220(13):2300066, 2023.

[79] P. Gallagher, K. Todd, and D. Goldhaber-Gordon. Disorder-induced gap behavior in graphene nanoribbons. *Phys. Rev. B*, 81:115409, Mar 2010.

[80] P. Galy, L. J. Camirand, P. Lemieux, F. Arnaud, D. Drouin, and M. Pioro-Ladrière. Cryogenic temperature characterization of a 28-nm FD-SOI dedicated structure for advanced CMOS and quantum technologies co-integration. *IEEE J. Electron Devices Society*, 6:594–600, 2018.

[81] W. Gannett, W. Regan, K. Watanabe, T. Taniguchi, M. F. Crommie, and A. Zettl. Boron nitride substrates for high mobility chemical vapor deposited graphene. *Appl. Phys. Lett.*, 98(24):242105, 2011.

[82] A. K. Geim and K. S. Novoselov. The rise of graphene. *Nat. Mater.*, 6:183–191, 2007.

[83] G. Ghibaudo, M. Aouad, M. Casse, S. Martinie, T. Poiroux, and F. Balestra. On the modelling of temperature dependence of subthreshold swing in MOSFETs down to cryogenic temperature. *Solid-State Electron.*, 170:107820, 2020.

[84] H. Ghoneim, J. Knoch, H. Riel, D. Webb, M. T. Bjoerk, E. Loertscher, S. Karg, H. Schmid, and W. Riess. Suppression of the ambipolar behavior in metal source/drain field-effect transistors. *Appl. Phys. Lett.*, 95:213504, 2009.

[85] E. Gnani, P. Maiorano, S. Reggiani, A. Gnudi, and G. Baccarani. Performance limits of superlattice-based steep-slope nanowire FETs. In *2011 International Electron Devices Meeting*, pages 5.1.1–5.1.4, 2011.

[86] E. Gnani, S. Reggiani, A. Gnudi, and G. Baccarani. Steep-slope nanowire FET with a superlattice in the source extension. *Solid-State Electron.*, 65–66:108–113, 2011. Selected Papers from the ESSDERC 2010 Conference.

[87] S. Gomes dos Santos Filho. A less critical cleaning procedure for silicon wafer using diluted HF dip and boiling in isopropyl alcohol as final steps. *J. Electrochem. Soc.*, 142(3):902, 1995.

[88] K. Gopalakrishnan, P. B. Griffin, and J. D. Plummer. I-mos: a novel semiconductor device with a subthreshold slope lower than kt/q. In *Digest. International Electron Devices Meeting*, pages 289–292, 2002.

[89] M. A. Gosalvez, Y. Zhou, Y. Zhang, G. Zhang, Y. Li, and Y. Xing. Simulation of microloading and ARDE in DRIE. In *2015 Transducers - 2015 18th International Conference on Solid-State Sensors, Actuators and Microsystems (TRANSDUCERS)*, pages 1255–1258, June 2015.

[90] D. Graf, F. Molitor, K. Ensslin, C. Stampfer, A. Jungen, C. Hierold, and L. Wirtz. Spatially resolved Raman spectroscopy of single- and few-layer graphene. *Nano Lett.*, 7(2):238–242, 2007. PMID: 17297984.

[91] N. E. Grant and J. D. Murphy. Temporary surface passivation for characterisation of bulk defects in silicon: A review. *Phys. Status Solidi RRL*, 11(11):1700243, 2017.

[92] H. Gross, S. Heidenreich, M.-A. Henn, G. Dai, F. Scholze, and M. Bär. Modelling line edge roughness in periodic line-space structures by Fourier optics to improve scatterometry. *J. Eur. Opt. Soc., Rapid Publ.*, 9(0), 2014.

[93] L. L. Guevel, G. Billiot, X. Jehl, S. De Franceschi, M. Zurita, Y. Thonnart, M. Vinet, M. Sanquer, R. Maurand, A. G. M. Jansen, and G. Pillonnet. 19.2 a 110 mk 295 μw 28 nm FDSOI CMOS quantum integrated circuit with a 2.8 GHz excitation and nA current sensing of an on-chip double quantum dot. In *2020 IEEE International Solid-State Circuits Conference - (ISSCC)*, pages 306–308, 2020.

[94] S. Guha and V. Narayanan. High-κ/metal gate science and technology. *Annu. Rev. Mater. Res.*, 39(1):181–202, 2009.

[95] M. H. D. Guimarães, H. Gao, Y. Han, K. Kang, S. Xie, C.-J. Kim, D. A. Muller, D. C. Ralph, and J. Park. Atomically thin ohmic edge contacts between two-dimensional materials. *ACS Nano*, 10(6):6392–6399, 2016. PMID: 27299957.

[96] Y. Guo, D. Liu, and J. Robertson. 3d behavior of Schottky barriers of 2d transition-metal dichalcogenides. *ACS Appl. Mater. Interfaces*, 7(46):25709–25715, 2015. PMID: 26523332.

[97] S. Gupta, P. P. Manik, R. K. Mishra, A. Nainani, M. C. Abraham, and S. Lodha. Contact resistivity reduction through interfacial layer doping in metal-interfacial layer-semiconductor contacts. *J. Appl. Phys.*, 113(23):234505, 06 2013.

[98] H.-C. Han, H.-L. Chiang, I. P. Radu, and C. Enz. Analytical modeling of source-to-drain tunneling current down to cryogenic temperatures. *IEEE Electron Device Lett.*, 44(5):717–720, 2023.

[99] M. Y. Han, J. C. Brant, and P. Kim. Electron transport in disordered graphene nanoribbons. *Phys. Rev. Lett.*, 104:056801, Feb 2010.

[100] Y. Han, J. Sun, J.-H. Bae, D. Grützmacher, J. Knoch, and Q.-T. Zhao. High performance 5 nm Si nanowire FETs with a record small SS = 2.3 mv/dec and high transconductance at 5.5 k enabled by dopant segregated silicide source/drain. In *2023 IEEE Symposium on VLSI Technology and Circuits (VLSI Technology and Circuits)*, pages 1–2, 2023.

[101] E. Hatta, J. Nagao, and K. Mukasa. Tunneling through a narrow-gap semiconductor with different conduction- and valence-band effective masses. *J. Appl. Phys.*, 79(3):1511–1514, 1996.

[102] C. Hedlund, H.-O. Blom, and S. Berg. Microloading effect in reactive ion etching. *J. Vac. Sci. Technol., A*, 12(4):1962–1965, 1994.

[103] A. Heinzig, T. Mikolajick, J. Trommer, D. Grimm, and W. M. Weber. Dually active silicon nanowire transistors and circuits with equal electron and hole transport. *Nano Lett.*, 13(9):4176–4181, 2013. PMID: 23919720.

[104] A. Heinzig, S. Slesazeck, F. Kreupl, T. Mikolajick, and W. M. Weber. Reconfigurable silicon nanowire transistors. *Nano Lett.*, 12(1):119–124, 2012. PMID: 22111808.

[105] M. M. Heyns, T. Bearda, I. Cornelissen, S. de Gendt, L. Loewenstein, P. W. Mertens, S. Mertens, M. Meuris, M. Schaekers, I. Teerlinck, R. Vos, and K. Wolke. Advanced cleaning strategies for ultraclean silicon surfaces. In *Proceedings of the 6th International Symposium of Electromechanical Society*, pages 3–16, 2000.

[106] G. Hills, C. Lau, A. Wright, S. Fuller, M. D. Bishop, T. Srimani, P. Kanhaiya, R. Ho, A. Amer, Y. Stein, D. Murphy, A. A. Chandrakasan, and M. M. Shulaker. Modern microprocessor built from complementary carbon nanotube transistors. *Nature*, 572:595–602, 2019.

[107] A. Hiraiwa and A. Nishida. Spectral analysis of line edge and line-width roughness with long-range correlation. *J. Appl. Phys.*, 108(3):034908, 2010.

[108] A. Hollmann, D. Jirovec, M. Kucharski, D. Kissinger, G. Fischer, and L. R. Schreiber. 30 GHz-voltage controlled oscillator operating at 4 K. *Rev. Sci. Instrum.*, 89(11):114701, 11 2018.

[109] http://hyperphysics.phy-astr.gsu.edu/hbase/Kinetic/menfre.html#c5.

[110] C. Hu, P. Patel, A. Bowonder, K. Jeon, S. H. Kim, W. Y. Loh, C. Y. Kang, J. Oh, P. Majhi, A. Javey, T. K. Liu, and R. Jammy. Prospect of tunneling green transistor for 0.1v CMOS. In *2010 International Electron Devices Meeting*, pages 16.1.1–16.1.4, 2010.

[111] K. Huang. *Statistical Mechanics*. John Wiley & Sons, 1987.

[112] P. Huang, T. Tanamoto, M. Goto, M. Takenaka, and S. Takagi. Investigation of electrical characteristics of vertical junction Si n-type tunnel FET. *IEEE Trans. Electron Devices*, 65(12):5511–5517, 2018.

[113] E. Icking, L. Banszerus, F. Wörtche, F. Volmer, P. Schmidt, C. Steiner, S. Engels, J. Hesselmann, M. Goldsche, K. Watanabe, T. Taniguchi, C. Volk, B. Beschoten, and C. Stampfer. Transport spectroscopy of ultraclean tunable band gaps in bilayer graphene. *Adv. Electron. Mater.*, 8(11):2200510, 2022.

[114] E. Icking, D. Emmerich, B. Beschoten, J. Knoch, M. C. Lemme, and C. Stampfer. Ultra steep slope cryogenic MOSFETs based on Bernal bilayer graphene. Submitted to *Nano Lett.*, 2023.

[115] T. Iffländer, S. Rolf-Pissarczyk, L. Winking, R. G. Ibrich, A. Al-Zubi, S. Blügel, and M. Wenderoth. Local density of states at metal-semiconductor interfaces: An atomic scale study. *Phys. Rev. Lett.*, 114:146804, Apr 2015.

[116] A. M. Ionescu and H. Riel. Tunnel field-effect transistors as energy-efficient electronic switches. *Nature*, 4793, 2011.

[117] M. Itano, F. W. Kern, M. Miyashita, and T. Ohmi. Particle removal from silicon wafer surface in wet cleaning process. *IEEE Trans. Semicond. Manuf.*, 6(3):258–267, Aug 1993.

[118] H. Jansen, H. Gardeniers, M. de Boer, M. Elwenspoek, and J. Fluitman. A survey on the reactive ion etching of silicon in microtechnology. *J. Micromech. Microeng.*, 6(1):14–28, Mar 1996.

[119] R. Jhaveri, V. Nagavarapu, and J. C. S. Woo. Asymmetric Schottky tunneling source SOI MOSFET design for mixed-mode applications. *IEEE Trans. Electron Devices*, 56(1):93–99, 2009.

[120] Y. Jia, X. Gong, P. Peng, Z. Wang, Z. Tian, L. Ren, Y. Fu, and H. Zhang. Toward high carrier mobility and low contact resistance: Laser cleaning of PMMA residues on graphene surfaces. *Nano-Micro Lett.*, 8:336–346, 2016.

[121] J. Joo, B. Y. Chow, and J. M. Jacobson. Nanoscale patterning on insulating substrates by critical energy electron beam lithography. *Nano Lett.*, 6(9):2021–2025, 2006. PMID: 16968019.

[122] F. Joucken, C. Bena, Z. Ge, E. A. Quezada-Lopez, F. Ducastelle, T. Tanagushi, K. Watanabe, and J. Velasco. Sublattice dependence and gate tunability of midgap and resonant states induced by native dopants in Bernal-stacked bilayer graphene. *Phys. Rev. Lett.*, 127:106401, Aug 2021.

[123] C. Jungemann, B. Richstein, T. Linn, and J. Knoch. Device modeling for admittance spectroscopy of pMOSFETs at cryogenic temperatures. Submitted to *IEEE Trans. Electron Dev.*, 2023.

[124] J. Kang, W. Liu, and K. Banerjee. High-performance MoS_2 transistors with low-resistance molybdenum contacts. *Appl. Phys. Lett.*, 104(9):093106, 2014.

[125] J. Kang, D. Shin, S. Bae, and B. H. Hong. Graphene transfer: key for applications. *Nanoscale*, 4:5527–5537, 2012.

[126] K.-H. Kao, T. R. Wu, H.-L. Chen, W.-J. Lee, N.-Y. Chen, W. C.-Y. Ma, C.-J. Su, and Y.-J. Lee. Subthreshold swing saturation of nanoscale MOSFETs due to source-to-drain tunneling at cryogenic temperatures. *IEEE Electron Device Lett.*, 41(9):1296–1299, 2020.

[127] T. Kato and S. Souma. Study of an application of non-parabolic complex band structures to the design for mid-infrared quantum cascade lasers. *J. Appl. Phys.*, 125(7):073101, 02 2019.

[128] W. Kern. The evolution of silicon wafer cleaning technology. *J. Electrochem. Soc.*, 137(6):1887–1892, 1990.

[129] W. Kern and D. Puotinen. Cleaning solutions based on hydrogen peroxide for use in silicon semiconductor technology. *RCA Rev.*, 31:187–206, 1970.

[130] M. A. Khayer and R. K. Lake. Effects of band-tails on the subthreshold characteristics of nanowire band-to-band tunneling transistors. *J. Appl. Phys.*, 110(7):074508, 10 2011.
[131] W.-B. Kim, T. Matsumoto, and H. Kobayashi. Ultrathin SiO_2 layer with an extremely low leakage current density formed in high concentration nitric acid. *J. Appl. Phys.*, 105(10):103709, 2009.
[132] W.-B. Kim, T. Matsumoto, and H. Kobayashi. Ultrathin SiO_2 layer with a low leakage current density formed with ~100 % nitric acid vapor. *Nanotechnology*, 21(11):115202, Feb 2010.
[133] C. Kittel and H. Kroemer. *Thermal Physics*. W.H. Freeman and Company, 1980.
[134] G. Klimeck, R. C. Bowen, T. B. Boykin, C. Salazar-Lazaro, T. A. Cwik, and A. Stoica. Si tight-binding parameters from genetic algorithm fitting. *Superlattices Microstruct.*, 27(2):77–88, 2000.
[135] J. Knoch. Chapter eight - nanowire tunneling field-effect transistors. In S. A. Dayeh, A. Fontcuberta i Morral and C. Jagadish, editors, *Semiconductor Nanowires II: Properties and Applications*, vol. 94 of Semiconductors and Semimetals, pages 273–295. Elsevier, 2016.
[136] J. Knoch and J. Appenzeller. Tunneling phenomena in carbon nanotube field-effect transistors. *Phys. Status Solidi A*, 205:679, 2008.
[137] J. Knoch and J. Appenzeller. Modeling of high-performance p-type iii–v heterojunction tunnel FETs. *IEEE Electron Device Lett.*, 31(4):305–307, 2010.
[138] J. Knoch, J. Appenzeller, B. Lengeler, R. Martel, P. Solomon, Ph. Avouris, Ch. Dieker, Y. Lu, K. L. Wang, J. Scholvin, and J. A. del Alamo. Technology for the fabrication of ultrashort channel metal-oxide-semiconductor field-effect transistors. *J. Vac. Sci Technol. A*, 19:1737–1741, 2001.
[139] J. Knoch, Z. Chen, and J. Appenzeller. Properties of metal graphene contacts. *IEEE Trans. Nanotechnol.*, 11:513–519, 2012.
[140] J. Knoch, B. Lengeler, and J. Appenzeller. Quantum simulations of an ultrashort channel single-gated n-MOSFET on SOI. *IEEE Trans. Electron Devices*, 49(7):1212–1218, 2002.
[141] J. Knoch, S. Mantl, and J. Appenzeller. Comparison of transport properties in carbon nanotube field-effect transistors with Schottky contacts and doped source/drain contacts. *Solid-State Electron.*, 49(1):73–76, 2005.
[142] J. Knoch, S. Mantl, and J. Appenzeller. Impact of the dimensionality on the performance of tunneling FETs: Bulk versus one-dimensional devices. *Solid-State Electron.*, 51:572–578, 2007.
[143] J. Knoch, S. Mantl, Y. Lin, Z. Chen, P. Avouris, and J. Appenzeller. An extended model for carbon nanotube field-effect transistors. In *Conference Digest [Includes 'Late News Papers' volume] Device Research Conference, 2004. 62nd DRC*, pages 135–136, vol. 1, 2004.
[144] J. Knoch and M. R. Müller. Electrostatic doping—controlling the properties of carbon-based FETs with gates. *IEEE Trans. Nanotechnol.*, 13(6):1044–1052, 2014.
[145] J. Knoch, B. Richstein, Y. Han, M. Frentzen, L. R. Schreiber, J. Klos, L. Raffauf, N. Wilck, D. König, and Q.-T. Zhao. Toward low-power cryogenic metal-oxide semiconductor field-effect transistors. *Phys. Status Solidi A*, 220(13):2300069, 2023.
[146] J. Knoch and B. Sun. Sub-linear current voltage characteristics of Schottky-barrier field-effect transistors. *IEEE Trans. Electron Devices*, 69(5):2243–2247, 2022.
[147] J. Knoch, M. Zhang, J. Appenzeller, and S. Mantl. Physics of ultrathin-body silicon-on-insulator Schottky-barrier field-effect transistors. *Appl. Phys. A*, 87:351–357, 2007.
[148] J. Knoch, M. Zhang, S. Feste, and S. Mantl. Dopant segregation in SOI Schottky-barrier MOSFETs. *Microelectron. Eng.*, 84:2563–2571, 2007.
[149] J. Knoch, M. Zhang, S. Mantl, and J. Appenzeller. On the performance of single-gated ultrathin-body SOI Schottky-barrier MOSFETs. *IEEE Trans. Electron Devices*, 53:1669–1674, 2006.
[150] L. Knoll, S. Richter, A. Nichau, A. Schäfer, K. K. Bourdelle, Q. T. Zhao, and S. Mantl. Gate-all-around Si nanowire array tunnelling FETs with high on-current of 75 μa/μm @ vdd=1.1v. In *2013 14th International Conference on Ultimate Integration on Silicon (ULIS)*, pages 97–100, 2013.
[151] H. Kobayashi Asuha, O. Maida, M. Takahashi, and H. Iwasa. Nitric acid oxidation of Si to form ultrathin silicon dioxide layers with a low leakage current density. *J. Appl. Phys.*, 94(11):7328–7335, 2003.

[152] R. Kohli and K. L. Mittal. *Developments in Surface Contamination and Cleaning, Volume 8: Cleaning Techniques*. Developments in Surface Contamination and Cleaning Series. Elsevier Science, 2015.

[153] D. Konig, D. Hiller, N. Wilck, B. Berghoff, M. Müller, S. Thakur, G. Di Santo, L. Petaccia, J. Mayer, S. Smith, and J. Knoch. Intrinsic ultrasmall nanoscale silicon turns n-/p-type with SiO_2/Si_3N_4-coating. *Beilstein J. Nanotechnol.*, 9:2255–2264, 8 2018.

[154] C. Kontis, M. R. Mueller, C. Kuechenmeister, K. T. Kallis, and J. Knoch. Optimizing the identification of mono- and bilayer graphene on multilayer substrates. *Appl. Opt.*, 51(3):385–389, Jan 2012.

[155] U. Kuenzelmann, M. R. Mueller, K. T. Kallis, F. Schuette, S. Menzel, S. Engels, J. Fong, C. Lin, J. Dysard, J. W. Bartha, and J. Knoch. Chemical-mechanical planarization of aluminium damascene structures. In *ICPT 2012 - International Conference on Planarization/CMP Technology*, pages 1–6, 2012.

[156] D. König, M. Frentzen, D. Hiller, N. Wilck, G. Di Santo, L. Petaccia, I. Píš, F. Bondino, E. Magnano, J. Mayer, J. Knoch, and S. C. Smith. Origin and quantitative description of the NESSIAS effect at Si nanostructures. *Adv. Phys. Res.*, 2(5):2200065, 2023.

[157] D. König, M. Frentzen, N. Wilck, B. Berghoff, I. Píš, S. Nappini, F. Bondino, M. Müller, S. Gonzalez, G. Di Santo, L. Petaccia, J. Mayer, S. Smith, and J. Knoch. Turning low-nanoscale intrinsic silicon highly electron-conductive by SiO_2 coating. *ACS Appl. Mater. Interfaces*, 13(17):20479–20488, 2021. PMID: 33878265.

[158] D. König, D. Hiller, S. Gutsch, and M. Zacharias. Energy offset between silicon quantum structures: Interface impact of embedding dielectrics as doping alternative. *Adv. Mater. Interfaces*, 1(9):1400359, 2014.

[159] D. König, D. Hiller, N. Wilck, B. Berghoff, M. Müller, S. Thakur, G. Di Santo, L. Petaccia, J. Mayer, S. Smith, and J. Knoch. Intrinsic ultrasmall nanoscale silicon turns n-/p-type with SiO_2/Si_3N_4-coating. *Beilstein J. Nanotechnol.*, 9:2255–2264, 2018.

[160] D. König, N. Wilck, D. Hiller, B. Berghoff, A. Meledin, G. Di Santo, L. Petaccia, J. Mayer, S. Smith, and J. Knoch. Electronic structure shift of deeply nanoscale silicon by SiO_2 versus Si_3N_4 embedding as an alternative to impurity doping. *Phys. Rev. Appl.*, 12:054050, Nov 2019.

[161] S. L. Lai, D. Johnson, and R. Westerman. Aspect ratio dependent etching lag reduction in deep silicon etch processes. *J. Vac. Sci. Technol., A*, 24(4):1283–1288, 2006.

[162] R. Lake, G. Klimeck, R. C. Bowen, and D. Jovanovic. Single and multiband modeling of quantum electron transport through layered semiconductor devices. *J. Appl. Phys.*, 81(12):7845–7869, 1997.

[163] V. Langrock, J. A. Krzywda, N. Focke, I. Seidler, L. R. Schreiber, and L. Cywiński. Blueprint of a scalable spin qubit shuttle device for coherent mid-range qubit transfer in disordered $Si/SiGe/SiO_2$. *PRX Quantum*, 4:020305, Apr 2023.

[164] F. Lanzerath, M. Goryll, D. Buca, M. Trinkaus, S. Mantl, J. Knoch, U. Breuer, W. Skorupa, and B. Gyhselen. Investigation of boron activation and diffusion in silicon-on-insulator and strained-silicon-on-insulator using rapid thermal annealing and flash lamp annealing. *J. Appl. Phys.*, 104:044908, 2008.

[165] C.-H. Lee, C.-L. Wang, H.-F. Lin, C.-Y. Chai, M.-Y. Hong, and C.-K. Ho. Toxicity of tetramethylammonium hydroxide: Review of two fatal cases of dermal exposure and development of an animal model. *Toxicol. Ind. Health*, 27(6):497–503, 2011.

[166] M. M. Lee and M. C. Wu. Thermal annealing in hydrogen for 3-d profile transformation on silicon-on-insulator and sidewall roughness reduction. *J. Microelectromech. Syst.*, 15(2):338–343, April 2006.

[167] A. F. M. Leenaars, J. A. M. Huethorst, and J. J. van Oekel. Marangoni drying: A new extremely clean drying process. *Langmuir*, 6:1701–1703, 1990.

[168] M. C. Lemme, T. J. Echtermeyer, M. Baus, and H. Kurz. A graphene field-effect device. *IEEE Electron Device Lett.*, 28(4):282–284, 03 2007.

[169] M. O. Li, D. Esseni, J. J. Nahas, D. Jena, and H. G. Xing. Two-dimensional heterojunction interlayer tunneling field effect transistors (thin-TFETs). *IEEE J. Electron Dev. Soc.*, 3(3):200–207, 2015.

[170] X. Li, X. Wang, L. Zhang, S. Lee, and H. Dai. Chemically derived, ultrasmooth graphene nanoribbon semiconductors. *Science*, 319:1229–1232, 2008.
[171] S. C. Lin and J. B. Kuo. Modeling the fringing electric field effect on the threshold voltage of FD SOI nMOS devices with the LDD/sidewall oxide spacer structure. *IEEE Trans. Electron Devices*, 50(12):2559–2564, Dec 2003.
[172] Y.-M. Lin, J. Appenzeller, J. Knoch, and P. Avouris. High-performance carbon nanotube field-effect transistor with tunable polarities. *IEEE Trans. Nanotechnol.*, 4(5):481–489, 2005.
[173] E. Lind, E. Memišević, A. W. Dey, and L. Wernersson. Iii-v heterostructure nanowire tunnel FETs. *IEEE J. Electron Dev. Soc.*, 3(3):96–102, 2015.
[174] G.-B. Liu, W.-Y. Shan, Y. Yao, W. Yao, and d. Xiao. Three-band tight-binding model for monolayers of group-vib transition metal dichalcogenides. *Phys. Rev. B*, 88:085433, Aug 2013.
[175] W. Liu, J. Kang, D. Sarkar, Y. Khatami, D. Jena, and K. Banerjee. Role of metal contacts in designing high-performance monolayer n-type WSe_2 field effect transistors. *Nano Lett.*, 13(5):1983–1990, 2013. PMID: 23527483.
[176] Y. Liu, J. Guo, E. Zhu, L. Liao, S.-J. Lee, M. Ding, I. Shakir, V. Gambin, Y. Huang, and X. Duan. Approaching the Schottky–Mott limit in van der Waals metal–semiconductor junctions. *Nature*, 557:696–712, 2018.
[177] P. Long, M. Povolotskyi, B. Novakovic, T. Kubis, G. Klimeck, and M. J. W. Rodwell. Design and simulation of two-dimensional superlattice steep transistors. *IEEE Electron Device Lett.*, 35(12):1212–1214, 2014.
[178] M. P. Lopez Sancho, J. M. Lopez Sancho, J. M. L. Sancho, and J. Rubio. Highly convergent schemes for the calculation of bulk and surface green functions. *J. Phys. F, Met. Phys.*, 15(4):851–858, Apr 1985.
[179] H. Lu and A. Seabaugh. Tunnel field-effect transistors: State-of-the-art. *IEEE J. Electron Dev. Soc.*, 2(4):44–49, 2014.
[180] H. C. Lu, T. Gustafsson, E. P. Gusev, and E. Garfunkel. An isotopic labeling study of the growth of thin oxide films on Si(100). *Appl. Phys. Lett.*, 67(12):1742–1744, 1995.
[181] M. Lundstrom. Elementary scattering theory of the Si MOSFET. *IEEE Electron Device Lett.*, 18(7):361–363, 1997.
[182] M. Lundstrom. *Fundamentals of Carrier Transport*. Cambridge University Press, 2nd edition, 2000.
[183] M. Lundstrom and C. Jeong. *Near-Equilibrium Transport: Fundamentals and Applications*. World Scientific, 2013.
[184] M. S. Lundstrom and D. A. Antoniadis. Compact models and the physics of nanoscale FETs. *IEEE Trans. Electron Devices*, 61(2):225–233, 2014.
[185] D. S. Macintyre, O. Ignatova, S. Thoms, and I. G. Thayne. Resist residues and transistor gate fabrication. *J. Vac. Sci. Technol., B Microelectron. Nanometer Struct. Process. Meas. Phenom.*, 27(6):2597–2601, 2009.
[186] M. T. Madzik, S. Asaad, A. Youssry, B. Joecker, K. M. Rudinger, E. Nielsen, K. C. Young, T. J. Proctor, A. D. Baczewski, A. Laucht, V. Schmitt, F. E. Hudson, K. M. Itoh, A. M. Jakob, B. C. Johnson, D. N. Jamieson, A. S. Dzurak, C. Ferrie, R. Blume-Kohout, and A. Morello. Precision tomography of a three-qubit donor quantum processor in silicon. *Nature*, 601:348–353, 2022.
[187] H. Z. Massoud. Thermal oxidation of silicon in dry oxygen growth-rate enhancement in the thin regime. *J. Electrochem. Soc.*, 132(11):2685, 1985.
[188] R. Maurand, X. Jehl, D. Kotekar-Patil, A. Corna, H. Bohuslavskyi, R. Lavieville, L. Hutin, S. Barraud, M. Vinet, M. Sanquer, and S. De Franceschi. A CMOS silicon spin qubit. *Nat. Commun.*, 7:13575, 2016.
[189] S. Mayo, K. F. Galloway, and T. F. Leedy. Radiation dose due to electron-gun metallization systems. *IEEE Trans. Nucl. Sci.*, 23(6):1875–1880, 1976.
[190] C. McConnell. *Particle Removal from Oxide, Nitride, and Bare Silicon Surfaces Using Direct-Displacement Isopropyl Alcohol (IPA) Drying*, pages 277–289. Springer US, Boston, MA, 1991.
[191] J. Meijer, B. Burchard, M. Domhan, C. Wittmann, T. Gaebel, I. Popa, F. Jelezko, and J. Wrachtrup. Generation of single color centers by focused nitrogen implantation. *Appl. Phys. Lett.*, 87(26):261909, 2005.

[192] M. Meuris and Pp. W. Mertens. The IMEC clean: A new concept for particle and metal removal on Si surfaces. *Solid State Technol.*, 38(7):109, 1995.

[193] T. Mikolajick, A. Heinzig, J. Trommer, T. Baldauf, and W. M. Weber. The RFET—a reconfigurable nanowire transistor and its application to novel electronic circuits and systems. *Semicond. Sci. Technol.*, 32(4):043001, Mar 2017.

[194] A. R. Mills, C. R. Guinn, M. J. Gullans, A. J. Sigillito, M. M. Feldman, E. Nielsen, and J. R. Petta. Two-qubit silicon quantum processor with operation fidelity exceeding 99 %. *Sci. Adv.*, 8:eabn5130, 2022.

[195] E. R. Mucciolo, A. H. Castro Neto, and C. H. Lewenkopf. Conductance quantization and transport gaps in disordered graphene nanoribbons. *Phys. Rev. B*, 79:075407, Feb 2009.

[196] M. R. Mueller, A. Gumprich, E. Ecik, K. T. Kallis, F. Winkler, B. Kardynal, I. Petrov, U. Kunze, and J. Knoch. Visibility of two-dimensional layered materials on various substrates. *J. Appl. Phys.*, 118(14):145305, 2015.

[197] M. R. Mueller, A. Gumprich, F. Schuette, K. Kallis, U. Kuenzelmann, S. Engels, C. Stampfer, N. Wilck, and J. Knoch. Buried triple-gate structures for advanced field-effect transistor devices. *Microelectron. Eng.*, 119:95–99, 2014. Micro/Nano Devices and Systems 2013.

[198] M. R. Mueller, R. Salazar, S. Fathipour, H. Xu, K. Kallis, U. Künzelmann, A. Seabaugh, J. Appenzeller, and J. Knoch. Gate-controlled WSe_2 transistors using a buried triple-gate structure. *Nanoscale Res. Lett.*, 11:512, 2016.

[199] W. Mönch. Barrier heights of real Schottky contacts explained by metal-induced gap states and lateral inhomogeneities. *J. Vac. Sci. Technol., B Microelectron. Nanometer Struct. Process. Meas. Phenom.*, 17(4):1867–1876, 1999.

[200] R. M. Y. Ng, T. Wang, F. Liu, X. Zuo, J. He, and M. Chan. Vertically stacked silicon nanowire transistors fabricated by inductive plasma etching and stress-limited oxidation. *IEEE Electron Device Lett.*, 30(5):520–522, 2009.

[201] S. V. Nguyen, D. Dobuzinsky, S. R. Stiffler, and G. Chrisman. Substrate trenching mechanism during plasma and magnetically enhanced polysilicon etching. *J. Electrochem. Soc.*, 138:1112, 1991.

[202] A. Noiri, K. Takeda, T. Nakajima, T. Kobayashi, A. Sammak, G. Scappucci, and S. Tarucha. Fast universal quantum gate above the fault-tolerance threshold in silicon. *Nature*, 601:338–342, 2022.

[203] K. S. Novoselov, A. K. Geim, S. V. Morozov, D. Jiang, Y. Zhang, S. V. Dubonos, I. V. Grigorieva, and A. A. Firsov. Electric field effect in atomically thin carbon films. *Science*, 306(5696):666–669, 2004.

[204] L. E. Ocola, M. Costales, and D. J. Gosztola. Development characteristics of polymethyl methacrylate in alcohol/water mixtures: a lithography and Raman spectroscopy study. *Nanotechnology*, 27(3):035302, Dec 2015.

[205] S. B. Orlinskii, J. Schmidt, E. J. J. Groenen, P. G. Baranov, C. de Mello Donegá, and A. Meijerink. Shallow donors in semiconductor nanoparticles: Limit of the effective mass approximation. *Phys. Rev. Lett.*, 94:097602, Mar 2005.

[206] G. Owen. Methods for proximity effect correction in electron lithography. *J. Vac. Sci. Technol., B Microelectron. Process. Phenom.*, 8(6):1889–1892, 1990.

[207] J. F. O'Hanlon. *A User's Guide to Vacuum Technology*. Wiley, New Jersey, 2004.

[208] J. L. Padilla, C. Medina-Bailon, C. Alper, F. Gamiz, and A. M. Ionescu. Confinement-induced InAs/GaSb heterojunction electron–hole bilayer tunneling field-effect transistor. *Appl. Phys. Lett.*, 112(18):182101, 2018.

[209] C.-S. Pang, S.-J. Han, and Z. Chen. Steep slope carbon nanotube tunneling field-effect transistor. *Carbon*, 180:237–243, 2021.

[210] Y. C. Pao, K. Tran, C. Shih, and N. Hardy. Solution to the e-beam gate resist blistering problem of 0.15 micron pHEMTs. *GaAs Mantech*, pages 1–3, 1999.

[211] H. M. Park. Control of ion energy in a capacitively coupled reactive ion etcher. *J. Electrochem. Soc.*, 145(12):4247, 1998.

[212] V. Passi, U. Sodervall, B. Nilsson, G. Petersson, M. Hagberg, C. Krzeminski, E. Dubois, B. Du Bois, and J.-P. Raskin. Anisotropic vapor HF etching of silicon dioxide for Si microstructure release. *Microelectron. Eng.*, 95:83–89, 2012.

[213] S. G. J. Philips, M. T. Madzik, S. V. Amitonov, S. L. de Snoo, M. Russ, N. Kalhor, C. Volk, W. I. L. Lawrie, D. Brousse, L. Tryputen, B. P. Wuetz, A. Sammak, M. Veldhorst, G. Scappucci, and L. M. K. Vandersypen. Universal control of a six-qubit quantum processor in silicon. *Nature*, 609:919–924, 2022.
[214] M. Pierre, R. Wacquez, X. Jehl, M. Sanquer, M. Vinet, and O. Cueto. Single-donor ionization energies in a nanoscale CMOS channel. *Nat. Nanotechnol.*, 5:133–137, 2010.
[215] J. D. Plummer, M. Deal, and P. D. Griffin. *Silicon VLSI Technology, Fundamentals, Practice, and Modeling*. Pearson, 2000.
[216] A. Plößl and G. Kräuter. Wafer direct bonding: tailoring adhesion between brittle materials. *Mater. Sci. Eng., R Rep.*, 25(1):1–88, 1999.
[217] C. Qiu, Z. Zhang, M. Xiao, Y. Yang, D. Zhong, and L.-M. Peng. Scaling carbon nanotube complementary transistors to 5-nm gate lengths. *Science*, 355:271–276, 2017.
[218] A. Rahman, J. Guo, S. Datta, and M. S. Lundstrom. Theory of ballistic nanotransistors. *IEEE Trans. Electron Devices*, 50(9):1853–1864, Sept 2003.
[219] S. Reich, J. Maultzsch, C. Thomsen, and P. Ordejón. Tight-binding description of graphene. *Phys. Rev. B*, 66:035412, Jul 2002.
[220] K. A. Reinhardt and R. F. Reidy, editors. *Handbook of Cleaning in Semiconductor Manufacturing: Fundamental and Applications*. John Wiley & Sons, Ltd, 2011.
[221] A. Revelant, A. Villalon, Y. Wu, A. Zaslavsky, C. Le Royer, H. Iwai, and S. Cristoloveanu. Electron-hole bilayer TFET: Experiments and comments. *IEEE Trans. Electron Devices*, 61(8):2674–2681, 2014.
[222] B. Richstein, Y. Han, Q. T. Zhao, L. Hellmich, J. Klos, S. Scholz, L. R. Schreiber, and J. Knoch. Interface engineering for steep slope cryogenic MOSFETs. *IEEE Electron Device Lett.*, 43(12):2149–2152, 2022.
[223] B. Richstein, L. Hellmich, and J. Knoch. Silicon nitride interface engineering for Fermi level depinning and realization of dopant-free MOSFETs. *Micro*, 1(2):228–241, 2021.
[224] S. Richter, C. Sandow, A. Nichau, S. Trellenkamp, M. Schmidt, R. Luptak, K. K. Bourdelle, Q. T. Zhao, and S. Mantl. ω-gated silicon and strained silicon nanowire array tunneling FETs. *IEEE Electron Device Lett.*, 33(11):1535–1537, 2012.
[225] F. Riederer, T. Grap, S. Fischer, M. R. Müller, D. Yamaoka, B. Sun, C. Gupta, K. T. Kallis, and J. Knoch. Alternatives for doping in nanoscale field-effect transistors. *Phys. Status Solidi A*, 215(7):1700969, 2018.
[226] A. L. P. Rotondaro, G. A. Hames, and T. Yocum. Use of H_2SO_4 for etch rate and selectivity control of boiling H_3PO_4. *ECS Proceedings*, 99-36:385, 1999.
[227] M. Sajjad, X. Yang, P. Altermatt, N. Singh, U. Schwingenschlögl, and S. De Wolf. Metal-induced gap states in passivating metal/silicon contacts. *Appl. Phys. Lett.*, 114(7):071601, 2019.
[228] R. N. Sajjad, W. Chern, J. L. Hoyt, and D. A. Antoniadis. Trap assisted tunneling and its effect on subthreshold swing of tunnel FETs. *IEEE Trans. Electron Devices*, 63(11):4380–4387, 2016.
[229] C. Sandow, J. Knoch, C. Urban, Q.-T. Zhao, and S. Mantl. Impact of electrostatics and doping concentration on the performance of SOI tunnel FETs. *Solid-State Electron.*, 53:1126, 2009.
[230] H. Schmid, M. T. Björk, J. Knoch, H. Riel, W. Riess, P. Rice, and T. Topuria. Patterned epitaxial vapor-liquid-solid growth of silicon nanowires on Si(111) using silane. *J. Appl. Phys.*, 103(2):024304, 2008.
[231] H. Schmid, D. Cutaia, J. Gooth, S. Wirths, N. Bologna, K. E. Moselund, and H. Riel. Monolithic integration of multiple iii-v semiconductors on Si for MOSFETs and TFETs. In *2016 IEEE International Electron Devices Meeting (IEDM)*, pages 3.6.1–3.6.4, 2016.
[232] D. S. Schulman, A. J. Arnold, and S. Das. Contact engineering for 2d materials and devices. *Chem. Soc. Rev.*, 47:3037–3058, 2018.
[233] E. Seo, B. K. Choi, and O. Kim. Determination of proximity effect parameters and the shape bias parameter in electron beam lithography. *Microelectron. Eng.*, 53(1):305–308, 2000.
[234] H. Shang, M. H. White, K. W. Guarini, P. Solomon, E. Cartier, F. R. McFeely, J. J. Yurkas, and W.-C. Lee. Interface studies of tungsten gate metal–oxide–silicon capacitors. *Appl. Phys. Lett.*, 78(20):3139–3141, 2001.

[235] J. C. Slater and G. F. Koster. Simplified LCAO method for the periodic potential problem. *Phys. Rev.*, 94:1498–1524, Jun 1954.

[236] Y.-W. Son, M. L. Cohen, and S. G. Louie. Energy gaps in graphene nanoribbons. *Phys. Rev. Lett.*, 97:216803, Nov 2006.

[237] W. G. Spitzer and H. Y. Fan. Determination of optical constants and carrier effective mass of semiconductors. *Phys. Rev.*, 106:882–890, Jun 1957.

[238] C. Stampfer, J. Güttinger, S. Hellmüller, F. Molitor, K. Ensslin, and T. Ihn. Energy gaps in etched graphene nanoribbons. *Phys. Rev. Lett.*, 102:056403, Feb 2009.

[239] W. Storm, H. A. Gerber, G.-F. Hohl, M. Naujok, and R. Schmolke. Determination of SC1 etch rates at low temperatures with microscope interferometry. In *Ultra Clean Processing of Silicon Surfaces IV*, vol. 65 of Solid State Phenomena, pages 275–278. Trans Tech Publications Ltd, 11 1998.

[240] T. Struck, J. Lindner, A. Hollmann, F. Schauer, A. Schmidbauer, D. Bougeard, and L. R. Schreiber. Robust and fast post-processing of single-shot spin qubit detection events with a neural network. *Sci. Rep.*, 11:16203, 2021.

[241] T. Struck, I. Seidler and, R. Xue, N. Focke, S. Trellenkamp, H. Bluhm, and L. R. Schreiber. Conveyor-mode single-electron shuttling in Si/SiGe for a scalable quantum computing architecture. *npj Quantum Inf.*, 8:100, 2022.

[242] B. Sun. *[Dissertation / PhD Thesis] Reconfigurable field-effect transistors based on wet-chemically etched silicon nanostructures*. PhD thesis, RWTH Aachen University, 2022.

[243] B. Sun, T. Grap, T. Frahm, S. Scholz, and J. Knoch. Role of electron and ion irradiation in a reliable lift-off process with electron beam evaporation and a bilayer PMMA resist system. *J. Vac. Sci. Technol., B*, 39(5):052601, 08 2021.

[244] B. Sun, B. Richstein, P. Liebisch, T. Frahm, S. Scholz, J. Trommer, T. Mikolajic, and J. Knoch. On the operation modes of dual-gate reconfigurable nanowire transistors. *IEEE Trans. Electron Devices*, 68(7):3684–3689, 2021.

[245] B. Sun, S. C. Scholz, A. Kemper, T. Grap, and J. Knoch. Modeling and prediction of hydrogen-assisted morphological evolution in silicon utilizing a level-set approach. *J. Microelectromech. Syst.*, 30(6):950–957, 2021.

[246] X. Sun and T.-J. K. Liu. Spacer gate lithography for reduced variability due to line edge roughness. *IEEE Trans. Semicond. Manuf.*, 23(2):311–315, 2010.

[247] S. M. Sze. *Physics of Semiconductor Devices*. Wiley, New York, 2nd edition, 1981.

[248] S. M. Sze. *VLSI Technology*. McGraw-Hill, 1983.

[249] J. Tersoff. Contact resistance of carbon nanotubes. *Appl. Phys. Lett.*, 74(15):2122–2124, 1999.

[250] M. Y. Toriyama, A. M. Ganose, M. Dylla, S. Anand, J. Park, M. K. Brod, J. Munro, K. A. Persson, A. Jain, and G. J. Snyder. Comparison of the tetrahedron method to smearing methods for the electronic density of states, 2021.

[251] M. Y. Toriyama, A. M. Ganose, M. Dylla, S. Anand, J. Park, M. K. Brod, J. M. Munro, K. A. Persson, A. Jain, and G. J. Snyder. How to analyse a density of states. *Mater. Today Electronics*, 1:100002, 2022.

[252] K. Uchida, J. Koga, and S. Takagi. Experimental study on electron mobility in ultrathin-body silicon-on-insulator metal-oxide-semiconductor field-effect transistors. *J. Appl. Phys.*, 102(7):074510, 10 2007.

[253] R. J. Umstattd, C. G. Carr, C. L. Frenzen, J. W. Luginsland, and Y. Y. Lau. A simple physical derivation of child–Langmuir space-charge-limited emission using vacuum capacitance. *Am. J. Phys.*, 73(2):160–163, 2005.

[254] F. Urbach. The long-wavelength edge of photographic sensitivity and of the electronic absorption of solids. *Phys. Rev.*, 92:1324, Dec 1953.

[255] C. Urban, C. Sandow, Q.-T. Zhao, J. Knoch, S. Lenk, and S. Mantl. Systematic study of Schottky barrier MOSFETs with dopant segregation on thin-body SOI. *Solid-State Electron.*, 54(2):185–190, 2010.

[256] W. G. Vandenberghe, A. S. Verhulst, B. Sorée, W. Magnus, G. Groeseneken, Q. Smets, M. Heyns, and M. V. Fischetti. Figure of merit for and identification of sub-60 mv/decade devices. *Appl. Phys. Lett.*, 102(1):013510, 2013.

[257] L. M. K. Vandersypen, H. Bluhm, J. S. Clarke, A. S. Dzurak, R. Ishihara, A. Morello, D. J. Reilly, L. R. Schreiber, and M. Veldhorst. Interfacing spin qubits in quantum dots and donors—hot, dense, and coherent. *npj Quantum Inf.*, 3:34, 2017.
[258] T. Vasen, P. Ramvall, A. Afzalian, G. Doornbos, M. Holland, C. Thelander, K. A. Dick, L.-E. Wernersson, and M. Passlack. Vertical gate-all-around nanowire GaSb-InAs core-shell n-type tunnel FETs. *Sci. Rep.*, 9:202, 2019.
[259] R. Venugopal, M. Paulsson, S. Goasguen, S. Datta, and M. S. Lundstrom. A simple quantum mechanical treatment of scattering in nanoscale transistors. *J. Appl. Phys.*, 93(9):5613–5625, 2003.
[260] A. Vetter. Resolution enhancement in mask aligner photolithography. PhD thesis, Karlsruhe Institute of Technology, 2019.
[261] P. Vogl, H. P. Hjalmarson, and J. D. Dow. A semi-empirical tight-binding theory of the electronic structure of semiconductors†. *J. Phys. Chem. Solids*, 44(5):365–378, 1983.
[262] R. Vos, M. Lux, K. Xu, W. Fyen, C. Kenens, T. Conard, P. Mertens, M. Heyns, Z. Hatcher, and M. Hoffman. Removal of submicrometer particles from silicon wafer surfaces using HF-based cleaning mixtures. *J. Electrochem. Soc.*, 148(12):G683, 2001.
[263] R. S. Wagner and W. C. Ellis. Vapor-liquid-solid mechanism of single crystal growth. *Appl. Phys. Lett.*, 4(5):89–90, 1964.
[264] J. Wan, C. Le Royer, A. Zaslavsky, and S. Cristoloveanu. A systematic study of the sharp-switching Z^2-FET device: From mechanism to modeling and compact memory applications. *Solid-State Electron.*, 90:2–11, 2013. Selected papers from EUROSOI 2012.
[265] J. Wang and M. Lundstrom. Channel material optimization for the ultimate planar and nanowire MOSFETs: a theoretical exploration. In *63rd Device Research Conference Digest, 2005. DRC '05*, vol. 1, pages 241–242, 2005.
[266] J. Wang, F. Ma, and M. Sun. Graphene, hexagonal boron nitride, and their heterostructures: properties and applications. *RSC Adv.*, 7:16801–16822, 2017.
[267] L. Wang, I. Meric, P. Y. Huang, Q. Gao, Y. Gao, H. Tran, T. Taniguchi, K. Watanabe, L. M. Campos, D. A. Muller, J. Guo, P. Kim, J. Hone, K. L. Shepard, and C. R. Dean. One-dimensional electrical contact to a two-dimensional material. *Science*, 342:614–617, 2013.
[268] X. Wang, Y. Ouyang, X. Li, H. Wang, J. Guo, and H. Dai. Room-temperature all-semiconducting sub-10-nm graphene nanoribbon field-effect transistors. *Phys. Rev. Lett.*, 100:206803, May 2008.
[269] T. Weichelt, U. Vogler, L. Stuerzebecher, R. Voelkel, and U. D. Zeitner. Resolution enhancement for advanced mask aligner lithography using phase-shifting photomasks. *Opt. Express*, 22(13):16310–16321, Jun 2014.
[270] E. Wigner and F. Seitz. On the constitution of metallic sodium. *Phys. Rev.*, 43:804–810, May 1933.
[271] B. Winstead and U. Ravaioli. Simulation of Schottky barrier MOSFETs with a coupled quantum injection/Monte Carlo technique. *IEEE Trans. Electron Devices*, 47(6):1241–1246, 2000.
[272] C.-L. Wu, S.-B. Su, J.-L. Chen, C.-P. Chang, and H.-R. Guo. Tetramethylammonium ion causes respiratory failure related mortality in a rat model. *Resuscitation*, 83(1):119–124, 2012.
[273] C.-L. Wu, S.-B. Su, J.-L. Chen, H.-J. Lin, and H.-R. Guo. Mortality from dermal exposure to tetramethylammonium hydroxide. *J. Occup. Health*, 50(2):99–102, 2008.
[274] C.-Y. Wu. Growth kinetics of silicon thermal nitridation. *J. Electrochem. Soc.*, 129(7):1559, 1982.
[275] P. Wu, A. Prakash, H. Ilatikhameneh, and J. Appenzeller. Understanding contact gating in Schottky barrier transistors from 2d channels. *Sci. Rep.*, 7:12596, 2017.
[276] S. Xiong, T.-J. King, and J. Bokor. Study of the extrinsic parasitics in nano-scale transistors. *Semicond. Sci. Technol.*, 20(6):652–657, May 2005.
[277] X. Xue, M. Russ, N. Samkharadze, B. Undseth, A. Sammak, G. Scappucci, and L. M. K. Vandersypen. Quantum logic with spin qubits crossing the surface code threshold. *Nature*, 601:343–347, 2022.
[278] R.-H. Yan, A. Ourmazd, and K. F. Lee. Scaling the Si MOSFET: From bulk to SOI to bulk. *IEEE Trans. Electron Devices*, 39:1704–1710, 1992.

[279] J. K. W. Yang and K. K. Berggren. Using high-contrast salty development of hydrogen silsesquioxane for sub-10-nm half-pitch lithography. *J. Vac. Sci. Technol., B Microelectron. Nanometer Struct. Process. Meas. Phenom.*, 25(6):2025–2029, 2007.

[280] Y. Yang and R. Murali. Impact of size effect on graphene nanoribbon transport. *IEEE Electron Device Lett.*, 31(3):237–239, March 2010.

[281] S. Yasin, D. G. Hasko, and H. Ahmed. Fabrication of <5 nm width lines in poly(methylmethacrylate) resist using a water: isopropyl alcohol developer and ultrasonically-assisted development. *Appl. Phys. Lett.*, 78(18):2760–2762, 2001.

[282] J. Yoneda, W. Huang, M. Feng, C. H. Yang, K. W. Chan, T. Tanttu, W. Gilbert, R. C. C. Leon, F. E. Hudson, K. M. Itoh, A. Morello, S. D. Bartlett, A. Laucht, A. Saraiva1, and A. S. Dzurak. Coherent spin qubit transport in silicon. *Nat. Commun.*, 12:4114, 2021.

[283] J. Yoneda, K. Takeda, T. Otsuka, T. Nakajima, M. R. Delbecq, G. Allison, T. Honda, T. Kodera, S. Oda, Y. Hoshi, N. Usami, K. M. Itoh, and S. Tarucha. A quantum-dot spin qubit with coherence limited by charge noise and fidelity higher than 99.9 %. *Nat. Nanotechnol.*, 13:102–106, 2018.

[284] K. K. Young. Short-channel effect in fully depleted SOI MOSFET's. *IEEE Trans. Electron Devices*, 36:399–402, 1989.

[285] P. Y. Yu and M. Cardona. *Fundamentals of Semiconductors*. Springer, 2010.

[286] N. F. Za'bah, K. S. K. Kwa, L. Bowen, B. Mendis, and A. O'Neill. Top-down fabrication of single crystal silicon nanowire using optical lithography. *J. Appl. Phys.*, 112(2):024309, 07 2012.

[287] M. Zhang, J. Knoch, S. Mantl, and J. Appenzeller. Improved carrier injection in SOI Schottky barrier MOSFETs. *IEEE Electron Device Lett.*, 28:223–225, 2007.

[288] M. Zhang, J. Knoch, Q. T. Zhao, A. Fox, S. Lenk, and S. Mantl. Low temperature measurements of Schottky-barrier SOI-MOSFETs with dopant segregation. *Electron. Lett.*, 41:1085–1086, 2005.

[289] M. Zhang, J. Knoch, Q. T. Zhao, S. Lenk, U. Breuer, and S. Mantl. Impact of dopant segregation on fully depleted Schottky-barrier SOI-MOSFETs. *Solid-State Electron.*, 50:594–600, 2006.

[290] Q. Zhao, S. Richter, C. Schulte-Braucks, L. Knoll, S. Blaeser, G. V. Luong, S. Trellenkamp, A. Schäfer, A. Tiedemann, J. Hartmann, K. Bourdelle, and S. Mantl. Strained Si and SiGe nanowire tunnel FETs for logic and analog applications. *IEEE J. Electron Dev. Soc.*, 3(3):103–114, 2015.

[291] I. Zubel and M. Kramkowska. The effect of isopropyl alcohol on etching rate and roughness of (100) Si surface etched in KOH and TMAH solutions. *Sens. Actuators A, Phys.*, 93(2):138–147, 2001.

Index

https://doi.org/10.1515/9783111054421-014